Nanotechnology: Materials, Techniques and Applications

Nanotechnology: Materials, Techniques and Applications

Editor: Peggy Rusk

NY RESEARCH PRESS
New York

Published by NY Research Press
118-35 Queens Blvd., Suite 400,
Forest Hills, NY 11375, USA
www.nyresearchpress.com

Nanotechnology: Materials, Techniques and Applications
Edited by Peggy Rusk

International Standard Book Number: 978-1-63238-570-3 (Hardback)

Cataloging-in-Publication Data

Nanotechnology : materials, techniques and applications / edited by Peggy Rusk.
p. cm.
Includes bibliographical references and index.
ISBN 978-1-63238-570-3
1. Nanotechnology. 2. Nanostructured materials. 3. Materials science. I. Rusk, Peggy.
T174.7 .N36 2018
620.5--dc23

Printed in China.

Contents

Preface

The main aim of this book is to educate learners and enhance their research focus by presenting diverse topics covering this vast field. This is an advanced book which compiles significant studies by distinguished experts in the area of analysis. This book addresses successive solutions to the challenges arising in the area of application, along with it; the book provides scope for future developments.

Nanotechnology is the science and engineering that is conducted at the nanoscale level. It studies the application of matter at the atomic, molecular and supramolecular scale. This book provides significant information of this discipline to help develop a good understanding of nanotechnology and its related fields. As this field is emerging at a fast pace, this book will help the readers to better understand the concepts of nanotechnology. Scientists and students actively engaged in nanotechnology will find this book full of crucial and unexplored concepts.

It was a great honour to edit this book, though there were challenges, as it involved a lot of communication and networking between me and the editorial team. However, the end result was this all-inclusive book covering diverse themes in the field.

Finally, it is important to acknowledge the efforts of the contributors for their excellent chapters, through which a wide variety of issues have been addressed. I would also like to thank my colleagues for their valuable feedback during the making of this book.

Editor

1

Observation of Zeeman effect in topological surface state with distinct material dependence

Ying-Shuang Fu[1,2], T. Hanaguri[2], K. Igarashi[3], M. Kawamura[2], M.S. Bahramy[2,4] & T. Sasagawa[3]

Manipulating the spins of the topological surface states represents an essential step towards exploring the exotic quantum states emerging from the time reversal symmetry breaking via magnetic doping or external magnetic fields. The latter case relies on the Zeeman effect and thereby we need to estimate the *g*-factor of the topological surface state precisely. Here, we report the direct observations of the Zeeman effect at the surfaces of Bi_2Se_3 and Sb_2Te_2Se by spectroscopic-imaging scanning tunnelling microscopy. The Zeeman shift of the zero mode Landau level is identified unambiguously by appropriately excluding the extrinsic effects arising from the nonlinearity in the band dispersion of the topological surface state and the spatially varying potential. Surprisingly, the *g*-factors of the topological surface states in Bi_2Se_3 and Sb_2Te_2Se are very different ($+18$ and -6, respectively). Such remarkable material dependence opens up a new route to control the spins of the topological surface states.

[1] School of Physics and Wuhan National High Magnetic Field Center, Huazhong University of Science and Technology, Wuhan 430074, China. [2] RIKEN Center for Emergent Matter Science, Wako, Saitama 351-0198, Japan. [3] Materials and Structures Laboratory, Tokyo Institute of Technology, Yokohama, Kanagawa 226-8503, Japan. [4] Department of Applied Physics, University of Tokyo, Bunkyo-ku, Tokyo 113-8656, Japan. Correspondence and requests for materials should be addressed to Y.-S.F. (email: yfu@hust.edu.cn) or to T.H. (email: hanaguri@riken.jp).

The helical Dirac fermions on the surfaces of topological insulators (TIs) host novel relativistic quantum phenomena in solids[1,2]. When the time reversal symmetry (TRS) of the topological surface state (TSS) is broken, a gap opens at the Dirac point. This brings about novel topological excitations, such as the magneto-electric effect[3,4], the quantum anomalous Hall effect[3,5] and the magnetic monopole effect[6]. Magnetic doping has proven to be an effective way to break the TRS via magnetic exchange interactions[7–9], thereby enabling the experimental observation of the quantum anomalous Hall state[10]. However, magnetic dopants may introduce charge inhomogeneity[11] and weaken the spin–orbit coupling of TI compounds[12].

The Zeeman effect, the coupling of spins with a magnetic field, offers an alternative way to break the TRS of the TSS without causing any of the above issues. The opened gap Δ is fully tunable by a perpendicular magnetic field B, that is, $\Delta = g_s\mu_B B$, where g_s is the electron g-factor of the TSS and μ_B is the Bohr magneton[13]. Consequently, to manipulate the TSS via the Zeeman effect, it is crucial to know its g_s.

Under B, Landau levels (LLs) are formed as a result of the cyclotron motion of the electrons. Inclusion of the Zeeman effect can lead to an energetic shift of the LLs as the corresponding TSS lack spin degeneracy (Fig. 1a). This is in stark contrast to the Zeeman splitting of the LLs observed in graphene[14] and conventional two-dimensional (2D) electron systems[15]. Such a Zeeman effect is most pronounced for the zeroth LL (LL_0), and decreases dramatically with increasing Landau index n (Supplementary Note 1).

Understanding the Zeeman shift behaviour and determining the g-factor of the TSS have entailed intense research investigations primarily by the means of quantum oscillation measurements[13,16,17]. However, the reported data are still largely controversial. The inconsistency is due to the fact that the host materials are inevitably doped by defects. This prevents the lower LLs, which exhibit prominent Zeeman shifts, from reaching the Fermi level and contributing to quantum oscillations in the accessible B (ref. 16). Furthermore, the sign of g_s cannot be determined from the Zeeman shift of the nonzero LLs[17]. A recent tunnelling spectroscopy study appears to be successfully probing the LL_0 of the TSS formed at an interface with a conventional semiconductor[18]. However, the method used depends on specific details of band-bending in such heterostructures, and therefore cannot be readily applied to other TI compounds. Spectroscopic imaging scanning tunnelling microscopy (SI-STM) can access electronic states in a wide energy range with high spatial and energy resolutions. Thus, it can be used to study any LL, regardless of the doping level of the TI compounds.

Using this technique, we have developed a methodology to unambiguously observe the Zeeman shift of LL_0. This allows us to determine the g-factors of the TSSs in Bi_2Se_3 and Sb_2Te_2Se precisely, which turn out to be very different.

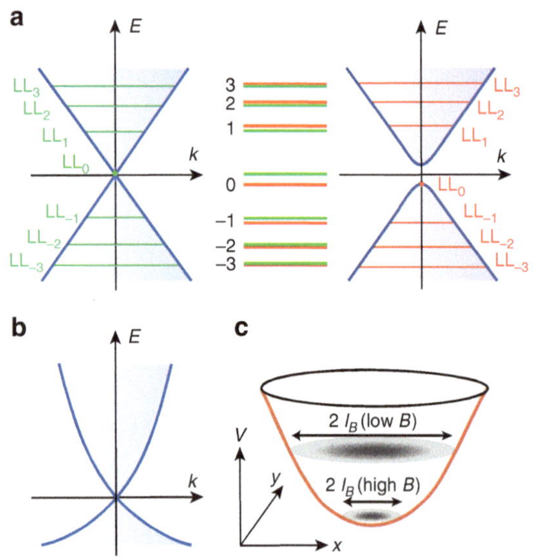

Figure 1 | Zeeman effect of TSS and extrinsic influences on its observation. (**a**) Schematic illustrating the Zeeman effect on the LLs of the TSS. When the Zeeman effect is absent, LLs (green lines) are formed in a perpendicular B. When the Zeeman effect is present, the TSS becomes massive and its LLs (red lines) exhibit an additional energy shift away from the Dirac point. The magnitude of the Zeeman shift decreases rapidly with increasing Landau index n. (**b**) Schematic showing the band structure of the actual TSS with a finite curvature superimposed on its linear dispersion. (**c**) Schematic of a 2D potential minimum and the spatial extension of the LL_0 wave function at different B. The dark (light) grey colour depicts high (low) intensity of the LL_0 wave function.

Results

Modelling the Zeeman effect of the TSS. In principle, the Zeeman shift of the LL_0 energy (E_0) of the TSS is linear with B, and its slope determines g_s. In practice, however, more factors are involved, hindering its direct observation. On one hand, a finite curvature is superimposed on the linear dispersion of the TSS in actual compounds (Fig. 1b). This induces an extra B-linear change in E_0 that is irrelevant to the Zeeman effect. On the other hand, there exists spatial potential variations in the TSS originating from the inhomogenously distributed charged defects[19]. This introduces an extrinsic B dependence of LL energies as the spatial extension of the LL wave functions shrinks with increasing B (Fig. 1c).

To quantify the effects of nonlinear dispersions and the potential variations, we consider the following model Hamiltonian[17].

$$H = \frac{1}{2m^* m_e}(\Pi_x^2 + \Pi_y^2) + v(\sigma_x \Pi_y - \sigma_y \Pi_x) + \frac{1}{2} g_s \mu_B B \sigma_z + V(x,y) \quad (1)$$

Here, $\boldsymbol{\sigma}$ are the Pauli matrices and $\boldsymbol{\Pi}$ are the canonical momenta. The first term denotes the nonlinearity of band dispersions, with m^* being the effective mass relative to that of the free electron (m_e). The second term represents the helical Dirac component of the TSS, with v being the electron velocity. The third term is the Zeeman term and the last term represents the potential variation. The energy of the n-th LL, E_n, has been given in ref. 20, albeit without contribution from the last term of equation (1).

To include the effect of this term, we consider a 2D parabolic potential model $V(x,y) = E_D + \alpha_x x^2 + \alpha_y y^2$ to approximate the shape and location of the potential extremes, where E_D is the Dirac-point energy. At the potential extreme, the B-dependence of E_0 can be calculated using a first-order approximation[21] (Supplementary Note 1). Including the effects of nonlinear dispersions and potential variations, the $E_0(B)$ is accordingly given as

$$E_0(B) = E_D + \frac{1}{2}\left(\frac{2}{m^*} - g_s\right)\mu_B B + (\alpha_x + \alpha_y)\frac{\hbar}{|e|B} \quad (2)$$

Note that m^* renormalizes g_s, and the potential variation introduces an additional $1/B$ dependence term. To determine the intrinsic g_s, we have developed an analysis scheme for our SI-STM data, which corrects these extrinsic factors, m^* and $\alpha_x + \alpha_y$, and applied it to two different TI materials, Bi_2Se_3 and Sb_2Te_2Se.

SI-STM study of the Zeeman effect of the TSS. As the first step, we evaluate m^* using momentum-resolved LL spectroscopy[22].

Figure 2a,b display the LL spectra of Bi_2Se_3 and Sb_2Te_2Se (Supplementary Note 2, Supplementary Figs 1and 2), respectively, as measured at a fixed location at various B. In contrast to the electron-doped Bi_2Se_3, Sb_2Te_2Se is hole-doped and thus its Dirac point is in the empty state. For each material, the corresponding E_n exhibits a quasi-linear scaling relation with a scaling variable $(nB)^{1/2}$ (Fig. 2c), which represents the energy–momentum dispersion of the TSS[22]. The potential effect on E_n is more significant for the LLs with small n exposed to high B than those with large n exposed to low B (ref. 21). In this regard, the observed scaling with $(nB)^{1/2}$ demonstrates that such potential effect has negligible influence on it at the measured location. For both compounds, a finite curvature is evidently seen in the dispersions indicating that m^* is finite, as also demonstrated by angle-resolved photoemission spectroscopy measurements and band calculations[7,23]. Remarkably, both compounds show a very

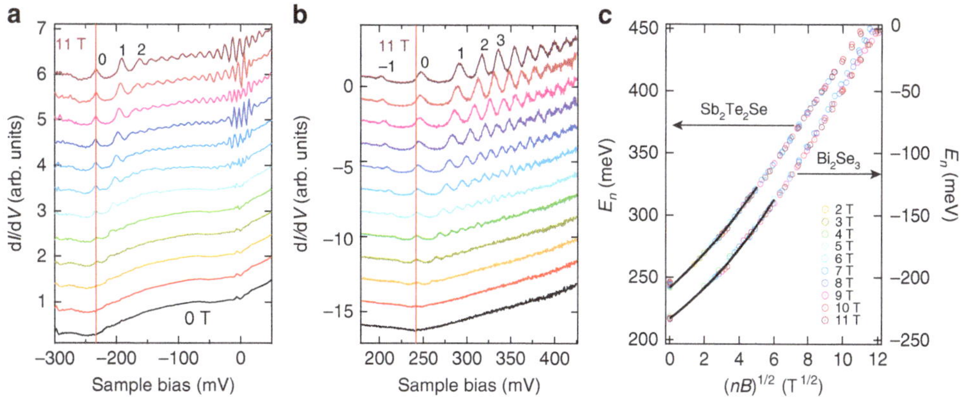

Figure 2 | LL spectroscopy of Bi_2Se_3 and Sb_2Te_2Se. Tunnelling spectroscopy showing the LLs of the TSS measured at a fixed location of (**a**) Bi_2Se_3 and (**b**) Sb_2Te_2Se surface at 1.5 K. The spectra were acquired at different B from 0 T to 11 T with an interval of 1 T and are offset vertically for clarity. The red lines mark the energies of the Dirac points. The data shown in **a** are the same data used in ref. 22. Measurement conditions of **b**: $V_s = -100$ mV, $I_t = 50$ pA and $V_{mod} = 1.4$ mV_{rms}. (**c**) Scaling analysis of $E_n(B)$ based on the data of Sb_2Te_2Se shown in **b** and a comparison with that of Bi_2Se_3. The E_n values were obtained by fitting the LL spectra with multiple Lorentz functions. The black curves depict the fitting to the low-energy parts of the scaling relations.

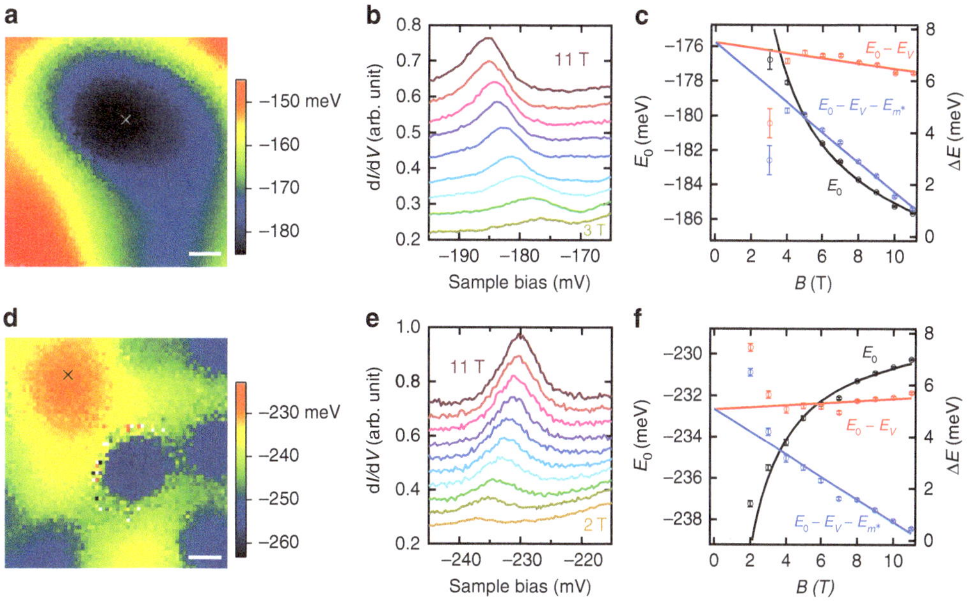

Figure 3 | Surface *g*-factor measurement on Bi_2Se_3. (**a**) Potential landscape of Bi_2Se_3 obtained by mapping E_0 at 11 T showing a potential minimum. The centre of the potential minimum is determined by the 2D parabolic potential fitting and is marked as a cross. The scale bar corresponds to 10 nm. Measurement conditions: $V_s = 50$ mV, $I_t = 50$ pA, $V_{mod} = 2.8$ mV_{rms} and $T = 1.5$ K. (**b**) Tunnelling spectra taken at the potential minimum centre at fields from 3 to 11 T with 1 T intervals. The spectra have been shifted for clarity. Measurement conditions: $V_s = -220$ mV, $I_t = 100$ pA, $V_{mod} = 1.4$ mV_{rms} and $T = 1.5$ K. (**c**) E_0 at different B obtained by fitting the data of **b** with a Lorentz line shape and plotted with black symbols (left axis). The error bars are the standard deviation of the fitting analysis. The effects of the potential and the non-ideal dispersions on the LL_0 energies are represented by $E_V = (\alpha_x + \alpha_y)\frac{\hbar}{|e|B}$ and $E_{m^*} = \frac{\mu_B B}{m^*}$, respectively. Their influences can be excluded by subtracting their contributions. The red symbols denote the LL_0 energies after subtracting the effect of the potential ($E_0 - E_V$). The blue symbols correspond to the LL_0 energies after subtracting the effects from both the potential and the non-ideal dispersions ($E_0 - E_V - E_{m^*}$) (right axis, in the relative energy scale). The black curve denotes the fitting to E_0 with B according to equation (2). The red and blue lines show the linear fitting of $E_0 - E_V$ and $E_0 - E_V - E_{m^*}$ with B, respectively. (**d**-**f**) Similar data and analysis as **a**-**c** for a potential maximum of Bi_2Se_3. Measurement conditions of **d** and **e**: $V_s = -200$ mV, $I_t = 165$ pA, $V_{mod} = 1.8$ mV_{rms} and $T = 1.5$ K.

similar band curvature, despite their different constituent elements. We thus expect m^* to be nearly the same for both the compounds. To evaluate this, m^* is extracted from the scaling function, while ignoring the potential variations. For each compound, we have performed three measurements at different locations/samples. On the basis of these measurements, m^* is found to be 0.12 ± 0.03 for Bi_2Se_3 and 0.13 ± 0.02 for Sb_2Te_2Se with negligible inhomogeneity.

Next, we assess the impact of potential variations on $E_0(B)$. The spatial variation of E_0 represents the potential landscape, although it is smeared out by the magnetic length l_B (refs 19,21). We start with Bi_2Se_3 and map out the potential landscape by performing a spectroscopic imaging of E_0 at a high B of 11 T, where $l_B \sim 7.7$ nm (Fig. 3a,d). We focus on areas around the potential extremes (Fig. 3a,d) and fit their shapes with the 2D parabolic potential model introduced above (Supplementary Fig. 3). After positioning the tip at the fitted potential centre (Fig. 3a,d, cross point), the B dependence of the LL_0 peak is measured (Fig. 3b,e). The measured $E_0(B)$ is plotted in Fig. 3c,f (black symbols). Intriguingly, $E_0(B)$ exhibits $\sim 1/B$ behaviours at both the potential minimum ($\alpha_x > 0, \alpha_y > 0$) and maximum ($\alpha_x < 0, \alpha_y < 0$), but shifts towards opposite directions. This is exactly expected from equation (2) and directly highlights the influence of potential variations on $E_0(B)$.

To eliminate the potential effect, we estimate the last term of equation (2) using the fitted values of α_x and α_y. After subtracting the estimated potential effect, we obtain a nearly B independent E_0 in the high B region at both the potential minimum and maximum (red symbols in Fig. 3c,f). This validates our methodology and indicates that the effect of m^* and the Zeeman effect contribute oppositely with an accidental cancellation. By further subtracting the contribution from m^*, we obtain the genuine Zeeman shift (Fig. 3c,f, blue symbols). Subsequently, g_s in Bi_2Se_3 is determined to be $+18 \pm 4$. (The error in g_s is the propagation of uncertainty evaluated from the errors of m^* and $\frac{2}{m^*} - g_s$, which are extracted above and measured at different potential extremes, respectively.) This differs significantly from its corresponding bulk value $+32$ as indicated by magneto-transport and nuclear magnetic resonance measurements[24,25].

The same methodology is applied to Sb_2Te_2Se (Fig. 4; Supplementary Figs 4 and 5). Distinct from Bi_2Se_3, the LL_0 state measured at the potential minimum (Fig. 4b) exhibits a nonmonotonic B dependence (Fig. 4c) and increases with B at high fields. Given that the effective masses of both compounds are almost the same, g_s in Sb_2Te_2Se must differ significantly from that of Bi_2Se_3. Indeed, we evaluate g_s in Sb_2Te_2Se to be -6 ± 2. While both systems share very similar band dispersions in their TSSs, their g-factors turn out to be strikingly different in both size and sign.

Discussion

We elucidate the above results under the framework of $\mathbf{k} \cdot \mathbf{p}$ theory. For narrow gap semiconductors, the conduction bands are coupled to the spin–orbit-split valence bands through a second-order perturbative term. Such a coupling substantially enhances the orbital sector of the g-factor. As a result, the total g-factor of such electrons/holes can significantly deviate from that of the free electrons in both magnitude and sign[26]. In the case of TIs, the strong atomic spin–orbit coupling creates a symmetry-inverted band gap. Thus, the atomic orbital characters of the wave function undergo a strong variation at and in the

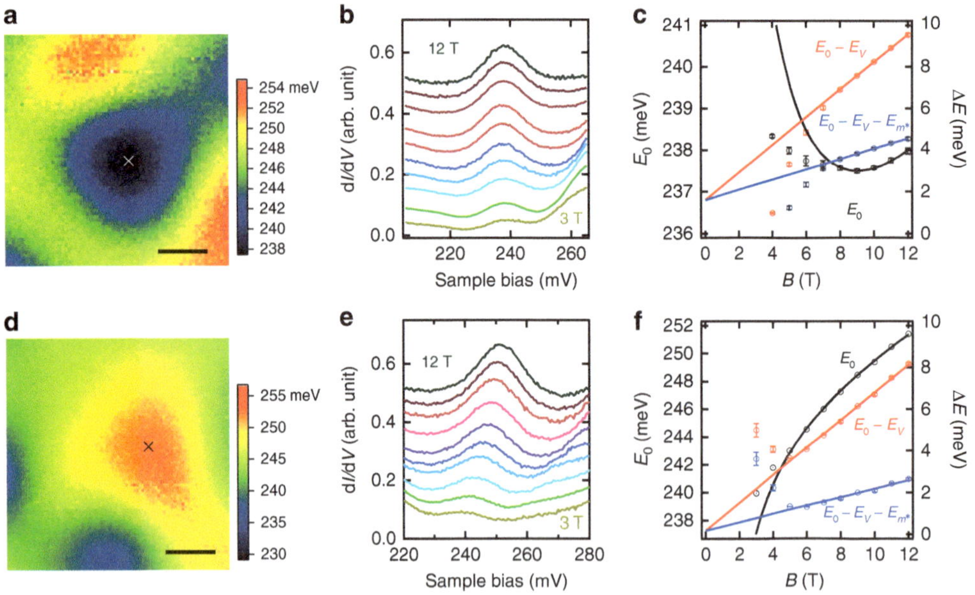

Figure 4 | Surface g-factor measurement on Sb_2Te_2Se. (**a**) Potential landscape of Sb_2Te_2Se obtained by mapping E_0 at 12 T showing a potential minimum. The centre of the potential minimum is determined by the 2D parabolic potential fitting and marked by a cross. The scale bar corresponds to 10 nm. (**b**) Tunnelling spectra taken at the potential minimum centre at fields from 3 to 12 T with 1 T intervals. The spectra have been shifted for clarity. (**c**) E_0 values at different B obtained by fitting the data of **b** with a Lorentz line shape and plotted with black symbols (left axis). The error bars are the standard deviation of the fitting analysis. The effects of the potential and the non-ideal dispersions on the LL_0 energies are represented by $E_V = (\alpha_x + \alpha_y)\frac{\hbar}{|e|B}$ and $E_{m^*} = \frac{\mu_B B}{m^*}$, respectively. Their influences can be excluded by subtracting their contributions. The red symbols denote the LL_0 energies after subtracting the effect of the potential ($E_0 - E_V$). The blue symbols correspond to the LL_0 energies after subtracting the effects from both the potential and the non-ideal dispersions ($E_0 - E_V - E_{m^*}$) (right axis, in the relative energy scale). The black curve denotes the fitting to E_0 with B according to equation (2). The red and blue lines represent the linear fitting of $E_0 - E_V$ and $E_0 - E_V - E_{m^*}$ with B, respectively. (**d**–**f**) Similar data and analysis as **a**–**c** for a potential maximum of Sb_2Te_2Se. Measurement conditions of **a** and **d**: $V_s = 215$ mV, $I_t = 50$ pA, $V_{mod} = 2.8$ mV$_{rms}$, $T = 4.4$ K and $B = 12$ T. Measurement conditions of **b** and **e**: $V_s = 210$ mV, $I_t = 50$ pA, $V_{mod} = 1.8$ mV$_{rms}$ and $T = 4.4$ K.

vicinity of the inverted band edges[27]. Such a variation manifests itself in the orbital character of TSS around the Dirac point as well[28]. As Sb_2Te_2Se and Bi_2Se_3 comprise different elements, their wave functions must have different orbital characters. Such a difference further implies that the g-factors of the two compounds should be different too. Note that a considerable energy dependence of the g-factor is expected, since the orbital character of the wave function changes notably in these systems. Our measurement is merely around the Dirac point, which is directly relevant to the gap opening of the TSS via B. In this regard, the g-factor of the TSS may be different from that of the bulk since they are measured at different energies. Further theoretical investigations regarding those factors are needed to develop a general theory that describes the g-factor.

Considering the significant material dependence of g_s, we envision an interesting possibility to tailor the g-factor of the TSS by controlling the chemical composition of the chalcogenide TI materials in the form of solid solutions[29,30]. This provides a new knob in manipulating the TSS for its spin-related applications.

Methods

Experiment description. The experiments were performed with a modified commercial UNISOKU low temperature STM at 4.4 or 1.5 K. Magnetic fields up to 12 T can be applied perpendicularly to the sample surface. Sb_2Te_2Se and Bi_2Se_3 crystals grown by a modified Bridgman technique were cleaved *in situ* under ultrahigh vacuum conditions at ~77 K. After cleaving, the crystals were transferred quickly to the low-temperature STM for subsequent measurements. Two Bi_2Se_3 and three Sb_2Te_2Se samples were measured. A tungsten tip was used as the STM probe, which had been cleaned and characterized with a field-ion microscope. The tunnelling spectra were obtained by lock-in detection of the tunnelling current with a modulation voltage at 617.3 Hz feeding into the sample bias. The tip was grounded to provide the reference voltage.

Evaluation of the finite curvature of LL scaling. We evaluated m^*, which characterizes the finite curvature, using the scaling analysis of the LL energies shown in Fig. 2c. Since only the Zeeman shift of the LL_0 is prominent, Supplementary equation (10) (Supplementary Note 1) can be approximated by $E_{n\neq 0} \cong \frac{\hbar|e|}{m^* m_e} nB \pm \sqrt{2\hbar v^2|e|nB}$. This indicates that the scaling of E_n with $(nB)^{1/2}$ still applies even in the case of finite m^*. Regarding E_0, its energy at 3 T was used for the scaling analysis, because its shift is negligible at low B. By fitting the low-energy part of the scaling relation, the m^* value can be obtained. It must be noted that the low-energy fitting considerably deteriorated when high-energy data were also included. Since the effect of m^* on E_0 is determined by the electronic states around the Dirac point, we only fitted the low-energy parts.

Fitting the potential extremes with 2D parabolic potential model. First, two sectional lines across the potential extreme were drawn and fitted with a one-dimensional parabolic potential to estimate its shape and location. The obtained parameters were subsequently input as the initial guess of the 2D parabolic fitting. The 2D fitting provided the shape and location of the fitted potential, whose equipotential lines were superimposed on the potential map (Supplementary Figs 3a,b and 4a,b). The generated fitting error was small, demonstrating that the 2D parabolic potential model fitted the measured potential well. This conclusion was further augmented by evaluating two sectional lines extracted along the major eclipse axes of both the fitted potential and the measured E_0 map (Supplementary Figs 3c–f and 4c–f). Since the 2D parabolic potential merely applies to potential extremes, it cannot be used at low B, where the LL_0 state spatially expands beyond the potential extremes. (The effect of the potential variations at low B is discussed in Supplementary Note 3, Supplementary Figs 6 and 7). Therefore, for the fitting of Figs 3c,f and 4c,f, we did not include the data below the critical B value, where the potential-corrected E_0 begins to deviate from the B-linear behaviour. The size of the LL_0 state ($2l_B$) at the critical B (Supplementary Figs 3a,b and 4a,b, dashed circles) has a good correspondence with the fitted area, which substantiates our model.

References

1. Hasan, M. Z. & Kane, C. L. Topological insulators. *Rev. Mod. Phys.* **82,** 3045–3067 (2010).
2. Qi, X. L. & Zhang, S. C. Topological insulators and superconductors. *Rev. Mod. Phys.* **83,** 105–1110 (2010).
3. Qi, X. L., Hughes, T. L. & Zhang, S. C. Topological field theory of time-reversal invariant insulators. *Phys. Rev. B* **78,** 195424 (2008).
4. Essin, A. M., Moore, J. E. & Vanderbilt, D. Magnetoelectric polarizability and axion electrodynamics in crystalline insulators. *Phys. Rev. Lett.* **102,** 146805 (2009).
5. Yu, R. *et al.* Quantized anomalous Hall effect in magnetic topological insulators. *Science* **329,** 61–64 (2010).
6. Qi, X. L., Li, R., Zang, J. & Zhang, S. C. Inducing a magnetic monopole with topological surface states. *Science* **323,** 1184–1187 (2009).
7. Chen, Y. L. *et al.* Massive Dirac fermion on the surface of a magnetically doped topological insulator. *Science* **329,** 659–662 (2010).
8. Checkelsky, J. G., Ye, J. T., Onose, Y., Iwasa, Y. & Tokura, Y. Dirac-fermion-mediated ferromagnetism in a topological insulator. *Nat. Phys.* **8,** 729–733 (2012).
9. Xu, S. Y. *et al.* Hedgehog spin texture and Berry's phase tuning in a magnetic topological insulator. *Nat. Phys.* **8,** 616–622 (2012).
10. Chang, C. Z. *et al.* Experimental observation of the quantum anomalous Hall effect in a magnetic topological insulator. *Science* **340,** 167–170 (2013).
11. Zhang, D. M. *et al.* Interplay between ferromagnetism, surface states, and quantum corrections in a magnetically doped topological insulator. *Phys. Rev. B* **86,** 205127 (2012).
12. Zhang, J. S. *et al.* Topology-driven magnetic quantum phase transition in topological insulators. *Science* **339,** 123905 (2013).
13. Analytis, J. G. *et al.* Two-dimensional surface state in the quantum limit of a topological insulator. *Nat. Phys.* **6,** 960–964 (2010).
14. Song, Y. J. *et al.* High-resolution tunnelling spectroscopy of a graphene quartet. *Nature* **467,** 185–189 (2010).
15. Hashimoto *et al.* Quantum Hall transition in real space: from localized to extended states. *Phys. Rev. Lett.* **101,** 256802 (2008).
16. Xiong, J. *et al.* High-field Shubnikov–de Haas oscillations in the topological insulator Bi_2Te_2Se. *Phys. Rev. B* **86,** 045314 (2012).
17. Taskin, A. A. & Ando, Y. Berry phase of nonideal Dirac fermions in topological insulators. *Phys. Rev. B* **84,** 035301 (2011).
18. Yoshimi, R. *et al.* Dirac electron states formed at the heterointerface between a topological insulator and a conventional semiconductor. *Nat. Mater.* **13,** 253–257 (2014).
19. Fu, Y. S. *et al.* Imaging the two-component nature of Dirac–Landau levels in the topological surface state of Bi_2Se_3. *Nat. Phys.* **10,** 815–819 (2014).
20. Reynoso, A., Usaj, G., Sánchez, M. J. & Balseiro, C. A. Theory of edge states in systems with Rashba spin-orbit coupling. *Phys. Rev. B* **70,** 235344 (2004).
21. Okada, Y. *et al.* Visualizing Landau levels of Dirac electrons in a one-dimensional potential. *Phys. Rev. Lett.* **109,** 166407 (2012).
22. Hanaguri, T., Igarashi, K., Kawamura, M., Takagi, H. & Sasagawa, T. Momentum-resolved Landau-level spectroscopy of Dirac surface state in Bi_2Se_3.. *Phys. Rev. B* **82,** 081305 (R) (2010).
23. Lin, H. *et al.* An isolated Dirac cone on the surface of ternary tetradymite-like topological insulators. *New J. Phys.* **13,** 095005 (2011).
24. Köhler, H. & Wöchner, E. The g-factor of the conduction electrons in Bi_2Se_3. *Phys. Stat. Sol. (b)* **67,** 675 (1975).
25. Mukhopadhyay, S. *et al.* Hyperfine coupling and spin polarization in the bulk of the topological insulator Bi_2Se_3. *Phys. Rev. B* **91,** 081105 (R) (2015).
26. Roth, L. M., Lax, B. & Zwerdling, S. Theory of optical magneto-absorption effects in semiconductors. *Phys. Rev.* **114,** 90–104 (1959).
27. Liu, C. X. *et al.* Model Hamiltonian for topological insulators. *Phys. Rev. B* **82,** 045122 (2010).
28. Cao, Y. *et al.* Mapping the orbital wavefunction of the surface states in three-dimensional topological insulators. *Nat. Phys.* **9,** 499–504 (2013).
29. Arakane, T. *et al.* Tunable Dirac cone in the topological insulator $Bi_{2-x}Sb_xTe_{3-y}Se_y$. *Nat. Commun.* **3,** 636 (2012).
30. Zhang, J. S. *et al.* Band structure engineering in $Bi_{2-x}Sb_xTe_3$ topological insulators. *Nat. Commun.* **2,** 574 (2011).

Acknowledgements

We thank Y. Tokura, X.-C. Xie, J.-H. Gao, P. Zhang, Z.-G. Wang, Y.-Y. Wang and X. Liu for helpful discussions. This work is funded by Grant-in-Aid for Scientific Research from the Ministry of Education, Culture, Sports, Science and Technology of Japan, as well as the National Science Foundation of China.

Author contributions

Y.-S.F. and T.H. performed the experiments and analysed the data. K.I. and T.S. grew single crystals of Sb_2Te_2Se and Bi_2Se_3. M.K. and M.S.B. contributed to the theoretical understanding. T.H. supervised the project. Y.-S.F. and T.H. wrote the manuscript with contributions from M.K. and M.S.B.

Additional information

Competing financial interests: The authors declare no competing financial interests.

2

Capillarity ion concentration polarization as spontaneous desalting mechanism

Sungmin Park[1,*], Yeonsu Jung[2,*], Seok Young Son[1], Inhee Cho[1], Youngrok Cho[1], Hyomin Lee[1,3], Ho-Young Kim[2,3,4] & Sung Jae Kim[1,4,5]

To overcome a world-wide water shortage problem, numerous desalination methods have been developed with state-of-the-art power efficiency. Here we propose a spontaneous desalting mechanism referred to as the capillarity ion concentration polarization. An ion-depletion zone is spontaneously formed near a nanoporous material by the permselective ion transportation driven by the capillarity of the material, in contrast to electrokinetic ion concentration polarization which achieves the same ion-depletion zone by an external d.c. bias. This capillarity ion concentration polarization device is shown to be capable of desalting an ambient electrolyte more than 90% without any external electrical power sources. Theoretical analysis for both static and transient conditions are conducted to characterize this phenomenon. These results indicate that the capillarity ion concentration polarization system can offer unique and economical approaches for a power-free water purification system.

[1] Department of Electrical and Computer Engineering, Seoul National University, Seoul 08826, South Korea. [2] Department of Mechanical and Aerospace Engineering, Seoul National University, Seoul 08826, South Korea. [3] Institute of Advanced Machines and Design, Seoul National University, Seoul 08826, South Korea. [4] Big Data Institute, Seoul National University, Seoul 08826, South Korea. [5] Inter-university Semiconductor Research Center, Seoul National University, Seoul 08826, South Korea. * These authors contributed equally to this work. Correspondence and requests for materials should be addressed to H.L. (email: nanofluidics@snu.ac.kr) or to H.-Y.K. (email: hyk@snu.ac.kr) or to S.J.K. (email: gates@snu.ac.kr).

Recent noteworthy investigations in micro/nanofluidic devices enable to develop not only various biomedical but also significant energy and environmental applications[1]. Among these groundbreaking technologies, micro/nanofluidic platforms have successfully served as the differential shells of macro-scale water purification/desalination systems[2–4]. One can microscopically investigate the salt separation mechanism occurring near nanoporous membranes or porous electrodes by flow field tracking or voltage–current measurements. Although state-of-the-art desalination methods, such as reverse-osmosis and flash distillation, can provide the gigantic amount of fresh water with high energy efficiency[5], low energy efficiency of small-scale platforms and the requirement of high-cost infrastructures have become acute nuisances, especially for remote/rural areas of the world. On the other hand, the micro/nanofluidic differential shells can be scalably integrated to a portable-scale system for individual usage or even to a mid-scale system for small village supply.

For such small-sized desalination/purification applications, ion concentration polarization (ICP) has been recently introduced[2,6–8]. ICP indicates the polarization of electrolytic concentrations at the anodic and cathodic side of a permselective membrane under d.c. bias. Typical behaviour is that the concentration becomes largely depleted at the anodic side (also known as an ion-depletion zone) and enriched at the cathodic side (also known as an ion-enrichment zone) in the case with a cation-selective membrane and vice versa[9]. In addition to nanoporous membranes[10–12], bipolar electrodes[13,14] and nanoporous particles[15,16] have been reported for a new practical platform to generate ICP as well. Intensive researches have been conducted to explore its fundamental aspects such as the source of over-limiting current[17–19], the vortex generation[20–22] and related instability issues[23] and its novel engineering applicability such as biological sample preparations[24,25], energy saving mechanisms[26] and electro-desalination applications[2,14]. Although ICP desalination method is unprecedentedly capable of extracting fresh water from highly saline, biologically infected or even heavy metal contaminated water, the necessity of a high external electrical power source[4,27] is still the most critical hurdle to be overcome. Therefore, a new desalination mechanism that consumes the minimum level of energy is highly demanded.

In nature, mangrove trees that grow in saline water along the tropical or subtropical coasts are well known for their unique physiological responses to survive in highly saline environments without any external supply of mechanical, chemical or electrical powers[28]. The phenomenological observations of mangrove[28] so far have found that the root is able to provide sap of 90% lower salinity than the external saline water by a reverse osmosis mechanism. Therefore, we are particularly interested in the ability of mangrove to exclude salt without any external energy. Motivated by this fact, we suggest a new spontaneous desalination mechanism by introducing the nanoscale electrokinetic and the hydrodynamic transportation of ions at the interfaces between the nanoporous structure and the saline environment. We visualize the ion movements that illustrate the new class of the ion-depletion phenomenon. By utilizing dry ionic hydrogel, the effective ionic flux through the hydrogel is induced by the capillarity of the hydrogel itself, instead of an external electrical bias, so that we name the currently exploited process as the capillarity ion concentration polarization (CICP). Osmo-poro-elastic process due to the swelling of the hydrogel[29,30] is excluded in this work by a mechanical confinement[31–33] within a microchannel to focus on the role of capillarity. Both analytical and numerical models are obtained to analyse the statics and dynamics of the spontaneous desalting mechanism. Since the mechanism is independent from any electrical power sources, it offers significant merits in terms of a power consumption and a stable operation compared with the conventional ICP mechanism. Therefore, CICP mechanism can be applied to power-free and small-scale (possibly portable) desalination/purification devices that would be particularly useful in remote/rural and disaster-stricken areas.

Results

Imbibition rates through the ionic hydrogel. The ionic hydrogel is capable of absorbing ambient aqueous solution by the capillarity. This phenomenon is critical to the presented CICP system, which utilizes the capillarity as a driving force for a liquid transportation. A hydrogel usually has an osmo-poro-elastic effect due to a significant swelling. However, the swelling has been reported to be minimized by a mechanical confinement[31–33]. In this work, the hydrogel swells within a microchannel and, thus, we are able to analyse the CICP phenomenon only by the capillarity effect. See Supplementary Note 1 for the swelling result either with or without mechanical confinements. As shown in Fig. 1a, there are two significant evidences of the imbibition by the capillarity. Micro-particles are injected near the hydrogel pad and their motions are captured at distinct time intervals. As indicated, their paths converge toward the hydrogel pad from both reservoirs in the centre-connection device, evidencing a mass sink at the centre. See Supplementary Video 1. In addition, the particle convergence in the end-connection device is also shown in Supplementary Note 2. The descriptions of the centre- and the end-connection device are given in the 'Methods' section. Since there are no external driving forces (such as gravity or external electric/magnetic/acoustic field) other than the capillary force, the total volume that has disappeared (V_{water}) is calculated by the speed of the particles. Since we confirm that there is no leakage at the bonding or hydrogel interfaces, the entire V_{water} should be transported into the hydrogel. Because the hydrogel has auto-fluorescence and it disappears when in contact with water molecules, the wetted volume of hydrogel (V_{wet}) is able to be measured by the progression of an imbibition front. Therefore, the fraction of these two volumes (V_{water}/V_{wet}) is the same as the porosity of the hydrogel and it is calculated to be around 0.1, which would be used as one of the important parameters in the theoretical calculations.

Furthermore, the dynamic measurements of the imbibition length reveal that it grows as the square root of time as shown in Fig. 1b. This result is consistent with the classical Washburn's equation that describes the temporal evolution of the imbibition length in a porous medium[34]. Thus, the imbibition rate shown in the inset of Fig. 1b indicates that the capability of absorbing solution decreases as a function of time. This fact significantly affects CICP phenomenon as a restoration phase and would be discussed later on as well.

The formation of the ion-depletion zone by CICP phenomenon. The conventional ICP phenomenon generated under an external d.c. bias results in significant concentration gradients at the anodic and cathodic sides of a permselective nanoporous membrane. Typical behaviour is that the ion concentration becomes extremely low at an anodic side (referred to as an ion-depletion zone) and enriches at a cathodic side (referred to as an ion-enrichment zone) in the case of a cation-selective membrane. Although the mechanism involves complex interactions of electric fields and electrokinetic flows[35–37], the eye-catching fact in terms of engineering applications is that ICP can play as an electrical filter. This virtual barrier rejects the entrance of charged species into the ion-depletion zone, so that it can be utilized as a water desalination/purification mechanism[2,38] and a biomolecular preconcentration mechanism[24,25]. Instead of an electrical bias, one can achieve the same ICP phenomena if another driving force pushes or pulls an electrolyte solution into the permselective membrane. Therefore, the capillarity is used in this work to initiate ICP phenomena. In this sense, the ion-depletion zone should also be formed in CICP system, which we confirm using a confocal microscope as shown in Fig. 1c. Since the driving force for a liquid flow due to the capillarity is much weaker than that due to an electrical bias in general, the depletion zone initially stays on the ionic hydrogel for a few hours (while the zone quickly expands in a conventional ICP system within a few seconds). A noticeable observation is that the fluorescent trackers are trapped at the top corners of a microchannel as shown in Fig. 1c. If there were no such ion-depletion zone, the trackers should have been accumulated at the bottom of the microchannel. In the meantime, the centre of the fluorescent cloud exhibits the brightest signal, since the flow converges into the hydrogel. Therefore, the confocal images with most of the fluorescent signals gathered at the corners of the microchannel provide strong evidences of CICP formation on the hydrogel pad.

CICP as the spontaneous desalting mechanism. Compared with the conventional ICP system, the formation of the ion-depletion zone in CICP system takes several hours since it relies only on the capillary force. The fluorescent microscopic images are captured as shown in Fig. 2a–c. Each snapshot is captured at different instances as indicated in the caption when the depletion zone expands in maximum. As shown in Fig. 2a, the fluorescent signal linearly decreases from the bulk solution as indicated at the line profile along A–A′. This is because the CICP is driven only by the capillarity, while the conventional ICP has a sudden drop and a flat concentration profile with fast electrokinetic vortices[18,19,36]. Before the measurement, the reference signals of various concentrations of fluorescent dye are obtained. See Supplementary Note 3. By comparing with the reference signal intensities, the dye molecules that infer the electrolyte concentration at the interface of hydrogel pad is expected to be removed with approximately 80%. Here, the ions are assumed to behave similar to fluorescent dyes. At least, the length of the depletion zone for ions and dye molecules are the same. This is confirmed by additional numerical simulations and experimental demonstrations. See Supplementary Note 4 for detailed

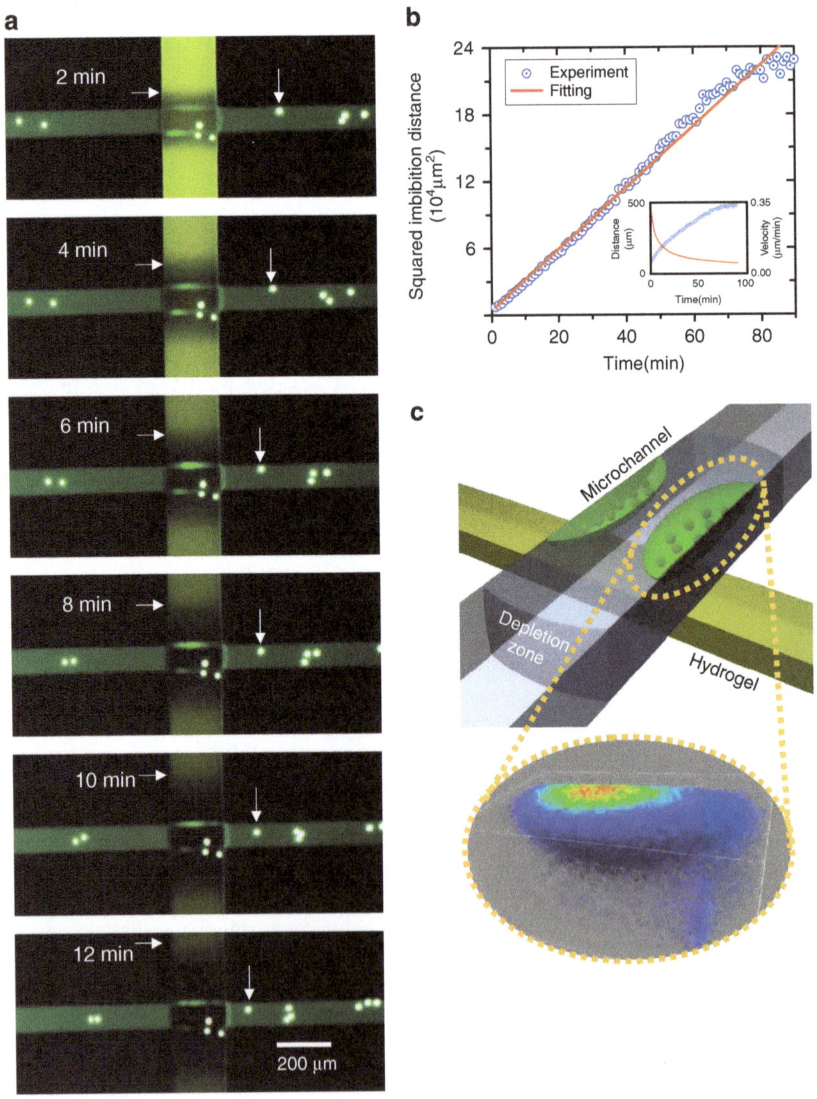

Figure 1 | The formation of CICP zone. (**a**) The snapshots of a fluid flow driven by nanoporous hydrogel (see Supplementary Video 1) and (**b**) the plot of measured imbibition lengths as a function of time. The imbibition velocity is also plotted in the inset. (**c**) The confocal microscopic image of accumulating fluorescent dyes at the top corner of microchannel, confirming the formation of an ion-depletion zone at the bottom of microchannel.

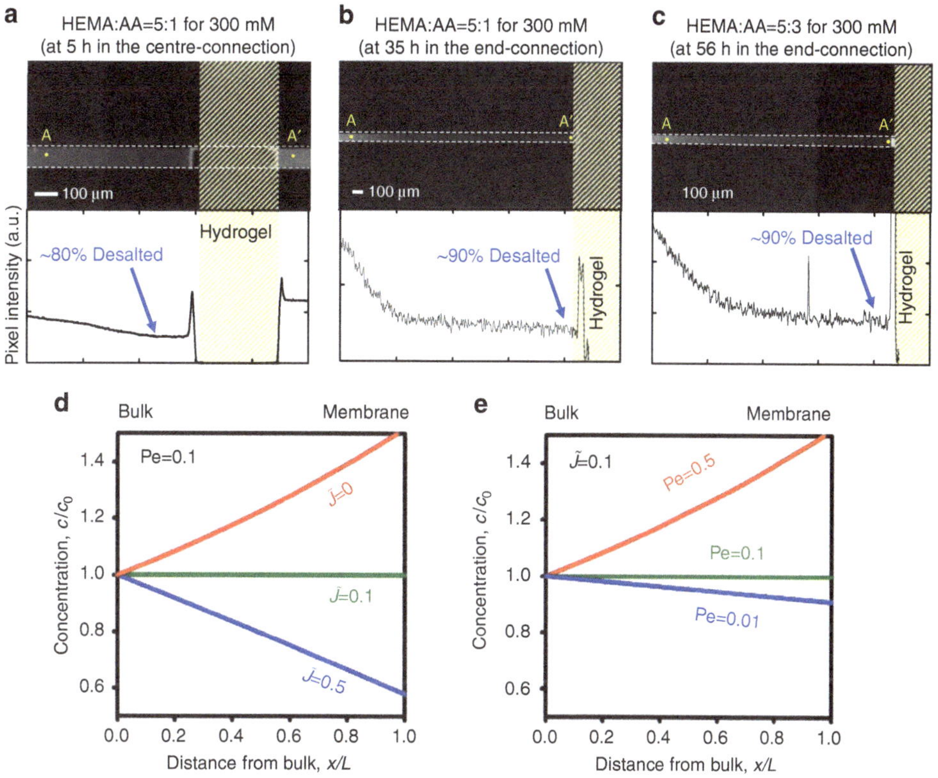

Figure 2 | Spontaneous desalting by CICP. A fluorescent signal tracking through the microchannel with different compositions of the ionic hydrogel, (**a**) HEMA: AA = 5:1 in the centre-connection device, (**b**) HEMA: AA = 5:1 and (**c**) HEMA: AA = 5:3 in the end-connection device. The plots of normalized concentration inside the ion-depletion zone with fixed (**d**) Pe at 0.1 and (**e**) ionic flux at 0.1.

explanation. Although Fig. 2a is conducted on the centre-connection device, the depletion zone asymmetrically expanded (that is, the depletion zone is rarely developed on the right side of the hydrogel). This is caused by an unavoidable residual flow from the level difference between the reservoirs. Detailed discussion would be given later on. To avoid the unwanted residual flow, the end-connection device is tested as shown in Fig. 2b,c. The dye concentration profiles have a similar linear drop and the removal efficiency is expected to be around 90%, which is similar to that of absorbed liquid by mangrove[28]. By comparing Fig. 2b with Fig. 2c, the duration of the depletion behaviour largely depends on the composition of the hydrogel. Increasing the volume fraction of charged group (acrylic acid (AA) in this work) provides the longer duration of CICP zone.

These behaviours, especially for a static case, are solved analytically. Under the assumption that the CICP system has an ideally cation-selective membrane with electroneutrality, a one-dimensional domain ignoring a surface effect and a quasi-equilibrium, the analytical solutions of the non-dimensional concentration ($\tilde{c}$) and electric potential ($\tilde{\psi}$) are given by

$$\tilde{c} = \frac{(\mathrm{Pe} - \tilde{J})\exp(\mathrm{Pe}\tilde{x}) + \tilde{J}}{\mathrm{Pe}} \tag{1}$$

and

$$\tilde{\psi} = -\mathrm{Pe}\,\tilde{x} + \ln\left[\frac{(\mathrm{Pe} - \tilde{J})\exp(\mathrm{Pe}\,\tilde{x}) + \tilde{J}}{\mathrm{Pe}}\right], \tag{2}$$

where $\tilde{x}$ is the dimensionless spatial coordinate whose origin coincided with the bulk reservoir and $\tilde{J}$ is the dimensionless ionic flux induced by the imbibition through the membrane.

Although CICP experiments are conducted without any electric power source, the electric interactions between ionic species should appear in terms of electro-migration. Pe is the Péclet number, which is the ratio of a convective transportation to a diffusive transportation (uL/D in this system). See the 'Methods' section for full governing equations, boundary conditions and detailed derivations. These simple analytical solutions imply the most important parameters for enhancing the desalination efficiency near the membrane. The experimental results show that the electrolyte concentration is depleted near the membrane (that is, the concentration gradient is negative, $dc/dx < 0$) so that $\tilde{J}$ is greater than Pe, since $d\tilde{c}/d\tilde{x}$ is equal to $(\mathrm{Pe} - \tilde{J})\exp[\mathrm{Pe}\,\tilde{x}]$. Figure 2d,e show the concentration profile plots at the different physicochemical parameters, while Pe and $\tilde{J}$ are fixed at 0.1 for each plot because the microfluidic device has Pe of ~0.1 with $D \sim O(10^{-9})\,\mathrm{m^2\,s^{-1}}$, L (the length of the microchannel) $\sim O(1)$ mm and u (a flow velocity due to an imbibition) $\sim O(10)\,\mu\mathrm{m\,min^{-1}}$. The physical intuitions from the plots are such that higher $\tilde{J}$ and lower Pe give a severer depletion. Since $\tilde{J}$ is proportional to the charge imbalance ($\tilde{J} = -\tilde{c}\frac{d\tilde{\psi}}{d\tilde{x}}$, See equations (13) and (16) in the 'Methods' section), the higher permselectivity would enhance the formation of the ion-depletion zone. This is also confirmed in Fig. 2b,c. By comparing the plots, the higher composition of the charged group (AA) leads to a stronger depletion behaviour which implies a longer duration. However, an excess amount of the charged group in the hydrogel results in a brittle structure so that one should tweak the composition as one's discretion. For the latter case, a diffusion limited system (that is, lower Pe) drives the depletion phenomenon more efficiently.

In Fig. 2a, the depletion zone propagates asymmetrically. An inevitable liquid level difference between each reservoir in the centre-connection device results in a residual flow over hydrogel pad. In the case of the residual flow from right to left (with the imbibition flow still converging to the centre from both reservoirs), the velocity of combined flow would increase at the right interface of the hydrogel and decrease at the left interface of the hydrogel. On the basis of the above theoretical analysis, low Pe would give a better depletion efficiency than high Pe. Since Pe is defined as a flow rate over a diffusion rate, there is a higher chance to have a stronger depletion zone at the left interface in Fig. 2a than the one at the right interface. See Supplementary Note 5 for a detailed description.

The restoration phase by a diminished imbibition. The aforementioned depletion zone is formed and sustained for a few hours, depending on the composition of the ionic hydrogel and the geometry of the microchannel. However, the ion-depletion zone collapses and the concentration restores to the bulk concentration as shown in Fig. 3. See Supplementary Videos 4–6. This has never been observed in a conventional ICP phenomenon, since it continuously applies an external power. Note that the snapshots are captured every 1, 7 and 8 h in Fig. 3a–c, respectively. One interesting feature other than the appearance of the restoration phase is that the rate of the restoration phase is much slower than the rate of the depletion phase.

The transient analysis of CICP provides a clue for understanding the restoration phase. In CICP, the imbibition rate gradually decreases as a function of time as discussed above using the Washburn's equation[34]. Without loss of generality, the imbibition rate is expressed as

$$u_{\mathrm{imb}} = \sqrt{\frac{S}{4t}} \qquad (3)$$

where u_{imb} is the imbibition velocity, S is the imbibition parameter and t is the time. As shown above, u_{imb} as the source of CICP is proportional to $t^{-0.5}$, implying a high initial imbibition rate, but it should decay with time. Detailed descriptions are given in the 'Methods' section.

Through the transient model, the dynamic changes of CICP layer are investigated. With chosen parameters in the 'Methods' section, the depletion phase as depicted in Fig. 3d lasts for $\tilde{t} \sim 0.4$ when the non-dimensional imbibition parameter $\tilde{S}$ is 100 (normalized by $D^2\tau_D/L^2$). However, beyond this point, the phase transits to the restoration phase as shown in Fig. 3e because of the decreased imbibition rate. See Supplementary Video 7. These simulation results largely correspond to the experimental results in Fig. 3a–c. Although the imbibition rate decreases gradually in both the depletion and the restoration phase, the contribution of the imbibition in the depletion phase is still larger than the dissipation of ionic species induced by a diffusion. Briefly, the depletion phase initially appears (that is, the imbibition rate > the dissipation) and changes into the restoration phase (that is, the imbibition rate < the dissipation).

In Fig. 3f, the minimum cation concentrations inside the depletion zone are plotted with varying imbibition parameters, $\tilde{S}$, as a function of time. The parameter could be tuned by the absorbing capability of the ionic hydrogel. The imbibition parameter is defined by $(\gamma\ d_{\mathrm{cap}})/(4D\eta)$ where γ is the surface tension, d_{cap} is the diameter of a capillary and η is the dynamic viscosity of the fluid. The increase of γ and d_{cap} or the decrease of η are the solutions to enhancing the imbibition rate. Although the Washburn's equation has been known to describe the imbibition through a single capillary tube, recent reports[39–41] suggested that the functional form of the imbibition in the case of a capillary network also follows the Washburn's equation. Nevertheless, the transient analysis with arbitrary $\tilde{S}$ provides the insight for understanding the transition from the depletion phase to the restoration phase. Larger $\tilde{S}$ generates a higher imbibition rate, which is enough to overcome the dissipation so that the depletion phase sustains longer. Moreover, the ionic hydrogel of a higher charges has more sustainable depletion phase by comparing Fig. 3b,c. The hydrogel used in Fig. 3c has the higher composition of the charged group, AA. In this theoretical model, the non-dimensional Donnan equilibrium concentration ($\tilde{N}$) is related to the intrinsic charge of the membrane. The membrane possessing a higher charge generates a higher ionic flux through the membrane, so that the depletion phase can last longer as shown in Fig. 3g with fixed $\tilde{S}$ at 100.

Discussion

In this study, we propose a new ICP phenomenon driven by capillarity. Instead of an external electric field, the inherent capillarity of nanoporous network drives a permselective ion transportation through the nanoporous hydrogel network, leading to the ion-depletion zone near the ionic hydrogel. Although the state-of-the-art desalination method utilizing a size exclusion, a thermal or electrical energy have remarkably advanced, the high level of energy requirements has hampered the development of a small-scale desalination/purification system. The fact that the CICP mechanism is working without any electrical power source offers significant merits in terms of the power consumption and the stable operation compared with the conventional electrokinetic ICP mechanism and other state-of-the-art large-scale desalination methods. Therefore, the present spontaneous desalting mechanism can lead to a power-free desalination system that is significantly useful in remote/rural and disaster-stricken areas. To achieve such a system, maintaining a quiescent fluid near the nanoporous membrane (that is, low Péclet number), enhancing the intrinsic surface charge of the membrane (that is, high ionic flux) and sustaining an imbibition at a certain level (that is, preventing restoration phase) are strongly required.

Methods

CICP device fabrication. The main building block of the CICP system is polydimethyl-siloxane (PDMS) material as schematically shown in Fig. 4. The final device consists of two layers of PDMS blocks. The bottom block is for a microchannel filled with the ionic hydrogel and the top block is for a microchannel to inject the saline water sample. Silicon masters are prepared using SU8 photoresist with the microchannel dimension of 400 μm (width) × 50 μm (depth) × 16 mm (length) for bottom PDMS block and 100 μm (width) × 15 μm (depth) × 10 mm (length) for top PDMS block. The PDMS blocks are fabricated as follows. The PDMS material (Sylgard 184 Silicone elastomer kit, Dow Corning, USA) mixed with a curing agent at 10:1 ratio is poured on the silicon masters that have desirable microchannel patterns. Then the polymer solution is cured in an oven for 4 h at 75 °C.

Detailed fabrication process from the first schematic is as follows. First, only the top PDMS for hydrogel mould is treated with oxygen plasma (CUTE-MP, Femto Science, Korea) for adhesion of the hydrogel and attached to a blank bottom PDMS piece. Then the microchannel is filled with a hydrogel precursor material consisting of 2-hydroxyethyl methacrylate (HEMA), acrylic acid (AA), a crosslinker (ethylene glycol dimethacrylate (EGDMA)) and a photoinitiator, (2,2-dimethoxy-2 phenyl-acetophenone (DMPA)) at the weight ratio of 32.51: 6: 0.41: 1.25 (named 5:1 hydrogel) or 32.51: 18: 0.41: 1.25 (named 5:3 hydrogel)[12]. The ratio of HEMA to AA would have a significant effect on CICP phenomenon as discussed. All the materials are purchased from Sigma-Aldrich. Ultraviolet irradiation for 8 min at the dose of 225 mW cm^{-2} (WUV-L50, Daihan Scientific, Korea) polymerizes the hydrogel inside the microchannel and then the top PDMS block is detached from the bottom blank PDMS piece. The hydrogel inside the microchannel is initially brittle and stiff (dry state) but the hydrogel absorbs a solution so that the portion of hydrogel becomes flexible and tough as the CICP proceeded. See Supplementary Video 8 for demonstrating only the water-absorbed portion is flexible and tough. Finally the flipped PDMS block for hydrogel and the PDMS for saline water samples are irreversibly bonded with oxygen plasma treatment. Assembled devices placed on glass slide are shown in Fig. 4b. There are two distinct designs, a 'centre-connection device' and an 'end-connection device'. The centre-connection device is designed for easiness of facilitating the experiment such as sample filling and flushing, while a residual flow is able to be completely eliminated in the end-connection device.

The microchannel inside top PDMS block is filled with KCl electrolyte solution at the concentration of 300 mM mixed with a fluorescent dye (Alexa 488, Sigma Aldrich, USA) as a tracker of the ion movements and negatively charged micro-

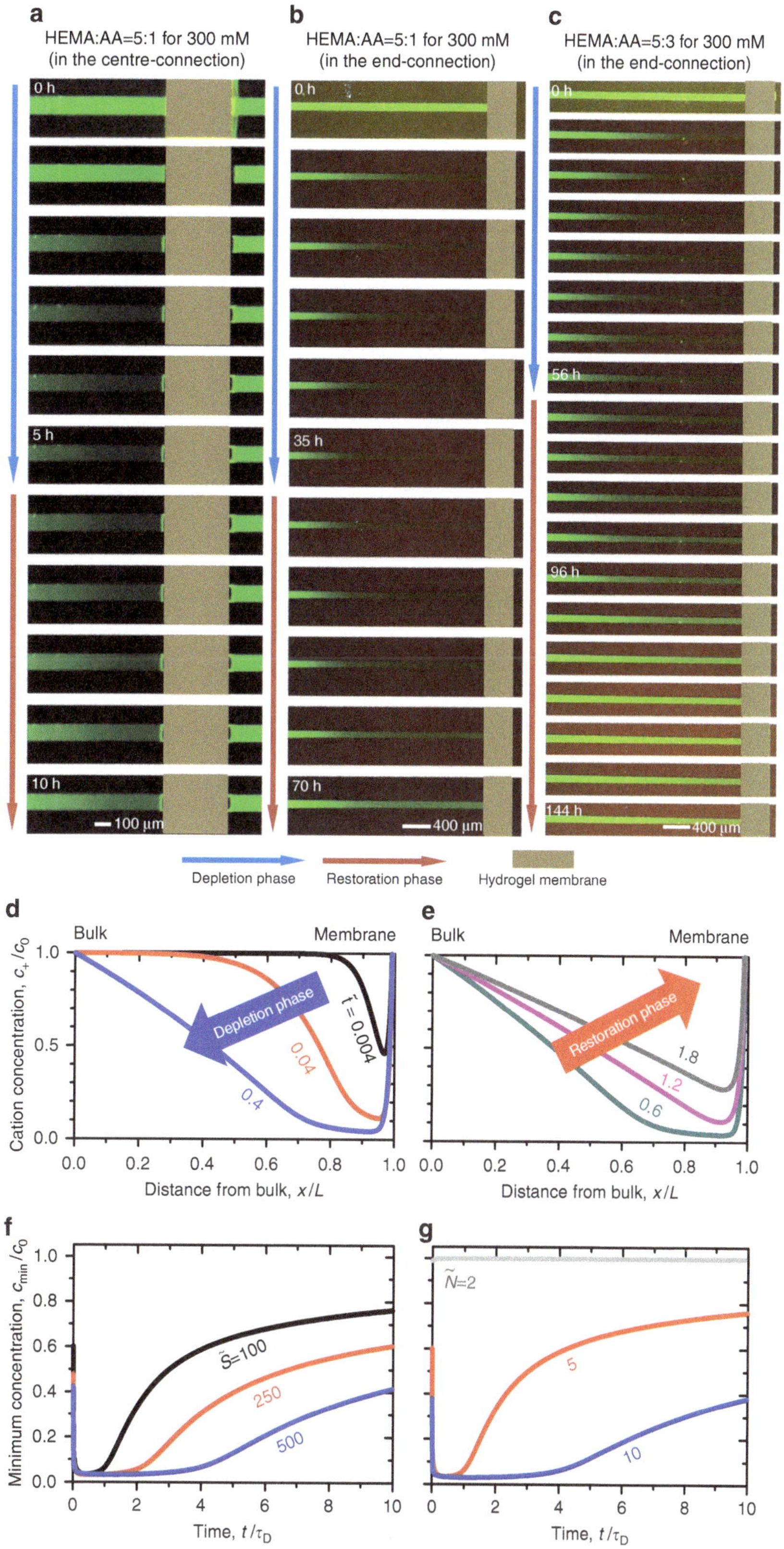

Figure 3 | The depletion and the restoration phase of CICP. The dynamics of CICP system showing the transition from the depletion phase to the restoration phase with the different composition of the ionic hydrogel. (**a**) HEMA: AA = 5:1 in the centre-connection device, (**b**) HEMA: AA = 5:1 and (**c**) HEMA: AA = 5:3 in the end-connection device. See Supplementary Videos 4-6 for each figure. Cation concentration profiles inside the CICP zone at (**d**) the depletion phase and (**e**) the restoration phase. See Supplementary Video 7. The minimum cation concentrations inside the ion-depletion zone with varying (**f**) imbibition parameters and (**g**) Donnan equilibrium concentrations.

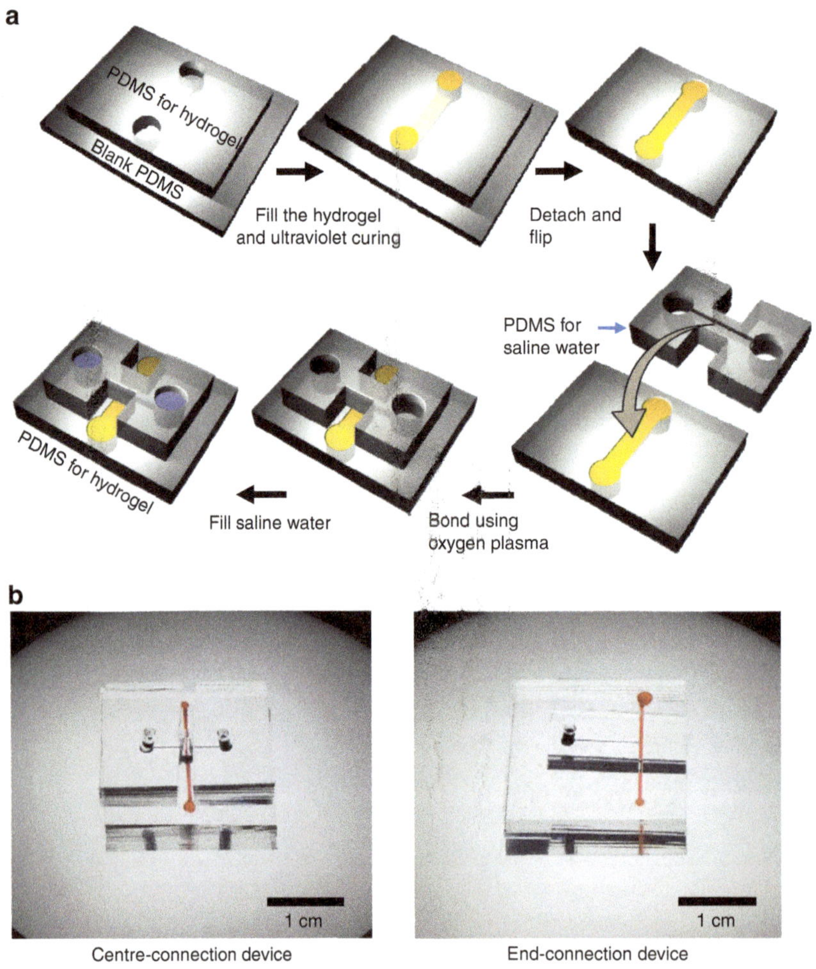

Figure 4 | A micro/nanofluidic CICP device. (**a**) The fabrication processes of CICP system. (**b**) The photo of assembled CICP devices of the centre-connection device and the end-connection device.

particles ($d = 5\,\mu m$, Invitrogen, USA) for flow field tracking. KCl is used for a simple theoretical interpretation by setting the same diffusivity of cation and anion and the concentration of 300 mM is the average salinity of brackish water.

Experimental setup. Compared with the conventional ICP process, any external voltage source is no longer required for the present CICP device. Instead, the total experimental time takes from a few hours to a few days depending on the device (for example, a conventional ICP occurs within a few seconds) so that the images of flow and concentration changes are captured every 2 min for a few days using a commercial computer software (CellSens, Olympus, Japan) and an inverted fluorescent microscope (IX 53, Olympus, Japan). The location where the hydrogel channel and the sample channel crossed is intensively analysed. A confocal microscope (FV1200, Olympus, Japan) is used for the three-dimensional concentration profiling.

Theoretical domain descriptions with governing equations. The numerical and analytical analyses are conducted for the following domain where CICP occurs. A microchannel with length L is filled with 1:1 electrolyte solution. One side of the microchannel opens to a well-mixed reservoir with bulk concentration, c_0 (at $x = 0$), while the other side contacts an ideal cation-selective surface (at $x = L$). Assuming that the ideal permselectivity allows us to avoid complexity in describing the transport phenomena inside the membrane[17,22,23,42,43], the domain from the reservoir to the membrane surface becomes the region of interest to be theoretically investigated. Since the streaming potential generated by the imbibition through the membrane is negligible, the effects of a surface conduction and an electro-osmotic flow on the CICP are excluded. Therefore, the domain is simplified as a one-dimensional configuration.

In general, the transport of ionic species along the x axis is governed by the Nernst–Planck equation,

$$\frac{\partial c_i}{\partial t} = -\frac{\partial J_i}{\partial x} \tag{4}$$

where t is the time, c_i is the concentration of i-th species and J_i is the ionic molar flux of the i-th species along the x axis. In accordance with transport mechanisms[44], the ionic flux, J_i is given by

$$J_i = -D_i\frac{\partial c_i}{\partial x} - \frac{z_iFD_i}{RT}c_i\frac{\partial \psi}{\partial x} + c_iu \tag{5}$$

where D_i is the diffusivity of i-th species, F is the Faraday constant, R is the gas constant, T is the absolute temperature, ψ is the electric potential and u is the flow velocity of x direction. In equation (5), each term on the right-hand side corresponds to the transport mechanism due to a diffusion, an electro-migration and a convection, respectively. Although CICP experiments are conducted without any electric power source, the term of the electro-migration should be considered to describe the electric interactions between ionic species. Thus, the interactions are explained by the Poisson equation,

$$-\varepsilon_f\frac{\partial^2 \psi}{\partial x^2} = F(c_+ - c_-) \tag{6}$$

where ε_f is the electrical permittivity of the electrolyte solution, and c_+ and c_- are the concentrations of cation and anion, respectively. The convective component of the ionic flux is usually obtained by solving the Stokes equations and the continuity equation with coupled partial differential equations (4)–(6). However, in this work, we adopt the simplified approach suggested by Dhopeshwarkar *et al.*[45] to avoid mathematical complication. In their work, the driving force to generate ICP phenomena was the difference of applied electric potentials through the microchannel–nanochannel–microchannel system. However, our system is a microchannel–nanochannel connected system so that the theoretical domain is different and the imbibition through the membrane drives the ICP phenomena. Thus, the domain is reduced to a microchannel part rather than a whole microchannel–nanochannel connected system, while the domain of Dhopeshwarkar *et al.* analysed a whole microchannel–nanochannel–microchannel system. Since the effective convective transport is induced by the imbibition into the membrane, the flow velocity in the microchannel (u) is related to the imbibition rate (u_{imb}) as

$$u = \phi_{\mathrm{p}} u_{\mathrm{imb}} \tag{7}$$

where ϕ_{p} is the porosity of the membrane, which is assumed to consist of a bundle of cylindrical capillaries. Equation (7) is derived from the fact that the flow rate in the whole system should be uniform. This imbibition is the driving force for CICP, while the driving force of conventional ICP platforms is usually the difference of applied electric potentials.

Detailed boundary conditions at the reservoir bulk ($x=0$) and the cation-selective surface ($x=L$) to describe the CICP would be discussed in the following sections.

The derivation of analytical solutions at the steady state. Since each ionic flux of the i-th species is constant over the domain in the steady state, the Nernst–Planck equation is reduced to simplified balance equations as

$$J_{+} = -D_{+}\frac{\mathrm{d}c_{+}}{\mathrm{d}x} - \frac{FD_{+}}{RT}c_{+}\frac{\mathrm{d}\psi}{\mathrm{d}x} + c_{+}u \tag{8}$$

and

$$J_{-} = -D_{-}\frac{\mathrm{d}c_{-}}{\mathrm{d}x} + \frac{FD_{-}}{RT}c_{-}\frac{\mathrm{d}\psi}{\mathrm{d}x} + c_{-}u \tag{9}$$

for 1:1 electrolyte solution. For simplicity, thin double layers and electroneutrality ($c_{+}=c_{-}=c$) are assumed in the microchannel as the domain. In addition, the diffusivity of each ionic species is fixed to the same value, ($D_{+}=D_{-}=D$). For the ideal cation-selective surface, no anion flux exists over the microchannel. As a result, the above equations are rewritten as

$$J_{+} = -D\frac{\mathrm{d}c}{\mathrm{d}x} - \frac{FD}{RT}c\frac{\mathrm{d}\psi}{\mathrm{d}x} + cu \tag{10}$$

and

$$0 = -D\frac{\mathrm{d}c}{\mathrm{d}x} + \frac{FD}{RT}c\frac{\mathrm{d}\psi}{\mathrm{d}x} + cu. \tag{11}$$

Combining equations (10) and (11) yields

$$J_{+} = -2D\frac{\mathrm{d}c}{\mathrm{d}x} + 2cu, \tag{12}$$

while subtracting equation (11) from equation (10) yields

$$J_{+} = -2\frac{DFc}{RT}\frac{\mathrm{d}\psi}{\mathrm{d}x}. \tag{13}$$

The equations (12) and (13) are further simplified by non-dimensionalization with

$$\tilde{x} \equiv \frac{x}{L},\ \tilde{c} \equiv \frac{c}{c_0},\ \tilde{\psi} \equiv \frac{F\psi}{RT},\ \tilde{J} \equiv \frac{LJ_{+}}{2Dc_0},\ \text{and Pe} \equiv \frac{uL}{D} \tag{14}$$

where Pe is known as the Péclet number, the ratio of a convective transport rate and a diffusive transport rate. Consequently, the non-dimensional forms of equations (12) and (13) are,

$$\tilde{J} = -\frac{\mathrm{d}\tilde{c}}{\mathrm{d}\tilde{x}} + \mathrm{Pe}\,\tilde{c} \tag{15}$$

and

$$\tilde{J} = -\tilde{c}\frac{\mathrm{d}\tilde{\psi}}{\mathrm{d}\tilde{x}}, \tag{16}$$

respectively. Because $\tilde{c}=1$ and $\tilde{\psi}=0$ at $\tilde{x}=0$, equations (15) and (16) can be integrated under constant $\tilde{J}$, which is induced by the imbibition through the membrane. Therefore, the following analytical solutions are obtained with respect to the concentration and the electric potential inside CICP layer as

$$\tilde{c} = \frac{(\mathrm{Pe}-\tilde{J})\exp(\mathrm{Pe}\,\tilde{x}) + \tilde{J}}{\mathrm{Pe}} \tag{17}$$

and

$$\tilde{\psi} = -\mathrm{Pe}\,\tilde{x} + \ln\left[\frac{(\mathrm{Pe}-\tilde{J})\exp(\mathrm{Pe}\,\tilde{x}) + \tilde{J}}{\mathrm{Pe}}\right]. \tag{18}$$

The formulation of transient model. The imbibition into a capillary tube is described by Washburn's equation[34] that is

$$L_{\mathrm{imb}}^{2} = \frac{\gamma\, d_{\mathrm{cap}}}{4\eta}t \tag{19}$$

where L_{imb} is imbibition length, γ is the surface tension of water/capillary wall/air interface and d_{cap} is the diameter of capillary. By differentiating equation (19), the imbibition rate is obtained as

$$u_{\mathrm{imb}} = \sqrt{\frac{S}{4t}} \tag{20}$$

where S is defined by $\gamma\, d_{\mathrm{cap}}/4\eta$. Because u_{imb} as the source of CICP is proportional to $t^{-0.5}$, the thickness of CICP layer induced by imbibition into the membrane would be decreased as time passed.

The spatiotemporal concentration profile inside the CICP layer is obtained by the following non-dimensional forms of equations (5) and (6) with appropriate boundary conditions,

$$\frac{\partial}{\partial\tilde{x}}\left(-\frac{\partial\tilde{\psi}}{\partial\tilde{x}}\right) = \frac{\tilde{c}_{+}-\tilde{c}_{-}}{2\tilde{\lambda}_{\mathrm{D}}} \tag{21}$$

and

$$\frac{\partial\tilde{c}_i}{\partial\tilde{t}} = -\frac{\partial}{\partial\tilde{x}}\left(-\frac{\partial\tilde{c}_i}{\partial\tilde{x}} - z_i\tilde{c}_i\frac{\partial\tilde{\psi}}{\partial\tilde{x}} + \phi_{\mathrm{p}}\tilde{c}_i\tilde{u}_{\mathrm{imb}}\right) \tag{22}$$

where the characteristic scales of c_i, ψ and x are denoted in equation (14), the characteristic time is chosen as the diffusion time scale, τ_{D} defined as L^2/D, and the characteristic velocity is set to be D/L. In equation (21), $\tilde{\lambda}_{\mathrm{D}}$ is the non-dimensional Debye length defined as

$$\tilde{\lambda}_{\mathrm{D}} = \frac{1}{L}\sqrt{\frac{\varepsilon_f RT}{2z^2F^2c_0}} \tag{23}$$

for z:z electrolyte solution. At the reservoir bulk ($\tilde{x}=0$), the following boundary conditions should be satisfied,

$$\tilde{\psi} = 0 \text{ and } \tilde{c}_i = 1. \tag{24}$$

At the ideally cation-selective surface ($\tilde{x}=1$), the cation flux through the ionic hydrogel pad is generated only by the imbibition,

$$-\frac{\partial\tilde{\psi}}{\partial\tilde{x}} = \frac{\partial\ln\tilde{c}_{+}}{\partial\tilde{x}}, \tag{25}$$

and the permselectivity is described by

$$\tilde{c}_{+} = \tilde{N} \tag{26}$$

and

$$-\frac{\partial\tilde{c}_{-}}{\partial\tilde{x}} + c_{-}\frac{\partial\tilde{\psi}}{\partial\tilde{x}} + \phi_{\mathrm{p}}\tilde{c}_i\tilde{u}_{\mathrm{imb}} = 0. \tag{27}$$

Equation (25) implies that the gradient of electrochemical potential ($\nabla(\tilde{\psi}+\ln\tilde{c}_{+})=0$) would be zero, so that only a convective transportation into the membrane contributes to form the CICP layer. The Donnan equilibrium concentration[17,22] inside the nanoporous membrane is described by equations (26) and (27), which impedes that anion flux into the membrane should be zero due to the ideal cation selectivity. Equations (21) and (22) with boundary conditions of equations (24)–(27) are solved numerically.

Numerical method. Numerical simulations are conducted by COMSOL Mutiphysics 4.4, a commercial finite element method tool, under the one-dimensional domain and the transient analysis mode. The domain is discretized into 1,000 elements, with finer meshes in the proximity to the membrane surface. To carry out a qualitative analysis for CICP system, we assume that the diffusion time scale (τ_{D}) and the porosity (ϕ_{p}) are 500 s and 0.1, respectively. In addition, the non-dimensional Debye length and Donnan equilibrium concentration are set to be 0.01 and 5. From Washburn's equation[34], the imbibition rate is infinite at $t=0$ s so that the initial conditions cannot be defined. To avoid this problem, a transient analysis is conducted from $t=1$ s rather than from $t=0$ s. The time step and the termination time are set to be 1 and 3,600 s, respectively.

References

1. van den Berg, A., Craighead, H. G. & Yang, P. From microfluidic applications to nanofluidic phenomena. *Chem. Soc. Rev.* **39,** 899–900 (2010).
2. Kim, S. J., Ko, S. H., Kang, K. H. & Han, J. Direct seawater desalination by ion concentration polarization. *Nat. Nanotechnol.* **5,** 297–301 (2010).
3. Kwak, R., Guan, G. F., Peng, W. K. & Han, J. Y. Microscale electrodialysis: concentration profiling and vortex visualization. *Desalination* **308,** 138–146 (2013).
4. Nikonenko, V. V. *et al.* Desalination at overlimiting currents: state-of-the-art and perspectives. *Desalination* **342,** 85–106 (2014).
5. Shannon, M. A. *et al.* Science and technology for water purification in the coming decades. *Nature* **452,** 301–310 (2008).
6. Kwak, R., Pham, V. S., Lim, K. M. & Han, J. Y. Shear flow of an electrically charged fluid by ion concentration polarization: scaling laws for electroconvective vortices. *Phys. Rev. Lett.* **110,** 114501 (2013).
7. Deng, D. *et al.* Overlimiting current and shock electrodialysis in porous media. *Langmuir* **29,** 16167–16177 (2013).
8. Deng, D. *et al.* Water purification by shock electrodialysis: deionization, filtration, separation, and disinfection. *Desalination* **357,** 77–83 (2015).

9. Pu, Q., Yun, J., Temkin, H. & Liu, S. Ion-enrichment and ion-depletion effect of nanochannel structures. *Nano. Lett.* **4**, 1099–1103 (2004).
10. Kim, S. J. & Han, J. Self-sealed vertical polymeric nanoporous junctions for high-throughput nanofluidic applications. *Anal. Chem.* **80**, 3507–3511 (2008).
11. Heo, J. *et al.* Ultra-high-aspect-orthogonal and tunable three dimensional polymeric nanochannel stack array for BioMEMS applications. *Nanoscale* **6**, 9681–9688 (2014).
12. Kim, P., Kim, S. J., Suh, K.-Y. & Han, J. Stabilization of ion concentration polarization using a heterogeneous nanoporous junction. *Nano Lett.* **10**, 16–23 (2010).
13. Fosdick, S. E., Knust, K. N., Scida, K. & Crooks, R. M. Bipolar electrochemistry. *Angew. Chem. Int. Ed.* **52**, 10438–10456 (2013).
14. Knust, K. N., Hlushkou, D., Anand, R. K., Tallarek, U. & Crooks, R. M. Electrochemically mediated seawater desalination. *Angew. Chem. Int. Ed.* **52**, 8107–8110 (2013).
15. Leinweber, F. C. & Tallarek, U. Nonequilibrium electrokinetic effects in beds of ion-permselective particles. *Langmuir.* **20**, 11637–11648 (2004).
16. Holtzel, A. & Tallarek, U. Ionic conductance of nanopores in microscale analysis systems: where microfluidics meets nanofluidics. *J. Sep. Sci.* **30**, 1398–1419 (2007).
17. Dydek, E. V. *et al.* Overlimiting current in a microchannel. *Phys. Rev. Lett.* **107**, 118301 (2011).
18. Nam, S. *et al.* Experimental verification of overlimiting current by surface conduction and electro-osmotic flow in microchannels. *Phys. Rev. Lett.* **114**, 114501 (2015).
19. Kim, S. J., Wang, Y.-C., Lee, J. H., Jang, H. & Han, J. Concentration polarization and nonlinear electrokinetic flow near nanofluidic channel. *Phys. Rev. Lett.* **99**, 044501 (2007).
20. Rubinstein, S. M. *et al.* Direct observation of a nonequilibrium electro-osmotic instability. *Phys. Rev. Lett.* **101**, 236101 (2008).
21. Pundik, T., Rubinstein, I. & Zaltzman, B. Bulk electroconvection in electrolyte. *Phys. Rev. E* **72** **(2005).**
22. Green, Y. & Yossifon, G. Dynamical trapping of colloids at the stagnation points of electro-osmotic vortices of the second kind. *Phys. Rev. E* **87** (2013).
23. Rubinstein, I. & Zaltzman, B. Electro-osmotically induced convection at a permselective membrane. *Phys. Rev. E* **62**, 2238–2251 (2000).
24. Kim, S. J., Song, Y.-A. & Han, J. Nanofluidic concentration devices for biomolecules utilizing ion concentration polarization: theory, fabrication, and application. *Chem. Soc. Rev.* **39**, 912–922 (2010).
25. Ko, S. H. *et al.* Massively-parallel concentration device for multiplexed immunoassays. *Lab Chip* **11**, 1351–1358 (2011).
26. Cho, I., Sung, G. & Kim, S. J. Overlimiting current through ion concentration polarization layer: hydrodynamic convection effects. *Nanoscale* **6**, 4620–4626 (2014).
27. Kim, S. J., Ko, S. H., Kang, K. H. & Han, J. Direct seawater desalination by ion concentration polarization. *Nat. Nanotechnol.* **8**, 609–609 (2013).
28. Parida, A. K. & Jha, B. Salt tolerance mechanisms in mangroves: a review. *Trees* **24**, 199–217 (2010).
29. Beebe, D. J. *et al.* Functional hydrogel structures for autonomous flow control inside microfluidic channels. *Nature* **404**, 588–590 (2000).
30. Lee, K. Y. & Mooney, D. J. Hydrogels for tissue engineering. *Chem. Rev.* **101**, 1869–1880 (2001).
31. Shengqiang, C. & Zhigang, S. Equations of state for ideal elastomeric gels. *Europhys. Lett.* **97**, 34009 (2012).
32. Marcombe, R. *et al.* A theory of constrained swelling of a pH-sensitive hydrogel. *Soft Matter* **6**, 784–793 (2010).
33. Suzuki, A. & Hara, T. Kinetics of one-dimensional swelling and shrinking of polymer gels under mechanical constraint. *J. Chem. Phys.* **114**, 5012–5015 (2001).
34. Dullien, F. A. L. & Brenner, H. *Porous Media: Fluid Transport and Pore Structure* (Elsevier Science, 2012).
35. Kim, S. J., Li, L. & Han, J. Amplified electrokinetic response by concentration polarization near nanofluidic channel. *Langmuir* **25**, 7759–7765 (2009).
36. Kim, S. J. *et al.* Multi-vortical flow inducing electrokinetic instability in ion concentration polarization layer. *Nanoscale* **4**, 7406–7410 (2012).
37. Lee, S.-H. *et al.* Sub-10 nm transparent all-around-gated ambipolar ionic field effect transistor. *Nanoscale* **7**, 936–946 (2015).
38. Kwak, R., Kim, S. J. & Han, J. Continuous-flow biomolecule and cell concentrator by ion concentration polarization. *Anal. Chem.* **83**, 7348–7355 (2011).
39. Elizalde, E., Urteaga, R. & Koropecki, R. R. Berli CLA. Inverse Problem of Capillary Filling. *Phys. Rev. Lett.* **112**, 134502 (2014).
40. Huang, Y.-R. *et al.* Polymer nanocomposite films with extremely high nanoparticle loadings via capillary rise infiltration (CaRI). *Nanoscale* **7**, 798–805 (2015).
41. Gruener, S., Hofmann, T., Wallacher, D., Kityk, A. V. & Huber, P. Capillary rise of water in hydrophilic nanopores. *Phys. Rev. E* **79**, 067301 (2009).
42. Rubinstein, I. & Shtilman, L. Voltage against current curves of cation exchange membranes. *J. Chem. Soc. Faraday Trans.* **75**, 231–246 (1979).
43. Druzgalski, C. L., Andersen, M. B. & Mani, A. Direct numerical simulation of electroconvective instability and hydrodynamic chaos near an ion-selective surface. *Phys. Fluids* **25**, 110804 (2013).
44. Masliyah, J. H. & Bhattacharjee, S. *Electrokinetic and Colloid Transport Phenomena* (Wiley, 2006).
45. Dhopeshwarkar, R., Crooks, R. M., Hlushkou, D. & Tallarek, U. Transient effects on microchannel electrokinetic filtering with an ion-permselective membrane. *Anal. Chem.* **80**, 1039–1048 (2008).

Acknowledgements

This work is mainly supported by Samsung Research Funding Center of Samsung Electronics under Project Number SRFC-MA1301-02. We also acknowledge the partial financial support to S.P. by the Center for Integrated Smart Sensor (Grant No. CISS-2011-0031870), H.L. by NRF (Grant Nos. 2012-0009563 and 2014-048162), I.C. by the Basic Science Research Program (Grant No. 2013R1A1A1008125) and Y.C., Y.J. by Korean Health Technology RND Project, Grants No. HI13C1468 and No. HI14C0559, respectively. All the authors gratefully acknowledge the financial support from BK21 plus program of Creative Research Engineer Development IT, Seoul National University.

Author contributions

S.P. performed most of the desalting experiments; Y.J. analysed the steady-state solution; S.Y.S. helped in designing the experiments; I.C. imaged the ion-depletion zone using the confocal microscope; Y.C. performed the fluorescent tracking; H.L. designed and analysed the transient model; H.L., H.-Y.K. and S.J.K. supervised the project, designed the experiments and wrote the paper.

Additional information

Competing financial interests: The authors declare no competing financial interests.

3

Light-enhanced liquid-phase exfoliation and current photoswitching in graphene–azobenzene composites

Markus Döbbelin[1], Artur Ciesielski[1], Sébastien Haar[1], Silvio Osella[2], Matteo Bruna[3], Andrea Minoia[2], Luca Grisanti[2], Thomas Mosciatti[1], Fanny Richard[1], Eko Adi Prasetyanto[4], Luisa De Cola[4], Vincenzo Palermo[5], Raffaello Mazzaro[6,7], Vittorio Morandi[6], Roberto Lazzaroni[2], Andrea C. Ferrari[3], David Beljonne[2] & Paolo Samorì[1]

Multifunctional materials can be engineered by combining multiple chemical components, each conferring a well-defined function to the ensemble. Graphene is at the centre of an ever-growing research effort due to its combination of unique properties. Here we show that the large conformational change associated with the *trans–cis* photochemical isomerization of alkyl-substituted azobenzenes can be used to improve the efficiency of liquid-phase exfoliation of graphite, with the photochromic molecules acting as dispersion-stabilizing agents. We also demonstrate reversible photo-modulated current in two-terminal devices based on graphene-azobenzene composites. We assign this tuneable electrical characteristics to the intercalation of the azobenzene between adjacent graphene layers and the resulting increase in the interlayer distance on (photo)switching from the linear *trans*-form to the bulky *cis*-form of the photochromes. These findings pave the way to the development of new optically controlled memories for light-assisted programming and high-sensitive photosensors.

[1] Nanochemistry Laboratory, ISIS & icFRC, Université de Strasbourg & CNRS, 8 allée Gaspard Monge, 67000 Strasbourg, France. [2] Laboratory for Chemistry of Novel Materials, University of Mons, Place du Parc 20, B-7000 Mons, Belgium. [3] Cambridge Graphene Centre, University of Cambridge, 9 JJ Thomson Avenue, Cambridge CB3 OFA, UK. [4] Laboratory of Supramolecular Biomaterials and Chemistry, ISIS & icFRC, Université de Strasbourg & CNRS, 8 allée Gaspard Monge, 67000 Strasbourg, France. [5] ISOF-CNR, via Gobetti 101, 40129 Bologna, Italy. [6] IMM-CNR Sezione di Bologna, via Gobetti, 101, 40129 Bologna, Italy. [7] Dipartimento di Chimica 'G. Ciamician', Università di Bologna, via Selmi 2, 40126 Bologna, Italy. Correspondence and requests for materials should be addressed to A.C. (email:ciesielski@unistra.fr) or to P.S. (email: samori@unistra.fr).

Graphene is a one-atom-thick two-dimensional material with unique mechanical, optical, thermal and electrical properties[1], promising future emerging technologies, including flexible and wearable electronics[2]. Two main approaches are being followed for graphene production: bottom-up and top-down[3]. The former relies on the assembly of suitably designed molecular building blocks, undergoing chemical reaction to form covalently linked networks[4]. The latter occurs via exfoliation of graphite into graphene[3,5]. Bottom-up techniques, in particular those based on organic syntheses starting from small molecular modules[6,7], when performed in liquid media, are both size limited, because macromolecules become less soluble with increasing size[6], and suffer from the occurrence of side reactions with increasing molecular weight[6]. The growth on solid (ideally catalytically active) surfaces allows one to circumvent these issues. Substrate-based growth can also be achieved by chemical vapour deposition (CVD)[8,9] or via silicon evaporation from silicon carbide[10], which rely on the ability to follow a narrow thermodynamic path. Top-down approaches can be accomplished under different environmental conditions. Among them, liquid-phase exfoliation (LPE)[2,3,5,11–15] is an attractive strategy, being extremely versatile and applicable to a variety of environments and on different substrates. While bottom-up methods, in particular CVD, can yield large size, LPE gives limited sheet sizes[11,16–18]. Nevertheless, LPE has several advantages. It is a viable low-cost process, which can be easily upscaled to mass-produced dispersions processable by well-established techniques such as spin-coating, drop-casting, screen-printing and ink-jet printing[17]. Increasing research efforts are being devoted to the production of graphene via LPE to improve the material's physicochemical and electrical properties.

Small organic molecules, such as surfactants or dispersion-stabilizing agents, can promote the exfoliation of graphite into graphene in organic solvents[5,11], in particular when such molecules have higher adsorption energies on the basal plane of graphene than those involved in solvent–graphene interaction[11]. We recently found that long alkanes are particularly suitable for enhancing the yield of exfoliation[19,20].

Photochromic molecules, in particular azobenzene-based molecules, covalently linked to reduced graphene oxide[21], physisorbed directly on graphene[22,23] or through pyrene anchoring groups[24] on the graphene surface, or even when graphene is adsorbed on an azobenzene self-assembled monolayer[25], can be used to reversibly modulate the graphene's electronic properties. Moreover, it was shown that covalent functionalization of gold surface[26] or carbon nanotubes[27] with azobenzene molecules can be used as a route for fabrication of light-driven electronic switches. Nonetheless, the use of additional responsive functions provided by a suitably designed molecule to assist the LPE of graphite into functional inks is still unexplored. This approach could lead to the production of functional hybrid materials and nanocomposites in a one-pot process.

Here we combine photochromic systems and LPE graphene exploiting the properties of these materials. We focus on a commercial alkoxy-substituted azobenzene, i.e., 4-(decyloxy) azobenzene. The presence of a long alkoxy side chain is expected to enhance the molecular interaction with graphene[28], and promote the exfoliation of graphite towards graphene. Here we demonstrate that the use of 4-(decyloxy)azobenzene has two major advantages: first, the amount of graphene dispersed in *N*-methyl-2-pyrrolidone (NMP) can be enhanced by exploiting the photoisomerization of 4-(decyloxy)azobenzene molecules during exfoliation. Second, the isomerization of 4-(decyloxy)azobenzene from *trans* to *cis* and vice versa, when physisorbed between adjacent graphene layers leads to a reversible modulation of the inter-flake distance on the sub-Ångström scale. This is reflected in a light response of the electrical properties of the hybrid material.

Results

Liquid-phase exfoliation. To test the ability of alkoxy-substituted azobenzenes to increase Y_W (%), which is the yield of graphene exfoliation, defined as the ratio between the weight of dispersed graphitic material and that of the starting graphite flakes, as well as to exploit the photochromic nature of such molecules when interacting with graphene, vials containing NMP, graphite powder and azobenzene are exposed to tip sonication for 3 h (see Methods and Supplementary Note 1 for details). This experiment is carried out either in dark or under ultraviolet irradiation, using a portable ultraviolet lamp, at two temperatures: 20 and 40 °C (see experimental set-up in Supplementary Fig. 1). Subsequently, the dispersions are allowed to settle for 15 min, then decanted and centrifuged for 1 h at 10,000 r.p.m. Control samples are also prepared, consisting of dispersions prepared in the absence of 4-(decyloxy)azobenzene both in dark and irradiated at 365 nm at either 20 or 40 °C. To quantify the concentration of graphene after centrifugation, a mixture of graphene dispersion and 2-propanol (IPA) is first heated to 50 °C for 30 min and then passed through polytetrafluoroethylene membrane filters. The remaining solvent and weakly interacting 4-(decyloxy) azobenzene molecules are washed out several times with diethyl ether and IPA. Measurements of the filtered mass are performed on a microbalance to infer the concentration of graphitic material in dispersion, needed to quantify Y_W (%) (see Fig. 1b). The presence of adsorbed molecules may affect the mass measurements and Y_W (%). Thus, the heating step is necessary to completely desorb the 4-(decyloxy)azobenzene molecules, as proven with X-ray photoelectron spectroscopy (Supplementary Fig. 2 and Supplementary Note 2). By analysing 20 independent experiments, we find that when the exfoliation is performed in the absence of 4-(decyloxy)azobenzene (control experiments), comparable Y_W are observed when dispersions are kept at 20 °C (0.64%) or heated at 40 °C (0.62%). The irradiation at 365 nm does not affect Y_W, neither at 20 nor at 40 °C (0.63% in both cases).

LPE in the presence of alkoxy-substituted azobenzene acting as dispersion-stabilizing agent can undergo different mechanisms depending on the isomeric form of the photochromic molecule (Fig. 1c). In our case, the presence of 4-(decyloxy)azobenzene has no major influence on Y_W of dispersions prepared in dark (0.71% and 0.72% at 20 and 40 °C, respectively). As previously reported[11], the use of small organic molecules, in particular those based on aromatic cores[11] and alkyl functionalization[19,20], can promote graphite exfoliation because of their high affinity for the basal plane of graphene hindering interflake stacking. The relation between interaction energy and exfoliation yield is not straightforward. In some cases, molecules with lower adsorption energy on graphene give higher exfoliation yield, as demonstrated for graphite exfoliation[29], as well as for exfoliation of a wide range of other layered materials[30]. This is because a critical step of the process is the adsorption of the molecule on the graphene surface and the displacement of the adsorbed solvent layer, overcoming an activation energy barrier that depends significantly on molecular structure and conformation. Therefore, even a minor increase of Y_W (0.06–0.09%) on addition of a small amount of 4-(decyloxy)azobenzene (5 wt%) highlights the importance of these molecules during LPE.

In contrast to the experiments performed in dark, irradiation with ultraviolet light at 365 nm increases the concentration of exfoliated material. At 20 °C we obtain $\sim 81 \pm 5\,\mu g\,ml^{-1}$,

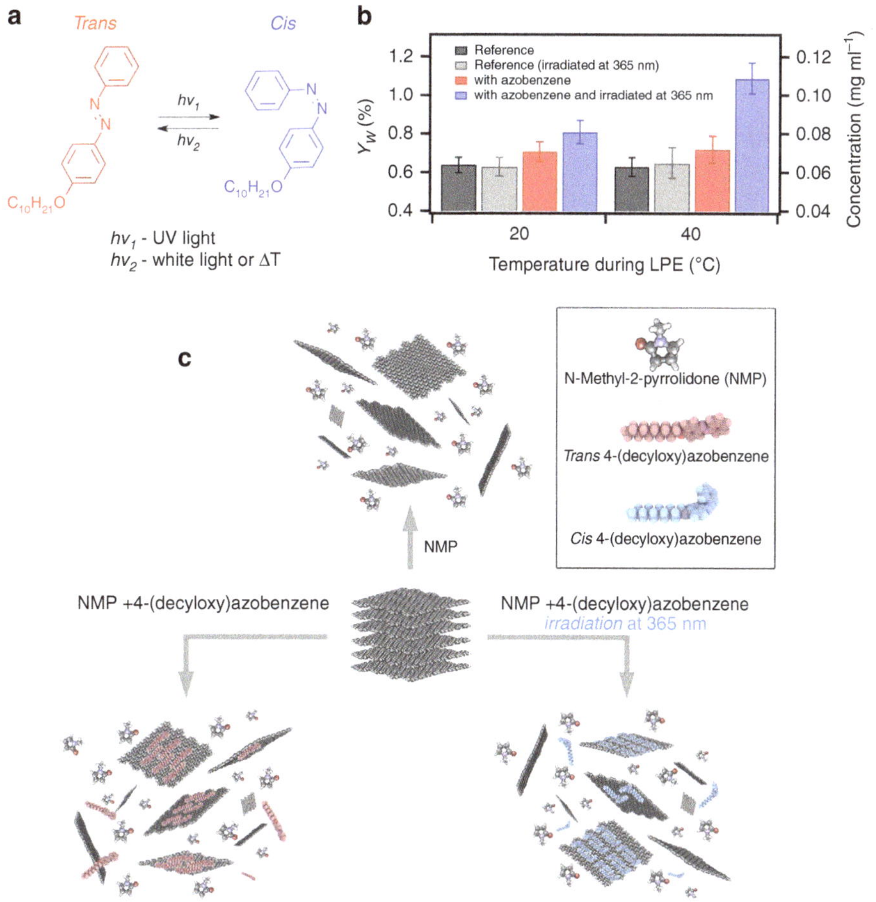

Figure 1 | Light-assisted graphene's LPE process in the presence of 4-(decyloxy)azobenzene and yield of exfoliation. (**a**) Chemical structure of 4-(decyloxy)azobenzene and reversible *trans-cis* photoisomerization under UV and visible light. (**b**) Y_W and concentration after filtration. The error bars reflect a statistical analysis on 20 independent experiments. (**c**) Schematic representation of the LPE process under different experimental conditions: in NMP (reference experiment); and in NMP with azobenzene in dark and under UV irradiation.

corresponding to ~30% increase in Y_W, when compared with reference samples (63 μg ml^{-1}). Yet, the largest increase in concentration is obtained by irradiating the dispersions heated at 40 °C, reaching 110 ± 8 μg ml^{-1}, which corresponds to a 75% increase in Y_W. These results demonstrate that the addition of 4-(decyloxy)azobenzene during LPE leads to a significant increase in Y_W only in samples containing 4-(decyloxy)azobenzene irradiated at 365 nm. Molecules rapidly switching conformation, such as azobenzenes on ultraviolet (UV) illumination, heating and sonication, can overcome the activation barriers to adsorption, acting as 'nano-spinners' that disrupt the adsorbed solvent layer more effectively than molecules having a static conformation. Therefore, the increase in concentration can be attributed to the motion resulting by the *trans*-to-*cis* photoisomerization. The Y_W temperature dependence may be due to differences in molecule–graphene interactions at 20 and 40 °C, as well as different kinetics of the *trans*-to-*cis* photoisomerization.

***Trans*-to-*cis* photoisomerization and *cis*-to-*trans* relaxation.** The transition from the thermodynamically stable *trans* to the metastable *cis* form can be induced by irradiation with UV light, and takes place on the picosecond timescale[31,32], whereas the *cis*-to-*trans* isomerization is typically triggered by irradiation with visible light or heat[33]. The latter can also be activated by the use of ultrasound-induced mechanical force[33]. Ref. 34 claimed that mechanoisomerization of azobenzene typically results in the cleavage of the N = N azo bond. However, more recent theoretical calculations[35] and experiments[36] have challenged this view.

To gain a quantitative understanding of the photoisomerization of 4-(decyloxy)azobenzene, we perform a kinetic analysis of both *trans*-to-*cis* and *cis*-to-*trans* isomerization. The UV–visible spectrum of 4-(decyloxy)azobenzene dissolved in NMP (Fig. 2a, black line) shows an absorption maximum at 353 nm, arising from the π–π^* transition[33], and a peak at 442 nm, due to n–π^* transitions[33]. The UV–visible spectrum of the above solution when irradiated at 365 nm shows a reduction of the π–π^* transition band at 353 nm, and an increase in absorbance of the 442 nm (n–π^* transition) peak, which is a characteristic feature of the *trans*-to-*cis* photoisomerization (Fig. 2a, red line; and Supplementary Fig. 3). The photostationary state can be reversibly converted back to the *trans* isomer on irradiation at 450 nm (see Methods for details). The yield of the conversion back to the thermodynamically stable *trans*-4-(decyloxy)azobenzene is higher than 95% (Fig. 2a, black dashed line).

To probe the *cis*-to-*trans* isomerization of the azobenzene molecules under experimental conditions corresponding to those used during the LPE process, i.e., at 20 and 40 °C, in the presence of ultrasound-induced mechanical forces, a stock solution of 4-(decyloxy)azobenzene in NMP (16.2 mM) is prepared, and then exposed to 365 nm light and sonicated for 3 h (see Supplementary Note 1 for details). Every 30 min

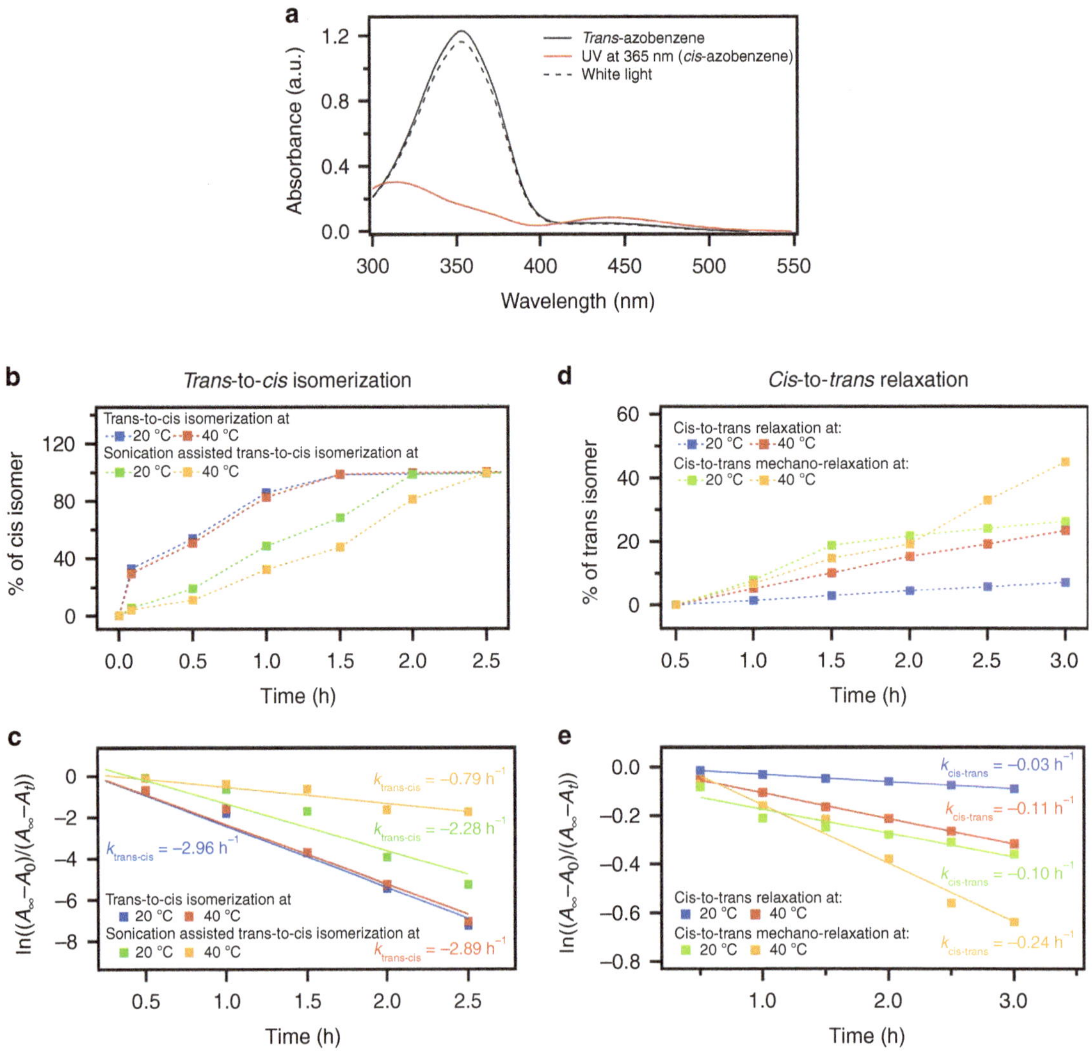

Figure 2 | Spectroscopic characterization of *trans*-to-*cis* isomerization and *cis*-to-*trans* relaxation. (**a**) UV-ndash;visible spectra of 4-(decyloxy) azobenzene in NMP (1.62 mM) showing the disappearance of the 353-nm band and an increase of the 442-nm band for 365-nm irradiation. The solid black line corresponds to the parent *trans*-4-(decyloxy)azobenzene spectrum, the red line represents a typical spectrum of a solution irradiated with UV light (*cis*-4-(decyloxy)azobenzene), whereas the black dashed line corresponds to a *cis*-4-(decyloxy)azobenzene solution irradiated with visible light (thus re-transformed into *trans*-4-(decyloxy)azobenzene). (**b**) Conversion percentage from *trans* to *cis* as a function of UV irradiation time at 20 and 40 °C. (**c**) First-order kinetic plots and rate constants ($k_{trans-cis}$) for *trans*-to-*cis* isomerization at 20 and 40 °C. (**d**) Conversion percentage from *cis* to *trans* as a function of time at 20 and 40 °C. (**e**) First-order kinetic plots and rate constants ($k_{cis-trans}$) for the *cis*-to-*trans* relaxation at 20 and 40 °C. (**b**–**e**) Reference samples prepared without sonication: blue (20 °C) and red (40 °C); sonicated samples: green (20 °C) and orange (40 °C).

(except for a first measurement performed after 5 min) 0.2 ml of UV-irradiated solution is transferred (in the dark) into a quartz cuvette, diluted 10 times with NMP. The optical absorption spectrum is then recorded (shown in Supplementary Fig. 4). Finally, to ensure the stability of the azo bond during sonication, a solution sonicated for 3 h is exposed to white light and an absorption spectrum is recorded. A similar procedure is applied for blank experiments, i.e., in the absence of ultrasound-induced mechanical forces.

Analysis of the absorption and proton nuclear magnetic resonance spectra (Supplementary Fig. 5) acquired for different experimental conditions allows us to quantify the percentage of conversion from *trans*- to *cis*-4-(decyloxy)azobenzene, as well as the rate constants for *trans*-to-*cis* isomerization. Both nuclear magnetic resonance and high-performance liquid chromatography data (Supplementary Figs 5 and 6, and Supplementary Note 3) show that the decyloxy chains are not being chopped from the azobenzene core. The slow kinetics of the *trans*-to-*cis* isomerization can be explained by the fact that both spectroscopic experiments and LPE process are carried out under unusual isomerization conditions, including a low UV light power density (0.34 mW cm^{-2}). A different time (30 min) is needed for reaching a photostationary state, in the reference and sonicated samples, as displayed in Fig. 2b. Previous studies[36,37] on the *trans*-to-*cis* isomerization of azobenzene in the presence of UV light showed that the reaction follows first-order kinetics and the rate constant k can be written as follows:

$$\ln\frac{(A_\infty - A_t)}{(A_\infty - A_0)} = -kt \tag{1}$$

where A_0, A_t and A_∞ are the absorbances before irradiation, at irradiation time t and after irradiation for a prolonged time (ca. 6 h). Applying equation (1), we find that the isomerization constant $k_{trans-cis}$ at 20 °C for the 4-(decyloxy)azobenzene linked to graphene is similar to $k_{trans-cis}$ at 40 °C: 2.96 and 2.89 h^{-1}, respectively. Significantly different $k_{trans-cis}$ kinetics are observed for the samples exposed to ultrasound-induced mechanical forces, i.e., 2.28 and 0.79 h^{-1} for dispersions prepared at 20 and 40 °C, respectively. To ensure the stability of the azo bond during sonication, the two samples are exposed to white light, and the absorption spectra recorded. The absorption at 365 nm is consistent with the *trans*-4-(decyloxy)azobenzene spectra in both cases (Supplementary Figs 3 and 4), indicating that the molecules, in particular the azo bond, are stable during sonication at various temperatures.

Since the *trans*-to-*cis* isomerization and the LPE processes are performed under UV light, the possible *cis*-to-*trans* back-conversion can be considered promoted by thermal or mechano-thermal relaxation, for experiments performed in the absence or presence of ultrasound-induced forces, respectively.

To probe both thermal and mechano-thermal isomerization, four solutions (16.2 mM) are prepared in NMP and the UV–visible spectra recorded. The solutions are then irradiated with UV light (365 nm) for 3 h to photoisomerize the 4-(decyloxy)azobenzene from its *trans* to *cis* configuration (Supplementary Fig. 3). The solutions are then kept for 3 h at 20 and/or 40 °C, in the presence and/or absence of ultrasound-induced mechanical forces. Every 30 min, 0.2 ml of solution is transferred (in dark) onto a quartz cuvette and diluted 10 times with NMP. Absorption spectra are then recorded. Owing to the lack of white light during the 3 h relaxation, 4-(decyloxy)azobenzene molecules do not convert fully from their *cis*- to *trans*-isomers. As shown in Fig. 2c, the percentage of *cis*-to-*trans* conversion in the blank samples (without sonication) at 20 and 40 °C is ~8.6% and 27.6%, respectively. In contrast, the ultrasound-induced conversion is much higher, i.e., 29.8% and 51.2% for samples kept at 20 and 40 °C, respectively. Similar to the *trans*-to-*cis* isomerization, the *cis*-to-*trans* relaxation of 4-(decyloxy)azobenzene follows first-order kinetics, and k can be obtained from equation (1). $k_{cis-trans}$ for 4-(decyloxy)azobenzene kept at 20 °C is lower than $k_{cis-trans}$ for 4-(decyloxy)azobenzene relaxed at 40 °C, i.e., 0.03 and 0.11 h^{-1}, respectively (Fig. 2d). Notably different $k_{cis-trans}$ are also observed for samples exposed to ultrasound-induced mechanical forces, i.e., 0.10 and 0.24 h^{-1} for dispersions kept at 20 and 40 °C, respectively.

These results indicate that the *trans*-to-*cis* isomerization of 4-(decyloxy)azobenzene strongly depends on the experimental conditions. In particular, it is hindered by ultrasound-induced mechanical forces, which most likely cause the occurrence of the competing *cis*-to-*trans* isomerization. Furthermore, comparison of $k_{trans-cis}$ and $k_{cis-trans}$ at 20 and 40 °C reveals that, at higher temperature, the process is more dynamic. As a result, the higher concentration of flakes (110 μg ml^{-1}) obtained at 40 °C can be attributed to the dynamic conformational *trans*-to-*cis* and *cis*-to-*trans* changes of 4-(decyloxy)azobenzene molecules triggered simultaneously by the competing effect activated by UV light, temperature and ultrasounds.

Analysis of dispersions. To fully characterize the exfoliated flakes, both qualitative and quantitative information are required. While quantitative insights can be assessed by providing the Y_W, a qualitative analysis must give more relevant details, such as the percentage of single-layer graphene (SLG) and multi-layer graphene (MLG) flakes, the lateral size of the flakes and the presence/absence of defects. The number of graphene layers (N) can be quantified using high resolution-transmission electron microscopy (HR-TEM)[3,5] and Raman spectroscopy[17,18]. Together with the information coming from electron diffraction patterns, in HR-TEM N can be directly counted by analysing the folded edges[38]. We focus on the samples prepared by LPE at 40 °C in the presence of azobenzene, while irradiating with UV light (highest increase in Y_W), and compare them with those exfoliated in NMP at 40 °C under UV irradiation (in the absence of azobenzene). The statistical analyses of N and the lateral flake size are reported in Supplementary Fig. 7a,b, respectively (Supplementary Note 4). In terms of N, minor differences are observed between the reference sample and flakes exfoliated in the presence of azobenzene molecules, where the percentage of SLG amounts to 20% and 14%, respectively. The lateral size of the flakes is not affected by the presence of azobenzene molecules during LPE (Supplementary Fig. 7b). Representative HR-TEM micrographs are reported in Supplementary Figs 8, 9 and 10.

The Raman spectrum of graphite and MLG consists of two fundamentally different sets of peaks. Those, such as D, G and 2D, present also in SLG, due to in-plane vibrations[39–41], and others such as the shear (C) modes[42] and the layer-breathing modes (LBM)[40,43], due to the relative motions of the planes themselves, are either perpendicular or parallel to their normal. The G peak corresponds to the high-frequency E_{2g} phonon at Γ.

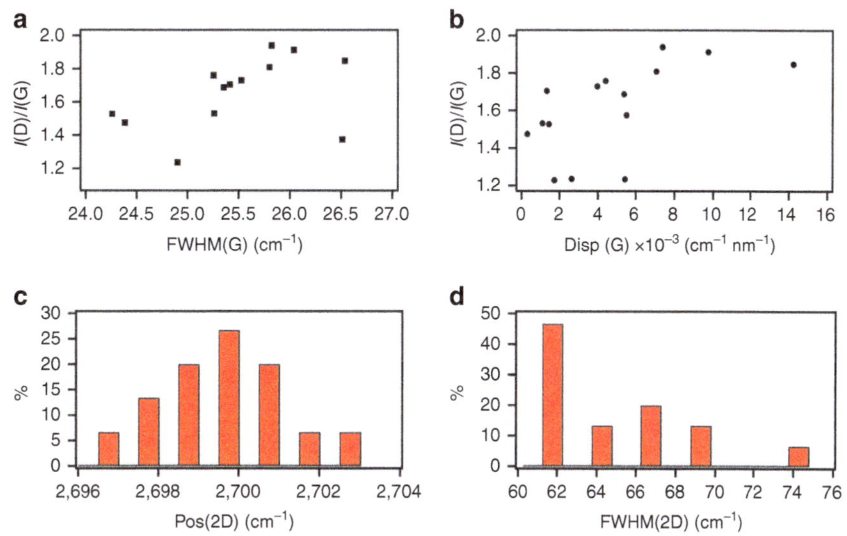

Figure 3 | Raman analysis of graphene/4-(decyloxy)azobenzene films. *I*(D)/*I*(G) as a function of (**a**) FWHM(G) and (**b**) Disp(G). Distribution of (**c**) Pos(2D) and (**d**) FWGM(2D).

The D peak is due to the breathing modes of six-atom rings and requires a defect for its activation[41,44,45]. It comes from transverse optical phonons around the Brillouin Zone edge **K** (refs 41,45), is active by double resonance[46] and is strongly dispersive with excitation energy due to a Kohn anomaly at **K** (ref. 47). Double resonance can also happen as intra-valley process, i.e., connecting two points belonging to the same cone around **K** or **K**′. This gives the so-called D′ peak. The 2D peak is the D peak overtone while the 2D′ peak is the D′ overtone. Since 2D and 2D′ originate from a process where momentum conservation is satisfied by two phonons with opposite wave vectors, no defects are required for their activation, and are thus always present[38]. The 2D peak is composed of a single Lorentzian in SLG, whereas it splits into several components as N increases, reflecting the evolution of the electronic band structure[38].

In disordered carbons, the G peak position, Pos(G), increases as the excitation wavelength decreases from infrared to ultraviolet[44]. Therefore, the dispersion of the G peak, Disp(G), i.e. the rate of change of Pos(G) with the laser excitation wavelength, increases with disorder[44]. Similar to Disp(G), also the full width at half maximum of the G peak, FWHM(G), always increases with disorder[44]. The analysis of the intensity ratio of the D and G peaks, I(D)/I(G), combined with that of FWHM(G) and Disp(G) allows us to discriminate between disorder localized at the edges and disorder in the bulk of the samples. In the latter case, a higher I(D)/I(G) would correspond to higher FWHM(G) and Disp(G).

Raman measurements are done on the same set of samples to characterize the quality of the graphitic material. Figure 3a,b show a small correlation between these parameters. This implies that the D peak is mostly due to edges, as well as the presence of some defects in the samples. FWHM(2D) (Fig. 3d) is larger with respect to that of graphene flakes produced by micromechanical cleavage, but still has a single Lorentzian line shape. This is consistent with the presence of MLGs composed of folded or juxtaposed SLGs, but still electronically decoupled[17].

Figure 4a plots the Raman spectra of a graphene–azobenzene film after visible and UV light illumination, acquired at 514.5 nm with power $<250\,\mu$W to avoid any Raman laser-induced isomerization of azobenzene. None of the graphene Raman parameters show significant change on UV or visible light irradiation indicating negligible change in the defects and doping.

The *trans*-to-*cis* isomerization can also be monitored by Raman spectroscopy[24,48]. Figure 4b plots the Raman spectrum of azobenzene after both visible and UV light illumination. The spectrum is composed of five main peaks, labelled C1–C5 (ref. 48). These can also be seen on the spectra of the graphene–azobenzene hybrids (Fig. 4c). The intensity ratio of peaks C3 and C4 can be used to monitor the isomerization of azobenzene as described in ref. 48. On illumination with visible light, the photochromic molecules fall into the *trans* configuration and I(C3)/I(C4) approaches ~1. When the sample is illuminated with UV light, azobenzene molecules undergo *trans*-to-*cis* isomerization and I(C3)/I(C4) decreases markedly. The effect is fully reversible and the *trans* form is recovered by a further visible light exposure as shown in Fig. 4d.

X-ray analysis. Powder X-ray diffraction is further used to characterize the structure of the graphene–azobenzene hybrid powder and to compare it with the control samples (Supplementary Note 5). Similarly to graphite, the powder prepared from LPE has a sharp peak at 26.7°, corresponding to an interlayer spacing of ~0.33 nm (Supplementary Fig. 11a). The X-ray diffraction patterns of powders prepared from graphene/*trans*-azobenzene and graphene/*cis*-azobenzene show new peaks at $2\theta = 12.3°$ ($d_{spacing}$ ~ 0.72 nm) and 9.9° ($d_{spacing}$ ~ 0.89 nm), respectively (Supplementary Fig. 11b,c). This confirms that *trans* and/or *cis*-azobenzenes can sandwich between graphene layers, thus increasing the overall spacing.

Electrical characterization. To probe the electrical properties of the graphene–azobenzene hybrid and, in particular, to explore the potential light-responsive nature of the material, we drop-cast a ~100-nm-thick graphene–azobenzene film on a n^{++} Si/SiO_2 substrates exposing pre-patterned interdigitated gold electrodes (channel length = 10 μm; Fig. 6a,b).

Typical I–V curves exhibit a nonlinear resistive behaviour, as shown in Fig. 5c. The conductivity of the graphene–azobenzene hybrid is lower than that measured in control devices, i.e.,

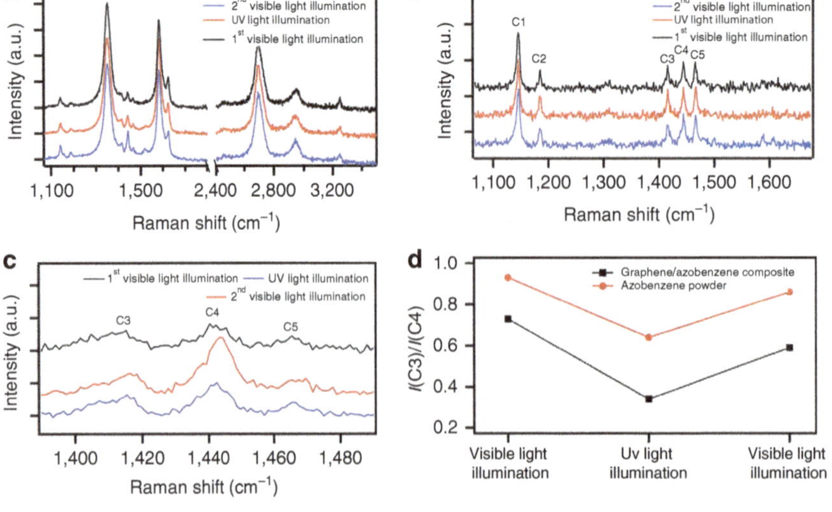

Figure 4 | Evolution of Raman spectra on illumination. (**a**) Comparison of the Raman spectrum of a graphene/4-(decyloxy) azobenzene hybrid film after 90-min illumination with visible light, UV light and visible light again to probe the effect of isomerization. (**b**) The same experiment on azobenzene powder. (**c**) 4-(decyloxy)azobenzene Raman peaks C3, C4 and C5 extracted from graphs in **a** showing the change of peaks ratio on illumination, which confirms the isomerization. (**d**) Dependence upon illumination of I(C3)/I(C4) for both hybrid material and powders.

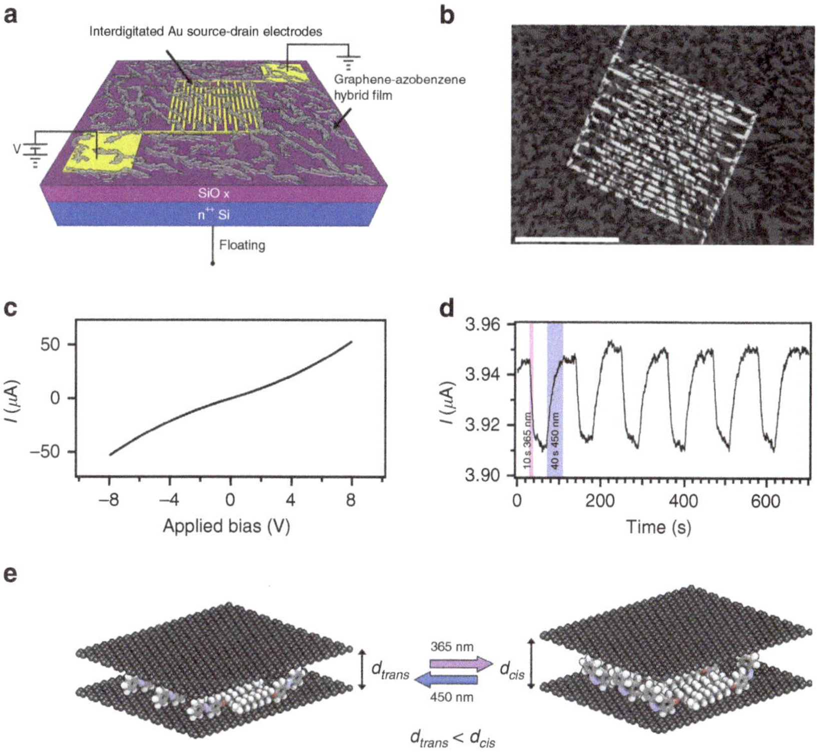

Figure 5 | Electrical characteristics of hybrid materials. (**a**) Scheme of the two-terminal device configuration, (**b**) scanning electron microscopy image of interdigitated Au electrodes covered with a hybrid film. Scale bar, 500 μm. (**c**) *I-V* characteristics, (**d**) optical modulation of current response for a static bias and dynamic alternative UV and visible light irradiation cycles. (**e**) Schematic of graphene-azobenzene hybrid undergoing UV and visible irradiation cycles.

comprising as active layer graphene exfoliated in NMP in the absence of azobenzene (Supplementary Fig. 12). Figure 6d shows the current modulation in a two-terminal device with graphene–azobenzene on cycles of UV and visible light irradiation, for 1 V applied between the two electrodes. One cycle consists of 10-s UV light (365 nm), 1-min rest in dark followed by 40-s visible light (450 nm) irradiation. Under UV exposure (magenta background) a current decrease is detected, which reversibly increases to about the initial value under visible light irradiation (violet background) over various cycles. No fatigue is observed during six cycles of photoisomerization. Conversely, the control devices show no response to light (Supplementary Fig. 12). Noteworthy, in some cases the increase of current in the hybrid films can be the result of π-electron interactions between the molecules and graphene[49]; however, such phenomenon, known as increased photoconductivity, is irreversible and can only be modulated by varying the ratio between graphene and *ad hoc* molecules.

In the azobenzene–graphene film (Supplementary Figs 13 and 14), the molecules are physisorbed on top and in between the graphene sheets. Under UV irradiation, the azobenzene molecules undergo a conformational change from the less-bulky *trans* isomer to the bulkier *cis* isomer. Such a process is therefore accompanied by an increase in the few-layer graphene (FLG) inter-sheet distance, thereby hindering the charge transport via hopping between the sheets, resulting in lower conductivity of the hybrid film. The conductivity of the hybrid material can be fully restored by irradiating with visible light, causing the *cis*-to-*trans* photoisomerization.

To gain greater insight into the conduction mechanism and the effect of isomerization, we then perform Kelvin probe measurements (Supplementary Note 6). We find that the work function is reversibly photo-modulated over various cycles (Supplementary Fig. 15), providing further evidence for the isomerization of the azobenzene molecules. This difference in work function has a direct effect on charge injection from the electrodes to the graphene/azobenzene conductive material, which is less favourable in the presence of the *cis* isomer. Light modulation can be applied multiple times to the systems, with no evidence of decrease in performance. We assign this to graphene's shielding of the azobenzene molecules from the outer environment. An ideal case consists of stacks of alternating graphene and azobenzenes, the latter having a thickness, which can be photo-modulated (Fig. 5e). In reality, the flakes (with average lateral sizes of 200 nm) form aggregates possessing a poor degree of order. Within these aggregates the azobenzenes are intercalated in between adjacent layers of graphene. The control samples show no change in work function on UV and visible light cycling (Supplementary Fig. 15).

Molecular modelling simulations. To gain a better understanding of the interactions of the azobenzene molecules with graphene, we perform molecular dynamics simulations (see Supplementary Note 7 for details). All calculations are done using the Groningen Machine for Chemical Simulations (GROMACS)[50] package and a modified version of the all-atom Optimized Potentials for Liquid Simulations (OPLSAA) force field developed for azobenzenes[51,52]. To assess the relative adsorption affinity of the *trans* versus *cis* conformations, we first compute the interaction energy for a single azobenzene molecule by running a 1-ns simulation in vacuum. The graphene layer is taken as an infinite rigid body in these calculations. The computed interaction energies of -130.1 ± 7.5 and $-115.7 \pm 8.0\,kJ\,mol^{-1}$ for *trans*- and *cis*-azobenzene, respectively, are consistent with the expected stronger adsorption of *trans*-azobenzene, resulting from its planar geometry and the concomitant maximized π–π interactions between the phenyl

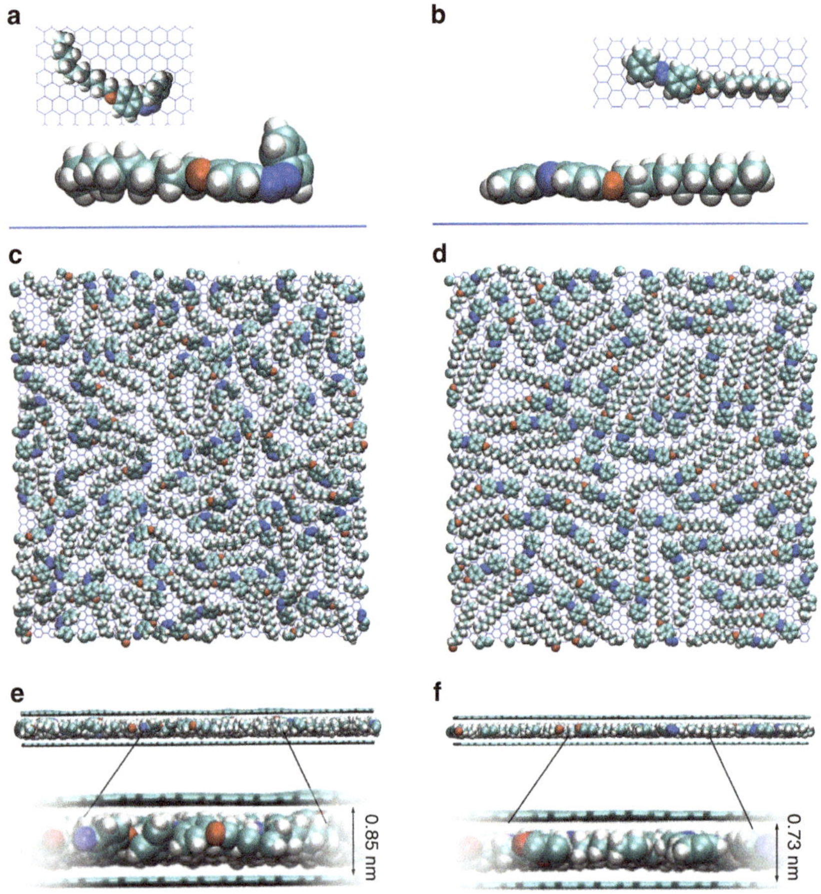

Figure 6 | Molecular modelling of graphene interaction with *cis* and *trans* azobenzene. Side and top views of a single (**a**) *cis*- and (**b**) *trans*-azobenzene molecule adsorbed on SLG. Top view of the (**c**) *cis* and (**d**) *trans*-monolayer, and side view of the molecular system consisting of the (**e**) *cis* and (**f**) *trans*-monolayer sandwiched between two SLGs, with the indication of the SLGs separation.

groups and graphene. The average adsorption distance for the *trans* molecule is ~0.07 nm shorter than *cis* (0.34 versus 0.41 nm). Figure 6a,b plots top and side views of the adsorbed *cis* and *trans* molecules from the last snapshots along the molecular dynamics trajectories.

It should be noted that the conformation giving the higher exfoliation yield is the *cis* one, with lower interaction energy with graphene (14.4 kJ mol^{-1} difference). This counterintuitive finding highlights the importance of kinetic factors on exfoliation yield rather than thermodynamic factors, in good agreement with previous results obtained with rigid molecules[29,30]. The 70% increase in exfoliation yield could be due to the rapid, dynamic change of conformation between *cis* and *trans* that the molecules undergo during the exfoliation, which is more rapid at high temperature and under sonication (Fig. 2).

We then study the supramolecular organization of *trans*- and *cis*-azobenzenes on SLG in vacuum. In the absence of experimental structural data, we consider highly regular self-assembled monolayers (Supplementary Fig. 16) and test their stability by performing simulations of the assemblies at room temperature. The regular assemblies are unstable at 300 K and disassemble after few tens of picoseconds, suggesting that the azobenzene molecules form disordered layers at the SLG surface. To investigate the formation and structure of such disordered layers, we use the following protocol. Azobenzene molecules are introduced one by one in the simulation box at a distance of 1.8 nm above the plane of the infinite SLG with a random orientation. The initial atomic velocities are assigned so that the resulting vectors point towards the SLG, thereby prompting physical adsorption on the surface. Room-temperature molecular dynamics simulations of a few hundreds of picoseconds are then performed to explore the configurational space after each molecular deposition, and the last saved configuration is used as input for the landing of the subsequent molecule. The scheme is iterated multiple times until the desired surface coverage is reached. A 5-ns simulation is then performed on the final structure to equilibrate the monolayer at room temperature. A more detailed description of the methodology is provided in Supplementary Note 8. Full coverage of the graphene periodic layer of 115.8 nm^2 is reached when 64 (72) molecules of *trans* (*cis*)-azobenzene are adsorbed, which translates into a surface contact area on graphene of ~1.8 nm^2 for the *trans* and ~1.6 nm^2 for the *cis* conformation. Figure 6c,d plots a top view of the azobenzene monolayers obtained for the two conformers in the case of full coverage. Supplementary Figs 17 and 18 show the formation of those monolayers as a function of the number of molecules adsorbed on SLG.

To validate the hypothesis that the insertion of a *trans-/cis*-monolayer sandwiched between two SLGs can result in different interlayer spacing according to the ideal model, a second SLG is placed on top of the azobenzene monolayers prepared using the deposition protocol described above. The 1-ns simulations are then done at room temperature. The results show that *cis*-azobenzenes maintain the SLG at a distance ~0.85 nm, while the layer separation decreases to ~0.73 nm for *trans* (Fig. 6e,f), in excellent agreement with the X-ray data. This supports the view that photoisomerization of the molecules from *trans* to *cis* can favour the exfoliation process. To investigate the effect of this switchable separation on charge transport, we perform quantum chemical calculations (at the intermediate neglect of differential overlap level[53]) on two square-shaped nanographenes ($\sim 20\,nm^2$) interacting either through-space or in the presence of the sandwiched (*trans* or *cis*) azobenzenes (Supplementary Fig. 19). In the first case (no sandwiched layer), the electronic couplings mediating positive and negative charge carrier tunnelling from one nanographene to the other, as obtained from the diabatization scheme reported in ref. 54, decrease by a factor ~100 when the interlayer distance is increased from 0.73 (corresponding to the azobenzenes in *trans* conformation) to 0.85 nm (*cis* conformation). In both cases, these interactions are very small due to the fast fall-off decay of the nanographene wavefunction overlap. Interestingly, the corresponding couplings calculated in the presence of the sandwiched azobenzenes are several orders of magnitude larger and show a less pronounced change (by a factor ~10) on photoisomerization, suggesting that the loss in electrical transmission associated with the larger inter-graphene distances is at least partly compensated by conformational-dependent superexchange effects involving the molecular orbitals of the azobenzenes. Our experimental current modulation on photoisomerization of azobenzenes (Fig. 5d) is much smaller than the calculated one (factor ~100). This can be understood considering that the simulations are run for two SLG intercalated by a tightly packed, physisorbed, self-assembled layer of azobenzene, whereas the experiments are done for hybrid films with a distribution of layer thicknesses (Supplementary Fig. 7a) separated by a small quantity of azobenzene (below 5% of the area of graphene).

As aforementioned the ideal model does not fully describe our real system that consists of poorly ordered aggregates comprising flakes intercalated with azobenzenes, and charges hopping between adjacent graphenes. The distance between the adjacent graphene sheets can be varied on photoisomerization of the azobenzenes.

Discussion

We demonstrated that alkoxy-substituted photochromic molecules can act as photo-addressable surfactant and as dispersion-stabilizing agents to enhance the yield of exfoliation in an upscalable molecule-assisted LPE-based method. The simultaneous use of UV light, promoting the *trans*-to-*cis* isomerization, as well as thermal annealing at 40 °C and mechanical forces generated by sonication, both favouring *cis*-to-*trans* isomerization, promotes the exfoliation in liquid media. The most effective exfoliation is obtained with azobenzene molecules irradiated with UV light in NMP at 40 °C, with a concentration of exfoliated graphene of $110\,\mu g\,ml^{-1}$. This corresponds to an ~80% increase in exfoliation yield when compared with pure NMP ($63\,\mu g\,ml^{-1}$). By depositing the hybrid film onto Au pre-patterned SiO_2 substrates, light-responsive thin hybrid films, formed in a one-step co-deposition process, can be realized, whose conductivity can be reversibly modulated by the *trans–cis* photoisomerization of the azobenzenes. By combining this approach with cost-effective techniques, such as ink-jet printing, more complex responsive device designs and architectures may be realized. This paves the way to future applications such as optically controllable memory switch elements for light-assisted programming and photosensors.

Methods

Materials. Graphite synthetic flakes (Product No. 332461), 4-(decyloxy) azobenzene (Product No. S931950) and NMP (Product No. 270458) are sourced from Sigma-Aldrich.

Liquid-phase exfoliation. Graphene dispersions are prepared by adding graphite powder (100 mg) to NMP (10 ml), and tip-sonication (Labsonic M, Platinum tip, diameter 2 mm, sound rating density $300\,W\,cm^{-2}$) for 3 h either in dark or under UV light irradiation, using a portable laboratory ultraviolet lamp (8 W, 365 nm, $0.34\,mW\,cm^{-2}$, Herolab GmbH), in glass vials (Pyrex), in the presence of azobenzene molecules (5.5 mg). The dispersions are then allowed to settle for 15 min, then decanted and centrifuged (Eppendorf, centrifuge 5804) for 1 hour at 10,000 r.p.m. From the centrifuged dispersions 70 vol% are pipetted off the top for characterization and film deposition. To quantify the concentration of flakes after centrifugation, the dispersions are passed through polytetrafluoroethylene membrane filters (pore size 100 nm). Measurements of the filtered mass are performed with a microbalance (Sartorius MSA2.75)

Device fabrication. n^{++}Si substrates with a thermally grown SiO_2 layer (230 nm) and pre-patterned interdigitated gold source and drain electrodes (IPMS Fraunhofer) with different channel length ($L = 2.5$, 5, 10 and 20 μm) and constant channel width ($W = 10$ mm) are used. The substrates are cleaned before device fabrication in an ultrasonic bath (FB 15047, Fisher Scientific) of acetone and isopropanol, 15 min in each solvent, and treated 5 min (+30 min incubation) with an UV surface decontamination system (PSD-UV, Novascan) to improve wetting of the solvent. The dispersions (50 μl) are drop-cast on clean substrates and dried for 48 h in vacuum at 30 °C.

Characterization. The electrical characterization is carried out in inert atmosphere (glovebox) with an electrometer (Keithley 2636A) interfaced with LabTracer software. For the light-induced switch, a Polychrome V monochromator (Till Photonics) is used as ultraviolet light (350 nm, $5.64\,mW\,cm^{-2}$) and visible light source (450 nm, $4.83\,mW\,cm^{-2}$). For the Raman experiments, graphene dispersions are drop-cast onto pre-cleaned n^{++}Si substrates with a thermally grown SiO_2 layer (300 nm) and dried for 48 h. Raman spectra are collected with a Renishaw InVia spectrometer at 457, 514.5 and 633 nm. The excitation power is kept below 1 mW to avoid the effects of local heating. The scattered light is collected with a ×100 objective. HR-TEM is done in a FEI Tecnai F20 equipped with a Schottky emitter and operated at 120 keV primary beam energy. Scanning electron microscopy images are recorded with a Quanta FEG 250 from FEI.

The thickness of the hybrid films is determined with an Alpha-Step IQ Surface Profiler from KLA Tencor. X-ray photoelectron spectroscopy analyses are carried out on a Thermo Scientific K-Alpha X-ray photoelectron spectrometer, with a basic chamber pressure $\sim 10^{-8}$ mbar and an Al anode as the X-ray source (X-ray radiation of 1,486 eV). Spot sizes of 400 μm, pass energies of 200 eV for survey scans and 50 eV for high-resolution scans are used. A volume of 150 μl of the dispersions are spin-coated on Au substrate for 1 min at 1,000 r.p.m. and substrates are annealed for 1 day at 100 °C in a oven under vacuum. Samples for X-ray diffraction are prepared by precipitating graphene and/or graphene/azobenzene by adding water:ethanol (1:1, vol:vol) into NMP. The collected precipitate is dried under vacuum for 24 h. The powder X-ray diffraction patterns are obtained using a Bruker AXS D2 Phaser (LYNXEYE detector) with Ni-filtrated Cu-Kα radiation ($\lambda = 1.5406$ Å) with a 1-mm air-scattering slit and 0.1-mm equatorial slit. Samples are deposited on the surface of a single crystal Si wafer (cut of (911)). X-ray diffraction patterns are collected with 0.016° steps and 10 s per step increments from 8° to 80°.

References

1. Geim, A. K. & Novoselov, K. S. The rise of graphene. *Nat. Mater.* **6,** 183–191 (2007).
2. Ferrari, A. C. *et al.* Science and technology roadmap for graphene, related two-dimensional crystals, and hybrid systems. *Nanoscale* **7,** 4598–4810 (2015).
3. Bonaccorso, F. *et al.* Production and processing of graphene and 2d crystals. *Mater. Today* **15,** 564–589 (2012).
4. Narita, A. *et al.* Synthesis of structurally well-defined and liquid-phase-processable graphene nanoribbons. *Nat. Chem.* **6,** 126–132 (2014).
5. Hernandez, Y. *et al.* High-yield production of graphene by liquid-phase exfoliation of graphite. *Nat. Nanotechnol.* **3,** 563–568 (2008).
6. Chen, L., Hernandez, Y., Feng, X. L. & Müllen, K. From nanographene and graphene nanoribbons to graphene sheets: chemical synthesis. *Angew. Chem. Int. Ed.* **51,** 7640–7654 (2012).
7. Palma, C.-A. & Samorì, P. Blueprinting macromolecular electronics. *Nat. Chem.* **3,** 431–436 (2011).
8. Kim, K. S. *et al.* Large-scale pattern growth of graphene films for stretchable transparent electrodes. *Nature* **457,** 706–710 (2009).
9. Li, X. S. *et al.* Large-area synthesis of high-quality and uniform graphene films on copper foils. *Science* **324,** 1312–1314 (2009).
10. Berger, C. *et al.* Electronic confinement and coherence in patterned epitaxial graphene. *Science* **312,** 1191–1196 (2006).
11. Ciesielski, A. & Samorì, P. Graphene via sonication assisted liquid-phase exfoliation. *Chem. Soc. Rev.* **43,** 381–398 (2014).

12. Parvez, K. *et al.* Electrochemically exfoliated graphene as solution-processable, highly conductive electrodes for organic electronics. *ACS Nano* **7,** 3598–3606 (2013).
13. Parvez, K. *et al.* Exfoliation of graphite into graphene in aqueous solutions of inorganic salts. *J. Am. Chem. Soc.* **136,** 6083–6091 (2014).
14. León, V. *et al.* Few-layer graphenes from ball-milling of graphite with melamine. *Chem. Commun.* **47,** 10936–10938 (2011).
15. Sampath, S. *et al.* Direct exfoliation of graphite to graphene in aqueous media with diazaperopyrenium dications. *Adv. Mater.* **25,** 2740–2745 (2013).
16. Coleman, J. N. Liquid exfoliation of defect-free graphene. *Acc. Chem. Res.* **46,** 14–22 (2013).
17. Torrisi, F. *et al.* Inkjet-printed graphene electronics. *ACS Nano* **6,** 2992–3006 (2012).
18. Marago, O. M. *et al.* Brownian motion of graphene. *ACS Nano* **4,** 7515–7523 (2010).
19. Ciesielski, A. *et al.* Liquid-phase exfoliation of graphene using intercalating compounds: a supramolecular approach. *Angew. Chem. Int. Ed.* **53,** 10355–10361 (2014).
20. Haar, S. *et al.* A supramolecular strategy to leverage the liquid-phase exfoliation of graphene in the presence of surfactants: unraveling the role of the length of fatty acids. *Small* **11,** 1691–1702 (2014).
21. Luo, W. *et al.* A high energy density azobenzene/graphene hybrid: a nano-templated platform for solar thermal storage. *J. Mater. Chem. A* **3,** 11787–11795 (2015).
22. Peimyoo, N. *et al.* Photocontrolled molecular structural transition and doping in graphene. *ACS Nano* **6,** 8878–8886 (2012).
23. Seo, S., Min, M., Lee, S. M. & Lee, H. Photo-switchable molecular monolayer anchored between highly transparent and flexible graphene electrodes. *Nat. Commun.* **4,** 1920 (2013).
24. Kim, M., Safron, N. S., Huang, C. H., Arnold, M. S. & Gopalan, P. Light-driven reversible modulation of doping in graphene. *Nano Lett.* **12,** 182–187 (2012).
25. Margapoti, E. *et al.* Emergence of photoswitchable states in a graphene–azobenzene–Au platform. *Nano Lett.* **14,** 6823–6827 (2014).
26. Mativetsky, J. M. *et al.* Azobenzenes as light-controlled molecular electronic switches in nanoscale metal-molecule-metal junctions. *J. Am. Chem. Soc.* **130,** 9192–9193 (2008).
27. Feng, Y. Y., Zhang, X. Q., Ding, X. S. & Feng, W. A light-driven reversible conductance switch based on a few-walled carbon nanotube/azobenzene hybrid linked by a flexible spacer. *Carbon* **48,** 3091–3096 (2010).
28. Rabe, J. P. & Buchholz, S. Commensurability and mobility in 2-dimensional molecular-patterns on graphite. *Science* **253,** 424–427 (1991).
29. Schlierf, A. *et al.* Nanoscale insight into the exfoliation mechanism of graphene with organic dyes: effect of charge, dipole and molecular structure. *Nanoscale* **5,** 4205–4216 (2013).
30. Yang, H. F. *et al.* Dielectric nanosheets made by liquid-phase exfoliation in water and their use in graphene-based electronics. *2D Mater.* **1,** 011012 (2014).
31. Griffiths, J. Photochemistry of azobenzene and its derivatives. *Chem. Soc. Rev.* **1,** 481–493 (1972).
32. Tamai, N. & Miyasaka, H. Ultrafast dynamics of photochromic systems. *Chem. Rev.* **100,** 1875–1890 (2000).
33. Bandara, H. M. D. & Burdette, S. C. Photoisomerization in different classes of azobenzene. *Chem. Soc. Rev.* **41,** 1809–1825 (2012).
34. Joseph, J. M., Destaillats, H., Hung, H. M. & Hoffmann, M. R. The sonochemical degradation of azobenzene and related azo dyes: rate enhancements via Fenton's reactions. *J. Phys. Chem. A* **104,** 301–307 (2000).
35. Turansky, R., Konopka, M., Doltsinis, N. L., Stich, I. & Marx, D. Switching of functionalized azobenzene suspended between gold tips by mechanochemical, photochemical, and opto-mechanical means. *Phys. Chem. Chem. Phys.* **12,** 13922–13932 (2010).
36. Surampudi, S. K., Patel, H. R., Nagarjuna, G. & Venkataraman, D. Mechano-isomerization of azobenzene. *Chem. Commun.* **49,** 7519–7521 (2013).
37. Shin, K. H. & Shin, E. J. Photoresponsive azobenzene-modified gold nanoparticle. *Bull. Korean Chem. Soc.* **29,** 1259–1262 (2008).
38. Ferrari, A. C. *et al.* Raman spectrum of graphene and graphene layers. *Phys. Rev. Lett.* **97,** 187401–187405 (2006).
39. Ferrari, A. C. & Robertson, J. Raman spectroscopy of amorphous, nanostructured, diamond-like carbon, and nanodiamond. *Philos. Trans. R. Soc. Lond. Ser. A* **362,** 2477–2512 (2004).
40. Ferrari, A. C. & Basko, D. M. Raman spectroscopy as a versatile tool for studying the properties of graphene. *Nat. Nanotechnol.* **8,** 235–246 (2013).
41. Tuinstra, F. & Koenig, J. L. Raman spectrum of graphite. *J. Chem. Phys.* **53,** 1126–1130 (1970).
42. Tan, P. H. *et al.* The shear mode of multilayer graphene. *Nat. Mater.* **11,** 294–300 (2012).
43. Lui, C. H. *et al.* Observation of layer-breathing mode vibrations in few-layer graphene through combination Raman scattering. *Nano Lett.* **12,** 5539–5544 (2012).
44. Ferrari, A. C. & Robertson, J. Resonant Raman spectroscopy of disordered, amorphous, and diamondlike carbon. *Phys. Rev. B* **64,** 075414–075426 (2001).
45. Ferrari, A. C. & Robertson, J. Interpretation of Raman spectra of disordered and amorphous carbon. *Phys. Rev. B* **61,** 14095–14107 (2000).
46. Thomsen, C. & Reich, S. Double resonant Raman scattering in graphite. *Phys. Rev. Lett.* **85,** 5214–5217 (2000).
47. Piscanec, S., Lazzeri, M., Mauri, F., Ferrari, A. C. & Robertson, J. Kohn anomalies and electron-phonon interactions in graphite. *Phys. Rev. Lett.* **93,** 185503 (2004).
48. Zheng, Y. B. *et al.* Surface-enhanced Raman spectroscopy to probe reversibly photoswitchable azobenzene in controlled nanoscale environments. *Nano Lett.* **11,** 3447–3452 (2011).
49. Chunder, A., Pal, T., Khondaker, S. I. & Zhai, L. Reduced graphene oxide/copper phthalocyanine composite and its optoelectrical properties. *J. Phys. Chem. C* **114,** 15129–15135 (2010).
50. Abraham, M. J. *et al.* GROMACS: high performance molecular simulations through multi-level parallelism from laptops to supercomputers. *SoftwareX* **1,** 19–25 (2015).
51. Schafer, L. V., Muller, E. M., Gaub, H. E. & Grubmuller, H. Elastic properties of photoswitchable azobenzene polymers from molecular dynamics simulations. *Angew. Chem. Int. Ed.* **46,** 2232–2237 (2007).
52. Pipolo, S. *et al.* First-principle-based MD description of azobenzene molecular rods. *Theor. Chem. Acc.* **131,** 1274–1287 (2012).
53. Ridley, J. & Zerner, M. An intermediate neglect of differential overlap technique for spectroscopy: pyrrole and the azines. *Theor. Chem. Acc.* **32,** 111–134 (1973).
54. Kondov, I., Cizek, M., Benesch, C., Wang, H. B. & Thoss, M. Quantum dynamics of photoinduced electron-transfer reactions in dye-semiconductor systems: first-principles description and application to coumarin 343-TiO_2. *J. Phys. Chem. C* **111,** 11970–11981 (2007).

Acknowledgements

We acknowledge funding from the European Commission through the Graphene Flagship (GA-604391), the Marie Curie project GREAT (PIEF-2011-GA-298246), the FP7-NMP-2012 project SACS (GA-310651), the Agence Nationale de la Recherche through the LabEx project Nanostructures in Interaction with their Environment (ANR-11-LABX-0058_NIE), the International Center for Frontier Research in Chemistry (icFRC), the Belgian National Fund for Scientific Research (FNRS), the FNRS/FRFC 'PHOTOGRAPH' PDR project, the Belgian Science Policy Office (IAP 7/05), the ERC synergy grant Hetero2D, ERC PoC HiGRAPHINK, and the Royal Society Wolfson Research Merit Award, EP/K01711X/1, EP/K017144/1 and EP/L016087/1. D.B. is a FNRS Research Director. M. Eredia, M. El Garah and A. Galanti are acknowledged for their assistance in X-ray photoelectron spectroscopy and high-performance liquid chromatography measurements.

Author contributions

P.S. and A.C. conceived the experiments and designed the study; M.D. and S.H. participated in the planning of the study and carried out the exfoliation (LPE) and characterization in solution (ultraviolet); M.D. performed the electrical characterization; S.H. and A.C. performed spectroscopic characterization; F.R. and S.H. performed the X-ray photoelectron spectroscopy experiments; M.D. and T.M. performed Kelvin probe measurements; M.B. and A.C.F. performed and analysed Raman experiments; S.O., A.M., R.L. and D.B. conducted theoretical studies; R.M. and V.M. analysed the samples with TEM; E.A.P. and L.D.C. performed and analysed X-ray diffraction experiments; all authors discussed results and contributed to the interpretation of data; D.B., A.C.F., A.C. and P.S. co-wrote the paper; all authors contributed to editing the manuscript.

Additional information

Competing financial interests: The authors declare no competing financial interests.

A low-cost hierarchical nanostructured beta-titanium alloy with high strength

Arun Devaraj[1], Vineet V. Joshi[2], Ankit Srivastava[3], Sandeep Manandhar[4], Vladimir Moxson[5], Volodymyr A. Duz[5] & Curt Lavender[2]

Lightweighting of automobiles by use of novel low-cost, high strength-to-weight ratio structural materials can reduce the consumption of fossil fuels and in turn CO_2 emission. Working towards this goal we achieved high strength in a low cost β-titanium alloy, Ti-1Al-8V-5Fe (Ti185), by hierarchical nanostructure consisting of homogenous distribution of micron-scale and nanoscale α-phase precipitates within the β-phase matrix. The sequence of phase transformation leading to this hierarchical nanostructure is explored using electron microscopy and atom probe tomography. Our results suggest that the high number density of nanoscale α-phase precipitates in the β-phase matrix is due to ω assisted nucleation of α resulting in high tensile strength, greater than any current commercial titanium alloy. Thus hierarchical nanostructured Ti185 serves as an excellent candidate for replacing costlier titanium alloys and other structural alloys for cost-effective lightweighting applications.

[1] Physical and Computational Sciences Directorate, Pacific Northwest National Laboratory, 902 Battelle Boulevard, Richland, Washington 99354, USA. [2] Energy and Environment Directorate, Pacific Northwest National Laboratory, 902 Battelle Boulevard, Richland, Washington 99354, USA. [3] Department of Materials Science and Engineering, Texas A&M University, 3003 College Station, Texas 77843, USA. [4] Environmental Molecular Sciences Laboratory, Pacific Northwest National Laboratory, 902 Battelle Boulevard, Richland, Washington 99354, USA. [5] Advance Materials Products Inc. (ADMA), 1890 Georgetown Road, Hudson, Ohio 44236, USA. Correspondence and requests for materials should be addressed to A.D. (email: Arun.Devaraj@pnnl.gov) or to V.V.J. (email: Vineet.Joshi@pnnl.gov).

With rapid global climate change and alarming levels of atmospheric CO_2 concentration, there is growing interest in and attention to reducing anthropogenic CO_2 emission[1,2]. Emissions from fossil fuel combustion in transportation vehicles accounts for a significant portion of the global CO_2 emissions[3]. This has recently led to development of number of government policies and mandates for global automobile manufacturers to meet increasingly stringent CO_2 emission standards and higher fuel efficiency targets for future years[4]. Lightweighting of transportation vehicles is the key to reducing fossil fuel consumptions and anthropogenic CO_2 emissions[1–3]. The primary approach to lightweighting is through increased use of structural materials with high specific strength (strength to weight ratio). In recent years, titanium alloys, more specifically β-titanium alloys, have been widely explored as candidate materials owing to their attractive specific strength, toughness and corrosion resistance[5–7], for example, the Boeing 787 Dreamliner, which is one of the most fuel efficient airplanes in its class, is utilizing about 15% titanium alloys, the largest percentage of titanium used in a passenger airplane till date[8]. However, wide scale adoption of β-titanium alloys in transportation applications has been limited due to its high cost.

The cost of β-titanium alloys can be lowered by replacing the expensive β stabilizers such as Mo, Cr and V (in full or in part) by low cost Fe (ref. 9) such as the case in Ti–1Al–8V–5Fe (Ti185) alloy introduced in 1960s (ref. 10). Ti185 alloy demonstrated high tensile and shear strength proving to be an excellent candidate for fastener applications. Despite the advantages, the production of Ti185 alloy in bulk by conventional ingot processing remained a challenge since addition of Fe in excess of 2.5 wt% in titanium alloys lead to segregation of Fe. The segregation of Fe results in the formation of inhomogeneous β structures also known as β flecks, which is detrimental to the mechanical performance of the alloy[11].

Recently the authors have demonstrated a novel route to develop engineering components of β-titanium alloys using a low-cost TiH_2 powder feedstock[12–15]. The mechanical properties of the Ti185 alloy developed through this low-cost process are found to be at par with ones developed via the conventional route. The strength of the low-cost powder processed Ti185 alloy is further improved to achieve strength level higher than all current commercial titanium alloys by achieving a hierarchical nanostructure. The sequence of phase transformation leading to the formation of this hierarchical nanostructure and its role in controlling the strength of the alloy is the subject of this publication.

Results

Tensile property measurement. The mechanical properties of the specimens characterized in terms of ultimate tensile strength (TS) and 0.2% offset yield strength (YS) subjected to the three solution treated and annealed (STA) conditions are shown in Fig. 1a. As shown in the figure, both the TS and YS of the specimens increased considerably with the increase in solution treatment temperature. In an average the TS increased from ~1,535 to 1,690 MPa and YS increased from 1,490 to 1,655 MPa for an increase in solution treatment temperature from 1,300 to 1,450 °F. The nominal strain to failure of all the specimens, regardless of the STA condition was approximately in the 4–6% range. The tensile strength of the STA treated low-cost Ti185 alloy produced here is ~15% greater than the currently available strongest commercial titanium alloy while the strain to failure (elongation) is at par, as shown in Fig. 1b (plotted using CES Selector software, Granta Design, Cambridge, UK). For comparison with a second database of mechanical properties of titanium alloys, as per 2012 ASM Titanium alloy database[16], the range of ultimate tensile strength of five different class of titanium alloys is as follows, alpha and near alpha alloys (581–1,240 MPa), alpha-beta alloys (827–1,580 MPa), alpha-beta/metastable beta (945 MPa), metastable beta alloys (545–1,573 MPa) and unalloyed and modified Ti (331–662 MPa). From the database, the highest mechanical properties are exhibited by Ti–6Al–2Sn–4Zr–6Mo alloy in STA heat treated condition with 1,580 MPa, the second highest strength was for Ti–15Mo–5Zr–3Al also in STA condition at 1,573 MPa and then the next strongest is Ti–13V–11Cr–3Al alloy with 1,483 MPa (ref. 16). The ultimate tensile strength value of 1,690 MPa achieved by the hierarchically nanostructured Ti185 alloy after 1,450-900-2 STA is higher than the ultimate strength of the strongest titanium alloys in this data base also. It is also noteworthy to mention that these alloys as compared with Ti185 have higher concentrations or additions of expensive elements.

Microstructure characterization of STA conditions. To explore the origin of increase in strength with increasing solution treatment temperature we analyzed the microstructure of the specimens subjected to three STA conditions. The representative scanning electron microscopy (SEM) backscattered electron (BSE) images of the three STA conditions 1,300-900-2, 1,375-900-2 and 1,450-900-2 are shown in Fig. 2. In general both grain boundary α-phase and intragranular α-phase precipitates are present in the specimens corresponding to the three STA conditions. The large intragranular α-phase precipitates seen in the SEM BSE images in Fig. 2 will be referred to as primary intragranular α in the remainder of this paper. The primary intragranular α precipitates in the 1,300-900-2 STA condition have an average length of 2.2 µm and width of 0.4 µm, with an inter-particle spacing of roughly 0.6 µm. In 1,375–900–2 STA condition the average length and width of primary α are 2.78 µm and 0.52 µm, respectively, with inter-particle spacing of roughly 0.89 µm. Finally in the 1,450-900-2 STA condition, the primary α precipitates on an average are 4.2 µm long, 0.8 µm wide with an inter-particle spacing of roughly 1.4 µm. To this point, for the three STA conditions, it can be noted that (i) the morphology of grain boundary α is roughly the same for all three STA conditions; (ii) the aspect ratio of the primary α precipitates are also roughly the same and are in the range 5.5–5.25 for all the three STA conditions; and (iii) the number density of primary α precipitates decreases significantly with increasing solution treatment temperature.

The decrease in the number density as well as increase in the inter-particle spacing should lead to decrease in the strength[17], which is in contradiction to the trend observed for tensile strength in Fig. 1b. Hence we carried out transmission electron microscopy (TEM) analysis of the microstructure of the specimens to check the presence of any smaller (nanometer) size scale precipitates. The bright field TEM images of STA condition 1,300-900-2 are shown in Fig. 3a,b, whereas the same for STA condition 1,450-900-2 are shown in Fig. 3c,d. The bright-field TEM images of STA 1,300-900-2 specimen show the presence of grain boundary α, primary intragranular α and additionally some fine scale α precipitates of an average width of 54 nm and length of 386 nm with an inter-particle spacing of roughly 80 nm. This fine scale α precipitate will be referred to as secondary α precipitate in the remainder of this paper. The secondary α precipitates seen in the TEM bright-field images of STA 1,450-900-2 (Fig. 3c,d) are clearly much smaller (34 nm wide and 66 nm long) and with an inter-particle spacing less than 45 nm, and hence much more densely distributed than the secondary α in STA 1,300-900-2.

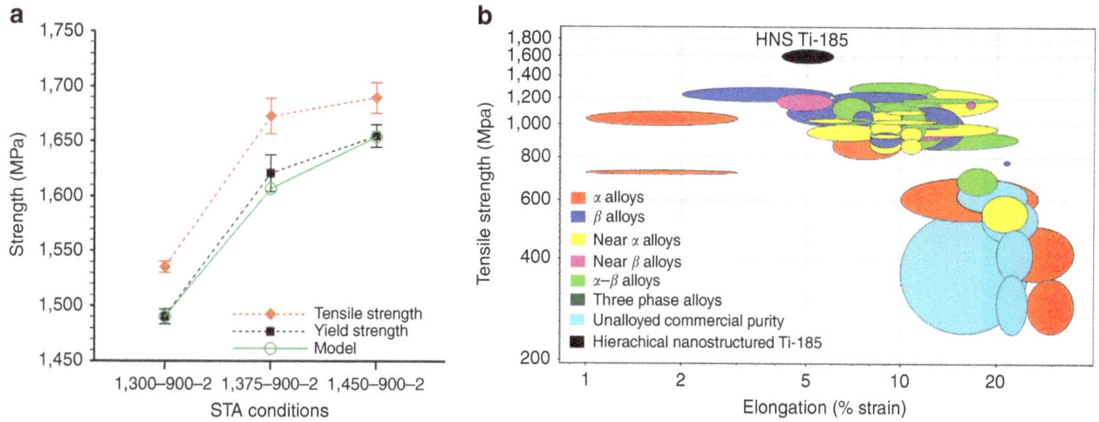

Figure 1 | Tensile strength comparison with model and other Ti alloys. (**a**). Ultimate tensile and yield strength of Ti185 for three STA conditions. Also shown in (**a**) is the prediction of empirical model in equation (2). (**b**) Comparison of tensile strength and strain to failure (elongation) of Ti185 alloy with other commercially available Titanium alloys. The graph in (**b**) is plotted using CES Selector software, Granta Design, Cambridge, UK.

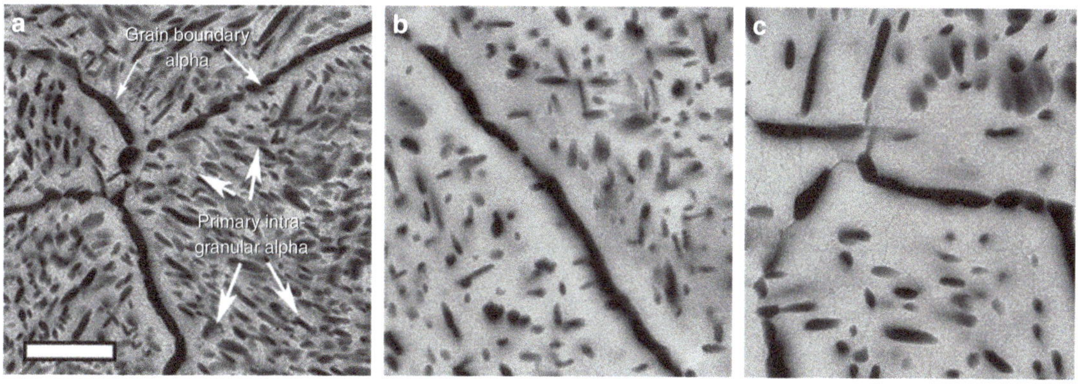

Figure 2 | Microstructure of STA conditions. Representative microstructure of the specimens subjected to three STA conditions (**a**) 1,300-900-2, (**b**) 1,375-900-2, and (**c**) 1,450-900-2. Length of scale bar is 5 μm.

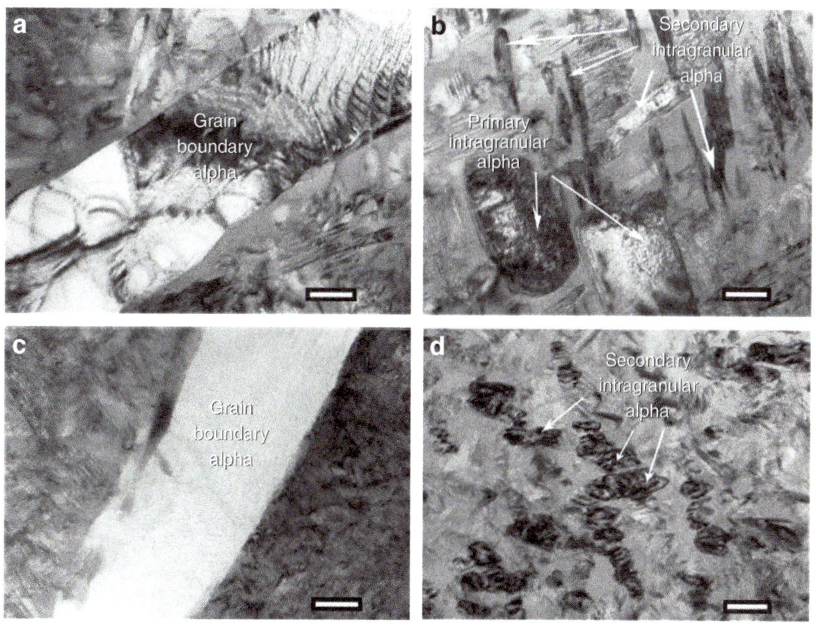

Figure 3 | TEM analysis of STA conditions. Bright-field TEM images showing (**a**) grain boundary α and intragranular α and (**b**) primary and secondary intragranular α in a STA 1,300-900-2 specimen. Scale bars are 200 nm in both (**a**-**c**) grain boundary α and nanoscale secondary intragranular α (scale bar is 200 nm) and (**d**) high-density nanoscale secondary intragranular α (scale bar is 100 nm) in a STA 1,450-900-2 specimen.

Subsequently atom probe tomography (APT) of these alloys was conducted to understand the solute portioning between the α- and the β-phase. The APT specimens were specifically lifted out from the intragranular regions of the specimens. The APT reconstructions from STA 1,300-900-2 specimen shown in Fig. 4a,b captured a portion of α-phase (enriched in Al (purple)) together with β-phase (enriched in V (red) and Fe (green)) on either side. The APT reconstruction shows a 63.1 × 63.3 × 63.1 nm^3 volume within the intragranular region of 1,300-900-2 specimen. From the APT reconstruction, α precipitate seems to have a thickness of 20 nm with length and width much larger than the reconstructed volume, pointing to a plate-like morphology. The compositional partitioning between α- and β-phases is plotted using a proximity histogram[18] across an 84 at% Ti isocomposition surface (Fig. 4c). The α precipitates in the APT reconstructions from STA 1,450-900-2 specimen shown in Fig. 4d,e have smaller size and a discrete morphology in contrast to the plate-like α precipitates observed in STA 1,300-900-2 specimen. The APT reconstruction from 1,450-900-2 specimen corresponds to 62.2 × 59.2 × 132.7 nm^3 volume from the intragranular region. The proximity histogram shown in Fig. 4f shows the corresponding compositional partitioning between α- and β-phases, and is very similar to elemental partitioning in STA 1,300-900-2 specimen. The steady-state concentration of α- and β-phases estimated from the steady state regions on either side of both the proximity histograms are given in Table 1. The steady-state compositions of α- and β-phases in both STA 1,300-900-2 and STA 1,450-900-2 specimens are comparable, which points to attainment of thermodynamically equilibrium solute partitioning during the 2 h aging heat treatment at 900 °F.

Microstructure characterization of ST conditions. To investigate the origin of these hierarchical nanostructures we conducted additional microstructural characterization of the Ti185 alloy subjected only to the solutionizing treatment at 1,300 and 1,450 °F for 1 h followed by water quenching. These conditions will be named as 1,300ST and 1,450ST conditions from here onwards. The SEM BSE image of 1,300ST shows the presence of grain boundary α and primary intragranular α (Fig. 5a). Bright field TEM image of 1,300ST condition is shown in Fig. 5b where primary α-phase precipitates and grain boundary α precipitates are marked. Additional distinct reflections arising from omega phase were observed in the [113] zone axis selected area electron diffraction pattern (SAD) of the β phase in between primary intragranular α phase precipitates (Fig. 5c inset). The ω reflections within the dotted circle in SAD were used to form a dark field TEM image highlighting the presence of nanoscale ω phase regions in the β phase matrix of 1,300ST (Fig. 5c). In comparison the SEM BSE image of 1,450ST clearly shows a much lower number density of intragranular primary α precipitates (Fig. 5d). The bright field TEM image of 1450ST only shows presence of

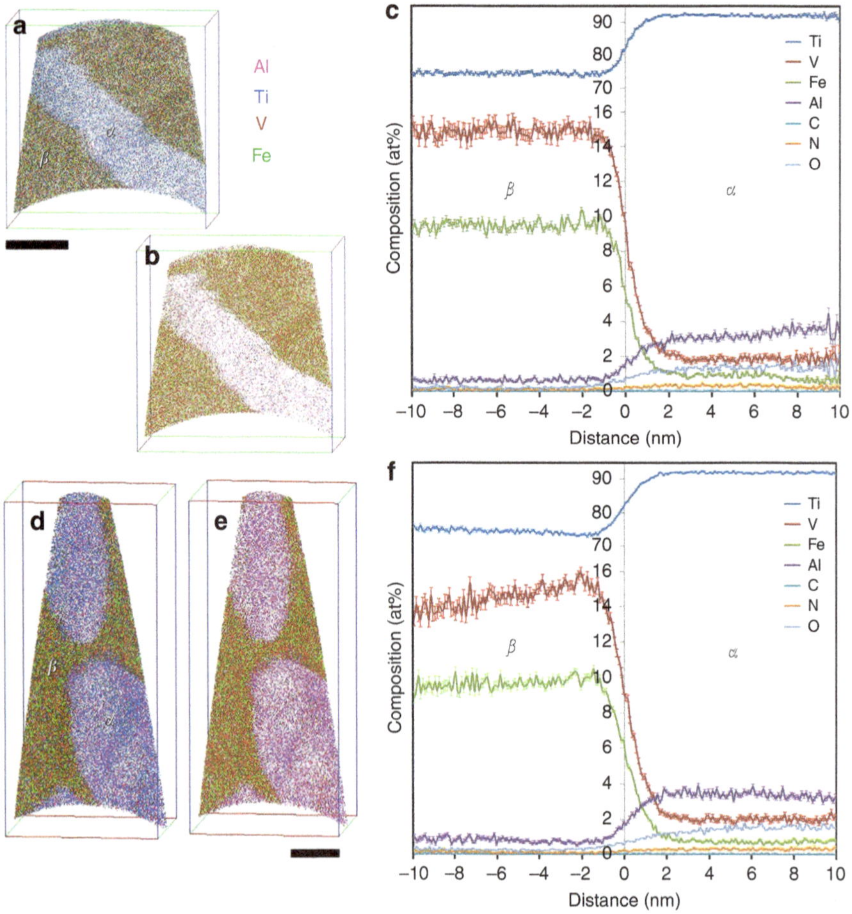

Figure 4 | APT analysis of STA conditions. (**a**) All-ionic view of the APT reconstruction showing Ti in blue, V in red, Al in purple and Fe in green, (**b**) ionic view of only Fe, V and Al ions, and (**c**) the solute partitioning between α and β phase showing V and Fe enrichment in the β phase and Ti and Al enrichment in the α phase in STA 1,300-900-2 specimen. (**d**) All-ionic view of the APT reconstruction showing the stubby α precipitates and the partitioning of Al (purple) from V (red) and Fe (green) is shown in (**e**). The compositional partitioning between α and β phases in STA 1,450-900-2 (**f**). Scale bars for (**a**,**b**,**d**,**e**) are 20 nm.

Table 1 | Phase composition analysis of STA conditions.

STA ID	Phase	Measured elemental composition from APT analysis (at %)						
		Ti	V	Fe	Al	C	N	O
1,300-900-2	α	91.95 ± 0.27	2.03 ± 0.14	0.75 ± 0.08	3.34 ± 0.18	0.05 ± 0.02	0.33 ± 0.06	1.56 ± 0.12
1,450-900-2	α	92.06 ± 0.39	1.91 ± 0.2	0.79 ± 0.12	3.48 ± 0.26	0.05 ± 0.03	0.29 ± 0.08	1.42 ± 0.17
1,300-900-2	β	74.93 ± 0.48	14.08 ± 0.39	9.55 ± 0.33	0.9 ± 0.10	0.07 ± 0.03	0.17 ± 0.05	0.3 ± 0.06
1,450-900-2	β	74.52 ± 0.46	14.90 ± 0.38	9.49 ± 0.31	0.67 ± 0.09	0.05 ± 0.02	0.14 ± 0.04	0.24 ± 0.05

A comparison of steady-state α and β-phase composition of STA 1,300-900-2 and STA 1,450-900-2 specimens.

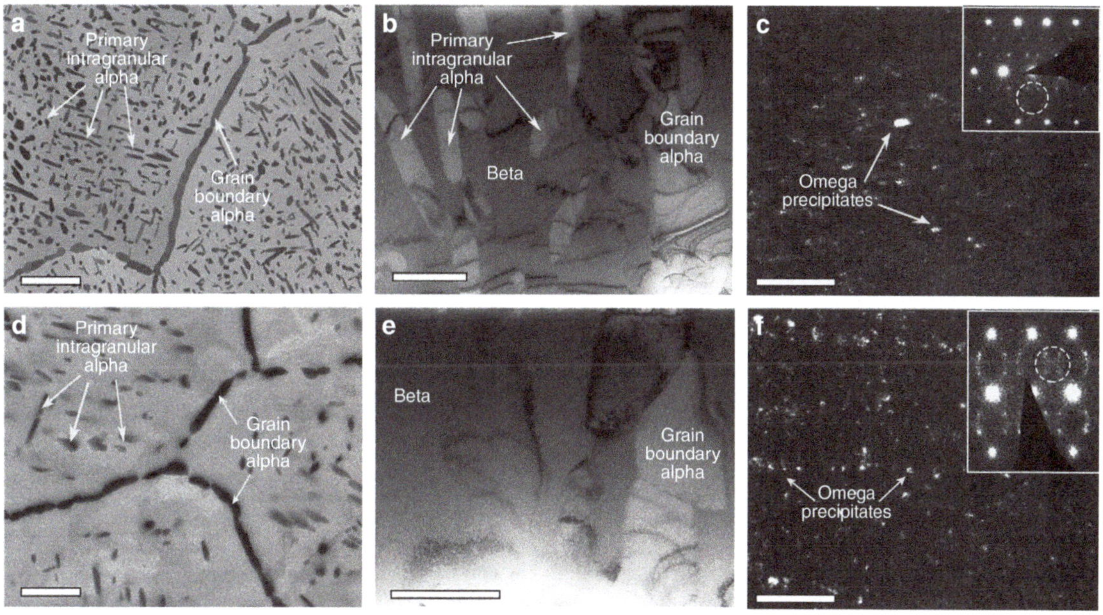

Figure 5 | Microstructure of ST conditions. (**a**) SEM BSE image of 1300ST showing grain boundary α and primary intragranular α (scale bar is 5 μm) (**b**) Bright-field TEM image showing grain boundary α and primary intragranular α with β matrix in between devoid of secondary α in a 1,300ST (scale bar is 1 μm); (**c**) dark field image of 1,300ST showing fine scale omega phase distributed within the β phase regions (scale bar is 100 nm). Dark field image was formed using omega phase reflections inside the dotted circle in the β[113] zone axis SAD shown as inset. (**d**) SEM BSE image of 1,450ST showing grain boundary α and primary intragranular α (scale bar is 5 μm) (**e**) Bright-field TEM image showing grain boundary α and β matrix devoid of secondary alpha in 1,450ST (scale bar is 1 μm) (**f**) dark field image showing distribution of fine scale omega phase within the β phase regions(scale bar is 100 nm). Dark field image was formed using omega phase reflections inside the dotted circle in the β[110]zone axis SAD shown as inset.

Table 2 | Beta phase composition analysis of ST conditions.

ST ID	Phase	Measured elemental composition from APT analysis (at %)						
		Ti	V	Fe	Al	C	N	O
1,300ST	β	79.90 ± 0.05	10.46 ± 0.04	6.52 ± 0.03	1.96 ± 0.02	0.06 ± 0.003	0.33 ± 0.01	0.78 ± 0.01
1,450ST	β	82.35 ± 0.04	8.87 ± 0.03	5.74 ± 0.02	2.04 ± 0.01	0.03 ± 0.002	0.15 ± 0.004	0.8 ± 0.01

Beta phase composition of only solutionized and water quenched samples measured from APT analysis.

grain boundary α and β phase regions (Fig. 5c). The [110] zone axis SAD of β phase, given in Fig. 5d also shows clear evidence of additional ω-phase reflections. The dark field TEM image formed using the ω-phase reflections shows the presence of qualitatively larger number density ω phase precipitates in the beta phase matrix of 1,450ST condition (Fig. 5d). From these TEM results, it is clear that ω phase precipitates are present after solutionizing treatment in 1,300ST and 1,450ST conditions with qualitatively different number densities. Subsequently APT analysis of both 1,300ST and 1,450ST conditions was conducted to quantify the composition of β phase regions in between the primary intragranular α precipitates. From the APT measured composition of β phase given in Table 2, it is clear that the β phase in 1,300ST has a higher concentration of β stabilizing elements (V and Fe concentration) than in 1,450ST condition.

Discussion

We now turn to explain the sequence of phase transformation leading to the formation of hierarchical nanostructure in the low-cost powder processed Ti185 specimens subjected to the three STA conditions. The increase in solution treatment temperature increases the stability of the β-phase, which explains the decrease in the volume fraction of intragranular primary α with increasing solution temperature. This in turn suggests that the large intragranular primary α in all three STA treated Ti185 specimens formed during the initial solution treatment stage, which is evident from the significant difference in intragranular primary α volume fraction observable in the SEM BSE images between 1,300ST and 1,450ST conditions given in Fig. 5a,d. The Ti185 alloy considered here has the same alloying elements but different concentration as in β-titanium alloy, Ti–10V–2Fe–3Al (Ti1023). A number of studies have been reported on varying the morphology and volume fraction of α-phase and metastable ω-phase precipitates by thermomechanical treatment of Ti1023 alloy[19–22]. The ω-phase was observed in the solution treated and water quenched Ti1023 specimens[23]. The formation of cuboidal shaped metastable ω-phase was also reported in another low misfit Ti–V-based alloy upon water quenching past solution treatment[24]. The formation of ω-phase in β-titanium alloys is in general associated with local rejection of β stabilizing elements resulting in local compositional instability[25–27]. This suggests that the microstructure of all three STA treated Ti185 specimens upon water quenching past solution treatment contained: grain boundary α, intragranular primary α, and retained β with ω-phase precipitates. The dark field TEM images of Ti185 specimens conditioned to 1,300ST and 1,450ST, Fig. 5c,f, clearly show that such is the case.

The primary α formed during the initial solution treatment stage would reject the β-stabilizing elements into the remaining β-phase hence increasing the solute concentration in the remaining β-phase. The high volume fraction of primary α precipitates in 1,300ST compared with 1,450ST explains the observed increase in beta stabilizing element concentration in the β-phase in 1,300ST (Table 2). The temperature for the start of ω-phase transformation during water quenching from β or α–β microstructure was found to be lowered with increasing solute content in the β-phase for Ti–V binary alloys[28], rendering β to ω-phase transformation unstable for higher solute concentrations in β-phase. Hence the β to ω-phase transformation is comparatively more stable during water quenching past solution treatment at 1,450 °F compared with 1,300 °F due to the lower solute concentration in the β-phase. This results in high number density (or say volume fraction) of ω-phase precipitates in the retained β-phase in the specimens water quenched past solution treatment at 1,450 °F compared with the specimens water quenched past solution treatment at 1,300 °F as qualitatively visible in the dark field TEM images in Fig. 5c,f.

The solution treated water quenched specimens were then subjected to ageing heat treatment at 900 °F for 2 h to complete the STA treatment. The long-term annealing at low temperatures leads to α-phase precipitation from existing ω-phase precipitates[28–32]. For Ti1023 alloy it was suggested in ref. 32 that fine-scale, blocky α-phase precipitation occurred from existing ω-phase precipitates. Hence the presence of high number density of ω-phase precipitates in the specimens solution treated at 1,450 °F resulted in high number density of fine-scale, blocky α-phase precipitates past aging treatment at 900 °F. However, the specimens solution treated at 1,300 °F had lower number density of the ω-phase precipitates in the retained β-phase, hence, comparatively low number density of secondary α-phase precipitates nucleated during the aging treatment.

In summary, the final microstructures formed in Ti185 specimens following the three STA conditions have: (i) grain boundary α, with roughly similar morphology irrespective of the solution treatment temperature; (ii) primary intragranular α, whose number density decreases with increasing solution treatment temperature; (iii) secondary intragranular α, whose number density increases with increasing solution treatment temperature; and (iv) α and β-phases with chemical compositions being independent of the solution treatment temperature. The detailed characterization of both STA and ST conditions permitted to clearly understand the microstructure evolution pathway leading to the final hierarchical nanostructures formed following the three STA conditions. The detailed understanding of the final microstructure for the three STA conditions underlies the understanding of increase in strength with increasing solution treatment temperature.

The roughly similar steady-state partitioning of alloying elements between α- and β-phase precipitates as well as presence of similar grain boundary α-phase precipitates irrespective of the STA condition, suggests that the observed hierarchical nanostructure (two size scales of α-phase precipitates) within the grain is the sole factor contributing to strengthening with increasing solution treatment temperature. The effect of bimodal distribution of α-phase precipitate on the critical resolved shear stress (CRSS) of the material can be empirically written as[33–36]:

$$\tau^k = \tau_p^k + \tau_s^k \tag{1}$$

where, τ is the overall CRSS of the material, τ_p is the contribution of primary α-phase precipitates,τ_s is the contribution of secondary α-phase precipitates and k is an adjustable parameter. The value of k lies in the range $1 \leq k \leq 2$ with $k = 1$ corresponding to linear superposition and $k = 2$ corresponding to the Pythagorean superposition. The STA conditioned Ti185 contains distribution of large primary α (more likely to be impenetrable that is, strong) in the midst of fine secondary α (more likely to be shearable that is, weak) and for such a microstructure linear superposition is more likely to be dominant[17,34]. Next, we know that the critical stress required for a dislocation to overcome obstacles is inversely proportional to the average spacing of the obstacles and the CRSS can be related to the yield strength of the material by the 'Taylor factor'[17]. Hence, equation (1) under the assumption of linear superposition can be rewritten as:

$$\sigma_y = \frac{K_p}{l_p} + \frac{K_s}{l_s} \tag{2}$$

where, σ_y is the yield strength of the material, l_p is the inter-particle spacing of the primary α-phase precipitates, l_s is the inter-particle spacing of the secondary α-phase precipitates and K_p and K_s are constants. Note, all the constants including the 'Taylor factor' are merged in K_p and K_s. The values of l_p for the three STA conditions 1,300-900-2, 1,375-900-2 and 1,450-900-2 in μm are roughly 0.6, 0.89 and 1.4, respectively while the values of l_s for the three STA conditions in μm are roughly 0.08, 0.054 and 0.045, respectively. The values of the constants K_P and $K_s$were found to be 442.6 MPa-μm and 60.2 MPa-μm, respectively. The prediction of the empirical model in equation (2) is compared with the yield strength of the three STA conditioned Ti185 alloy in Fig. 1a and good agreement is noted.

Hence following equation (2) it can be now concluded that the decrease in the number density of primary α (increase in inter-particle spacing) with increase in solution treatment temperature decreases the strength of the alloy, which is not only compensated but is surpassed by the increase in the number density (decrease in inter-particle spacing) of fine-scale secondary α precipitates. The existence of these two competing changes in the microstructure also explains the relatively small increment in strength by increasing the solution treatment temperature from 1,375 to 1,400 °F compared with the large increment in strength by increasing the solution treatment temperature from 1,300 to 1,375 °F. On the other hand, since intergranular fracture is the dominant failure mechanism in these alloys, the roughly similar strain to failure irrespective of the STA conditions can be dedicated to the uniformity of grain boundary α-phase precipitates[37].

In conclusion, the increase in uniformly distributed, fine-scale secondary α precipitates led to the observed increase in strength with increase in solution treatment temperature. Also due to the uniformity of grain boundary α precipitates the % elongation of the alloy remained at 4–6% irrespective of the solution treatment temperature. The authors believe that the excellent mechanical properties, when additionally complemented by the low cost of this alloy manufactured by TiH_2 based powder processing can indeed provide an attractive alternative for replacing more expensive titanium alloys as well as other heavier engineering alloys. We anticipate that this approach can also be used to fabricate other β-titanium alloys with hierarchical nanostructure and superior strength.

Methods

Ti185 Alloy production and heat treatment. Sintered Ti185 billets produced using low cost TiH2-based powder processes were obtained from Advance Materials Products, Inc., (ADMA) Hudson, Ohio. The billets were then thermo-mechanically processed (hot rod-rolled) to achieve a reduction ratio of 31:1 to form 16 mm diameter rods, for further details see ref. 15. The samples were later solution treated (below the β transus temperature, which is 1,526 °F for this alloy) at 1,300, 1,375 and 1,450 °F for 1 h and aged at 900 °F for 2 h. (Note: for convenience degree Fahrenheit is used in the current work instead of degree Celsius). Hereafter this heat treatment condition is abbreviated as STA and the three STA conditions are designated as 1,300-900-2, 1,375-900-2 and 1,400-900-2.

Tensile testing. The mechanical behavior of 4–5 specimens corresponding to each STA condition were tested under uniaxial tension in a MTS Model 312.21 test frame with a 45kN Lebow 3116 load cell at a constant nominal strain rate of $10^{-3}s^{-1}$. The extension in the gauge length was measured using a 12.7 mm MTS Model 634.32E-24 extensometer.

Microstructural characterization. The microstructure of all three STA conditions were characterized using SEM BSE imaging in an FEI Quanta dual beam focused ion beam (FIB) system. The TEM samples and APT samples were also prepared using the FEI Quanta dual beam FIB. TEM imaging was performed using JEOL 2010F transmission electron microscope. The APT specimen preparation method by site specific FIB lift-out and annular milling[38] aided in selecting specific regions within the intragranular region of the Ti185 alloys. A CAMECA LEAP 4000XHR system equipped with pulsed UV laser (355 nm wavelength) was used to perform APT experiments using 20 pJ laser pulse energy and 0.005 atoms/pulse evaporation rate at specimen temperature of 60 K. The APT results were reconstructed and analyzed using Interactive Visualization and Analysis Software (IVAS) 3.6.8 using standard reconstruction procedure[39]. The reconstruction and composition measurement of all APT data was done using x–y–z voxels of $1 \times 1 \times 1$ nm size with delocalization of $3 \times 3 \times 1.5$ nm, respectively. For composition measurements, the mass-to-charge spectra peak at 27 Da has an overlap of Al^{+1} and the 54 isotope of Fe^{+2}. This peak overlap was de-convoluted using the count of non-overlapping Fe^{+2} isotope counts at 28, 28.5 and 29 Da and estimating corresponding Fe^{+2} counts at 27 Da using the natural abundance of 5.845%. The error bars for composition measurements were estimated based on statistical error for measured atom count as per the equation $\sigma = \sqrt{\frac{C^i*(1-C^i)}{N}}$, where C^i corresponds to measured atomic concentration fraction of the element and N is the total atom counts in the bin.

References

1. Canadell, J. G. *et al.* Contributions to accelerating atmospheric CO_2 growth from economic activity, carbon intensity, and efficiency of natural sinks. *Proc. Natl Acad. Sci. USA* **104,** 18866–18870 (2007).
2. Raupach, M. R. *et al.* Global and regional drivers of accelerating CO_2 emissions. *Proc. Natl Acad. Sci. USA* **104,** 10288–10293 (2007).
3. DeCicco, J. & Fung, F. (ed). in *Global Warming On the Road: The Climate Impact of Americas Automobiles* (Environmental Defense, 2006).
4. Executive Office of the President, US. *Driving Efficiency Cutting Costs for Families at the Pump and Slashing Dependence on Oil* (Executive Office of the President, 2011).
5. Boyer, R. R. & Briggs, R. D. The use of beta titanium alloys in the aerospace industry. *J. Mater. Eng. Perform.* **14,** 681–685 (2005).
6. Nyakana, S. L., Fanning, J. C. & Boyer, R. R. Quick reference guide for beta titanium alloys in the 00s. *J. Mater. Eng. Perform.* **14,** 799–811 (2005).
7. Boyer, R. R. Attributes, characteristics, and applications of titanium and its alloys. *Jom-Us* **62,** 21–24 (2010).
8. Lu, K. The future of metals. *Science* **328,** 319–320 (2010).
9. Esteban, P. G., Ruiz-Navas, E. M., Bolzoni, L. & Gordo, E. Low-cost titanium alloys? Iron may hold the answers. *Met. Powder Rep.* **63,** 24–27 (2008).
10. Boyer, R., Welsch, G. & Collings, E. W. *Materials Properties Handbook - Titanium Alloys* (ASM International, 1998).
11. Morrissey, R. *Frequency and Mean Stress Effects in High Cycle Fatigue of Ti-6Al-4V.* Master's Thesis 1997) http://www.dtic.mil/dtic/tr/fulltext/u2/a330696.pdf.
12. Ivasishin, O. M., Savvakin, D. G., Bondareva, X. O. & Dekhtyar, O. I. in Synthesis of PM Titanium Alloys Using Titanium Hydride Powder: Mechanism of Densification *Proceedings of the 10th World Conference on Titanium Held at the CCH-Congress Center Hamburg, Germany 13–18 July, 2003* 495–502 (Wiley-VCH Verlag GmbH & Co. KGaA, 2003).
13. Ivasishin, O. M., Eylon, D., Bondarchuk, V. I. & Savvakin, D. G. in Diffusion during powder metallurgy synthesis of titanium alloys *Diffusion and Diffusional Phase Transformations in Alloys* 277, 177–185 (Trans Tech Publications Ltd, 2008).
14. Ivasishin, O. M. & Savvakin, D. G. in The impact of diffusion on synthesis of high-strength titanium alloys from elemental powder blends *TMS 2010 Spring Symposium on Cost-Affordable Titanium III, February 14, 2010; 18 February 2010* 113–121 (Trans Tech Publications Ltd, 2010).
15. Joshi, V. V. *et al.* Development of Ti-6Al-4V and Ti-1Al-8V-5Fe alloys using low-cost TiH 2 powder feedstock. *J. Mater. Eng. Perform.* **22,** 995–1003 (2013).
16. Kaufman, J. G. Knovel http://app.knovel.com/hotlink/toc/id:kpTAD00001/titanium-alloy-database/titanium-alloy-database.
17. Martin, J. W. *Precipitation Hardening: Theory and Applications* (Butterworth-Heinemann, 2012).
18. Hellman, O. C. *et al.* Analysis of three-dimensional atom-probe data by the proximity histogram. *Microsc. Microanal.* **6,** 437–444 (2000).
19. Boyer, R. R. & Kuhlman, G. W. Processing properties relationships of Ti-10V-2Fe-3Al. *Metall. Trans. A* **18,** 2095–2103 (1987).
20. Boyer, R. R. & Kuhlman, G. W. Processing properties relationships of Ti-10V-2Fe-3Al forgings. *Jom-J Min. Met. Mat. S* **37,** A75–A75 (1985).
21. Jha, S. K. & Ravichandran, K. S. Effect of mean stress (stress ratio) and aging on fatigue-crack growth in a metastable beta titanium alloy, Ti-10V-2Fe-3Al. *Metall. Mater. Trans. A* **31,** 703–714 (2000).
22. Srinivasu, G., Natraj, Y., Bhattacharjee, A., Nandy, T. K. & Rao, G. V. S. N. Tensile and fracture toughness of high strength beta titanium alloy, Ti-10V-2Fe-3Al, as a function of rolling and solution treatment temperatures. *Mater. Design* **47,** 323–330 (2013).
23. Bhattacharjee, A., Varma, V. K., Kamat, S. V., Gogia, A. K. & Bhargava, S. Influence of beta grain size on tensile behavior and ductile fracture toughness of titanium alloy Ti-10V-2Fe-3Al. *Metall. Mater. Trans. A* **37A,** 1423–1433 (2006).
24. Ng, H. P. *et al.* Phase separation and formation of omega phase in the beta matrix of a Ti-V-Cu alloy. *Acta Mater.* **59,** 2981–2991 (2011).
25. Devaraj, A. *et al.* Experimental evidence of concurrent compositional and structural instabilities leading to omega precipitation in titanium-molybdenum alloys. *Acta Mater.* **60,** 596–609 (2012).
26. Devaraj, A. *et al.* Three-dimensional morphology and composition of omega precipitates in a binary titanium-molybdenum alloy. *Scripta Mater.* **61,** 701–704 (2009).
27. Nag, S. *et al.* Novel mixed-mode phase transition involving a composition-dependent displacive component. *Phys. Rev. Lett.* **106** (2011).
28. Menon, E. S. K. & Krishnan, R. Phase-transformations in Ti-V alloys .2. Alpha-and-omega-formation. *J. Mater. Sci.* **18,** 375–382 (1983).
29. Prima, F., Vermaut, P., Texier, G., Ansel, D. & Gloriant, T. Evidence of alpha-nanophase heterogeneous nucleation from omega particles in a beta-metastable Ti-based alloy by high-resolution electron microscopy. *Scripta Mater.* **54,** 645–648 (2006).
30. Nag, S. *et al.* Omega-assisted nucleation and growth of alpha precipitates in the Ti-5Al-5Mo-5V-3Cr-0.5Fe beta titanium alloy. *Acta Mater.* **57,** 2136–2147 (2009).
31. Xu, Z. *et al.* Simulation of heterogeneous atom probe tip shapes evolution during field evaporation using a level set method and different evaporation models. *Comput. Phys. Commun.* **189,** 106–113 (2015).
32. Terlinde, G. T., Duerig, T. W. & Williams, J. C. Microstructure, tensile deformation, and fracture in aged Ti 10V-2Fe-3Al. *Metall. Trans. A* **14,** 2101–2115 (1983).
33. Brown, L. M. & Ham, R. K. in *Crystals* (eds Kelly, A. & Nicholson, R. B.), 9 (Applied Science Publishers, London, 1971).
34. Kocks, U. The theory of an obstacle-controlled yield strength—report after an international workshop. *Mater. Sci. Eng.* **27,** 291–298 (1977).
35. Ardell, A. Precipitation hardening. *Metall. Trans. A* **16,** 2131–2165 (1985).
36. Nembach, E. *Particle Strengthening of Metals and Alloys* (J. Wiley, 1997).
37. Osovski, S., Srivastava, A., Williams, J. C. & Needleman, A. Grain boundary crack growth in metastable titanium beta alloys. *Acta Mater.* **82,** 167–178 (2015).

38. Thompson, K. *et al. In situ* site-specific specimen preparation for atom probe tomography. *Ultramicroscopy* **107**, 131–139 (2007).
39. Larson, D. J. *et al. Local Electrode Atom Probe Tomography: A User's Guide* (Springer, 2013).

Acknowledgements

The authors would like to thank the US Department of Energy Vehicle Technologies Office (DOE/VTO) Propulsion materials program for the financial support provided for this work. This work was also partially supported by the Laboratory Directed Research and Development (LDRD) program of Pacific Northwest National Laboratory (PNNL) as a part of the Chemical Imaging Initiative. Pacific Northwest National Laboratory is operated by Battelle Memorial Institute for the US Department of Energy under contract DE-AC05-76RLO1830. The research was performed using the Environmental Molecular Sciences Laboratory, a national scientific user facility sponsored by the Department of Energy's Office of Biological and Environmental Research and located at PNNL. Authors would also like to thank Stuart Dyer for providing the graph from CES Selector software, Granta Design, Cambridge, UK, www.grantadesign.com.

Author contributions

A.D., V.J. and C.L. conceived the research. V.M. and V.A.D. fabricated the Ti185 alloy through low cost TiH_2 based powder processing. V.J. performed the STA heat treatments and tensile testing. A.D. executed all the microstructural characterization by SEM, TEM and APT including specimen preparation, characterization experiments and analysed the results. S.M. assisted in APT specimen preparation and APT experiments for ST conditions. A.S. performed the mathematical model for linking mechanical property with precipitate size and interparticle spacing. A.D., V.J. and A.S. wrote the manuscript with input from all the other authors.

Additional information

Competing financial interests: The authors declare no competing financial interests.

Balancing surface adsorption and diffusion of lithium-polysulfides on nonconductive oxides for lithium–sulfur battery design

Xinyong Tao[1,2,*], Jianguo Wang[3,*], Chong Liu[2], Haotian Wang[2], Hongbin Yao[2], Guangyuan Zheng[2], Zhi Wei Seh[2], Qiuxia Cai[3], Weiyang Li[2], Guangmin Zhou[2], Chenxi Zu[2] & Yi Cui[2,4]

Lithium-sulfur batteries have attracted attention due to their six-fold specific energy compared with conventional lithium-ion batteries. Dissolution of lithium polysulfides, volume expansion of sulfur and uncontrollable deposition of lithium sulfide are three of the main challenges for this technology. State-of-the-art sulfur cathodes based on metal-oxide nanostructures can suppress the shuttle-effect and enable controlled lithium sulfide deposition. However, a clear mechanistic understanding and corresponding selection criteria for the oxides are still lacking. Herein, various nonconductive metal-oxide nanoparticle-decorated carbon flakes are synthesized via a facile biotemplating method. The cathodes based on magnesium oxide, cerium oxide and lanthanum oxide show enhanced cycling performance. Adsorption experiments and theoretical calculations reveal that polysulfide capture by the oxides is via monolayered chemisorption. Moreover, we show that better surface diffusion leads to higher deposition efficiency of sulfide species on electrodes. Hence, oxide selection is proposed to balance optimization between sulfide-adsorption and diffusion on the oxides.

[1] College of Materials Science and Engineering, Zhejiang University of Technology, Hangzhou 310014, China. [2] Department of Materials Science and Engineering, Stanford University, Stanford, California 94305, USA. [3] College of Chemical Engineering, Zhejiang University of Technology, Hangzhou 310014, China. [4] Stanford Institute for Materials and Energy Science, SLAC National Accelerator Laboratory, Menlo Park, California 94025, USA. * These authors contributed equally to this work. Correspondence and requests for materials should be addressed to Y.C. (email: yicui@stanford.edu).

Rechargeable lithium-sulfur (Li-S) batteries have recently become one of the more exciting energy storage systems due to the low-cost and high-specific energy of sulfur cathodes[1-27]. Although there have been significant developments for designing state-of-the-art Li-S batteries in the past two decades, the practical application is still hindered by many material challenges, including dissolution of intermediate lithium polysulfides (Li_2S_x, $x>3$) in the electrolyte[28], large volumetric expansion (80%) of sulfur upon lithiation[6], and poor electronic/ionic conductivity of sulfur and lithium sulfide (Li_2S) (ref. 6). To date, tremendous efforts have been made to solve the above problems by constructing advanced composite cathode materials. One effective strategy is the encapsulation of sulfur to prevent the leakage of active materials and suppress the shuttle effect of high-order Li_2S_x (refs 3,6). Oxides[6], carbon[3,29], polymers[30] and metals[31] are proved to be good matrix materials for the encapsulation of sulfur[19]. The second approach is the controllable deposition of the discharge product Li_2S, which is an ionic and electronic insulator[4]. The detaching and irreversible phase transformation of Li_2S is considered as the main reason for capacity fading[4,32]. The third strategy is using Li_2S as a starting cathode material, which undergoes volumetric contraction instead of the expansion in the case of sulfur[20]. In addition, Li_2S-based cathodes can be paired with lithium metal-free anodes such as graphite, silicon and alloys[33], thus suppressing the dendrite growth and the corresponding safety concerns of lithium-metal anodes.

All the previous research reveals the importance of understanding the sulfide species interaction with the matrix materials. Our earlier work pointed out that the usual carbon substrates interact with Li_2S_x weakly but the polar group enabled strong interaction with Li_2S_x, which can facilitate the Li_2S_x trapping and promote the attachment of solid Li_2S_2 and Li_2S and improve the cycling stability of Li-S batteries[34]. Many similar examples followed-up using polymers[35,36], oxides[4,19], sulfides[20], functionalized graphene[18,26,27,37,38], metal organic framework[39] and nitrogen doped carbon[36,40], which all have polar surfaces to adsorb Li_2S_x species. Results from this study seem to suggest that the stronger the binding, the better the Li-S batteries. In addition, it was recognized that using conducting materials such as indium tin oxide[4] and Ti_4O_7 (ref. 19) is preferable due to the electron transfer needed to induce electrochemical reaction. Our recent work showed that indium tin oxide decorated carbon nanofibres can enhance the redox kinetics of Li_2S_x, realize the controllable deposition of Li_2S and improve the electrochemical performance of Li-S batteries[4].

Besides the conductive matrix material, there are indeed abundant insulating materials to trap Li_2S_x. But there is a fundamental problem here: insulated materials cannot transport electrons. Trapping Li_2S_x on insulating materials would cause them to be accumulated in electronically inactive areas and reduce the capacity retention. However, some studies have shown improvements of the electrochemical properties after the decoration of the electrode with poor conductive oxides such as MnO_2 (ref. 16), $Mg_{0.6}Ni_{0.4}O$ (ref. 12), Al_2O_3 (ref. 41) and La_2O_3 (ref. 42).

The above background research has motivated us to hypothesize that surface diffusion of Li_2S_x species on solid substrates can play an important role in Li-S battery electrochemical performance. This is particularly important for insulating solid materials with strong adsorption of Li_2S_x. The competition between the adsorption and diffusion of the Li_2S_x adsorbates on solid substrates can be very important, yet has been overlooked for Li-S batteries.

Usually, the nonconductive metal oxides work together with the carbon matrix to improve the conductivity of sulfur cathodes. For the modified carbon matrix with some nonconductive metal oxides nanostructures (Fig. 1), there is no direct electron transfer between these oxides and Li_2S_x species. Because of the poor conductivity of the metal oxide, the absorbed Li_2S_x should be transferred from the surface of the oxide to the conductive carbon substrate to undergo the electrochemical reaction. Therefore the competitive surface diffusion and adsorption of sulfur species must play key roles in the Li-S batteries. If the metal oxide has weak Li_2S_x capture capability (Fig. 1a), a large amount of Li_2S_x can diffuse away from the carbon matrix, resulting in serious shuttle effect and uncontrollable deposition of Li_2S. When the diffusion of sulfur species from the surface of oxide to carbon is difficult (Fig. 1c), the electrochemical reaction of Li_2S_x and the corresponding growth of Li_2S on the oxide/carbon is impeded. Therefore, the balance optimization between Li_2S_x adsorption and diffusion on the metal oxides surface is necessary (Fig. 1b).

In this work, various nonconductive metal oxide (MgO, Al_2O_3, CeO_2, La_2O_3 and CaO) nanoparticles-decorated carbon flakes are synthesized via a facile and generic biotemplating method using Kapok trees fibres as both the template and the carbon source. The sulfur cathodes based on MgO, CeO_2 and La_2O_3 show higher capacity and better cycling stability. Adsorption test, microstructure analysis and electrochemical performance evaluation combined with density functional theory (DFT) calculations reveal that better surface diffusion leads to higher deposition efficiency of sulfide species. A comprehensive-oxide-selection criteria referring to the strong binding, high surface area and good surface diffusion properties is proposed.

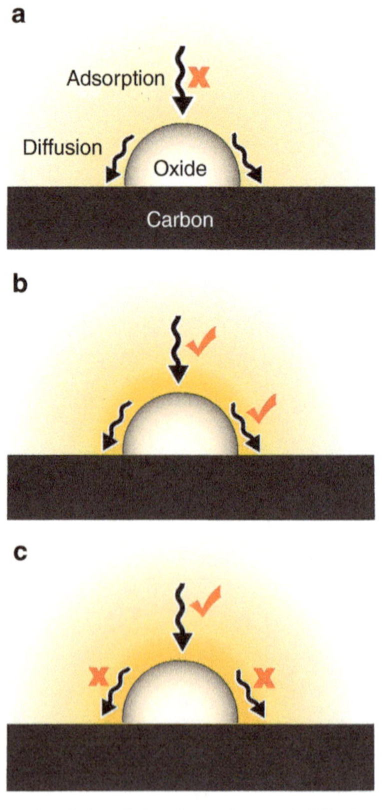

Figure 1 | Schematic of the Li_2S_x adsorption and diffusion on the surface of various nonconductive metal oxides. **(a)** The metal oxide with weak Li_2S_x adsorption capability; only few Li_2S_x can be captured by the oxide; **(b)** the metal oxide with both strong adsorption and good diffusion, which is favourable for the electrochemical reaction and the controllable deposition of sulfur species; **(c)** the metal oxide with strong bonding but without good diffusion; the growth of Li_2S and the electrochemical reaction on the oxide/C surface is impeded.

Results

Sulfide capture by metal oxides. To reveal the role of metal oxides in Li–S batteries, five kinds of pure metal oxide-nanoparticles were prepared by a generic Pechini sol–gel method[13]. 1,3-dioxolane (DOL) and dimethoxyethane (DME) are commonly used solvent in the Li–S battery electrolyte[4]. Therefore, 0.005 M Li_2S_8 in DOL/DME (1:1, v-v) was prepared for the adsorption test of sulfides. Figure 2a shows the camera image of the adsorption test using different mass of oxide samples with the same total surface area. The colour of the solution containing Al_2O_3 and CeO_2 is lighter than the others, indicating better adsorption of these two metal oxide nanoparticles. Inductively coupled plasma-optical emission spectroscopy (ICP-OES) results reveal that Al_2O_3 and CaO show the biggest and the smallest absorption capability, respectively (Fig. 2b). In addition, it was found that the adsorption capacity increases slightly with the rise of temperature (Fig. 2b and Supplementary Table 1), which is one essential characteristic of chemisorption. The measured Li_2S_8 adsorption quantity of metal oxides is in the range of 2.78–4.94 μmol m^{-2}, close to the simulated monolayer adsorption capacity ranging from 2.76 to 4.88 μmol m^{-2} (Fig. 2b). Therefore, the Li_2S_8 capture involves monolayer adsorption, which is another well-known characteristic of chemisorption.

In order to better understand the absorption mechanism, DFT calculation was performed to reveal the corresponding adsorption energy and sites (Fig. 2c,d and Supplementary Figs 2–6). Considering both low-order and high-order, Li_2S_x are important discharge products of Li–S batteries, we choose both Li_2S and Li_2S_8 as the prototype for modelling. Figure 2c shows the optimized geometries of the most stable Li_2S on CeO_2(111), Al_2O_3(110), La_2O_3(001), MgO(100) and CaO(100) surfaces. On the entire surface, the most favourable binding site of Li_2S is two Li atoms bonding with the oxygen atom of metal oxide (Fig. 2c). On Al_2O_3(110), the Li of Li_2S is the bridge site of two oxygen atoms, on other four metal oxides, the Li is on the atop site of oxygen. On MgO(100) CaO(100) and La_2O_3(001) surfaces, the sulfur of Li_2S is away from the metal oxides surface. Sulfur is bonding with oxygen on CeO_2(111), in which the sulfur–oxygen distance is 1.70 Å while with Al on Al_2O_3(110), in which the Al–S distance is 2.21 Å. The adsorption energy of Li_2S on CeO_2(111), Al_2O_3(110), La_2O_3(001), MgO(100) and CaO(100) surfaces is − 6.33, − 7.12, − 0.5.85, − 5.71 and − 5.49 eV, respectively. The optimized stable configurations of Li_2S_8 on five different surfaces are also shown in Fig. 2d and Supplementary Figs 2–6. Although the Li of Li_2S_8 has similar bonding sites on the metal oxide surface as that of Li_2S, the optimized most stable configuration of Li_2S_8 on each metal oxide is quite different. A lot of initial geometries of Li_2S_8 on each metal oxide have been considered in our calculations. It is found that after optimization, the structures have big change due to the interaction between Li_2S_8 and metal oxides. For both Li_2S and Li_2S_8 on metal oxides, the bonding between Li and oxygen plays a major role. In addition, only two or three sulfur atoms of Li_2S_8 are bonding with the oxide surface. The adsorption energies of Li_2S_8 have similar trend on these metal surfaces with Li_2S, while are much weaker than Li_2S. The calculated adsorption energies of both Li_2S and Li_2S_8 are in agreement with the experimental adsorption test results of Li_2S_8 on the oxide nanoparticles (Fig. 2b and Supplementary Table 1).

Biotemplated fabrication of oxides/carbon nanostructures. Although most of these oxides have remarkable adsorption behaviour for sulfide species, they are poor electronic conductors. Conductivity is one of the most important factors affecting the performance of Li–S batteries. Therefore, we fabricated metal-oxides (MgO, Al_2O_3, CeO_2, La_2O_3 and CaO) nanoparticles anchored on porous carbon nanoflakes to form an electronic conductive oxide/carbon nanocomposite. To fabricate the nano-composite, Kapok fibres (KFs)[43] were used as both the template and the carbon source (Fig. 3a). KFs are low-cost and high-yield agriculture products derived from the fruits of Kapok tree, which is chemically composed of 64% cellulose, 13% lignin and 23% pentosan[43]. In addition, the KFs have unique hollow

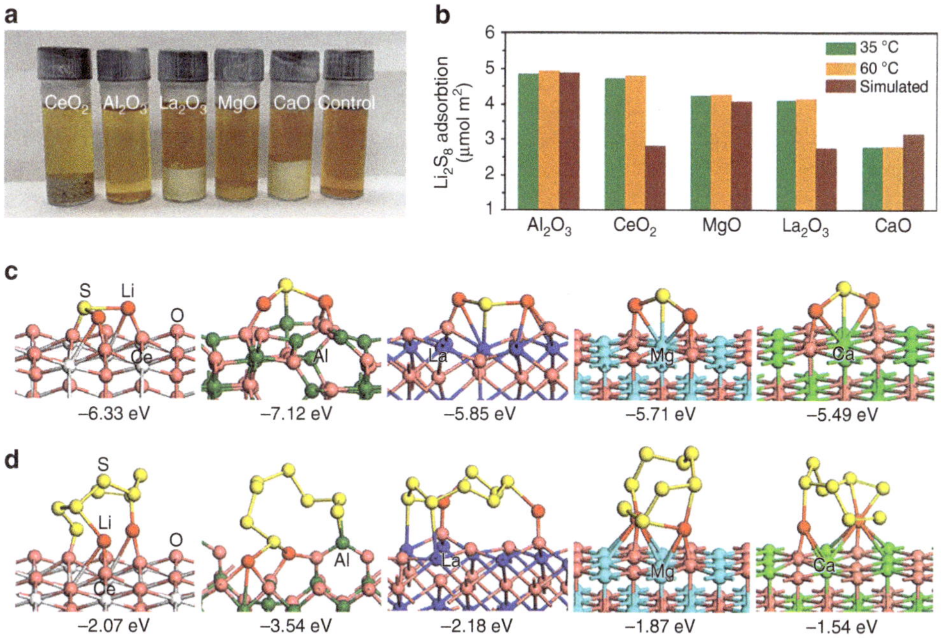

Figure 2 | Adsorption test and relative models of sulfide species on the surface of metal oxides. (**a**) Digital images of the Li_2S_8 trapping by the metal oxide nanoparticles in DOL/DME (1:1, v-v) solution. (**b**) Experimental and simulated adsorption amount of Li_2S_8 on different metal oxides. The simulated adsorption was based on the monolayer adsorption model. (**c**) Optimized geometries of the most stable Li_2S on CeO_2(111), Al_2O_3(110), La_2O_3(001), MgO(100) and CaO(100) surfaces. (**d**) Optimized geometries of most stable Li_2S_8 on the metal oxide surface.

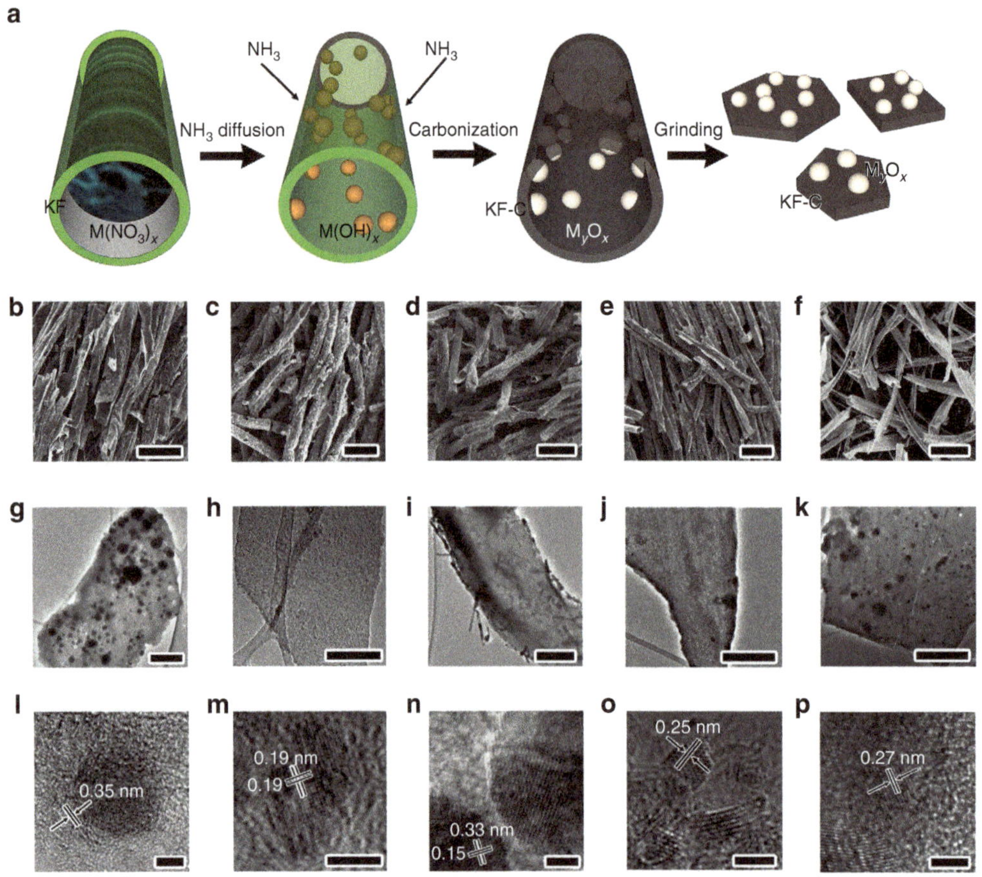

Figure 3 | Fabrication and microstructures of oxides/carbon nanostructures. (**a**) Schematic illustration of synthesis of oxides/carbon using the kapok tree fibres (KF) as both template and carbon sources. (**b-f**) SEM images of Al_2O_3/C, CeO_2/C, La_2O_3/C, MgO/C and CaO/C composites, respectively (scale bar = 50 μm). (**g-k**) The corresponding TEM images of Al_2O_3/C, CeO_2/C, La_2O_3/C, MgO/C and CaO/C, respectively (scale bar = 300 nm). (**l-p**) The corresponding HRTEM images of Al_2O_3/C, CeO_2/C, La_2O_3/C, MgO/C and CaO/C, respectively (scale bar = 3 nm).

lumens with a thin wall thickness ≤1 μm, which enables good sorption capacity through capillary force. Therefore, metal nitrate solution can be easily absorbed into the KFs (Fig. 3a). When the NH_3 gas diffuses into the KFs through the porous cell wall and the open end, metal hydroxides nanoparticles will be formed on the surface of the cell wall due to the confinement effect of KF template. After drying at 90 °C for 12 h, carbonization at 850 °C for 1.5 h and the following facile grinding, carbon nanoflakes decorated with metal oxide nanoparticles can be obtained (Fig. 3a). Figure 3b–f shows the scanning electron microscopy (SEM) images of Al_2O_3/C, CeO_2/C, La_2O_3/C, MgO/C and CaO/C nanocomposites, respectively. The composite remains the macromorphology of the original Kapok tree fibres, which have a unique hollow structure with a large lumen and a thin fibre wall. After simple grinding, the delicate carbon microtubes are converted to carbon nanoflakes (Fig. 3g–k and Supplementary Figs 7–9).

Moreover, transmission electron microscopy (TEM) images show that abundant oxide nanostructures are located on the carbon matrix (Fig. 3g–k). Figure 3l is the representative high resolution TEM (HRTEM) image of the Al_2O_3/C sample, showing the amorphous structure of Al_2O_3 and partially graphitized structure of C. The interlayer spacing is about 0.35 nm, corresponding to (002) planes of graphite. Figure 3m shows the [001] zone axis HRTEM image of a typical CeO_2 nanoparticle in CeO_2/C composite. The lattice spacing of 0.19 nm in Fig. 3m can be attributed to {220} crystal planes of the face centred cubic (fcc) phase CeO_2. The TEM image of La_2O_3/C is showed in Fig. 3i, suggesting that abundant rod-shaped nanoparticles are distributed in the carbon matrix. The corresponding HRTEM image in Fig. 3n shows that the La_2O_3 particle is single crystalline. The lattice spacings (0.15 and 0.33 nm) and the interplanar angle match the finger print of (004) and (100) planes of hexagonal phase La_2O_3. HRTEM image of MgO/C sample in Fig. 3o indicates lattice fringes with regular spacing of 0.25 nm, which can be indexed to (111) planes of fcc MgO. Figure 3p shows the HRTEM image of CaO/C sample. The fringes with spacing of 0.27 nm is corresponding to (111) planes of fcc CaO. These TEM results indicate that the Kapok tree fibre can act as ideal and general template for the synthesis of oxide nanostructures due to the confinement effect of fibre substrate.

Electrochemical performance of composite cathodes. Thermal diffusion method was used to fabricate the sulfur/M_xO_y/C composite[19]. The mass loading of the electrode ranges from 0.7 to 1.2 mg cm^{-2}. The electrolyte was 1 M Lithium bis (trifluoromethanesulphonyl)imide in DOL and DME, with $LiNO_3$ as additive to passivate the lithium anode. Figure 4a shows the representative charge–discharge curves of the composite electrode based on different oxide/carbon nanostructures at a current rate of 0.1 C (1 C = 1,672 mA g^{-1}). All the discharge curves show two typical discharge plateaus at 2.35 and 2.10 V, which can be assigned to the formation of high-order and low-order Li_2S_x, respectively[9]. No obvious difference can be found for the potential of the discharge plateaus. However, the CaO/C and

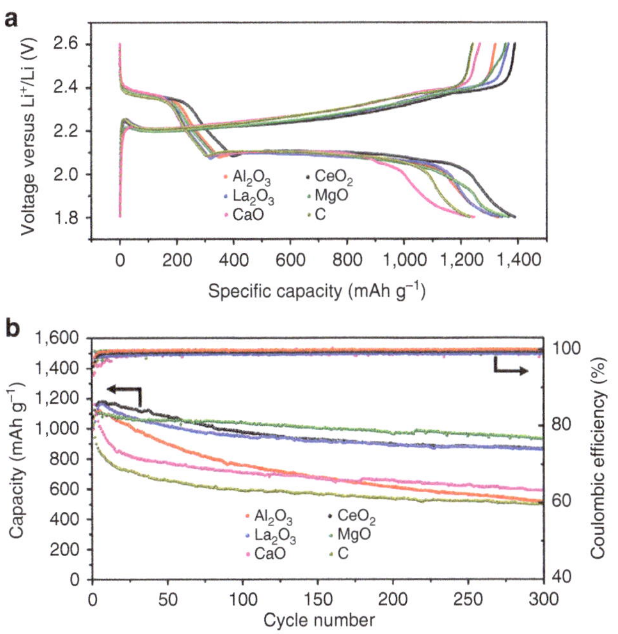

Figure 4 | Charge-discharge curves and cycling performance of the sulfur composite electrodes. (**a**) Representative charge-discharge profiles of the composite electrodes based on different oxide/carbon nanostructures at 0.1 C. (**b**) Specific capacity and the corresponding Coulombic efficiency of the composite electrodes upon prolonged 300 charge-discharge cycles at 0.5 C.

C composite cathodes show higher charge over-potentials than those of CeO_2/C, La_2O_3/C, Al_2O_3/C and MgO/C composite electrodes (Fig. 4a). The specific discharge capacities of Al_2O_3/C, CeO_2/C, La_2O_3/C, MgO/C, CaO/C and C at 0.1 C rate are measured to be 1,330, 1,388, 1,345, 1,368, 1,246 and 1,230 mAh g^{-1}, respectively (Fig. 4a). It can be found that the CaO/C and C composite cathodes show relatively lower discharge capacity. The high over-potential and low discharge capacity may result from the serious dissolution of Li_2S_x in electrolyte (shuttle effect), which causes the active material loss and the increase of electrolyte viscosity. ICP-OES test (Supplementary Table 2) based on the same mass of Al_2O_3/C, CeO_2/C, La_2O_3/C, MgO/C, CaO/C and C samples reveal that both CaO/C and C have poorer Li_2S_x capture capability compared with other samples.

Besides the specific capacity, cycling performance is one of the most important characteristics for Li–S batteries. Therefore, all the cathodes were subject to prolonged cycling. Figure 4b shows the discharge capacity and the corresponding Coulombic efficiency of the cathodes upon prolonged 300 cycles at 0.5 C. The representative charge/discharge curves (Fig. 4a) and the cycling performance (Fig. 4b) indicate that Al_2O_3/C, CeO_2/C, La_2O_3/C and MgO/C composite electrode show high specific capacity in the first several cycles. However, the composite electrodes show distinct capacity retention capability. The Al_2O_3/C, CaO/C and C cathodes exhibit obvious capacity fading especially in the first 100 cycles. Compared with Al_2O_3/C, CaO/C and C cathodes, the electrode based on CeO_2/C, La_2O_3/C and MgO/C show better cycling performance and the MgO/C cathode is the best among all the samples. The capacity decay per cycle is 0.171, 0.066, 0.047, 0.034, 0.136 and 0.170% for Al_2O_3/C, CeO_2/C, La_2O_3/C, MgO/C, CaO/C and C, respectively. Considering the serious capacity, decay mainly happens in the first 50 cycles, the average Columbic efficiency in the first 100 cycles was calculated (Fig. 4b). Al_2O_3/C, CeO_2/C, La_2O_3/C, MgO/C, CaO/C and C cathodes show 99.6, 99.1, 98.7, 99.4, 98.3 and 98.8% columbic efficiency. Lower Columbic efficiency of Li–S batteries resulted from the significant Li_2S_xdissolution, which cause the loss of S material and shuttle effect[19]. This can be also supported by the ICP-OES results, revealing that Al_2O_3/C has the best capture capability and CaO/C and C have poorer capture capability for Li_2S_8(Supplementary Table 2). Although Al_2O_3/C cathodes possess high initial discharge capacity and good Columbic efficiency, the rate of capacity decay is higher than those of CeO_2/C, La_2O_3/C and MgO/C cathodes. The distinct Columbic efficiency for Al_2O_3 (99.6%) cathode from both CaO/C (98.3%) and C (98.8%) cathodes may imply different capacity decay mechanism.

Analysis of capacity failure mechanism of composite cathodes. In order to further reveal the detailed capacity decay mechanism, some batteries were disassembled after 100 cycles at 0.5 C to observe the morphology evolution of the cathode materials by SEM. Figure 5a–f shows low magnification SEM images of the cycled electrodes based on Al_2O_3/C, CeO_2/C, La_2O_3/C, MgO/C, CaO/C and C nanostructures, respectively. In contrast to CeO_2/C (Fig. 5b), La_2O_3/C (Fig. 5c) and MgO/C (Fig. 5d) electrodes with uniform and flat surface, the Al_2O_3/C (Fig. 5a), CaO/C (Fig. 5e) and C (Fig. 5f) cathodes show high surface roughness. Some carbon nanoflakes with well-defined profile can be observed on the surface of the Al_2O_3/C (Fig. 5a), CaO/C (Fig. 5e) and C (Fig. 5f) cathodes after 100 cycles. In addition, some cracks and pinholes can be found in the CaO/C and carbon cathodes, which may result from the significant dissolution and loss of sulfur. Figure 5g–l and m–r shows the top-view and cross-sectional SEM images of cycled Al_2O_3/C, CeO_2/C, La_2O_3/C, MgO/C, CaO/C and C cathodes, respectively. Only a thin, uniform and dense Li_2S film can be found on the surface of Al_2O_3/C (Fig. 5g,m). Abundant Li_2S particles with irregular shape were formed between the Al_2O_3/C nanoflakes (Supplementary Fig. 1). Similar phenomenon can also be found in the CaO/C (Fig. 5k,q) and C (Fig. 5l,r) cathodes. Some characteristic stripes (Fig. 5k,l) and distinct fracture surface (Fig. 5q,r) indicate that the Li_2S particles are detached from the oxides/C matrix and may become electrochemically inactive. This is due to the uncontrollable precipitation of Li_2S on the non-polar or weakly polar surface[6], which leads the further decay of capacity. In contrast, it is not easy to identify the oxide/carbon nanoflakes from the CeO_2/C (Fig. 5h), La_2O_3/C (Fig. 5i) and MgO/C (Fig. 5j) cathodes. To obtain the cross-sectional morphology of the cathodes, the electrode materials were scraped from the Al foil current collector and mounted on a copper foil with rough surface for SEM observation. CeO_2/C (Fig. 5n), La_2O_3/C (Fig. 5o) and MgO/C (Fig. 5p) nanoflakes were wrapped by thick Li_2S layer. Supposing that the average thickness of carbon nanoflakers is 450 nm, the thickness of Li_2S layer deposited on the surface of CeO_2/C, La_2O_3/C and MgO/C will be 1,400, 1,677 and 1,573 nm, respectively. After mechanical scrapping and mounting processes, no detachment can be observed, indicating that there is good cohesion between the Li_2S layer and the oxide/C nanoflakes. These results are consistent with our DFT calculation results in Fig. 2. Because the Li_2S is a poor ionic and electronic conductor, the good combination of Li_2S with the conductive matrix must be favourable for the reversible electrochemical reaction in the following charging process[4]. Therefore, the better cycling performance of CeO_2/C, La_2O_3/C and MgO/C cathodes can be attributed to the controllable precipitation of Li_2S on the polar surface of carbon matrix.

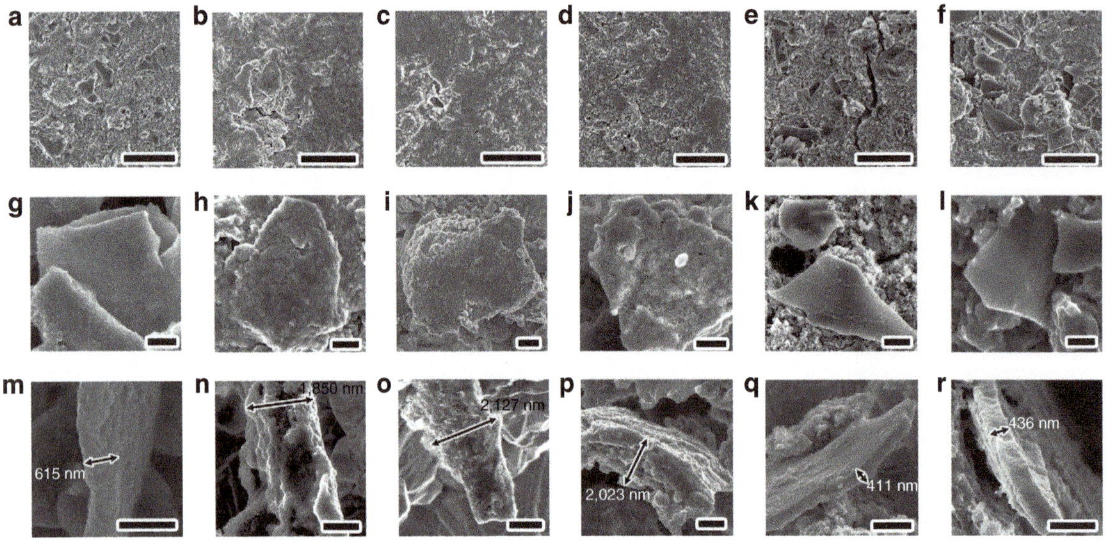

Figure 5 | Morphology of the discharged composite electrodes after cycling. (**a**-**f**) SEM images of the cycled composite electrodes based on Al_2O_3/C, CeO_2/C, La_2O_3/C, MgO/C, CaO/C and carbon nanostructures, respectively. (**g**-**l**) Top view and (**m**-**r**) cross-sectional SEM images show typical Al_2O_3/C, CeO_2/C, La_2O_3/C, MgO/C, CaO/C and C nanostructures after cycling, respectively. Scale bars = 10μm (**a**-**f**) and 1 μm (**g**-**r**).

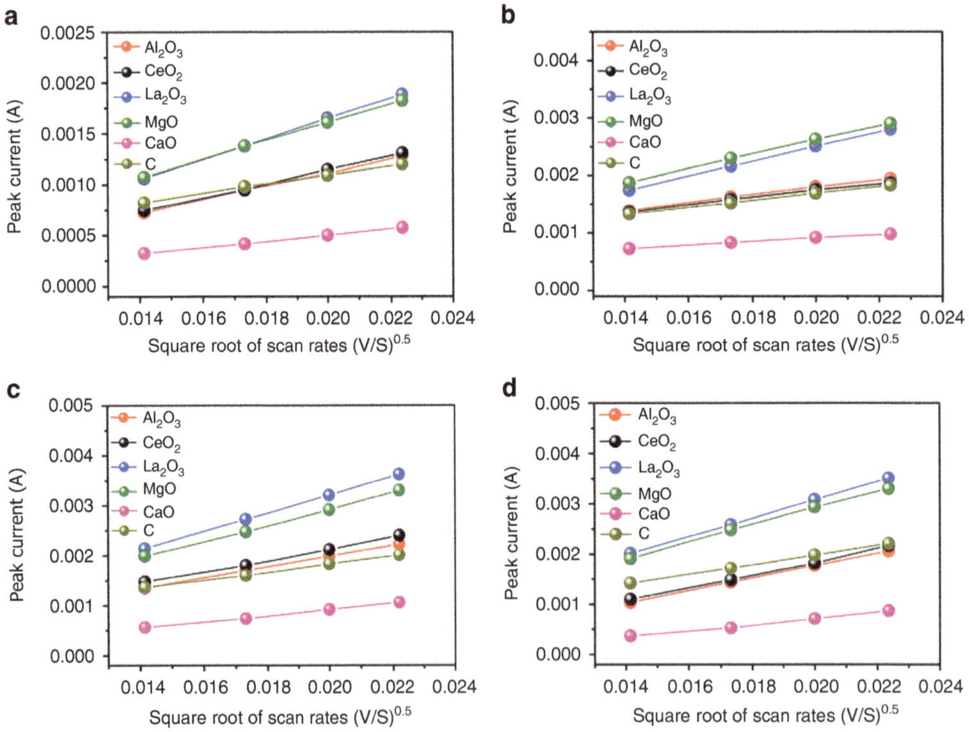

Figure 6 | Lithium ion diffusion properties of the electrode at various voltage scan rates. Plot of CV peak current of (**a**) the cathodic reaction 1 ($S_8 \rightarrow Li_2S_4$), (**b**) the cathodic reaction 2 ($Li_2S_4 \rightarrow Li_2S$), (**c**) the anodic reaction 1 ($Li_2S \rightarrow Li_2S_4$) and (**d**) the anodic reaction 2 ($Li_2S_4 \rightarrow S_8$) versus the square root of scan rates.

By now, some important questions arise: why Li_2S shows different deposition behaviour on various oxide/C surfaces? Which kind of oxide/C surface is favourable for the controllable precipitation of Li_2S? Because the nucleation and initial growth sites of Li_2S are located on the surface of oxide/C, the growth behaviour of Li_2S must be related to surface chemical properties of oxide/C matrix. First, the oxide/C surface should absorb high-order Li_2S_x owing to its strong adsorption ability (Fig. 2d), which acts as the sulfur source for the growth of Li_2S. The absorbed Li_2S_x must be transferred from the oxide surface to conductive carbon surface to enable the electrochemical reactions due to the insulating properties of these metal oxides. Therefore, the distribution and structure of Li_2S will be affected by the surface diffusion properties of Li_2S_x species on the oxide/C substrate.

Lithium ion diffusion properties and mechanism of the cathodes. Although it is very difficult to get the diffusivity of Li_2S_x on the oxide surface from the electrochemical measurement, the lithium diffusivity in the whole Li-S batteries can offer the important information about Li_2S_x surface diffusion because of that the most favourable binding site of sulfides is two Li atoms bonding with metal oxide (Fig. 2c,d). In order to explore the lithium diffusion properties, we performed cyclic voltammetry (CV) measurements under different scanning rates ranging from 0.2 to 0.5 mV s^{-1}. As shown in Fig. 6, all cathodic and anodic peak currents are linear with the square root of scan rates, from which the lithium diffusion performance can be estimated using the classical Randles Sevcik equation[44]:

$$I_p = (2.69\times10^5)n^{1.5}aD^{0.5}Cv^{0.5}\Delta C_o, \qquad (1)$$

where, I_p is the peak current, n is the number of electrons per reaction species, a is the active electrode area, D is the lithium ion diffusion coefficient, ΔC_o is the Li concentration change corresponding to the electrochemical reaction. The n, a and ΔC_o are constant in our battery system. The slopes of curves in Fig. 6a–d are positively correlated to the corresponding lithium ion diffusion, which indicates that the sulfur composite cathode based on both CaO/C and C have lower diffusivity. Abundant high viscosity Li_2S_x dissolved in the electrolyte and the poor Li_2S_x capture capability of both CaO/C and C is believed to be the main reason for their low diffusivity. Compared with CaO/C, CeO_2/C, Al_2O_3/C and C electrodes, La_2O_3/C and MgO/C samples show better diffusion properties and the measured diffusivity of Al_2O_3/C cathode is comparable with the CeO_2/C.

The diffusion of lithium on the surface of various metal oxides has been investigated by the DFT calculation (Fig. 7). Because the most favourable binding site of sulfides is two Li atoms bonding with metal oxide (Fig. 2c,d), the calculated Li ion diffusion can also indicate the diffusivity of sulfides species on the surface of the oxide. On MgO(100), CaO(100) and La_2O_3(001) surfaces, the diffusions of Li in different dimensions can be realized on three equivalent adsorption sites (Fig. 7). Among the three kinds of surfaces, the diffusion barrier of Li on CaO(100) is largest. The space group of MgO is same with CaO, however, the diffusion barrier of Li on MgO(100) is about 0.45 eV lower than that on CaO(100). The suitable adsorption energies of sulfide species and small diffusion barriers of Li on MgO will lead to the formation of abundant Li_2S particles on MgO/C surfaces, which are responsible for the best cycling performance of MgO/C cathodes. On CeO_2(111) surfaces, the large diffusion barrier is 0.66 eV, which is similar with that on La_2O_3(001) surfaces. This may explain why CeO_2/C and La_2O_3/C cathodes show similar cycling performance. Among the five kinds of metal oxides surfaces, the largest diffusion barrier (1.22 eV) of Li is found to be on Al_2O_3(110), which is about three times of that on MgO(100). It is seen that sulfide species can strongly adsorb, however, difficult to diffuse on Al_2O_3. Although Al_2O_3 has the strongest Li_2S_8 adsorption (Fig. 2), the slow diffusion of Li_2S_x indicated that Al_2O_3 may not be a good additive for sulfur cathode.

Discussion

Based on our experimental results, we clarify three functions of these oxides. The first basic function of these metal oxides is the Li_2S_x adsorption. Although many literatures have reported the Li_2S_x capture capacities, the detailed absorption mechanism is still unclear. Our DFT calculation and temperature swing adsorption experiments (Fig. 2) confirm that the monolayer chemisorption is dominant during the Li_2S_x capture. The second function of these metal oxides, especially some nonconductive oxides, is the Li_2S_x transfer station, which transports the Li_2S_x from the poorly conductive oxide surface to high conductive carbon matrix to ensure the full electrochemical conversion. The third function is to induce the controlled growth of Li_2S species on the surface of the composite instead of random deposition. Many reports proved that the uncontrolled deposition will result in electrochemically inactive large agglomerations of Li_2S. The subsequent detachment of Li_2S from the oxide/carbon matrix is the main capacity decay mechanism, which can be supported by the SEM observation in Fig. 5. The SEM observations, diffusion test and DFT calculations revealed that the deposition of Li_2S on the surface of oxide/carbon matrix may be influenced by the lithium ion diffusion properties on the surface of metal oxides, which has not yet been identified. Surface diffusion properties will affect the distribution and growth of Li_2S.

Based on these functions of nonconductive metals oxides, we can propose an oxide selection criterion for the Li–S batteries.

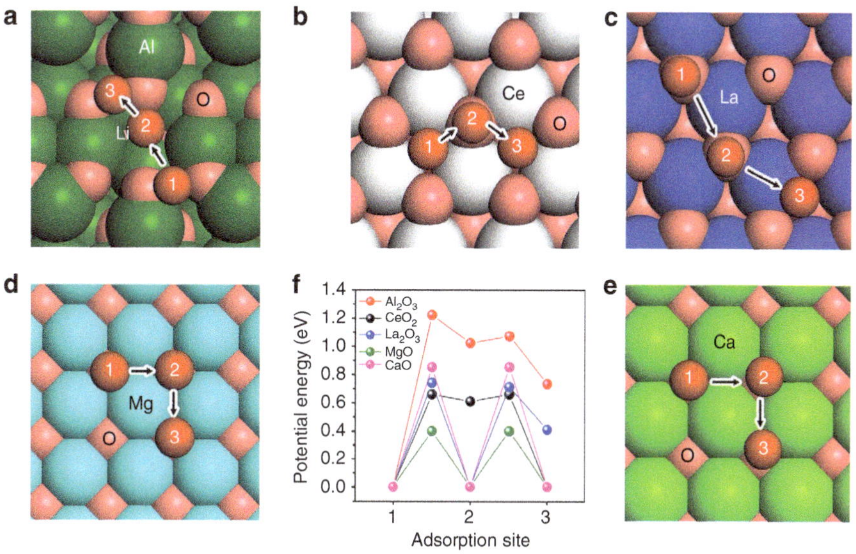

Figure 7 | Lithium diffusion mechanism on the surface of various metal oxides. (**a–e**) Minimum energy path for lithium ion diffusion on Al_2O_3(110), CeO_2(111), La_2O_3(001), MgO(100) and CaO(100) surfaces, respectively. (**f**) Potential energy profiles for Li^+ diffusion along different adsorption sites on the oxide surface.

Because the first role of oxides is adsorption, the binding between the sulfides species and the matrix should be strong, which can both suppress the shuttle effect and enable the full utilization of active materials. Considering that the Li_2S_x capture is the monolayer chemisorption and the adsorption amount will depend on the surface area of oxides, uniformly distributed oxides nanostructures with high surface area are essential. Although strong binding and high surface area are preconditions, the surface diffusion properties of oxides are also very important, which affect the distribution and structure of Li_2S. An optimized balance between Li_2S_x adsorption and surface diffusion is favourable for the sulfide species to deposit on the surface of oxide/carbon matrix, keep active during the cycling and ensure the final good cycling performance of batteries. In addition, some other factors such as electric conductivities, chemical stability and lithiation/delithiation of the oxides also need to be considered.

In conclusion, a series of nonconductive metal oxides (MgO, Al_2O_3, CeO_2, La_2O_3 and CaO) nanoparticles anchored on porous carbon nanoflakes have been synthesized successfully via a facile and generic biotemplating method using Kapok trees fibres as both the template and the carbon source. The composite cathode materials based on the MgO/C, La_2O_3/C and CeO_2/C nanoflakes show higher capacity and better cycling performance. Moreover, the working mechanisms of these oxides were revealed by adsorption test, microstructure analysis, electrochemical performance evaluation and DFT calculations. In addition, the comprehensive oxide selection criteria referring to the strong binding, high surface area and good surface diffusion properties were proposed for the first time. We believe that our proposed selection criteria can be generalized to other matrix materials for high performance Li–S batteries such as metal sulfides, metal nitrides, metal chlorides, and so on.

Methods

Preparation of metal oxide nanoparticles. A generic Pechini sol–gel method[13] was used to synthesize the pure metal oxides. 0.01 mol metal nitrates of the desired metals were dissolved in 5 ml of deionized water under stirring. 0.015 mol citric acid was then added in to the prepared solution to chelate the metal ions and heated to 200 °C to form a dry and porous gel. After the calcination of the gel at 850 °C for 1 h, pure metal oxide (Al_2O_3, CeO_2, La_2O_3, MgO and CaO) nanoparticles were obtained.

Biotemplated fabrication of metal oxide/carbon nanoflakes. Owing to the poor electric conductivity of Al_2O_3, CeO_2, La_2O_3, MgO and CaO, a series or metal oxide decorated porous carbon flakes were fabricated using the Kapok tree fibres as both the template and the carbon sources. To start with, 0.01 mol metal nitrates of the desired metals were dissolved in 40 ml of deionized water to form transparent solution. Commercial Kapok tree fibre was washed with acetone and ethanol to remove the surface wax and dipped in to the nitrate solution. After 2 h soaking, the nitrate loaded fibres were separated from the solution and introduced into ammonia atmosphere to ensure the formation of the metal hydroxides, followed by the drying at 100 °C for 10 h to remove the water. Then the dried fibres were inserted into the tube furnace and calcined at 850 °C for 1 h with 100 sccm continuous flow of argon. After cooling, the obtained carbon microtubes remaining the macromorphology of original Kapok tree fibres were easily grinded into carbon nanoflakes decorated with metal oxide nanoparticles.

Adsorption test. To prepare Li_2S_8 solution, stoichiometric Li_2S and sulfur were dissolved in 1,3-dioxolane/1,2-dimethoxyethane solution (1:1 in volume) and stirred at 80 °C for 24 h. Then the Li_2S_8 solution was diluted to 0.005 M for the capture test. Before the adsorption test, all the oxide and oxide/carbon samples were dried under vacuum at 80 °C for 24 h. After the immersion of the samples in 0.005 M Li_2S_8 at different temperatures, 20 μl of the solution was transferred for the ICP-OES test[45].

Characterization. The morphology and the microstructure were studied using SEM (FEI, XL30 Sirion) and TEM (FEI, Tecnai G2 F20 X-TWIN). The specific surface area was characterized from nitrogen adsorption–desorption measurement (Micromeritics, ASAP 2020). ICP-OES was conducted using a Thermo Scientific ICAP 6300 Duo View Spectrometer.

Electrochemical measurements. The sulfur cathode materials were prepared via a facile thermal diffusion method[19]. First, sulfur and grinded oxide/carbon nanoflakes with a weight ratio of 1:3 were mixed with appropriate amount of CS_2 solution. After the evaporation of CS_2, the mixture were pressed and heated at 155 °C for 12 h under vacuum to obtain the composite cathode materials. To fabricate the 2,023 type coin cells for the electrochemical measurements, the synthesized composites were grinded again and mixed with conductive carbon black and polyvinylidene fluoride binder in N-methyl-2-pyrrolidinone (70:20:10 by weight) to form a slurry. After a 12 h magnetic stirring, the slurry was coated onto the aluminium foil and dried overnight at 60 °C under vacuum. The sulfur content was in the range of 63–70 wt% and the mass loading of the electrodes ranges from 0.7 to 1.2 mg cm^{-2}. 1 M lithium bis(trifluoromethanesulphonyl)imide dissolved in a mixture of 1,3-dioxolaneand dimethoxymethane (1:1 by volume) with 0.1 M $LiNO_3$ additive was used as the electrolyte. Galvanostatic cycling was performed on Arbin or MTI testers with the potential range of 1.8–2.6 V versus Li/Li^+ at ambient temperature.

DFT calculations. All calculations were performed using the Vienna *ab initio* Simulation Package code[46] based on density functional theory. The projector augmented wave potentials were used to describe the interaction between ions and electrons[47,48]. Nonlocal exchange correlation energy was evaluated using the Perdew-Burke-Ernzehof functional. The electron wave function was expanded using plane waves with an energy cutoff of 400 eV. All structures were optimized with force convergence criterion of 10 meV $Å^{-1}$. All oxide surfaces were created based on the corresponding optimized bulk unit cell, which were in good agreement with the experimental values. A 4 × 4 × 1 Monkhorst–Pack k-point mesh and a vacuum slab of about 15 Å was inserted between the surface slabs for all the metal oxide models. The cell parameter is 8.40 × 8.07 × 25.00 Å for Al_2O_3(110), 11.67 × 11.67 × 22.94 Å for CeO_2(111), 11.82 × 11.82 × 25.00 Å for La_2O_3(001), 8.99 × 8.99 × 20.00 Å for MgO(100) and 10.25 × 10.25 × 20.00 Å for CaO(100). DFT + U approach, where U is an empirical parameter for on site electronic correlations, was used in the calculation of CeO_2(111) with a U value of 5.0 eV (ref. 49). The adsorption energies (E_a) for S and Li_2S on the metal oxide surfaces are defined as $E_a = E_{total} - E_{ads} - E_{suf}$, where E_{total} is the total energy of the adsorbed system, E_{ads} is the energy of the adsorbate in vacuum and E_{suf} is the energy of the optimized clean surface slab.

References

1. Bruce, P. G., Freunberger, S. A., Hardwick, L. J. & Tarascon, J. M. Li-O_2 and Li-S batteries with high energy storage. *Nat. Mater.* **11,** 19–29 (2012).
2. Yamin, H., Gorenshtein, A., Penciner, J., Sternberg, Y. & Peled, E. Lithium sulfur battery-oxidation reduction-mechanisms of polysulfides in the solutions. *J. Electrochem. Soc.* **135,** 1045–1048 (1988).
3. Ji, X. L., Lee, K. T. & Nazar, L. F. A highly ordered nanostructured carbon-sulphur cathode for lithium-sulphur batteries. *Nat. Mater.* **8,** 500–506 (2009).
4. Yao, H. B. *et al.* Improving lithium–sulphur batteries through spatial control of sulphur species deposition on a hybrid electrode surface. *Nat. Commun.* **5,** 3943 (2014).
5. Su, Y. S. & Manthiram, A. Lithium-sulphur batteries with a microporous carbon paper as a bifunctional interlayer. *Nat. Commun.* **3,** 1166 (2012).
6. Seh, Z. W. *et al.* Sulphur-TiO_2 yolk-shell nanoarchitecture with internal void space for long-cycle lithium-sulphur batteries. *Nat. Commun.* **4,** 1331 (2013).
7. Guo, J. C., Xu, Y. H. & Wang, C. S. Sulfur-impregnated disordered carbon nanotubes cathode for lithium-sulfur batteries. *Nano Lett.* **11,** 4288–4294 (2011).
8. Demir-Cakan, R. *et al.* Li-S batteries: simple approaches for superior performance. *Energy Environ. Sci.* **6,** 176–182 (2013).
9. Yang, Y., Zheng, G. Y. & Cui, Y. Nanostructured sulfur cathodes. *Chem. Soc. Rev.* **42,** 3018–3032 (2013).
10. Zhou, G. M. *et al.* A graphene-pure-sulfur sandwich structure for ultrafast, long-life lithium-sulfur batteries. *Adv. Mater.* **26,** 625–631 (2014).
11. Kim, J. S., Hwang, T. H., Kim, B. G., Min, J. & Choi, J. W. A lithium-sulfur battery with a high areal energy density. *Adv. Funct. Mater.* **24,** 5359 (2014).
12. Zhang, Y. G., Zhao, Y., Yermukhambetova, A., Bakenov, Z. & Chen, P. Ternary sulfur/polyacrylonitrile/$Mg_{0.6}Ni_{0.4}O$ composite cathodes for high performance lithium/sulfur batteries. *J. Mater. Chem. A* **1,** 295–301 (2013).
13. Tao, X. Y. *et al.* Highly mesoporous carbon foams synthesized by a facile, cost-effective and template-free Pechini method for advanced lithium-sulfur batteries. *J. Mater. Chem. A* **1,** 3295–3301 (2013).
14. Zhou, G. M. *et al.* A flexible sulfur-graphene-polypropylene separator integrated electrode for advanced Li-S batteries. *Adv. Mater.* **27,** 641–647 (2015).
15. Wang, J., He, Y.-S. & Yang, J. Sulfur-based composite cathode materials for high-energy rechargeable lithium batteries. *Adv. Mater.* **27,** 569–575 (2015).
16. Liang, X. *et al.* A highly efficient polysulfide mediator for lithium-sulfur batteries. *Nat. Commun.* **6,** 5682 (2015).
17. Zhao, M.-Q. *et al.* Unstacked double-layer templated graphene for high-rate lithium-sulphur batteries. *Nat. Commun.* **5,** 3410 (2014).

18. Wang, Z. *et al.* Enhancing lithium-sulphur battery performance by strongly binding the discharge products on amino-functionalized reduced graphene oxide. *Nat. Commun.* **5**, 5002 (2014).
19. Tao, X. Y. *et al.* Strong sulfur binding with conducting Magneli-phase Ti_nO_{2n-1} nanomaterials for improving lithium-sulfur batteries. *Nano Lett.* **14**, 5288–5294 (2014).
20. Seh, Z. W. *et al.* Two-dimensional layered transition metal disulphides for effective encapsulation of high-capacity lithium sulphide cathodes. *Nat. Commun.* **5**, 5017 (2014).
21. Qiu, Y. C. *et al.* High-rate, ultra long cycle-life lithium/sulfur batteries enabled by nitrogen-doped graphene. *Nano Lett.* **14**, 4821–4827 (2014).
22. Huang, C. *et al.* Manipulating surface reactions in lithium-sulphur batteries using hybrid anode structures. *Nat. Commun.* **5**, 3015 (2014).
23. Suo, L., Hu, Y.-S., Li, H., Armand, M. & Chen, L. A new class of Solvent-in-Salt electrolyte for high-energy rechargeable metallic lithium batteries. *Nat. Commun.* **4**, 1481 (2013).
24. Kim, H., Lee, J., Ahn, H., Kim, O. & Park, M. J. Synthesis of three-dimensionally interconnected sulfur-rich polymers for cathode materials of high-rate lithium-sulfur batteries. *Nat. Commun.* **6**, 7278 (2015).
25. Chen, H. W. *et al.* Monodispersed sulfur nanoparticles for lithium sulfur batteries with theoretical performance. *Nano Lett.* **15**, 798–802 (2015).
26. Chen, H. W. *et al.* Rational design of cathode structure for high rate performance lithium-sulfur batteries. *Nano Lett.* **15**, 5443–5448 (2015).
27. Wang, B., Alhassan, S. M. & Pantelides, S. T. Formation of large polysulfide complexes during the lithium-sulfur battery discharge. *Phys. Rev. Appl.* **2**, 034004 (2014).
28. Mikhaylik, Y. V. & Akridge, J. R. Polysulfide shuttle study in the Li/S battery system. *J. Electrochem. Soc.* **151**, A1969–A1976 (2004).
29. Wang, H. L. *et al.* Graphene-wrapped sulfur particles as a rechargeable lithium-sulfur battery cathode material with high capacity and cycling stability. *Nano Lett.* **11**, 2644–2647 (2011).
30. Yang, Y. *et al.* Improving the performance of lithium-sulfur batteries by conductive polymer coating. *ACS Nano.* **5**, 9187–9193 (2011).
31. Tao, X. Y. *et al.* Decoration of sulfur with porous metal nanostructures: an alternative strategy for improving the cyclability of sulfur cathode materials for advanced lithium-sulfur batteries. *Chem. Commun.* **49**, 4513–4515 (2013).
32. Cheon, S. E. *et al.* Rechargeable lithium sulfur battery—II. Rate capability and cycle characteristics. *J. Electrochem. Soc.* **150**, A800–A805 (2003).
33. Yang, Y. *et al.* New nanostructured Li_2S/silicon rechargeable battery with high specific energy. *Nano Lett.* **10**, 1486–1491 (2010).
34. Zheng, G. Y. *et al.* Amphiphilic surface modification of hollow carbon nanofibers for improved cycle life of lithium sulfur batteries. *Nano Lett.* **13**, 1265–1270 (2013).
35. Li, W. Y. *et al.* Understanding the role of different conductive polymers in improving the nanostructured sulfur cathode performance. *Nano Lett.* **13**, 5534–5540 (2013).
36. Zhou, W. D., Xiao, X. C., Cai, M. & Yang, L. Polydopamine-coated, nitrogen-doped, hollow carbon sulfur double-layered core-shell structure for improving lithium sulfur batteries. *Nano Lett.* **14**, 5250–5256 (2014).
37. Wang, C. *et al.* Macroporous free-standing nano-sulfur/reduced graphene oxide paper as stable cathode for lithium-sulfur battery. *Nano Energy* **11**, 678–686 (2015).
38. Zhou, G. M., Paek, E., Hwang, G. S. & Manthiram, A. Long-life Li/polysulphide batteries with high sulphur loading enabled by lightweight three-dimensional nitrogen/sulphur-codoped graphene sponge. *Nat. Commun.* **6**, 7760 (2015).
39. Zheng, J. M. *et al.* Lewis acid-base interactions between polysulfides and metal organic framework in lithium sulfur batteries. *Nano Lett.* **14**, 2345–2352 (2014).
40. Zhou, G. M., Zhao, Y. B. & Manthiram, A. Dual-confined flexible sulfur cathodes encapsulated in nitrogen-doped double-shelled hollow carbon spheres and wrapped with graphene for Li-S batteries. *Adv. Energy Mater.* **5**, 1402263 (2015).
41. Dong, K., Wang, S. P., Zhang, H. Y. & Wu, J. P. Preparation and electrochemical performance of sulfur-alumina cathode material for lithium-sulfur batteries. *Mater. Res. Bull.* **48**, 2079–2083 (2013).
42. Sun, F. G. *et al.* A high-rate lithium-sulfur battery assisted by nitrogen-enriched mesoporous carbons decorated with ultrafine La_2O_3 nanoparticles. *J. Mater. Chem. A* **1**, 13283–13289 (2013).
43. Chaiarrekij, S., Apirakchaiskul, A., Suvarnakich, K. & Kiatkamjornwong, S. Kapok I: characteristcs of Kapok fiber as a potential pulp source for papermaking. *Bioresources* **7**, 475–488 (2012).
44. Li, P. *et al.* Lithium sodium vanadium phosphate and its phase transition as cathode material for lithium ion batteries. *Electrochim. Acta* **180**, 120–128 (2015).
45. Diao, Y., Xie, K., Xiong, S. & Hong, X. Analysis of polysulfide dissolved in electrolyte in discharge-charge process of Li-S battery. *J. Electrochem. Soc.* **159**, A421–A425 (2012).
46. Kresse, G. & Furthmuller, J. Efficiency of ab-initio total energy calculations for metals and semiconductors using a plane-wave basis set. *Comput. Mater. Sci.* **6**, 15–50 (1996).
47. Blochl, P. E. Projector augmented-wave method. *Phys. Rev. B* **50**, 17953–17979 (1994).
48. Kresse, G. & Joubert, D. From ultrasoft pseudopotentials to the projector augmented-wave method. *Phys. Rev. B* **59**, 1758–1775 (1999).
49. Nolan, M., Parker, S. C. & Watson, G. W. The electronic structure of oxygen vacancy defects at the low index surfaces of ceria. *Surf. Sci.* **595**, 223–232 (2005).

Acknowledgements

Y.C. acknowledges the support from the Assistant Secretary for Energy Efficiency and Renewable Energy, Office of Vehicle Technologies of the US Department of Energy under the Battery Materials Research (BMR) Program. X.Y.T. and J.G.W. acknowledge support from the Natural Science Foundation of Zhejiang Province (grants LR13E020002) and the National Natural Science Foundation of China (51002138 and 21136001).

Author contributions

Y.C. and X.Y.T. conceived the idea. X.Y.T. designed the experiments, synthesized the materials and performed electrochemical test. J.G.W. and Q.X.C. carried out the DFT calculation and analyzed the result. C.L., H.T.W., G.Y.Z. and other authors conducted materials characterization. All authors discussed the electrochemical results and the whole paper.

Additional information

Competing financial interests: The authors declare no competing financial interests.

6

Effective energy storage from a triboelectric nanogenerator

Yunlong Zi[1,*], Jie Wang[1,*], Sihong Wang[1,*], Shengming Li[1], Zhen Wen[1], Hengyu Guo[1] & Zhong Lin Wang[1,2]

To sustainably power electronics by harvesting mechanical energy using nanogenerators, energy storage is essential to supply a regulated and stable electric output, which is traditionally realized by a direct connection between the two components through a rectifier. However, this may lead to low energy-storage efficiency. Here, we rationally design a charging cycle to maximize energy-storage efficiency by modulating the charge flow in the system, which is demonstrated on a triboelectric nanogenerator by adding a motion-triggered switch. Both theoretical and experimental comparisons show that the designed charging cycle can enhance the charging rate, improve the maximum energy-storage efficiency by up to 50% and promote the saturation voltage by at least a factor of two. This represents a progress to effectively store the energy harvested by nanogenerators with the aim to utilize ambient mechanical energy to drive portable/wearable/implantable electronics.

[1]School of Materials Science and Engineering, Georgia Institute of Technology, Atlanta, Georgia 30332, USA. [2]Beijing Institute of Nanoenergy and Nanosystems, Chinese Academy of Sciences, Beijing 100083, China. * These authors contributed equally to this work. Correspondence and requests for materials should be addressed to Z.L.W. (email: zhong.wang@mse.gatech.edu).

With the rapid development of mobile and portable electronics, more and more efforts have been dedicated to looking for sustainable mobile energy source at the power levels of micro-to-milli watts for these electronics. Currently, the major approach for powering these electronics is by using energy storage units such as batteries[1-3] and capacitors[4,5]. However, the main drawback of these energy storage units is the limited lifetime, so that they cannot drive the electronics sustainably. Alternatively, to drive small electronics sustainably through harvesting ubiquitous ambient small-scale energy, nanogenerators[6] based on piezoelectric, pyroelectric and triboelectric effects have been developed, which are dramatically different from traditional generators. Among nanogenerators, triboelectric nanogenerators (TENG)[7-10] have attracted attention due to their high output and high energy conversion efficiency. Hence, our study here mainly focuses on TENGs. The fundamental working mechanism of a TENG is based on coupling of triboelectrification and electrostatic induction[10-12]. First, at least one pair of triboelectric layers made from materials with distinct electron affinities get into physical contact to create triboelectric charges. Second, as triggered by external mechanical force, the relative motion between the triboelectric layers breaks the balanced electrostatic charge distribution on electrodes. As a result, the potential difference between electrodes is built and free electrons flow through external circuits to achieve new equilibrium. When the triboelectric layers move back, the free electrons flow back to return to the original electrostatic equilibrium. Under periodical mechanical motions such as vibrations[13-15], human walking[16,17] and ocean waves[18,19], pulsed alternating current (AC) output is delivered via the TENG to external circuit. Due to the nature of variable frequency and irregular amplitude of the pulsed AC output, TENGs cannot be directly used to drive most electronic devices. An energy storage unit is required to store the energy harvested by nanogenerators and to provide a regulated and manageable output. Consequently, self-charging power systems[20-22] have been developed by hybridizing a nanogenerator and an energy storage unit, and the latter is charged by the former through a full-wave bridge rectifier.

However, most of the previous research on TENGs mainly focused on output performance under external load resistances[12,17,23-26]; while only a few papers[20,22,27,28] explored the process of using a TENG to charge an energy storage unit, from which we obtained initial understandings on the charging characteristics. From these studies, it has been demonstrated that the charging rate decays quickly after several charging cycles. The saturation voltage, which is the highest achievable voltage of the energy storage unit, is much smaller than the open-circuit voltage of the TENG, resulting in a low energy-storage efficiency regardless of the energy conversion efficiency of the TENG. Therefore, studies are required to further understand the charging process and to improve the energy storage performances for TENGs. According to a recent study on the standards of TENGs, the average output power is mainly determined by the encircled area of its operation cycle in the built-up voltage V - transferred charge Q plot[12]. Therefore, the corresponding operation cycle for charging the energy storage units should be studied by considering the charging process of the TENG to achieve the maximized energy-storage efficiency.

In this work, we first analysed the operation cycle of using a TENG to directly charge a battery/capacitor through a bridge rectifier by our recently proposed V–Q plot[12]. A sliding freestanding-triboelectric-layer (SFT) mode TENG was fabricated to experimentally measure the V–Q plots of the direct charging cycle. Then a rationally designed cycle was proposed to improve the charging rate and the saturation voltage. The voltage of the energy storage unit was demonstrated to be a key parameter for optimizing the stored energy per cycle. As a proof of concept, a motion-triggered switch was added to the fabricated TENG to achieve this designed charging cycle. Furthermore, the enhanced performances including the enhanced charging rate, the improved energy-storage efficiency and the promoted saturation voltage were achieved experimentally. Finally, the self-charging power system operated under this designed charging cycle was used to drive a commercial calculator in the sustainable mode. Our work represents a paradigm shift in strategies and experimental methods to achieve effective energy harvesting and storage by TENGs as well as other nanogenerators.

Results

Theoretical analysis of the direct charging cycle. Conventional integration of a TENG and an energy storage device was achieved through a full-wave bridge rectifier, as shown in the inset of Fig. 1a. We utilize the V–Q plot as the analytical tool for TENG energy storage, as shown in Fig. 1, in which V is the potential difference between the two electrodes, and Q is the amount of charge transferred between the electrodes. Several important parameters of the TENG are defined in Table 1. The parameters for TENG energy storage are given in Supplementary Table 1. We use the most commonly utilized minimum achievable charge reference state[29], so both Q and V at $x = 0$ position are set to be 0 initially, where x is the relative displacement between triboelectric layers, which is used to specify the operation of the TENG. The cycle for maximized energy output with infinite load resistance, which has the largest possible energy output per cycle E_m, is also plotted in Fig. 1a,c using the dashed lines (with equations given in Supplementary Note 1)[12]. Here we assume that the charging voltage V_C, which is the voltage of the energy storage unit, does not change significantly during one cycle of the TENG operation (as stated in Supplementary Note 2).

Therefore, starting from $(Q, V) = (0, 0)$, the direct charging cycles of the TENG can be described with the following steps (Fig. 1 a,b). For the first cycle, the first step is to change from status I to II. The initial status I of the system is at $(Q, V) = (0, 0)$ and $x = 0$. Because the initial voltage V supplied by the TENG is lower than the charging voltage V_C, all of the diodes in the rectifier are at 'off' state, which makes the external circuit condition close to the open-circuit condition. When the operation of the TENG starts, V begins to increase with x. No charge transfer occurs during this step, until V reaches V_C as shown in status II.

The second step is to change from status II to III. Two of the four diodes in the rectifier are turned on to enable the charging process. Then x continues to increase until $x = x_{max}$, during which the TENG charges the battery/capacitor at $V = V_C$ (See Supplementary Note 3 and Supplementary Fig. 1 for the detailed process) until status III is achieved. Please note that in status III the charges cannot be fully transferred to the other electrode ($Q < Q_{SC,max}$), since $V = V_C$ should be maintained during the charging process in this step.

The third step is to change from status III to IV: x starts to decrease. Since the reversed-direction current is forbidden, V decreases without charge transfers, until $V = -V_C$ as shown in status IV. During this step all of the diodes in the rectifier are off because the net voltage $|V|$ is lower than the threshold voltage at V_C.

The fourth step is to change from status IV to V. The other two diodes in the rectifier are turned on to enable the charging process again. x continues to decrease until $x = 0$, as shown in status V, during which, the TENG charges the battery/capacitor with $V = -V_C$. Similar to status III, in status V the charge cannot be fully transferred back to the original electrode ($Q > 0$), since $V = -V_C$ should be maintained during the charging process in this step. Here the first charging cycle is completed.

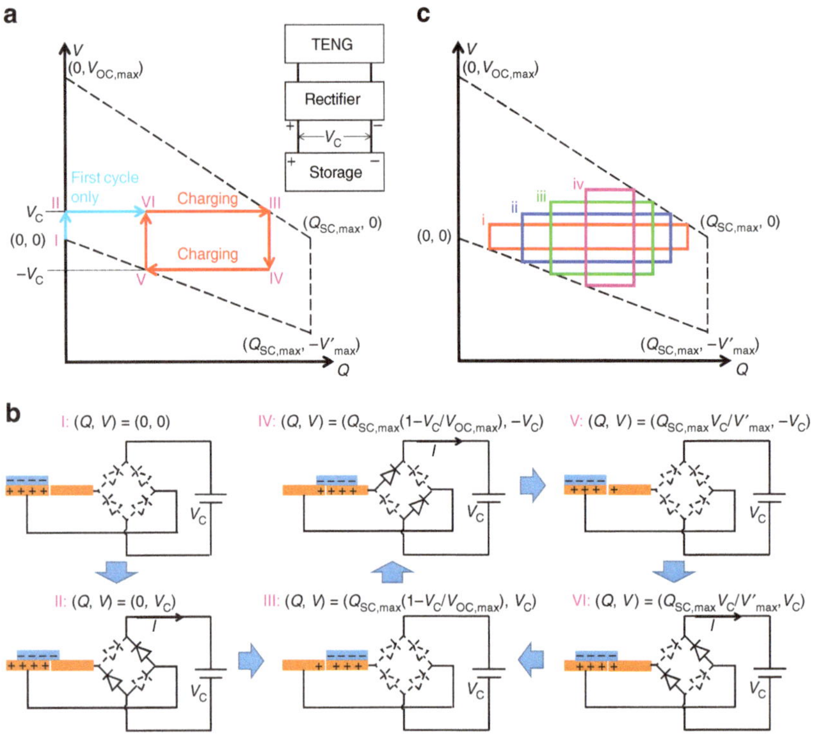

Figure 1 | The *V-Q* plot and the physical process of the direct charging cycle. (**a**) The *V-Q* plot. The inset of **a** shows the circuit diagram of the system. The dashed lines in **a** represent the cycle with the maximized energy output with infinite resistance. (**b**) The schematic diagrams of the physical process. In **b**, the diodes drawn by solid and dashed lines are for the on and off states, respectively. (**c**) The direct charging cycles (the second cycle and thereafter) for different charging voltages (voltage iv > iii > ii > i).

Table 1 | The definitions of important parameters of the TENG.

Symbol	Definitions
$Q_{SC,max}$	The maximum short-circuit transferred charge
$V_{OC,max}$	The maximum open-circuit voltage at $Q=0$
V'_{max}	The maximum achievable absolute voltage at $Q=Q_{SC,max}$
E_m	The largest possible energy output per cycle

For the second cycle and thereafter (starting from status V), x starts to increase, all of the diodes in the rectifier are off (since $|V| < V_C$) and V increases without charge transfers, until $V = V_C$ as shown in status VI. x continues to increase and the corresponding two diodes are turned on to allow the charging process continue until $x = x_{max}$, the same as status III. The process to complete one cycle is the same as steps 3 and 4 of the first cycle, which comes back to status V. Therefore, the second cycle and thereafter follow the sequence of V→VI→III→IV→V, which is the steady-state charging cycle.

The corresponding V–Q cycle is plotted in Fig. 1a, with the (Q, V) coordinates corresponding to statuses calculated in Supplementary Note 4. The stored energy per cycle $E_{C,direct}$, which is proportional to the average output power, can be calculated as the encircled area of each cycle[12]:

$$E_{C,direct} = \begin{cases} V_C Q_{SC,max}\left(2 - 2V_C/V_{OC,max} - V_C/V'_{max}\right), & \text{for the first cycle} \\ 2V_C Q_{SC,max}\left(1 - V_C/V_{OC,max} - V_C/V'_{max}\right), & \text{for the second cycle and thereafter} \end{cases} \quad (1)$$

Since the charging of an energy storage device requires many cycles of operation of the TENG, we will only consider the equation for the steady-state cycles. The amount of charge ($Q_{C,direct}$) flowing into the energy storage unit per cycle can be calculated as $E_{C,direct}/V_C$, which equals to the total length of the sides that are parallel to the Q axis in the V–Q plot. The direct charging cycles corresponding to different charging voltages V_C are plotted in Fig. 1c, which shows the decrease of $Q_{C,direct}$ due to the increase of V_C (which is further discussed in Supplementary Note 5).

From equation (1), with varied V_C, we can derive that the maximum value of $E_{C,direct}$ is achieved when $V_C = \frac{1}{2}\frac{V_{OC,max}V'_{max}}{V_{OC,max}+V'_{max}}$, which is:

$$E_{C,direct,max} = \frac{Q_{SC,max}}{2}\frac{V_{OC,max}V'_{max}}{V_{OC,max}+V'_{max}} \quad (2)$$

We define the energy storage efficiency η as the percentage of the stored energy per cycle in the largest possible energy output per cycle E_m. Therefore, for the direct charging cycle, the maximum energy-storage efficiency is:

$$\eta_{direct} = \frac{E_{C,direct,max}}{E_m} = \frac{V_{OC,max}V'_{max}}{\left(V_{OC,max}+V'_{max}\right)^2} \leq 25\% \quad (3)$$

And the largest possible V_C to operate the direct charging cycle is at the saturation voltage:

$$V_{Sat,direct} = \frac{V_{OC,max}V'_{max}}{V_{OC,max}+V'_{max}} \quad (4)$$

Experimental demonstration of the direct charging cycle. A SFT mode TENG is fabricated to measure the direct charging cycle experimentally. The TENG is schematically illustrated in the inset of Fig. 2a. The top static part of the TENG is made by attaching two parallel aluminium (Al) electrodes on the bottom surface of an acrylic board. The moving part of the TENG is made by attaching a piece of fluorinated ethylene propylene (FEP) film on the top surface of another acrylic board. The FEP

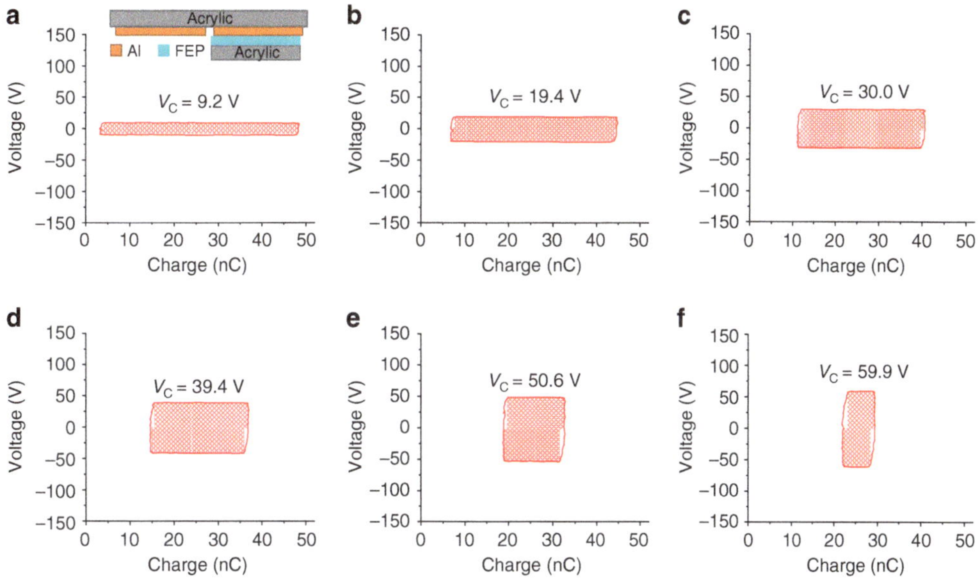

Figure 2 | The V-Q plots of the direct charging cycle for a commercial capacitor with charging voltage V_C increased from 9.2 V to 59.9 V. For these plots, V_C are (**a**) 9.2 V, (**b**) 19.4 V, (**c**) 30.0 V, (**d**) 39.4 V, (**e**) 50.6 V and (**f**) 59.9 V. The inset of **a** shows the TENG structure. The shaded areas are the stored energy per cycle $E_{C,direct}$.

film and Al electrodes are the triboelectric layers. The fabrication process is described in detail in the Methods section. In this structure, $V_{OC,max} = V'_{max}$ since the capacitance between the two electrodes is kept identical[29,30]. The measured cycle for maximized energy output with infinite load resistance is shown in Supplementary Fig. 2b, where $V_{OC,max} = V'_{max} \approx 140$ V, and $Q_{SC,max} \approx 50$ nC. The circuit diagram for this measurement is shown in Supplementary Fig. 2a. A commercial capacitor (0.73 μF) and fabricated lithium-ion batteries connected in series are used as the energy storage devices, respectively. For the capacitor, the charging voltage V_C increases gradually during the charging process, and the V–Q curves of direct charging cycle with different V_C are plotted in Fig. 2. For the batteries, each battery can supply a voltage of about 3–4 V, and different number of batteries are connected to achieve different V_C. The fabrication method of the batteries and the measurement set-up are described in the Methods section and Supplementary Fig. 3, respectively. The V–Q curves of the direct charging cycle for batteries are plotted in Supplementary Fig. 4. From these V–Q curves, we notice that all of the curves in the direct charging cycle for both batteries and the capacitor have a rectangular shape. The total length of the two sides parallel to the Q axis can be directly used as $Q_{C,direct}$, as demonstrated in Supplementary Note 6 and Supplementary Fig. 5. $Q_{C,direct}$ decreases with increasing V_C, and the saturation voltage $V_{Sat,direct}$ is measured at almost 70 V, which are consistent with the calculated results from equations (1) and (4).

Theoretical analysis of the designed charging cycle. From the equations and experimental results presented above, we can make three observations on the direct charging cycle. First, the charge flowing into the energy storage unit per cycle $Q_{C,direct}$, which is proportional to the charging rate, quickly decreases with the increased V_C. Second, the maximum value of the energy storage efficiency η is only 25%. Finally, the saturation voltage $V_{Sat,direct}$ is much smaller than $V_{OC,max}$ and V'_{max}. To improve these charging performances, we rationally design a charging cycle. A switch is added in parallel to the TENG, which is controlled to instantaneously turn on only at $x = 0$ and $x = x_{max}$, and turn off at all the other x. Starting from $(Q, V) = (0, 0)$, the process of the designed charging cycle can be described as follows (Fig. 3a,b):

Steps 1 and 2 (status I–III) are the same as those in the first cycle of the direct charging cycle. Similarly, in status III the charge cannot be fully transferred to the other electrode ($Q < Q_{SC,max}$).

Step 3 (status III–IV): while in status III, x reaches x_{max}, and then the switch is turned on. The charge transfers in short-circuit condition to achieve $(Q, V) = (Q_{SC,max}, 0)$, which is status IV.

Step 4 (status IV–V): the switch is turned off as x begins to decrease; at the same time all of the diodes in the rectifier are off since $|V| < V_C$. Therefore, V decreases without any charge transfers until it reaches $-V_C$, which is status V.

Step 5 (status V–VI): the second half-cycle of the charging starts as x continues decreasing, until $x = 0$ at status VI. Also, in status VI the charge cannot be fully transferred back to the original electrode ($Q > 0$).

Step 6 (status VI to I): the switch is turned on again to make (Q, V) return to (0, 0), that is, status I.

Therefore, the repeated process for the designed charging cycle is through I→II→III→IV→V→VI→I. As concluded from steps 3 and 6, the function of the switch is to enable the complete charge transfer by creating an instantaneous short-circuit condition. Consequently, in the next half-cycle, there are more charges available to charge the energy storage unit, which is the key to enhance the charging performances by using the designed charging cycle.

By marking the coordinates of corresponding statuses (as calculated in Supplementary Note 4) in the V–Q plot (Fig. 3a), we observe that there are three regions of energy in the encircled area in the designed charging cycle. Area 1 represents the energy that can be stored in both the direct and the designed charging cycles; area 3 represents the energy released through the switch; and the energy of area 2 is the part that can only be stored in the designed charging cycle. Thus, the stored energy per cycle in the designed charging cycle should only include energies in areas 1 and 2. During steps 1–3 and steps 4–6, the energies stored are derived as $V_C Q_{SC,max}(1 - V_C/V_{OC,max})$ and $V_C Q_{SC,max}(1 - V_C/V'_{max})$, respectively. Therefore, the total stored energy per cycle is:

$$E_{C,designed} = V_C Q_{SC,max}(2 - V_C/V_{OC,max} - V_C/V'_{max})$$

Note that this equation is only valid for $V_C \leq \min\{V_{OC,max},$

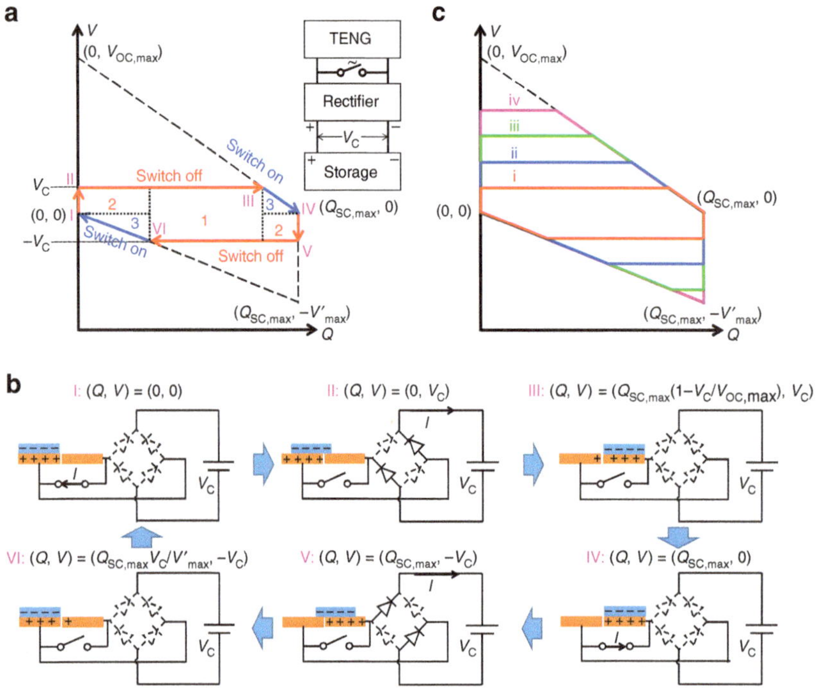

Figure 3 | The *V-Q* plot and the physical process of the rationally designed charging cycle. (**a**) The *V-Q* plot. The inset of **a** shows the circuit for the designed charging cycle. The dashed lines in **a** represent the cycle of maximized energy output with infinite resistance. (**b**) The schematic diagrams of the physical process. In **b**, the diodes drawn by solid and dashed lines are for the on and off states, respectively. (**c**) The designed charging cycle for different charging voltages, with charging voltage intensity iv > iii > ii > i.

$V'_{max}\}$. Considering the symmetric roles of $V_{OC,max}$ and V'_{max} (as explained in Supplementary Note 7), we can simply assume $V_{OC,max} \geq V'_{max}$. If $V'_{max} < V_C \leq V_{OC,max}$, since during step 4 the voltage cannot achieve $-V_C$, there is no energy stored during steps 4–6. Then the equation becomes:

$$E_{C,designed} = V_C Q_{SC,max}\left(1 - V_C/V_{OC,max}\right)$$

Therefore:

$$E_{C,designed} = \begin{cases} V_C Q_{SC,max}\left(2 - V_C/V_{OC,max} - V_C/V'_{max}\right), V_C \leq V'_{max} \leq V_{OC,max} \\ V_C Q_{SC,max}\left(1 - V_C/V_{OC,max}\right), V'_{max} < V_C \leq V_{OC,max} \end{cases} \quad (5)$$

Similarly, the other characteristic parameters of the designed charging cycle can be calculated as follows:

The maximum stored energy per cycle (achieved at $V_C = \frac{V_{OC,max} V'_{max}}{V_{OC,max} + V'_{max}}$ or $V_C = \frac{V_{OC,max}}{2}$):

$$E_{C,designed,max} = \max\left\{ Q_{SC,max}\frac{V_{OC,max} V'_{max}}{V_{OC,max} + V'_{max}}, \frac{Q_{SC,max} V_{OC,max}}{4} \right\} \quad (6)$$

The maximum energy storage efficiency:

$$\eta_{designed} = \frac{E_{C,designed,max}}{E_m} = \max\left\{ \frac{2V_{OC,max} V'_{max}}{\left(V_{OC,max} + V'_{max}\right)^2}, \frac{V_{OC,max}}{2\left(V_{OC,max} + V'_{max}\right)} \right\} \leq 50\% \quad (7)$$

And the saturation voltage is:

$$V_{Sat,designed} = \max\left\{ V_{OC,max}, V'_{max} \right\} \quad (8)$$

As derived by these equations of the designed charging cycle: the amount of charge flowing to the battery/capacitor per cycle of the designed charging cycle ($Q_{C,designed}$) is larger than that of the direct charging cycle ($Q_{C,direct}$), corresponding to the same V_C (Supplementary Note 5); the maximum energy storage efficiency has been increased up to 50%; and the saturation voltage $V_{Sat,designed}$ is the same as the larger one in $V_{OC,max}$ and V'_{max}, which is at least twice the value of $V_{Sat,direct}$. Therefore, this designed cycle represents an effective strategy to greatly enhance the performances of using TENG to charge an energy storage unit.

Experimental demonstration of the designed charging cycle. We demonstrate this designed charging cycle experimentally by adding a motion-triggered switch on the fabricated SFT mode TENG, which is similar to the design of the pulsed nanogenerator as reported previously[31], as shown in Fig. 4. The top static part is kept the same (Fig. 4a). Two parallel titanium (Ti) ribbons are attached on the acrylic board of the moving part (Fig. 4b). These two Ti ribbons are connected to the two Al electrodes through copper wires, respectively. Two Ti bars (Fig. 4c) are fixed in the positions of $x = 0$ and $x = x_{max}$. The configuration of the integrated structure is shown in Fig. 4d and the detailed fabrication process is described in the Methods section. When $x = 0$ or $x = x_{max}$, both of the two Ti ribbons contact with one of the Ti bars, and then the two electrodes are connected, which represents the switch on. Otherwise, the electrodes are not connected, which represents the switch off.

The *V–Q* curves of these TENGs operated in the designed charging cycle are measured. Figure 5 presents the *V–Q* plots of the designed charging cycle for the same capacitor used in the experiments for the direct charging cycle, and the plots for the batteries in series are shown in Supplementary Fig. 6. Similarly, the total length of the two sides that are parallel to *Q* axis in *V–Q* plot can be used directly as $Q_{C,designed}$. The stored energy per cycle $E_{C,designed}$ is marked as the shaded area in all the plots in Fig. 5, in which the red and grey shaded areas represent areas 1 and 2 in Fig. 3a, respectively. Compared with the direct charging cycle, the designed charging cycle can store more energy per cycle for the

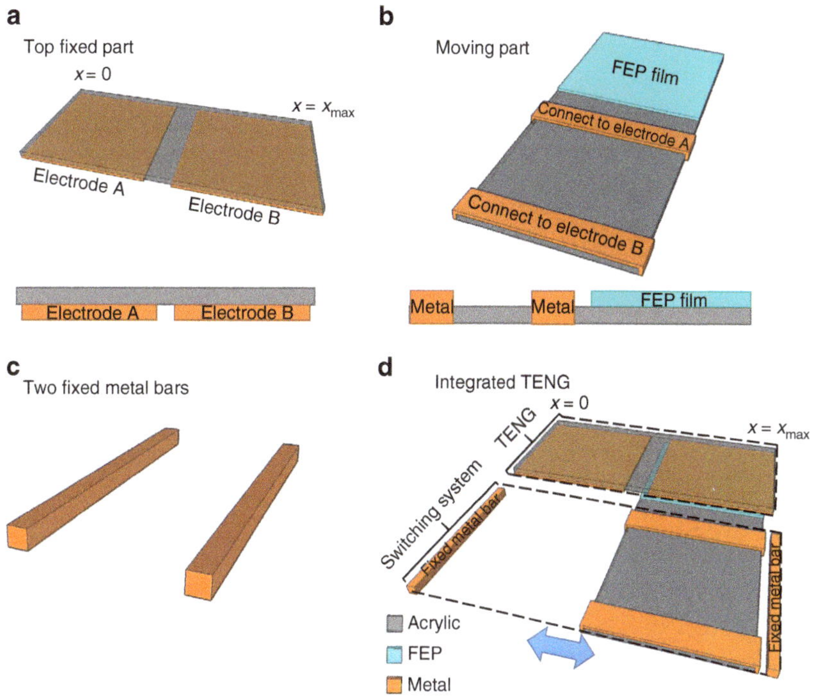

Figure 4 | The TENG with the motion-triggered switch fabricated for the designed charging cycle. This TENG is composed of **a** the static part, (**b**) the moving part and (**c**) the two fixed metal bars. (**d**) The TENG as integrated by the three parts in **a**-**c**. The insets in **a**,**b** show the front view of the static part and the right-side view of the moving part.

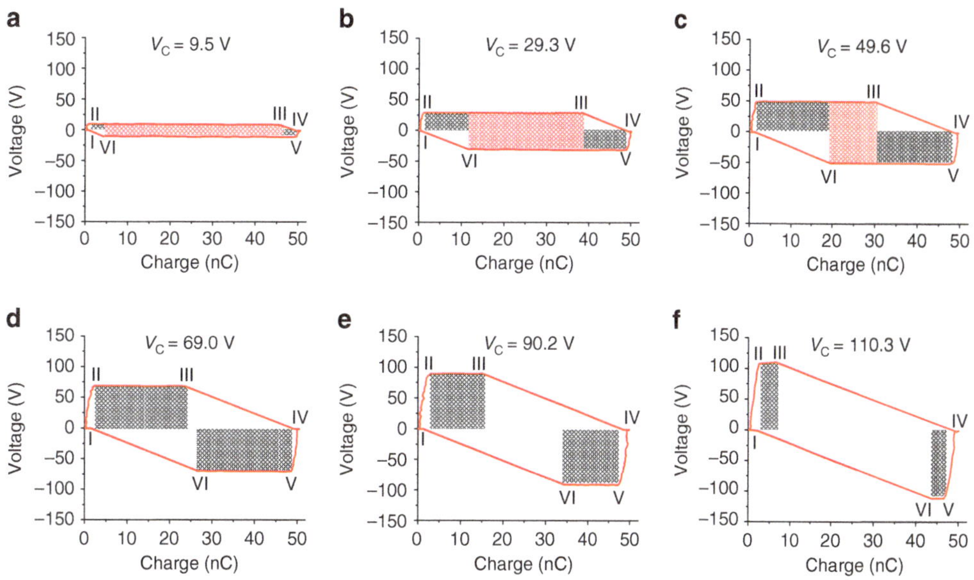

Figure 5 | The *V*-*Q* plots of the designed charging cycle for a commercial capacitor with charging voltage increased from 9.5 V to 110.3 V. For these plots, V_C are (**a**) 9.5 V, (**b**) 29.3 V, (**c**) 49.6 V, (**d**) 69.0 V, (**e**) 90.2 V and (**f**) 110.3 V. The six statuses in the designed charging cycle are marked in the plots. The shaded areas are the stored energy per cycle by the designed charging cycle $E_{C,designed}$, in which the red shaded areas are the part of the energy that can also be stored in the direct charging cycle with the same V_C, while the grey shaded areas are the 'extra' energy that can only be stored in the designed charging cycle.

same V_C, and V_{Sat} can be enhanced to be close to 140 V. To further understand the V–Q plots in Fig. 5d,f when V_C is larger than $V_{Sat,direct}$ but smaller than $V_{Sat,designed}$, the corresponding charging cycles are discussed in Supplementary Note 8, and the theoretically derived V–Q plots are displayed in Supplementary Fig. 7. All of these experimental results are consistent with our theoretical derivations.

Quantitative comparisons between the charging cycles. To further reveal the advantages of the designed charging cycle in energy storage and verify our theoretical derivations above, quantitative comparisons between experimental results of the direct and the designed charging cycles are performed, as shown in Fig. 6. Here the plots for the capacitor are used since it can easily achieve different V_C during the charging process.

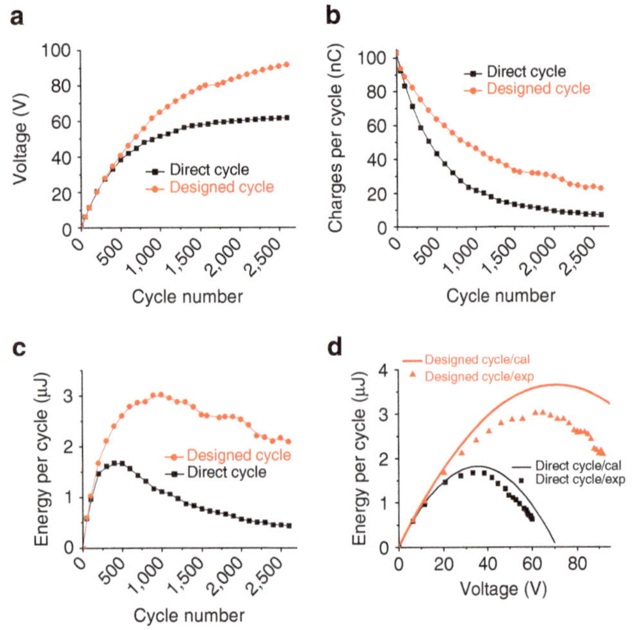

Figure 6 | Quantitative comparisons between the charging cycles. (**a**) The charging voltage V_C, (**b**) the amount of charge Q_C flowing to the capacitor per cycle and (**c**) the stored energy per cycle E_C changes versus the number of the charging cycles for both the direct and the designed charging cycles. (**d**) The experimental (dots) and calculated (lines) results (as calculated from equations (1) and (5)) of the stored energy per cycle E_C versus the charging voltage V_C for both the direct and the designed charging cycles.

Figure 6a–c show the changes of the charging voltage V_C, the charge Q_C flowing to the capacitor per cycle and the stored energy per cycle E_C versus the number of the charging cycles, respectively. For both types of cycles, with the increase of the number of the charging cycles, V_C increases since more and more charges flow into the capacitor; Q_C decreases due to the increases of V_C, as discussed in Supplementary Note 5. Compared with the direct charging cycle, V_C increases faster, Q_C decreases slower and hence E_C is significantly promoted in the designed charging cycle. The reason of slower decay of Q_C in the designed charging cycle is during the switch-on operations (statuses III to IV and VI to I), the charges are fully transferred to $Q_{SC,max}$ or 0, so that in the next half-cycle there are more charges available to flow into the capacitor (as further discussed in Supplementary Note 5). We also notice that the higher saturation voltage V_{Sat} can be achieved in the designed charging cycle, since for the direct charging cycle there is no energy stored once V_C approaches $V_{Sat,direct}$, as discussed in Supplementary Note 8. The curves of E_C versus V_C for both the direct and the designed charging cycles are also plotted in Fig. 6d for both experimental and calculated results (as calculated from equations (1) and (5)). From this we can conclude that, the designed charging cycle can supply a larger E_C under the same V_C, and a higher V_{Sat} can be achieved.

Application of the designed charging cycle. To further demonstrate the promoted energy storage in the designed charging cycle, we use a TENG working in the designed charging cycle to power a commercial calculator (working voltage 1.5 V) in sustainable mode using the circuit in Fig. 7a. To provide a constant voltage to the calculator, the average input power from the TENG needs to equal to the power consumption in the output circuit[20]. The capacitor is charged to let $V_C = 19.2$ V initially. As indicated by the dashed frame in Fig. 7a, the output circuit, which is connected to the capacitor in parallel, includes a power

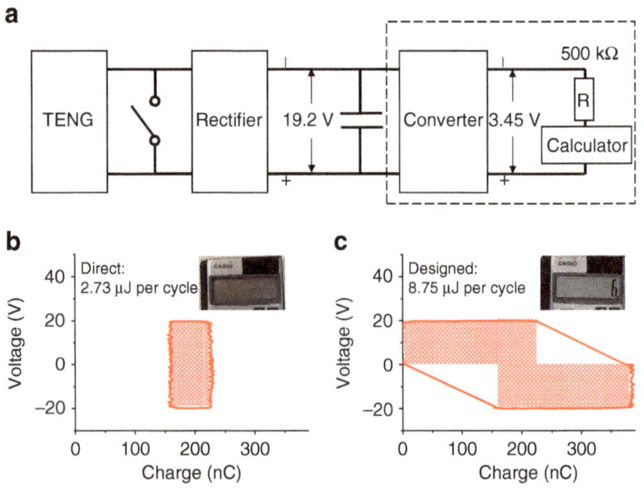

Figure 7 | Demonstration of a calculator sustainably powered by the self-charging power system through the designed charging cycle. (**a**) Schematic diagram of the operation circuit; (**b**,**c**) the *V*–*Q* plots of the direct and the designed charging cycle for the TENG. The shaded areas represent the energy stored per cycle for both cycles. The power required by units in dashed frame in **a** is satisfied by that supplied by TENG through the designed cycle. The inset pictures show the calculator's display screen as charged by the direct and designed charging cycles, respectively.

converter (from Linear Technology) to regulate the voltage to 3.45 V, the calculator and a divider resistor of 500 kΩ to provide the calculator a voltage of ~1.5 V. The measured current required to drive the output circuits is ~1.5 μA (Supplementary Fig. 8), which indicates that an average output power of ~28.8 μW needs to be provided by the TENG. If we set the working period as 0.3 s for the TENG, the required stored energy per cycle is ~8.65 μJ. To achieve that, we fabricate a TENG that can generate ~780 nC per cycle in maximum ($Q_{SC,max} \approx 390$ nC). The measured *V*–*Q* plots of both the direct and the designed charging cycles at $V_C = 19.2$ V are shown in Fig. 7b,c. The measured E_C for the direct charging cycle is 2.73 μJ, which is much smaller than the required energy per cycle of 8.65 μJ. Consequently, the capacitor's energy operated in the direct charging cycle is consumed quickly, and the calculator turned off quickly, as shown in the inset of Fig. 7b. As a comparison, the measured E_C for the designed charging cycle is 8.75 μJ, which is larger than the required energy per cycle of 8.65 μJ. Therefore, as shown in the insets of Fig. 7c, with the same $V_C = 19.2$ V, the TENG operating in the designed charging cycle can drive the calculator sustainably. The calculator working in sustainable mode powered by the self-charging power system through the designed charging cycle is also shown in Supplementary Movie 1.

Discussion

In summary, a rationally designed route for more effective energy storage of the random-pulsed output power generated by TENGs was developed theoretically and experimentally. The step-by-step charging processes of both the direct and the designed charging cycles were illustrated by *V*–*Q* plots, which revealed the advantages of the designed charging cycle in enhanced charging rate, improved energy-storage efficiency (up to 50%), and promoted saturation voltage (by at least a factor of two). As a proof of concept, we designed a TENG with motion-triggered switch to achieve the designed charging cycle. The results from both approaches were compared quantitatively, and the results confirmed the advantages of the designed charging cycle, as predicted by theoretical derivations. A commercial calculator was demonstrated to be sustainably powered by the TENG-based self-charging power system through the designed charging cycle. This designed charging cycle provides a route to enhance the energy harvesting and storage for TENGs as well as other nanogenerators, which represents a solid progress towards effectively utilizing ambient mechanical energy as a sustainable power source for electronics.

Methods

Fabrication and operation of the TENG. The static part of the TENG was fabricated by attaching Al foils on a 10.1 × 7 cm acrylic board to form two rectangular shape electrodes each with a size of 5 × 7 cm and the spacing of 1 mm. A 5 × 7-cm 50-μm FEP film attached along one side of a 5 × 15-cm acrylic board was used as the moving part. For the designed charging cycle, two Ti metal ribbons (1 cm in width for each, 5 cm in spacing) were attached on the other side of the acrylic board in parallel in the moving part, which were connected to two electrodes, respectively. Two Ti bars (7 cm in length) were also prepared. To operate the TENG in the designed charging cycle, the moving part was mounted on a linear motor and the static part and the two Ti bars were mounted on 3 three-dimensional stages, respectively, and the FEP surface and one Al electrode were placed face to face, as shown in Fig. 4d. The linear motor was controlled to move periodically between two Al electrodes with a displacement of 5.1 cm. For the direct charging cycle, the two Ti bars were removed. The TENG used to demonstrate the sustainable mode was fabricated in the same procedure with twice of all the sizes. The voltage and transferred charge were measured by two Keithley 6,514 system electrometers simultaneously. The measurement circuits for *V–Q* plots are illustrated in Supplementary Fig. 3.

Fabrication of the lithium-ion battery. Lithium-ion rechargeable coin cell batteries were fabricated by using $LiCoO_2$/carbon black/binder mixture on Al foil (1 cm in diameter) as the anode, polyethylene (PE) as separator (2 cm in diameter), the graphite/carbon black/binder mixture on Al foil (1.5 cm in diameter) as the cathode. The electrolyte (1 M $LiPF_6$ in 1:1:1 ethylene carbonate/dimethyl carbonate/diethyl carbonate) was injected inside between the anode and cathode before the coin cell was pressed firmly. The charging-discharging curves of these batteries were tested as shown in Supplementary Fig. 9, which exhibits a plateau voltage of ~3.8 V. In our experiments, each battery can supply voltage of ~3–4 V.

References

1. Dunn, B., Kamath, H. & Tarascon, J. M. Electrical energy storage for the grid: a battery of choices. *Science* **334,** 928–935 (2011).
2. Huskinson, B. *et al.* A metal-free organic-inorganic aqueous flow battery. *Nature* **505,** 195–198 (2014).
3. Peng, Z. Q., Freunberger, S. A., Chen, Y. H. & Bruce, P. G. A reversible and higher-rate $Li-O_2$ battery. *Science* **337,** 563–566 (2012).
4. El-Kady, M. F., Strong, V., Dubin, S. & Kaner, R. B. Laser scribing of high-performance and flexible graphene-based electrochemical capacitors. *Science* **335,** 1326–1330 (2012).
5. Zhu, Y. W. *et al.* Carbon-based supercapacitors produced by activation of graphene. *Science* **332,** 1537–1541 (2011).
6. Wang, Z. L. & Song, J. Piezoelectric nanogenerators based on zinc oxide nanowire arrays. *Science* **312,** 242–246 (2006).
7. Fan, F.-R., Tian, Z.-Q. & Lin Wang, Z. Flexible triboelectric generator. *Nano Energy* **1,** 328–334 (2012).
8. Zhu, G. *et al.* Linear-grating triboelectric generator based on sliding electrification. *Nano. Lett.* **13,** 2282–2289 (2013).
9. Wang, Z. L., Chen, J. & Lin, L. Progress in triboelectric nanogenerators as a new energy technology and self-powered sensors. *Energy Environ. Sci.* **8,** 2250–2282 (2015).
10. Wang, Z. L. Triboelectric nanogenerators as new energy technology and self-powered sensors - Principles, problems and perspectives. *Faraday Discuss.* **176,** 447–458 (2014).
11. Wang, Z. L. Triboelectric nanogenerators as new energy technology for self-powered systems and as active mechanical and chemical sensors. *ACS Nano* **7,** 9533–9557 (2013).
12. Zi, Y. *et al.* Standards and figure-of-merits for quantifying the performance of triboelectric nanogenerators. *Nat. Commun.* **6,** 8376 (2015).
13. Chen, J. *et al.* Harmonic-resonator-based triboelectric nanogenerator as a sustainable power source and a self-powered active vibration sensor. *Adv. Mater.* **25,** 6094–6099 (2013).
14. Yang, J. *et al.* Triboelectrification-based organic film nanogenerator for acoustic energy harvesting and self-powered active acoustic sensing. *ACS Nano* **8,** 2649–2657 (2014).
15. Fan, X. *et al.* Ultrathin, rollable, paper-based triboelectric nanogenerator for acoustic energy harvesting and self-powered sound recording. *ACS Nano* **9,** 4236–4243 (2015).
16. Yang, W. *et al.* Harvesting energy from the natural vibration of human walking. *ACS Nano* **7,** 11317–11324 (2013).
17. Xie, Y. *et al.* Grating-structured freestanding triboelectric-layer nanogenerator for harvesting mechanical energy at 85% total conversion efficiency. *Adv. Mater.* **26,** 6599–6607 (2014).
18. Chen, J. *et al.* Networks of triboelectric nanogenerators for harvesting water wave energy: a potential approach toward blue energy. *ACS Nano* **9,** 3324–3331 (2015).
19. Wang, X. *et al.* Triboelectric Nanogenerator based on fully enclosed rolling spherical structure for harvesting low-frequency water wave energy. *Adv. Energy Mater.* **5,** 15011467 (2015).
20. Wang, S. *et al.* Motion charged battery as sustainable flexible-power-unit. *ACS Nano* **7,** 11263–11271 (2013).
21. Wang, J. *et al.* A flexible fiber-based supercapacitor–triboelectric-nanogenerator power system for wearable electronics. *Adv. Mater.* **27,** 4830–4836 (2015).
22. Luo, J. *et al.* Integration of micro-supercapacitors with triboelectric nanogenerators for a flexible self-charging power unit. *Nano Res.* **8,** 3934–3943 (2015).
23. Zhou, Y. S. *et al.* Manipulating nanoscale contact electrification by an applied electric field. *Nano. Lett.* **14,** 1567–1572 (2014).
24. Zi, Y. *et al.* Triboelectric–pyroelectric–piezoelectric hybrid cell for high-efficiency energy-harvesting and self-powered sensing. *Adv. Mater.* **27,** 2340–2347 (2015).
25. Han, M. *et al.* Magnetic-assisted triboelectric nanogenerators as self-powered visualized omnidirectional tilt sensing system. *Sci. Rep.* **4,** 4811 (2014).
26. Seung, W. *et al.* Nanopatterned textile-based wearable triboelectric nanogenerator. *ACS Nano* **9,** 3501–3509 (2015).
27. Niu, S. *et al.* Optimization of triboelectric nanogenerator charging systems for efficient energy harvesting and storage. *IEEE Trans. Electron Devices* **62,** 641–647 (2015).
28. Pu, X. *et al.* Efficient charging of Li-ion batteries with pulsed output current of triboelectric nanogenerators. *Adv. Sci.* **3,** 1500255 (2015).
29. Niu, S. *et al.* Theory of freestanding triboelectric-layer-based nanogenerators. *Nano Energy* **12,** 760–774 (2015).
30. Wang, S., Xie, Y., Niu, S., Lin, L. & Wang, Z. L. Freestanding triboelectric-layer-based nanogenerators for harvesting energy from a moving object or human motion in contact and non-contact modes. *Adv. Mater.* **26,** 2818–2824 (2014).
31. Cheng, G., Lin, Z.-H., Lin, L., Du, Z.-l. & Wang, Z. L. Pulsed nanogenerator with huge instantaneous output power density. *ACS Nano* **7,** 7383–7391 (2013).

Acknowledgements

Y.Z., J.W. and S.W. contributed equally to this work. This research was supported by the National Science Foundation (DMR-1505319), the Hightower Chair foundation and the 'thousands talents' programme for pioneer researcher and his innovation team, China, National Natural Science Foundation of China (Grant No. 51432005). We would like to thank Dr Ken C. Pradel for his help in revising the manuscript.

Author contributions

Y.Z., J.W., S.W. and Z.L.W. conceived the idea, discussed the data and prepared the manuscript. Y.Z. and S.W. did the theoretical derivations and calculations on the direct charging cycle and the designed charging cycle. Y.Z. and J.W. did the experiments and analysed the data. S.L., Z.W. and H.G. helped the experiments. All the authors helped revise the manuscript.

Additional information

Competing financial interests: The authors declare no competing financial interest.

7

Linking stress-driven microstructural evolution in nanocrystalline aluminium with grain boundary doping of oxygen

Mo-Rigen He[1], Saritha K. Samudrala[2], Gyuseok Kim[1], Peter J. Felfer[2], Andrew J. Breen[2], Julie M. Cairney[2] & Daniel S. Gianola[1,3]

The large fraction of material residing at grain boundaries in nanocrystalline metals and alloys is responsible for their ultrahigh strength, but also undesirable microstructural instability under thermal and mechanical loads. However, the underlying mechanism of stress-driven microstructural evolution is still poorly understood and precludes rational alloy design. Here we combine quantitative *in situ* electron microscopy with three-dimensional atom-probe tomography to directly link the mechanics and kinetics of grain boundary migration in nanocrystalline Al films with the excess of O atoms at the boundaries. Site-specific nanoindentation leads to grain growth that is retarded by impurities, and enables quantification of the critical stress for the onset of grain boundary migration. Our results show that a critical excess of impurities is required to stabilize interfaces in nanocrystalline materials against mechanical driving forces, providing new insights to guide control of deformation mechanisms and tailoring of mechanical properties apart from grain size alone.

[1] Department of Materials Science and Engineering, University of Pennsylvania, Philadelphia, Pennsylvania 19104, USA. [2] Australian Centre for Microscopy and Microanalysis, School of Aerospace, Mechanical and Mechatronic Engineering, University of Sydney, Sydney, New South Wales 2006, Australia.
[3] Department of Materials, University of California Santa Barbara, Santa Barbara, California 93106, USA. Correspondence and requests for materials should be addressed to D.S.G. (email: gianola@engr.ucsb.edu).

Nanocrystalline (NC) metals and alloys, polycrystals with grain sizes less than ~100 nm, offer a suite of appealing mechanical properties for structural applications[1–4], including high strength and hardness[5,6], enhanced fatigue resistance[7,8] and unprecedented tribological robustness[9,10]. These virtues derive from the large fraction of materials that reside at grain boundaries (GBs). At the same time, this high fraction of interfacial materials is responsible for deleterious effects such as thermal instability[11,12] and relatively poor damage tolerance[13,14] in nominally pure NC metals. Despite recent efforts[15–18] to synthesize thermally stable NC metals via segregation of alloying species to the GBs, many of these materials, when subjected to mechanical loads, still exhibit grain growth[8,10,19,20] that can lead to marked changes in mechanical response[21–23]. However, the relationship between thermal and mechanical stability is still not obvious. Many present and future applications of NC metals and alloys such as robust coatings, electrical interconnects, micro- and nano-electro-mechanical systems, and soft magnets involve extreme mechanical duress that can activate microstructural evolution and in turn dynamically change material properties during the lifetime of the device. Characterization and control of mechanical stability in these materials will therefore play an essential role in prediction and optimization of their performances.

While the occurrence of stress-driven grain growth can lead to decreases in strength (via Hall–Petch scaling), it has been shown to markedly enhance tensile ductility via crack-tip blunting[24]. Atomistic simulations have pointed to a critical shear stress needed to promote coupled GB migration[25,26], speculated to be the mechanism underlying stress-driven grain growth as experimentally observed[27,28]. Thus, if this threshold can be raised above the macroscopic yield stress, microstructural evolution should only occur in the vicinity of stress concentrators such as crack tips. While manipulation of GB chemistry has been shown in atomistic simulations to stabilize the NC microstructure and control the threshold for stress-driven GB migration[29,30], no experiments have yet quantitatively determined this critical stress, let alone its dependence on the GB excess of doping solutes. Indeed, mapping of the character dependence of GB migration in realistic NC ensembles subjected to far-field stress and linking post-mortem observations of grain growth to the incipient GB migration has yielded limited success. Nevertheless, the successful characterization and control of the GB dopant concentration and thus the thermal and mechanical stability could lead to a new paradigm of alloy design, wherein the critical stress for GB migration could be tailored to impart an unprecedented combination of high strength (from a stable microstructure under ordinary loads) and triggered damage tolerance (from a microstructure that may evolve under stresses concentrated at potential failure sites).

Here we employ quantitative *in situ* nanoindentation inside a transmission electron microscope (TEM) to impart localized mechanical stresses on targeted individual GBs within NC Al films and demonstrate a positive correlation between the critical shear stress required for the onset of GB migration and the GB excess of clustered O atoms, as determined from three-dimensional atom-probe tomography (APT). Measurements of a variety of film compositions and distinct GBs uncover the crossover of the governing deformation mechanisms between dislocation motion and stress-driven GB migration, both directly observed in TEM, thereby providing an estimate for the critical GB excess of impurities necessary for retaining mechanical stability of the NC microstructures.

Results

Grain growth induced by *ex situ* nanoindentation. NC Al films with thicknesses (t) around 160 nm were deposited on Si(100) substrates and micro-machined Si wedges[31,32]. The global content of O impurities (C_O) in these films varied from 0.7 to 2.1 at.%, as controlled by confocal co-sputtering of pure Al and α-Al_2O_3 targets. More details of film deposition and characterization are described in Supplementary Figs 1–5 and our previous work[33]. The microstructural instability of these films under mechanical stresses was first investigated by instrumented nanoindentation on the films deposited on Si(100) substrates. Cross-sectional specimens across indents were then lifted out inside a focused ion beam/scanning electron microscope (FIB/SEM) from the films with C_O of 0.7 and 2.1 at.%, as shown in Fig. 1a,b, respectively. In the regions far from indents, which resembled the as-deposited microstructure, both films showed a predominance of columnar grains with the major axis aligned out of plane (Fig. 1c,e). In contrast, the grains beneath the indent showed extraordinary in-plane expansion in the purest film (Fig. 1d), whereas such grain growth was suppressed in the most impure film (Fig. 1f). For further quantification of the grain size evolution in post-deformed regions, each cross-sectional specimen was laterally divided into separate regions using a bin size of $2t$. The maximum grain size measured in each bin (normalized by the average grain size of the film) was used as a metric for discontinuous grain growth[28] and plotted versus the distance to the indent (normalized by t). As Fig. 1g shows, grain growth was driven by the higher stress near the indent and suppressed due to the increase of C_O, indicating a higher critical stress required for microstructural evolution in more impure films in view of equivalent maximum loads (and thus similar intensity of the stress field) applied to the two films (see inset of Fig. 1g). More details of this quantification are provided in Supplementary Figs 6 and 7. These results are consistent with observations that showed suppression of stress-driven grain growth by O impurities near the fracture ends of NC Al films following micro-tensile tests[21,33].

GB-targeted, quantitative *in situ* TEM nanoindentation. While post-deformed measurements highlighted the pinning effect of O impurities, the real-time mechanical and microstructural response of individual nanograins (and GBs) subjected to the local stress field near an indent has yet to be elucidated, especially on a quantitative level. To this end, indentation tests were performed *in situ* inside a TEM (JEOL 2010F) using a Hysitron PicoIndenter (PI 95)[34] on the NC Al films deposited on Si wedges. As shown in Fig. 2a,b, the film on the apex of wedge was transparent to the electron beam[31,32], and was aligned to the axis of a diamond cube-corner tip (see more details of experimental set-up in Supplementary Fig. 8). Indentation was carried out in a displacement-controlled manner with simultaneous force measurement and dark-field TEM image acquisition, enabling a direct correlation between mechanical response and microstructural evolution beneath the indenter.

Figure 3a–d represent a series of snapshots recorded during *in situ* indentation test of NC Al film with C_O of 0.7 at.%. A columnar grain is highlighted in dark-field images using individual diffraction spots in Al(111) and Al(200) rings (see inset of Fig. 3a), and Fig. 3i shows shape evolution of this grain and the corresponding force–time response, from which a few conclusions can be drawn. First, grain growth was only initiated when the applied force reached a critical value, and the onset of grain growth, occurring between frames 3a and b,

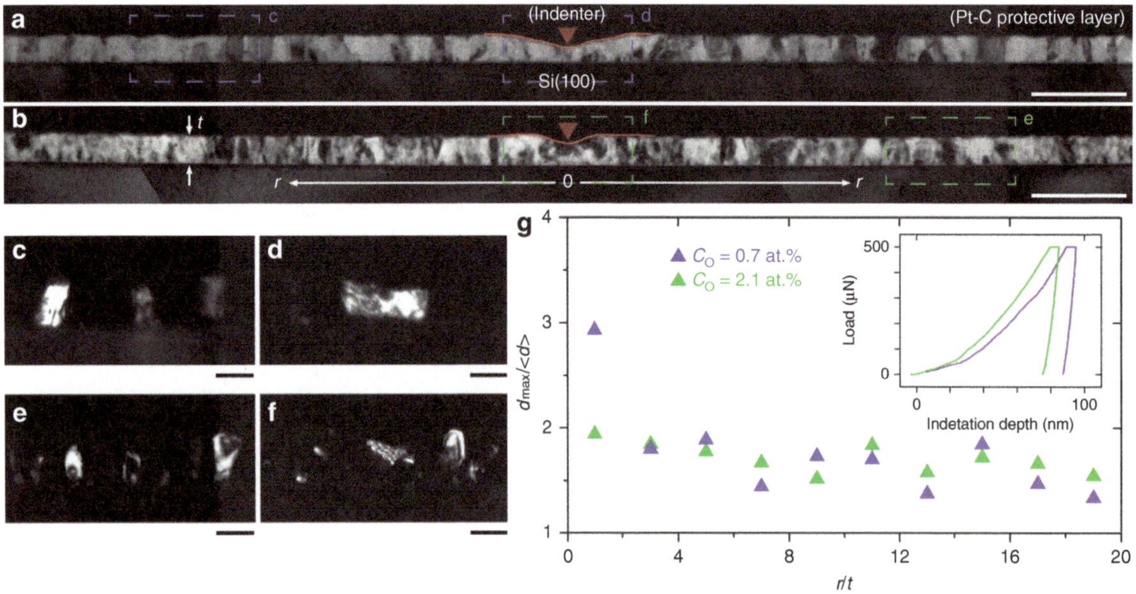

Figure 1 | Post-mortem observation of grain growth induced by nanoindentation and its suppression by impurities. (**a**,**b**) Bright-field TEM images of cross-sectional specimens extracted across indents in the NC Al films with global O content (C_O) of (**a**) 0.7 and (**b**) 2.1 at.%. Red lines show surface profiles of indents. Scale bars, 500 nm. (**c**–**f**) Dark-field TEM images of individual grains (**c**,**e**) far from and (**d**,**f**) just beneath indents show occurrence and suppression of grain growth in the film with (**c**,**d**) low and (**e**,**f**) high C_O, respectively. The images correspond to the outlined boxes in (**a**,**b**). Scale bars, 100 nm. (**g**) Quantification of mechanical grain growth. The maximum grain size (d_{max}) in each lateral bin normalized by averaged grain size of the film ($\langle d \rangle$) is plotted versus the distance to indent (r) normalized by film thickness (t). Bin size is $2t$. Inset: indentation load-depth curves of the two films.

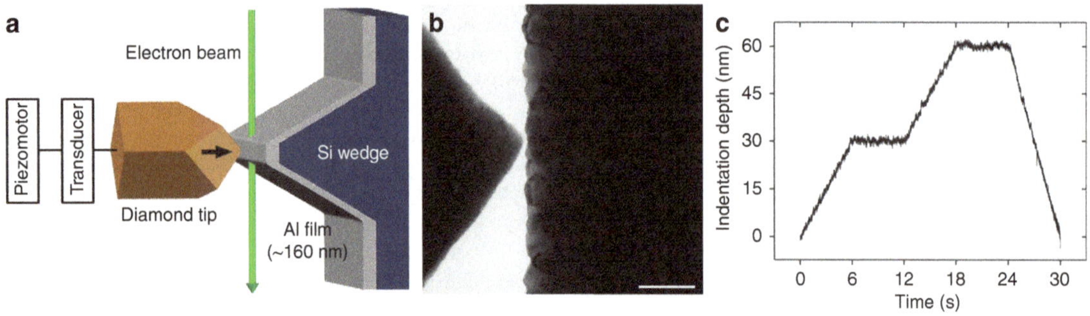

Figure 2 | Layout of quantitative *in situ* TEM indentation tests. (**a**) Schematic of experimental set-up showing thin film deposited on a wedge-shaped substrate to enable electron transparency at the wedge apex, as well as provisions for instrumented measurements and alignment. (**b**) Bright-filed TEM image of diamond tip and NC Al film on Si wedge substrate. Scale bar, 200 nm. (**c**) Loading and unloading profiles for a representative test.

corresponded to a measurable plateau in the real-time force signal (Fig. 3i). This correlation between the change in mechanical response and the onset of microstructural evolution is indicative of the stress-driven character of grain growth and its contribution to the incipient plastic deformation of NC metals. Second, grain growth was unambiguously observed with applied force beyond the critical value, although in some instances this occurred in conjunction with[35,36] or subsequent to[37,38] intragranular dislocation activities (as discussed later). Importantly, only the GB on the right side of the grain, which was closer to the indenter and thus undergoing higher stress, was found to be mobile and was subsequently driven towards the indenter, that is, the region with even higher stress. No detectable changes to the grain shape opposite to the indenter tip owing to GB migration or dislocation-mediated offsets were observed. We note that mobile GBs were always driven towards regions with higher shear stress, which is consistent with the shear-coupled mechanisms of GB migration[25,39]. However, one notable difference of our indentation tests with the models based on uniform shear stress is the presence of large stress gradients, highlighting the role of localized inhomogeneity of driving force in governing the progression from individual GB migration to discontinuous grain growth[40,41]. Finally, we observed that GB migration was not uniformly driven along the entire GB, but always initiated at some local hotspot, followed by bowing out of the neighbouring GB segment (see inset of Fig. 3i). Thus, the critical stress for GB migration can be defined according to the stress field at the hotspot, which reflects the local propensity for GB migration associated with the effectiveness of GB pinning, for example, by O impurities.

Addition of O impurities in the films mediated the onset of GB migration. Figure 3e–h,j represents another test of a NC Al film with C_O of 2.1 at.%. Similar characteristics of stress-driven grain growth and GB migration were observed, including (1) a critical force and concomitant yielding, (2) GB migration driven by (and towards) higher shear stress and consequent grain growth and (3) spatially inhomogeneous GB migration. However, the extent of GB migration was much less pronounced due to the increase of C_O, which qualitatively supports the post-deformed measurements shown above.

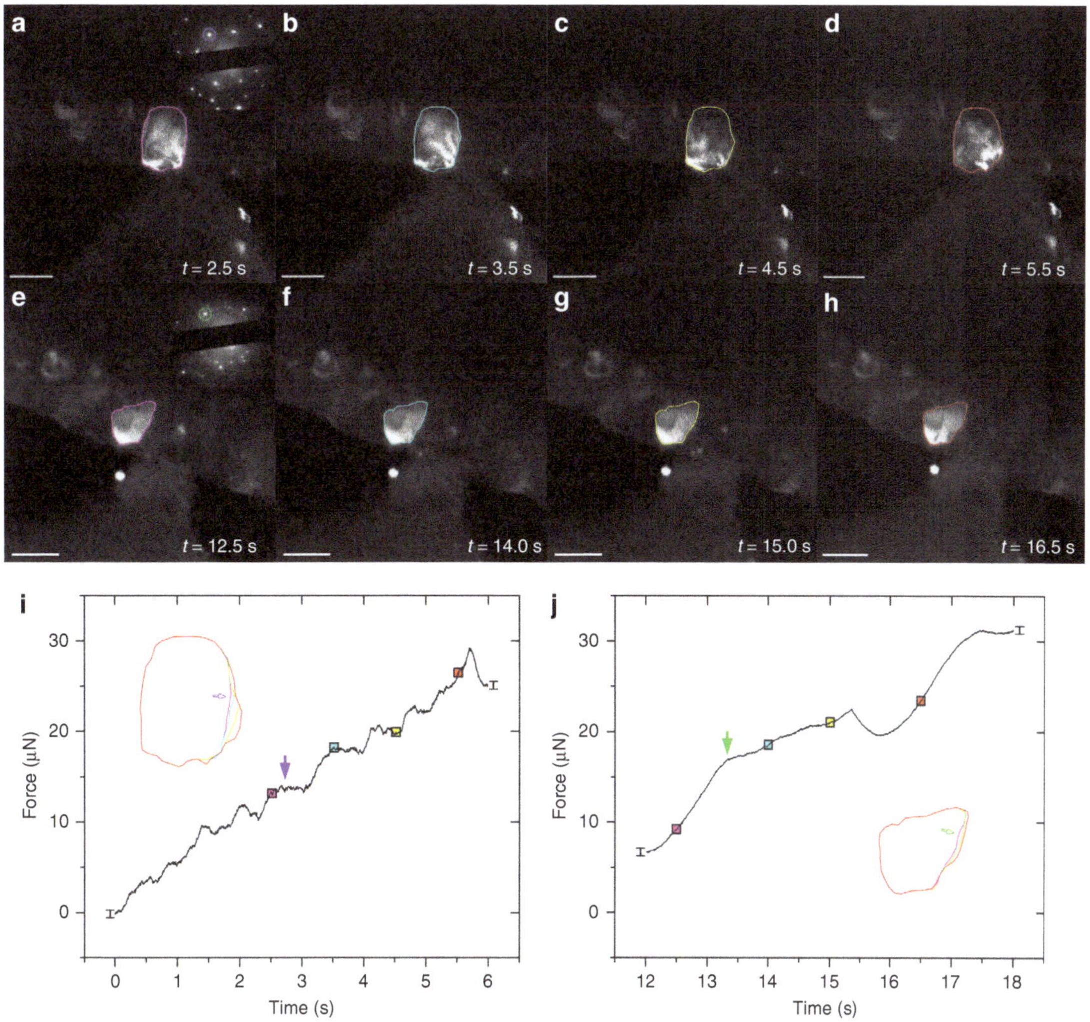

Figure 3 | *In situ* TEM indentation tests of NC Al films showing mechanically driven GB migration. (**a–d**) Time-series of snapshots recorded during test of the film with C_O of 0.7 at.%. Inset: selected-area diffraction pattern of the film, with the outlined Al(111) spot used for dark-field TEM imaging. Scale bars, 100 nm. (**i**) Force–time curve. Vertical error bars show the peak-to-peak noise (~1 μN) in force. Solid squares correspond to the snapshots shown in **a–d**. Solid arrow shows the onset of stress-driven GB migration. Inset: shape evolution of the grain outlined in **a–d**. GB migration only occurs in regions near the indenter. Open arrow shows the 'hotspot' where GB migration was initiated. (**e–h,j**) Corresponding results of the film with C_O of 2.1 at.%. GB migration is still observed albeit to a lesser extent and initiated at higher forces. Scale bars, 100 nm.

GB kinetics mediated by impurities. These *in situ* indentation tests then enable GB-specific quantification of the characteristic parameters describing the mechanics and kinetics of stress-driven GB migration. We first quantify the velocity of GB migration by tracking the shape evolution of grains with applied force beyond the critical value. For simplicity, the effective GB displacement at each recorded snapshot (for example, Fig. 3a–h) was estimated as

$$\langle x \rangle = \frac{\Delta A}{l_{GB}}, \qquad (1)$$

where l_{GB} is the original projected length of the mobile segment of GB and ΔA is the projected area swept by GB (Fig. 4a). The mean velocity of GB migration between each consecutive snapshots was thus calculated, and Fig. 4a shows GB migration to proceed in a jerky manner, which could be attributed to the spatially inhomogeneous de-pinning from impurities[29]. While quantitative modelling of GB kinetics and evaluation of GB mobility under various driving forces needs further theoretical efforts, a pronounced drag effect by (segregated) impurities is nonetheless revealed for the GB mean velocities averaged over each indentation test. As shown in Fig. 4b, GB mean velocity was decreased by ~1 order of magnitude with C_O increasing from 0.7 to 2.1 at.%. The very fast GB migration in the purest films, with mean velocities up to 5–10 nm s^{-1}, agreed well with previous *in situ* TEM tests of pure NC Al films[24,42], in which grain growth and GB migration were activated by the high stress near crack tips. On the other hand, even the most retarded GB migration in the relatively impure films, with mean velocities of 0.7–1 nm s^{-1}, were still significantly higher than the upper limit of GB velocity in thermal-driven GB migration, typically 0.2 nm s^{-1} for high-angle GBs and as small as 2×10^{-5} nm s^{-1} for low-angle GBs[24,42,43]. Therefore, we infer that the GB migration observed in our experiments was not (dominantly) carried out by atomic diffusion, but instead by the stress-driven collective motion of GB structural units, consistent again with the shear-coupled mechanisms[25,26].

Quantification of the critical shear stress for GB migration. Now we turn our focus to the critical driving force (that is, shear stress) for GB migration. As a first-order approximation neglecting elastic anisotropy (expected to be small in Al (ref. 21))

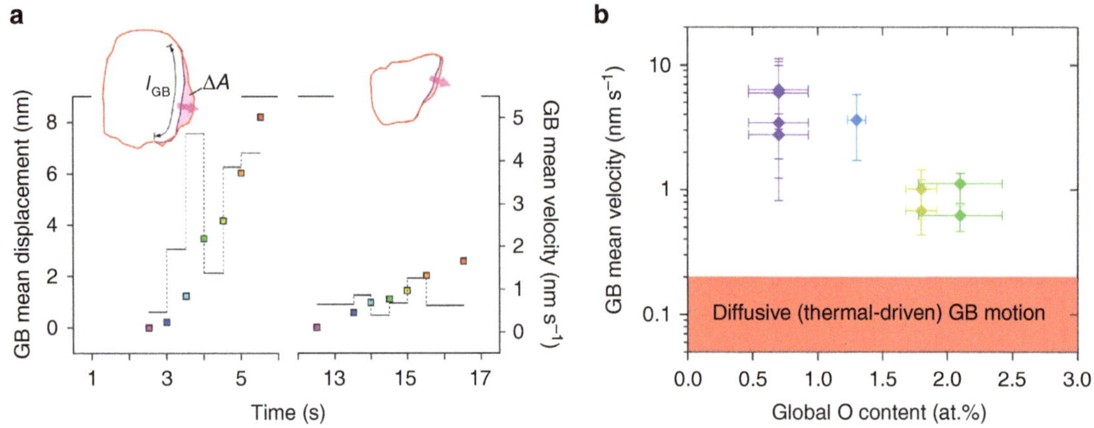

Figure 4 | Retardation of GB kinetics by impurities. (**a**) Direct measurements of GB displacements (colour squares, left axis) and velocities (black lines, right axis) over time during *in situ* TEM indentation tests of the films with $C_O = 0.7$ at.% (left panel) and 2.1 at.% (right panel). The effective displacements at each snapshot relative to the pre-deformed state are approximated by the area swept by the moving GB (ΔA) normalized by the original GB length (l_{GB}), and the mean velocities between consecutive snapshots reveal GB migration in a jerky manner. (**b**) Dependence of GB mean velocities (averaged over each indentation test) on C_O showing a significant retardation of GB migration owing to impurity pinning. Error bars of GB mean velocities for each test are determined as the s.d. of GB velocities shown in **a**. GB velocities representative of thermal-driven GB migration that is of a diffusive nature are shown in shaded region[24], and are substantially lower than our measurements of stress-driven GB migration (over two orders of magnitude in the pure films).

and any plasticity before GB migration, the stress field in the vicinity of the indenter was estimated using the Hertzian model of elastic contact[44], which described shallow indentation on a homogeneous half-space. The magnitude of shear stress at coordinate (r, z), as mapped in Fig. 5a,b, was calculated based on[45]:

$$\tau_{rz} = \frac{3P}{2\pi a^2}\frac{ruz^2}{u^4 + a^2z^2}\frac{a^2}{u^2 + a^2}, \qquad (2)$$

where $u = \sqrt{\frac{(r^2+z^2-a^2)+\sqrt{(r^2+z^2-a^2)^2+4a^2z^2}}{2}}$, $a = \left(\frac{3PR}{4E_r}\right)^{\frac{1}{3}}$ is contact radius, P is the critical force at the onset of GB migration, R is radius of curvature of the indenter as directly measured from TEM images and $E_r = \left(\frac{1-v_i^2}{E_i} + \frac{1-v_f^2}{E_f}\right)^{-1}$ is the reduced modulus derived from the modulus (E) and Poisson's ratio (v) of indenter (i) and Al film (f). The influence of the wedge-shaped Si substrate on the stress distribution in the Al film was further examined using finite-element modelling. More details of modelling and error analysis of shear stresses are described in Supplementary Figs 9–12. Taken as a whole, the Hertzian model was proved to be a reasonable approximation for our experimental configuration to evaluate the stress field leading to the onset of GB migration. We note, though, that the modelling of the stress field during GB migration remains a challenge owing to limited knowledge of the stress relaxation mechanisms at dynamic GBs.

Role of GB chemistry on mechanical stability. The critical shear stress (τ_{CR}) for GB migration was consequently determined as the resolved shear stress along the mobile GBs, calculated from full stress fields (τ_{rz}, τ_{rr} and τ_{zz}) given by the Hertzian model[45], at the hotspot of GB migration (Fig. 5a,b). Interestingly, GB migration was not necessarily initiated at the GB segment experiencing the highest resolved shear stress. This suggests either microstructural constraint (for example, immobile GB triple junctions[46,47] and α-Al_2O_3 precipitates[33]) or a spatial variation of pinning strength within individual GBs associated with the inhomogeneous segregation of impurities. Herein, we focus attention on the O-rich clusters (with averaged sizes of 25–35 O atoms) segregated along GBs, which have been observed in APT characterization of the as-deposited films[33] and are expected to dominate the GB pinning strength, though details of their inhomogeneous distribution have yet to be quantified. Figure 6a,b shows the spatial distribution of O-rich clusters in the films with C_O of 2.1 and 0.7 at.%, respectively. While the O-rich clusters were generally associated with GB regions, their area density, that is, the GB excess of O impurities (Γ_O), varied not only globally (between different GBs) but also locally (within individual GBs), as quantified in Fig. 6c,d. More details of the 'cylinder method' to calculate Γ_O are described in Supplementary Fig. 2.

Interestingly, two groups of GBs with distinct distributions of Γ_O were revealed in the impure films (Fig. 6c and Supplementary Fig. 3): one with a large difference between the minimum and maximum Γ_O, and another with a relatively narrow distribution of Γ_O. These distinct distributions of Γ_O presumably reflect the result of different GB structures that accommodate different amounts of impurities[48], and can be rationalized by considering that the diameter of analysed regions (6 nm) was on the same order of the mean effective size of O-rich clusters (4–5 nm), but considerably smaller than the mean distance between clusters (10–15 nm), as revealed by previous APT analysis[33]. As a consequence, Γ_O would be insensitive to the position of analysed regions for the GBs only containing randomly distributed O solutes (such as low-angle GBs and ideal coincident site lattice GBs[48]), whereas GBs containing segregated O-rich clusters (such as random high-angle GBs and coincident site lattice GBs with larger deviations[48]) would show a wider range of Γ_O depending on the position of analysed regions relative to the discrete clusters. The inhomogeneity of Γ_O within such cluster-segregated GBs can be striking, in the extreme case ranging from ~0.1 to ~0.7 atom per nm^2. By contrast, the distribution of Γ_O in the purest films (Fig. 6d), both between different GBs and within individual GBs, was found to be much narrower. Whereas τ_{CR} required for the onset of GB migration is governed by the interplay between the inhomogeneous distribution of Γ_O and the localized driving force, we note that confinement of microstructural features (for example, triple junctions[46,47]) may also contribute to the spatial inhomogeneity of GB pinning strength, while a higher extent of impurity segregation was expected near these features[23]. On the

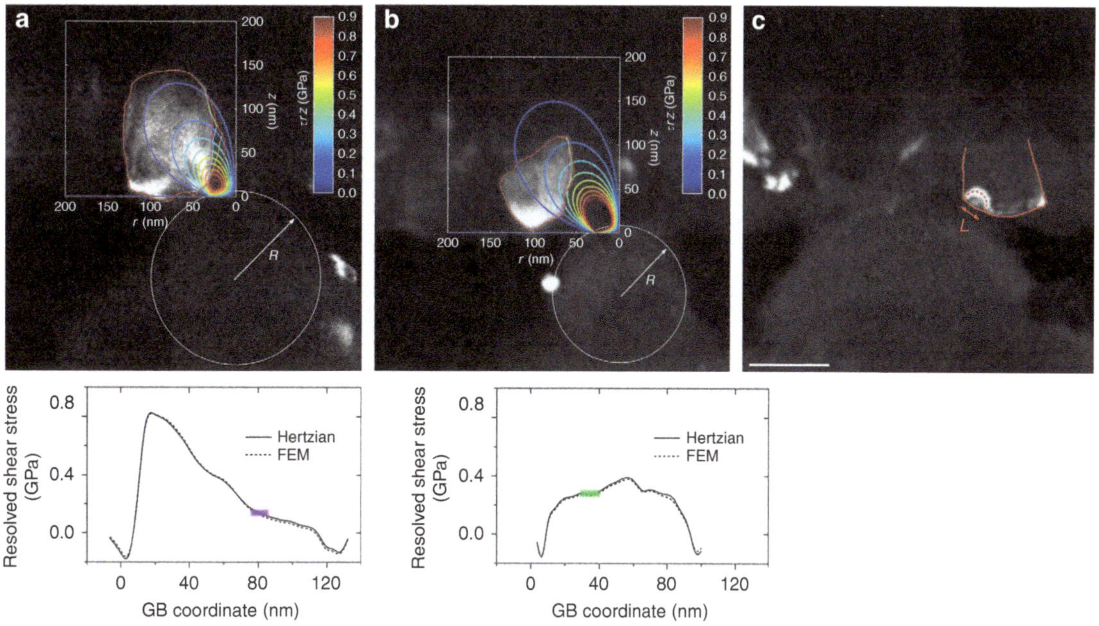

Figure 5 | Quantification of impurity-mediated critical shear stress for GB migration and other deformation mechanisms. (**a**,**b**) Snapshots at the onset of GB migration in the films with C_O of (**a**) 0.7 and (**b**) 2.1 at.%, superimposed with subsequent shape evolution of the grains and contour maps of shear stress (τ_{rz}) calculated using a Hertzian model (see text for details). Lower panels show corresponding profiles of resolved shear stress along the mobile GBs (both occurring on the right side of grains nearest to the indenter tip). Shaded bars denote the location of the onset of GB migration and estimated range of critical shear stress. (**c**) Snapshot of a different experiment showing the emission of dislocation (dash line) from a GB or triple junction (solid line). Scale bar, 100 nm. GBs were generally found to be immobile in cases where clear evidence of dislocation propagation was observed, indicating an alternative relaxation mechanism.

other hand, the effect of grain size alone is not considered to be prevalent, since the characteristic length scale of the spatial variation of Γ_O was remarkably smaller than grain sizes in the studied NC Al films. Taken as a whole, it is reasonable to examine the correlation between τ_{CR} and the full statistics of Γ_O.

As shown in Fig. 6e, we find a positive correlation between our measurements of τ_{CR} (from *in situ* TEM indentation) and Γ_O (from APT analysis) in several NC Al films with C_O ranging from 0.7 to 2.1 at.%. This trend is robust to the specific quantity of Γ_O that was extracted from the statistical analysis of a variety of GBs, as both the minimum and maximum Γ_O within individual GBs generally increased with C_O. Despite relatively dilute O concentrations in the films studied (and correspondingly small Γ_O values below 1 atom per nm^2), we measure nearly a fourfold increase in τ_{CR} over the range of C_O. Our results show qualitatively similar trends with atomistic simulations of a Σ75 $(\bar{7}51)[112]$ symmetric-tilt Al GB (ref. 29) showing a linear dependence of τ_{CR} for shear-coupled GB migration on the GB excess of O solute atoms (dotted line in Fig. 6e), despite the large differences in temperature, applied strain rate and GB characters. We also recognize the fact that the mechanical response of GBs vary with their crystallographic characters, as well as local chemical, structural and topological environment[48]. Nonetheless, the vast majority of GBs in our NC Al films, irrespective of C_O, were found to be random high-angle GBs (see Supplementary Figs 4 and 5 and Supplementary Table 1 for more detailed characterization and statistics of GB crystallography), which preclude any simple correlation between GB excess and GB characters. This suggests that the detailed crystallography of GBs may serve only a secondary role in controlling GB migration in NC alloys, in contrast with pure materials. Thereby, the predominant role of segregated impurities in pinning GB migration is unequivocally revealed in our experiments, which represent a real NC system with complex GB networks.

Discussion

Our experiments demonstrate that the critical stress for mechanically driven GB migration and its associated kinetics are primarily controlled by impurities, in our case O-rich clusters, that segregate to GBs. In light of this, we propose a transition in the governing deformation mechanisms in NC metals at a critical value of GB excess (Γ_{CR}), which ultimately determines whether GBs (and thus NC microstructures) are stable under mechanical loads. Since τ_{CR} for stress-driven GB migration steadily increase with Γ_O (Fig. 6f, mean GB excess values were used for simplicity), it is likely to surpass the driving force for conventional dislocation activities beyond Γ_{CR}. In these instances, the GBs are strongly pinned (in analogy to a Cottrell pinning atmosphere for dislocations) and plastic deformation would thereby need to be accommodated by propagation (that is, pinning and de-pinning) of discrete dislocations along GB ledges, a mechanism known to be operative and rate-limiting in 'static' NC materials where intragranular dislocation interactions are abated[49,50]. Indeed, such dislocation activity in mechanically stable grains was observed on multiple occasions in our experiments (for example, Fig. 5c) and over the full range of C_O from 0.7 to 2.1 at.%. The shear stress to propagate a dislocation along a GB was then estimated using the simple relationship:

$$\tau_{CR} = Gb/L, \tag{3}$$

where G is shear modulus, b is Burgers vector and L is the minimum diameter of the bowing dislocation segment observed before de-pinned from GB pinning points[24]. We note that intragranular dislocation activities did occur in mechanically unstable grains, nonetheless, the onset of stress-driven microstructural evolution was unambiguously defined by the τ_{CR} for GB migration. Combining both the mechanically dynamic and static GBs thereby provides a full map of τ_{CR} for incipient

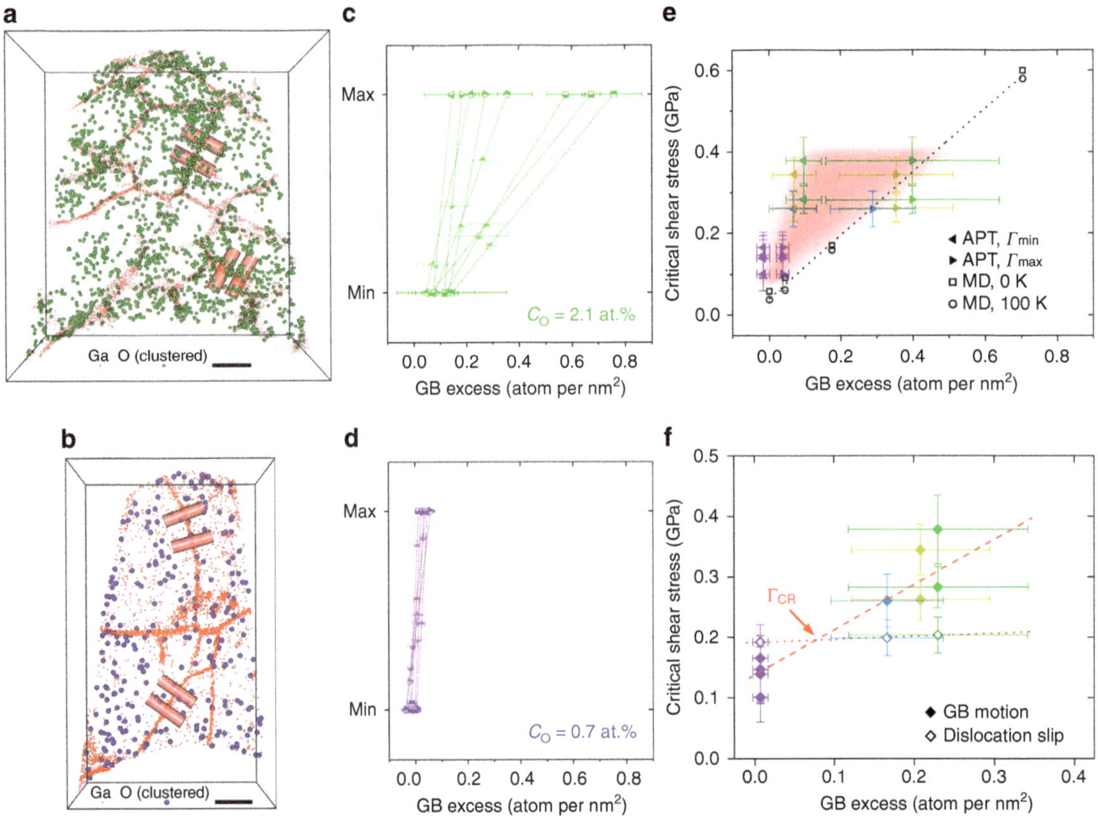

Figure 6 | Quantification of GB excess of O atoms (Γ_O) using APT and its role on mechanical stability of NC Al films. (**a**,**b**) APT reconstructions of films with C_O of (**a**) 2.1 and (**b**) 0.7 at.% showing the spatial distribution of O-rich clusters relative to GBs, highlighted by segregated Ga implanted during atom-probe sample preparation. Scale bars, 20 nm. (**c**) Variation of the areal density of O-rich clusters (Γ_O) in the films with $C_O = 2.1$ at.% showing large inhomogeneity within a given GB (shown by a single line bounded by the minimum and maximum Γ_O values), as well as between different GBs (shown by different lines). Error bars of Γ_O are determined with the 'cylinder method' (Supplementary Fig. 2). (**d**) A much narrower distribution of Γ_O was measured in the film with $C_O = 0.7$ at.%. Four representative analysed regions, that is, 'cylinders' for such quantification are indicated in **a** and **b**. (**e**) Correlation between critical shear stress (τ_{CR}) of GB migration and Γ_O averaged over all GBs measured by APT. Error bars of τ_{CR} are determined as described in Supplementary Figs 11 and 12. Error bars of Γ_O reflect the inhomogeneity between different GBs, and the ranges between local minima (left triangles) and local maxima (right triangles) of Γ_O reflect the inhomogeneity within individual GBs. Atomistic simulations[29] of an Al bicrystal doped with O is shown (open symbols, dotted line) for comparison. (**f**) Full map of correlation between averaged GB excess and τ_{CR} for GB migration (solid symbols, linear fitting shown as dashed line) and dislocation propagation (open symbols, linear fitting shown as dotted line). The crossover of these two lines denotes a critical GB excess (Γ_{CR}) for a transition of deformation mechanism that maintain NC microstructural stability under mechanical loads.

plastic deformation versus Γ_O (Fig. 6f), revealing an apparent crossover at $\Gamma_{CR} \sim 0.1$ atom per nm^2.

This delineation between mechanically dynamic and static NC microstructures, as controlled by the impurity content (and GB excess), qualitatively agrees with previous micro-tension tests of NC Al films that showed suppression of mechanical grain growth owing to GB impurities, yet where only global information was available[23]. Observing that the macroscopic response of NC materials is determined by averaging over the ensemble of grains with widespread GB characters and orientations of slip systems, the crossover of τ_{CR} with increasing Γ_O indicates a transition of the most probable deformation mechanism from GB migration to dislocation propagation. This new insight helps address the reported transition of micro-tensile behaviour of NC Al films from 'ductile and moderately strong' to 'very strong but brittle'[23]. More importantly, the highly localized stress field generated by *in situ* indentation tests enabled us to discover all branches of deformation mechanisms shown in Fig. 6f provided that the individual grain and GB were of proper orientation, even though dislocation slip in pure films and GB migration in impure films were statistically unfavourable.

The striking implication from our *in situ* measurements and analysis is twofold. First, the critical stress for the promotion of GB migration, the precursor for widespread mechanical grain growth in NC materials, can be measured at the individual GB level, and the value of this stress can be modulated by processing with alloying species that segregate to GBs. Comparing the degree of alloying required for thermal stability with that of mechanical stability, and examining the atomic mechanisms responsible for thermodynamic stability and retarded kinetics would be an interesting avenue for future pursuits. Second, our results suggest a crossover between dynamic and static deformation mechanisms, and consequently a tunable mechanical stability of NC materials, with only a dilute concentration of alloying species. This enables the wide spectrum of materials design with minimal constraints of materials costs and abundance. Most tantalizing, though, is the notion of a new structural materials design strategy that offers high strength paired with thermal and mechanical stability, juxtaposed with adaptive deformation mechanisms that can respond to locally high stresses in extreme conditions in a self-healing fashion.

Methods

Synthesis of NC Al films. NC Al films studied in our *ex situ* and *in situ* indentation tests were synthesized by magnetron co-sputtering of a 99.999% pure Al target and a 99.995% pure α-Al_2O_3 target with a confocal geometry[33]. The direct-current (d.c.) power of Al target was varied from 200 to 400 W, and the radio-frequency (RF) power of α-Al_2O_3 target was varied from 0 to 50 W. The as-deposited films manifested a composite-like microstructure[33] containing O impurities with distinct morphologies, that is, α-Al_2O_3 precipitates, O-rich clusters and O solute atoms.

***Ex situ* and *in situ* nanoindentation experiments.** The *ex situ* nanoindentation experiments were performed using a Hysitron TriboIndenter (TI 950) employing a diamond Berkovich tip, with an indentation strain rate of 0.05 s^{-1} and maximum indentation depth of $\sim 0.5t$. Additional indentation tests were performed *in situ* inside a TEM (JEOL 2010F) using a Hysitron PicoIndenter (PI 95)[34] on the NC Al films deposited on Si wedges. Indentation was carried out with a constant rate of 5 nm s^{-1} applied in step-wise loading segments (Fig. 2c), while force was simultaneously measured with a peak-to-peak noise level of 1 μN.

Atom-probe tomography and data analysis. The same atom-probe data sets as reported in ref. 33 were utilized for the calculation of GB excess using 'cylinder method'. Briefly, atom-probe specimens were prepared inside a Zeiss Auriga FIB/SEM by a lift-out approach[51]. Atom-probe data were acquired using a Cameca LEAP 4000X Si operated in voltage-pulsing mode at 40 K, with pulsing voltage being 20% of d.c. bias voltage, and evaporation rate kept at 1% of the pulsing rate of 200 kHz (ref. 33). Data reconstruction and visualization were performed using IVAS software (Cameca). O^+, AlO^+ and minor amount of O_2^+, O^{2+}, AlO^{2+}, AlO_2^+ and AlO_2^{2+} were identified in the mass spectrum as O-containing species. Analysis of O-rich clusters was facilitated by means of Voronoi volume distribution through custom MATLAB (Mathworks Inc.) programs.[52]

References

1. Kumar, K. S., Van Swygenhoven, H. & Suresh, S. Mechanical behavior of nanocrystalline metals and alloys. *Acta Mater.* **51,** 5743–5774 (2003).
2. Dao, M., Lu, L., Asaro, R., De Hosson, J. T. M. & Ma, E. Toward a quantitative understanding of mechanical behavior of nanocrystalline metals. *Acta Mater.* **55,** 4041–4065 (2007).
3. Meyers, M. A., Mishra, A. & Benson, D. J. Mechanical properties of nanocrystalline materials. *Prog. Mater. Sci.* **51,** 427–556 (2006).
4. Greer, J. R. & De Hosson, J. T. M. Plasticity in small-sized metallic systems: Intrinsic versus extrinsic size effect. *Prog. Mater. Sci.* **56,** 654–724 (2011).
5. Lu, L., Chen, X., Huang, X. & Lu, K. Revealing the maximum strength in nanotwinned copper. *Science* **323,** 607–610 (2009).
6. Wu, X. L., Zhu, Y. T., Wei, Y. G. & Wei, Q. Strong strain hardening in nanocrystalline nickel. *Phys. Rev. Lett.* **103,** 205504 (2009).
7. Hanlon, T., Tabachnikova, E. & Suresh, S. Fatigue behavior of nanocrystalline metals and alloys. *Int. J. Fatigue* **27,** 1147–1158 (2005).
8. Padilla, H. A. & Boyce, B. L. A review of fatigue behavior in nanocrystalline metals. *Exp. Mech.* **50,** 5–23 (2009).
9. Hanlon, T., Chokshi, A., Manoharan, M. & Suresh, S. Effects of grain refinement and strength on friction and damage evolution under repeated sliding contact in nanostructured metals. *Int. J. Fatigue* **27,** 1159–1163 (2005).
10. Rupert, T. J. & Schuh, C. A. Sliding wear of nanocrystalline Ni–W: Structural evolution and the apparent breakdown of Archard scaling. *Acta Mater.* **58,** 4137–4148 (2010).
11. Ames, M. *et al.* Unraveling the nature of room temperature grain growth in nanocrystalline materials. *Acta Mater.* **56,** 4255–4266 (2008).
12. Kacher, J., Robertson, I. M., Nowell, M., Knapp, J. & Hattar, K. Study of rapid grain boundary migration in a nanocrystalline Ni thin film. *Mater. Sci. Eng. A* **528,** 1628–1635 (2011).
13. Kumar, K. S., Suresh, S., Chisholm, M. F., Horton, J. A. & Wang, P. Deformation of electrodeposited nanocrystalline nickel. *Acta Mater.* **51,** 387–405 (2003).
14. Sharon, J. A., Padilla, H. A. & Boyce, B. L. Interpreting the ductility of nanocrystalline metals. *J. Mater. Res.* **28,** 1539–1552 (2013).
15. Weissmüller, J. Alloy effects in nanostructures. *Nanostruct. Mater.* **3,** 261–272 (1993).
16. Liu, F. & Kirchheim, R. Nano-scale grain growth inhibited by reducing grain boundary energy through solute segregation. *J. Crystal Growth* **264,** 385–391 (2004).
17. Chookajorn, T., Murdoch, H. A. & Schuh, C. A. Design of stable nanocrystalline alloys. *Science* **337,** 951–954 (2012).
18. Koch, C. C., Scattergood, R. O., Saber, M. & Kotan, H. High temperature stabilization of nanocrystalline grain size: thermodynamic versus kinetic strategies. *J. Mater. Res.* **28,** 1785–1791 (2013).
19. Zhang, K., Weertman, J. R. & Eastman, J. A. Rapid stress-driven grain coarsening in nanocrystalline Cu at ambient and cryogenic temperatures. *Appl. Phys. Lett.* **87,** 061921 (2005).
20. Fan, G. J., Fu, L. F., Choo, H., Liaw, P. K. & Browning, N. D. Uniaxial tensile plastic deformation and grain growth of bulk nanocrystalline alloys. *Acta Mater.* **54,** 4781–4792 (2006).
21. Gianola, D. S. *et al.* Stress-assisted discontinuous grain growth and its effect on the deformation behavior of nanocrystalline aluminum thin films. *Acta Mater.* **54,** 2253–2263 (2006).
22. Gianola, D. S., Mendis, B. G., Cheng, X. & Hemker, K. J. Grain-size stabilization by impurities and effect on stress-coupled grain growth in nanocrystalline Al thin films. *Mater. Sci. Eng. A* **483-484,** 637–640 (2008).
23. Tang, F., Gianola, D. S., Moody, M. P., Hemker, K. J. & Cairney, J. M. Observations of grain boundary impurities in nanocrystalline Al and their influence on microstructural stability and mechanical behaviour. *Acta Mater.* **60,** 1038–1047 (2012).
24. Legros, M., Gianola, D. S. & Hemker, K. J. *In situ* TEM observations of fast grain-boundary motion in stressed nanocrystalline aluminum films. *Acta Mater.* **56,** 3380–3393 (2008).
25. Cahn, J. W., Mishin, Y. & Suzuki, A. Coupling grain boundary motion to shear deformation. *Acta Mater.* **54,** 4953–4975 (2006).
26. Rajabzadeh, A., Mompiou, F., Legros, M. & Combe, N. Elementary mechanisms of shear-coupled grain boundary migration. *Phys. Rev. Lett.* **110,** 265507 (2013).
27. Gianola, D. S., Eberl, C., Cheng, X. M. & Hemker, K. J. Stress-driven surface topography evolution in nanocrystalline Al thin films. *Adv. Mater.* **20,** 303–308 (2008).
28. Rupert, T. J., Gianola, D. S., Gan, Y. & Hemker, K. J. Experimental observations of stress-driven grain boundary migration. *Science* **326,** 1686–1690 (2009).
29. Elsener, A., Politano, O., Derlet, P. M. & Van Swygenhoven, H. Variable-charge method applied to study coupled grain boundary migration in the presence of oxygen. *Acta Mater.* **57,** 1988–2001 (2009).
30. Schäfer, J. & Albe, K. Competing deformation mechanisms in nanocrystalline metals and alloys: Coupled motion versus grain boundary sliding. *Acta Mater.* **60,** 6076–6085 (2012).
31. Jin, M., Minor, A. M., Stach, E. A. & Morris, J. W. Direct observation of deformation-induced grain growth during the nanoindentation of ultrafine-grained Al at room temperature. *Acta Mater.* **52,** 5381–5387 (2004).
32. Soer, W. A., De Hosson, J. T. M., Minor, A. M., Morris, J. W. & Stach, E. A. Effects of solute Mg on grain boundary and dislocation dynamics during nanoindentation of Al–Mg thin films. *Acta Mater.* **52,** 5783–5790 (2004).
33. He, M. R. *et al.* Understanding the mechanical behavior of nanocrystalline Al-O thin films with complex microstructures. *Acta Mater.* **77,** 269–283 (2014).
34. Minor, A. M. *et al.* A new view of the onset of plasticity during the nanoindentation of aluminum. *Nat. Mater.* **5,** 697–702 (2006).
35. Lohmiller, J., Baumbusch, R., Kraft, O. & Gruber, P. A. Differentiation of deformation modes in nanocrystalline Pd films inferred from peak asymmetry evolution using in situ X-ray diffraction. *Phys. Rev. Lett.* **110,** 066101 (2013).
36. Farkas, D., Frøseth, A. & Van Swygenhoven, H. Grain boundary migration during room temperature deformation of nanocrystalline Ni. *Scripta Mater.* **55,** 695–698 (2006).
37. Mompiou, F. *et al.* Inter- and intragranular plasticity mechanisms in ultrafine-grained Al thin films: An in situ TEM study. *Acta Mater.* **61,** 205–216 (2013).
38. Rajabzadeh, A. *et al.* The role of disconnections in deformation-coupled grain boundary migration. *Acta Mater.* **77,** 223–235 (2014).
39. Caillard, D., Mompiou, F. & Legros, M. Grain-boundary shear-migration coupling. II. Geometrical model for general boundaries. *Acta Mater.* **57,** 2390–2402 (2009).
40. Tucker, G. J. & Foiles, S. M. Molecular dynamics simulations of rate-dependent grain growth during the surface indentation of nanocrystalline nickel. *Mater. Sci. Eng. A* **571,** 207–214 (2013).
41. Panzarino, J. F., Ramos, J. J. & Rupert, T. J. Quantitative tracking of grain structure evolution in a nanocrystalline metal during cyclic loading. *Model. Simul. Mater. Sci. Eng.* **23,** 025005 (2015).
42. Mompiou, F. *et al.* In situ TEM observation of grain annihilation in tricrystalline aluminum films. *Acta Mater.* **60,** 2209–2218 (2012).
43. Winning, M. & Rollett, A. D. Transition between low and high angle grain boundaries. *Acta Mater.* **53,** 2901–2907 (2005).
44. Johnson, K. L. *Contact Mechanics* (Cambridge Univ. Press, 1987).
45. Feng, G., Qu, S., Huang, Y. & Nix, W. D. An analytical expression for the stress field around an elastoplastic indentation/contact. *Acta Mater.* **55,** 2929–2938 (2007).
46. Gianola, D. S., Farkas, D., Gamarra, M. & He, M. The role of confinement on stress-driven grain boundary motion in nanocrystalline aluminum thin films The role of confinement on stress-driven grain boundary motion in nanocrystalline aluminum thin films. *J. Appl. Phys.* **112,** 124313 (2012).
47. Velasco, M., Van Swygenhoven, H. & Brandl, C. Coupled grain boundary motion in a nanocrystalline grain boundary network. *Scripta Mater.* **65,** 151–154 (2011).
48. Herbig, M. *et al.* Atomic-scale quantification of grain boundary segregation in nanocrystalline material. *Phys. Rev. Lett.* **112,** 126103 (2014).

49. Yamakov, V., Wolf, D., Phillpot, S. R., Mukherjee, A. K. & Gleiter, H. Dislocation processes in the deformation of nanocrystalline aluminium by molecular-dynamics simulation. *Nat. Mater.* **1,** 45–48 (2002).
50. Van Swygenhoven, H., Derlet, P. M. & Frøseth, A. G. Nucleation and propagation of dislocations in nanocrystalline fcc metals. *Acta Mater.* **54,** 1975–1983 (2006).
51. Felfer, P., Alam, T., Ringer, S. P. & Cairney, J. M. A reproducible method for damage-free site-specific preparation of atom probe tips from interfaces. *Microsc. Res. Tech.* **75,** 484–491 (2012).
52. Felfer, P., Ceguerra, A. V., Ringer, S. P. & Cairney, J. M. Detecting and extracting clusters in atom probe data: a simple, automated method using Voronoi cells. *Ultramicroscopy* **150,** 30–36 (2015).

Acknowledgements

This work is supported by the National Science Foundation Materials Network Program (grant number DMR-1008222) and the Australian Research Council. We are grateful for scientific and technical input and support from the Penn Nanoscale Characterization Facility and the Australian Microscopy & Microanalysis Research Facility node at the University of Sydney.

Author contributions

D.S.G. and J.M.C. designed and directed the project; M.-R.H. designed the experiments, produced the samples, performed *ex situ* and *in situ* TEM measurements, and analysed all results; S.K.S., P.J.F. and A.J.B. carried out APT experiments and analyses; G.K. carried out FEM simulations; all authors discussed the results; M.-R.H. and D.S.G. wrote the manuscript with input from all authors.

Additional information

Competing financial interests: The authors declare no competing financial interests.

8

Chirality of nanophotonic waveguide with embedded quantum emitter for unidirectional spin transfer

R.J. Coles[1], D.M. Price[1], J.E. Dixon[1], B. Royall[1], E. Clarke[2], P. Kok[1], M.S. Skolnick[1], A.M. Fox[1] & M.N. Makhonin[1]

Scalable quantum technologies may be achieved by faithful conversion between matter qubits and photonic qubits in integrated circuit geometries. Within this context, quantum dots possess well-defined spin states (matter qubits), which couple efficiently to photons. By embedding them in nanophotonic waveguides, they provide a promising platform for quantum technology implementations. In this paper, we demonstrate that the naturally occurring electromagnetic field chirality that arises in nanobeam waveguides leads to unidirectional photon emission from quantum dot spin states, with resultant in-plane transfer of matter-qubit information. The chiral behaviour occurs despite the non-chiral geometry and material of the waveguides. Using dot registration techniques, we achieve a quantum emitter deterministically positioned at a chiral point and realize spin-path conversion by design. We further show that the chiral phenomena are much more tolerant to dot position than in standard photonic crystal waveguides, exhibit spin-path readout up to $95 \pm 5\%$ and have potential to serve as the basis of spin-logic and network implementations.

[1]Department of Physics and Astronomy, University of Sheffield, Hicks Building, Sheffield S3 7RH, UK. [2]EPSRC National Centre for III-V Technologies, Department of Electronic and Electrical Engineering, University of Sheffield, Sheffield S1 3JD, UK. Correspondence and requests for materials should be addressed to M.N.M. (email: m.makhonin@sheffield.ac.uk).

Quantum information processing promises a dramatically increased performance in computing, secure communications and simulations of quantum systems. In addition, a network of quantum nodes enables distributed quantum computing[1], and may facilitate the quantum internet[2]. Such a distributed architecture requires the conversion between matter qubits for local quantum memories and photonic qubits for quantum communication between nodes[3–5]. Furthermore, any scalable quantum technology must solve the miniaturization and fabrication problem, which likely demands that the quantum nodes must be implemented on integrated circuits and waveguides[6]. Recent developments of passive components[7] with embedded quantum dots (QDs) using advanced semiconductor technologies and enhanced-coherence using resonant techniques on-chip[8] contribute to potential solutions. The spin of an electron or hole in a QD is a promising candidate to serve as the matter-qubit in a quantum network, but its implementation requires an efficient on-chip spin–photon interface.

The coupling of spin to the direction of photon emission in nanophotonic waveguides provides a potential solution. The subject of unidirectional light propagation in nanophotonic structures under circularly polarized laser excitation has been advancing rapidly since the first reports in 2013 for surface plasmons and atomic dipoles[9,10]. The chiral effect is understood to arise from the longitudinal component of evanescent fields in nanophotonic structures, which enables circularly polarized states to propagate with the field rotating within the longitudinal plane defined by the geometry of the waveguide[9–16]. An important step forward was made very recently in which directional emission was demonstrated for a quantum-dot emitter embedded within a specially engineered glide-plane waveguide[14], confirming theoretical predictions for chiral emission in photonic crystal waveguides (PhC WGs)[14,15].

In this work, we demonstrate the internal intrinsic chirality of the electromagnetic field in a system with no specially engineered chirality using the QD as an internal probe. We demonstrate the importance of the position of the quantum emitter, in such a way that spin-dependent directional emission is achieved. We achieve efficient coupling of QD exciton spin to the direction of photon emission in nanophotonic waveguides. This chiral behaviour occurs even though the waveguide geometry is completely symmetric, and the dielectric material is non-chiral. We use numerical simulations to demonstrate that the chiral emission originates from the chirality of the electromagnetic field inside the waveguide and exploit it for efficient coupling of the QD emitter by lateral positioning within the waveguide. The dots located at the centre of the waveguide exhibit non-chiral emission, but those displaced from the centre show a varying degree of chiral emission with distance from the centre. The maximal spin-dependent directionality occurs for QDs located at chiral points shifted from the waveguide centre by 32% of the waveguide width. We first confirm this hypothesis by measuring the variation of the spin-dependent directionality of a large number of dots, with good agreement found between the experimental data and the predictions of simulations for randomly positioned emitters within the waveguide. We compare results for nanobeam and PhC WGs, and find that the simpler nanobeam structure provides substantially better spin-dependent directionality because of the lower sensitivity of the chirality to the exact dot position. We then demonstrate designable spin-path coupling by using enhanced registration techniques[17,18] (see the Methods for details) to achieve highly directional emission for a QD deterministically positioned at a chiral point. We conclude that directionality factors approaching 100% ($95 \pm 5\%$) are achievable for the exciton spin of an embedded dot to a single photon state propagating in a nanophotonic waveguide. We also confirm the expected bi-directional emission for a dot deterministically positioned at the centre of the waveguide. We further note that our results provide a natural explanation for the bi-directional emission reported but not explained in cross-waveguide structures[19]. Taken together, the results establish a route towards the transfer and entanglement of the spin state of an emitter to a photonic qubit in an integrated network.

Results

Directional emission in dielectric nanobeam waveguides. The system we investigate is a QD in a nanophotonic waveguide. The dot serves as an integrated quantum emitter with addressable spin-exciton eigenstates, generating spin- and position-dependent directional emission of single photons in the waveguide by exciton recombination. Nanophotonic structures support both longitudinal and transverse field components (E_x and E_y) and, as we show, permit the transfer of in-plane circular polarization ($E_x \pm iE_y$; see Fig. 1a,b for definition of axes). Despite its non-chiral geometry and material, the waveguide exhibits a chiral electromagnetic field mode. As a result, a well-positioned QD will emit in-plane circularly polarized single photons unidirectionally, with a high correlation between the spin state of the QD and the which-path information of the photon.

A schematic of the experimental geometry is given in Fig. 1a. A cross-section of the device including the layer of QDs is shown schematically in Fig. 1b. The left- and right-circularly polarized photons arising from recombination of up and down exciton spin states from a QD embedded within a suspended nanobeam waveguide (NWG) are coupled to NWG modes, and then diffracted by out-coupler gratings at opposite ends towards external detectors. As shown in Fig. 1c–e, for all three charge states of the dots, exciton recombination leads to circularly polarized photon emission. In a non-chiral structure, the emission probability for a circularly polarized dipole is identical in both directions, and this behaviour is indeed observed when the dot is located at the centre of the waveguide (see Fig. 1f obtained from finite difference time domain (FDTD) simulations, see the Methods). By contrast, when the dot is displaced from the centre of the waveguide, the emission direction depends on the spin, with σ^+ photons emitted in one direction and σ^- in the other. The unidirectional emission is shown in Fig. 1g,h (the results of FDTD simulations), and occurs through the coupling of the σ^+/σ^- dipole emitters to direction-dependent modes at the chiral points of the NWG.

Electric field distributions in NWG. The origin of the unidirectional emission can be understood by consideration of the electromagnetic field distribution within the laterally confined nanophotonic geometry. Figure 2a shows a schematic representation of an unterminated NWG with the field distribution in the xy-plane at $z = 0$ (position of the QDs layer) calculated by solving Maxwell's equations (Supplementary Note 1). The amplitudes of the E_y and E_x field components, together with their relative phases, are shown in Fig. 2b. The $|E_y|$ component has a maximum at the centre of the NWG, whereas $|E_x|$ has two maxima closer to the edges (Supplementary Note 2). The relative phase between E_x and E_y is constant at either $\pm\pi/2$ and changes sign when crossing $y = 0$. It is this asymmetry in the phase that enables chiral behaviour: the field is right- or left-circularly polarized (that is, $\sigma^\pm$ corresponding to $E_x \pm iE_y$ fields) at points where $|E_y| = |E_x|$ and the phase is $\pm\pi/2$. This is shown more clearly in Fig. 2c, where the electric field is plotted vectorially at an instant in time. The chiral points are positions where the electric field rotates in time during propagation of the waveguide mode. The rotation direction depends on the propagation

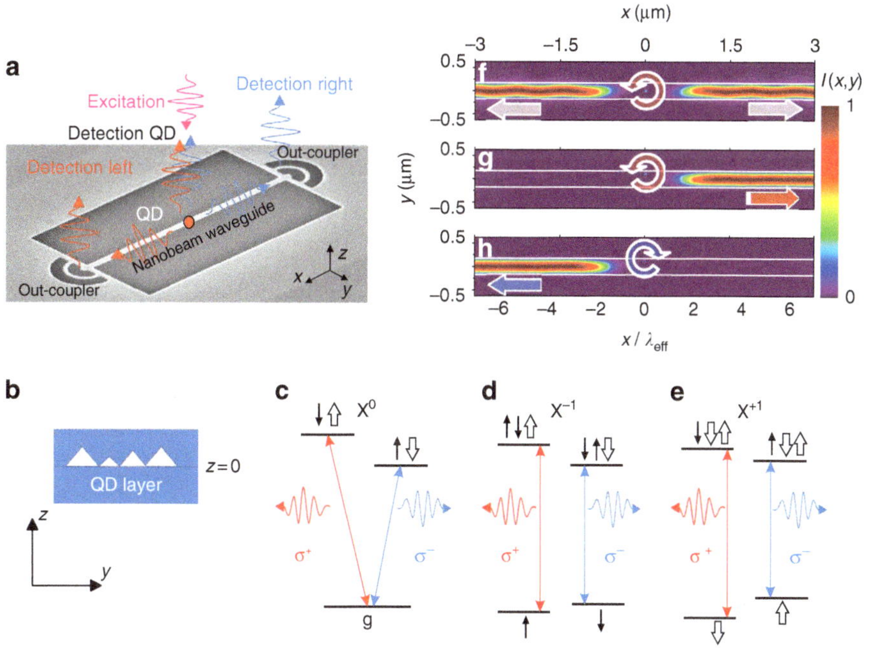

Figure 1 | Directional emission in dielectric nanobeam waveguides. (**a**) Schematic representation of the experiments on directional emission from quantum dots embedded inside a single-mode nanobeam waveguide (NWG) with out-couplers for photon collection. Red (blue) arrows correspond to photons originating from right (left) circularly polarized dipoles in the quantum dot. Scheme labels are plotted on top of an SEM image of the real device. (**b**) A schematic of *yz* cross-section of the NWG with the QD layer at $z=0$. Level diagrams and optical transitions for (**c**) neutral (X^0) and (**d**,**e**) charged (X^{-1}) and (X^{+1}) exciton QD states in $\mathbf{B}\neq 0$. (**f**) Time averaged intensity distribution of the emission $I(x,y)$ (colour scale) in the *xy*-plane at $z=0$ from circularly polarized σ^+ dipole at the centre of the NWG ($y=0$). (**g**,**h**) Emission for a circularly polarized dipole at the chiral point, displaced from the centre of the waveguide by 32% of the waveguide width: σ^+ dipole; σ^- dipole. The undulation arises from periodic coupling with time of the rotating circular dipole at the non-chiral point. The intensity distributions in (**f**-**h**) are calculated after the dipole source is switched off. The *x* axis is normalized to the NWG mode effective wavelength λ_{eff}. The white horizontal lines show the nanostructure boundaries. Fully animated versions of the simulations may be found in the Supplementary Notes 1-6.

direction as indicated in Fig. 2d, so that $\sigma^\pm$ emitters couple preferentially to modes propagating in opposite directions. By contrast, the point at the centre of the waveguide ($y=0$) has no longitudinal component ($E_x=0$) and is therefore linearly polarized (Fig. 2c,d). Translation along *y* from the centre to the chiral points thus transforms the field polarization from linear, through elliptical, to circular.

Spin readout in a NWG and PhC WG. To demonstrate the chiral effects experimentally, we use a single-mode NWG with randomly distributed self-assembled QDs, positioned by growth in the *xy*-plane at $z=0$. The guided photoluminescence (PL) of single QDs is detected from out-couplers at opposite ends of the NWG (Fig. 1a). The inset in Fig. 3c shows a typical auto-correlation function $g^{(2)}(\tau)$ confirming the single-photon nature of the QD emission. The sample was placed in an out-of-*xy*-plane magnetic field B_z to quantize the QD spins into $S_z=\pm 1$ states (spin up/down). This enables identification of exciton Zeeman spin components from the energy of the circularly polarized (σ^+/σ^-) photons emitted: $h\nu_x(S_z)=h\nu_0+\mu_B g_x B_z S_z$, where g_X is an excitonic g-factor that varies from dot to dot with a typical value of $\sim$1.2. $B_Z=1$ T for the data in Figs 3 and 4. Self-assembled growth leads to random positions of QDs, enabling the observation of emission from both chiral and non-chiral areas of the waveguide. Figure 3 contrasts the behaviour for two dots selected from the random distribution, with one showing chiral behaviour (Fig. 3a), and the other not (Fig. 3b). For the QD in the chiral point in Fig. 3a, both Zeeman-split components are seen when detecting directly from the dot ('det QD'), but only one component is observed for detection from the left and right out-couplers ('det L' and 'det R'). The two Zeeman components emit in opposite directions, implying spin-dependent directional emission. By contrast, non-directional emission is observed for the dot in centre of the waveguide (Fig. 3b), with both Zeeman components detected from both out-couplers. The degree of contrast in spin-readout (the directionality factor) is defined as:

$$C_{\text{LEFT/RIGHT}}=\frac{I^{\text{L/R}}_{\sigma+}-I^{\text{L/R}}_{\sigma-}}{I^{\text{L/R}}_{\sigma+}+I^{\text{L/R}}_{\sigma-}} \quad (1)$$

where the superscripts L and R refer to the left and right out-couplers. The results for 50 QDs are shown in Fig. 3c. The diagonal line across the figure is the expectation for a random distribution of dots in the *xy*-plane at $z=0$ whose circular dipoles couple to the confined electromagnetic fields. QDs at the linear point at the centre of the waveguide correspond to zero contrast, whereas dots at the chiral points give rise to contrasts of ± 1. Agreement between the trends in contrast between experiment and simulation is seen (the experimental scatter around the diagonal is discussed below).

The data from Fig. 3c are plotted as a histogram of numbers of dots of absolute contrast $C=(|C_{\text{LEFT}}|+|C_{\text{RIGHT}}|)/2$ in Fig. 3d and compared with binned data of the FDTD simulations of contrast (see Supplementary Fig. 7 for details) for out-coupler-terminated waveguide in Fig. 3e. Details of the comparison between infinite and terminated waveguides are given in the Supplementary Notes 2 and 3. The out-coupler gratings introduce back-reflections that lead to modulation of the field distributions; they do not change the mechanisms responsible for unidirectionality, but result in reduction in the probability for dots with high

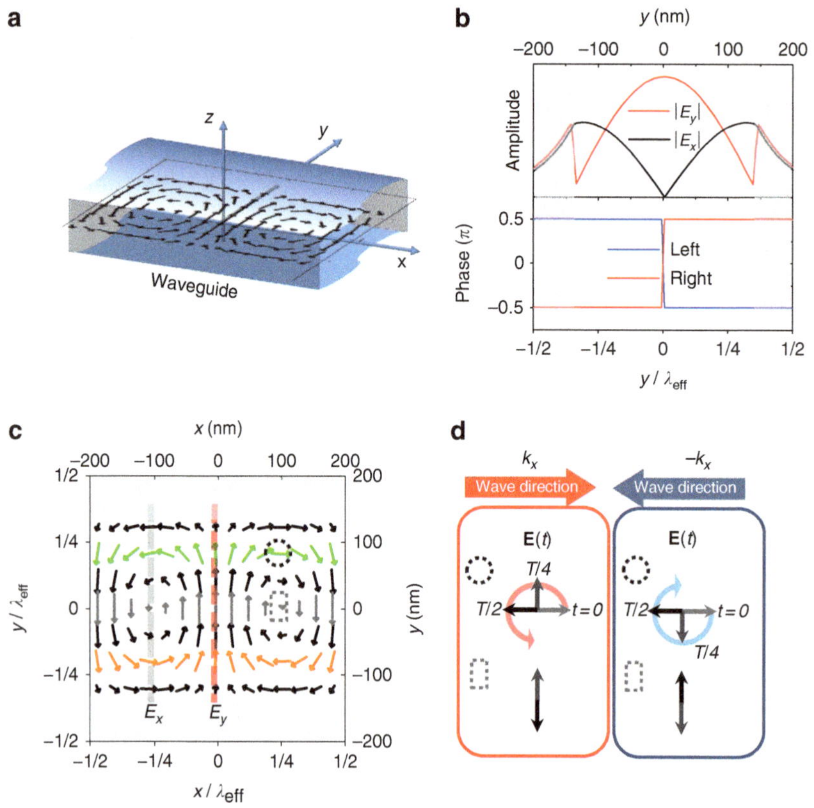

Figure 2 | Electric field distributions in nanobeam waveguide. (**a**) Schematic of an infinite length NWG showing the orientation of the axes and the field distribution in the *xy*-plane at $z=0$, as defined by the central layer of the waveguide containing QDs. (**b**) Electric field amplitudes (top) for the *x* and *y* components and their relative phase (bottom) for left/right propagating modes. (**c**) Simulated distribution of the electric field vector for the first fundamental mode at a fixed moment in time. Arrows show the direction of the field and the length of the magnitude. Colour is used as a guide to the eye: green/orange for opposite-helicity chiral points, and grey for central points with linear polarization. The left/bottom scales are normalized to the NWG mode effective wavelength λ_{eff}. (**d**) Electric field time evolution in space from **c** for propagating modes at chiral (circular dotted area) and non-chiral (rectangular dotted area) points in the NWG. The red and blue arrows show the k_x and $-k_x$ mode propagation direction, respectively. The rotation of the electric field vectors exhibiting circular right/left polarization is shown by red/blue circular arrows. The grey areas in **b** and **c** correspond to air cladding regions at the NWG edges.

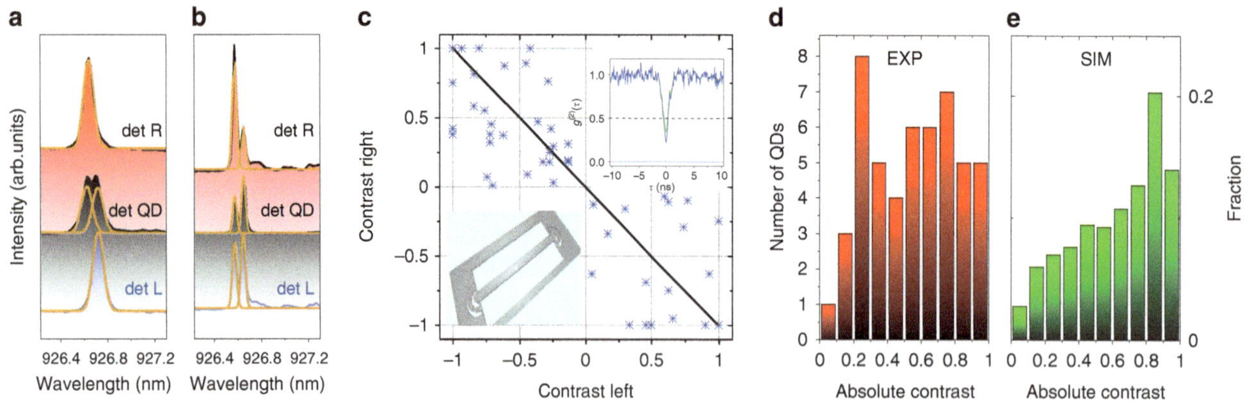

Figure 3 | Experimental demonstration of spin readout in nanobeam waveguides. (**a**) Photoluminescence spectra of QDs with high contrast in NWGs. (**b**) Experimental photoluminescence spectra of QDs with low contrast in NWGs. Fitted peaks in **a** and **b** are shown with orange curves. (**c**) Readout contrast $C=(I_{\sigma+}-I_{\sigma-})/(I_{\sigma+}+I_{\sigma-})$ for left and right out-couplers for randomly distributed self-assembled QDs in NWGs (50 QDs). The points correspond to experimental data, and the black diagonal line shows the expected distribution for ideal circularly polarized dipoles. (**d**) Statistical distributions of the absolute average contrasts $C=(|C_{LEFT}|+|C_{RIGHT}|)/2$ of QD spin readout taken from the experimental data in **c** for NWGs. All experimental data are recorded at $B_Z=1$ T. (**e**) Simulated statistical distributions of absolute contrasts calculated from the coupling of circularly polarized dipoles distributed across the WGs in the *xy*-plane ($z=0$) taking into account the electric field distributions (Supplementary Note 3) due to the out-couplers for NWG. The top inset in **c** shows a typical QD autocorrelation function $g^{(2)}(\tau)$ (blue curve with background subtraction $g^{(2)}(0)=0.22$, green without $g^{(2)}(0)=0.34$). The inset in **c** shows schematic of the NWG.

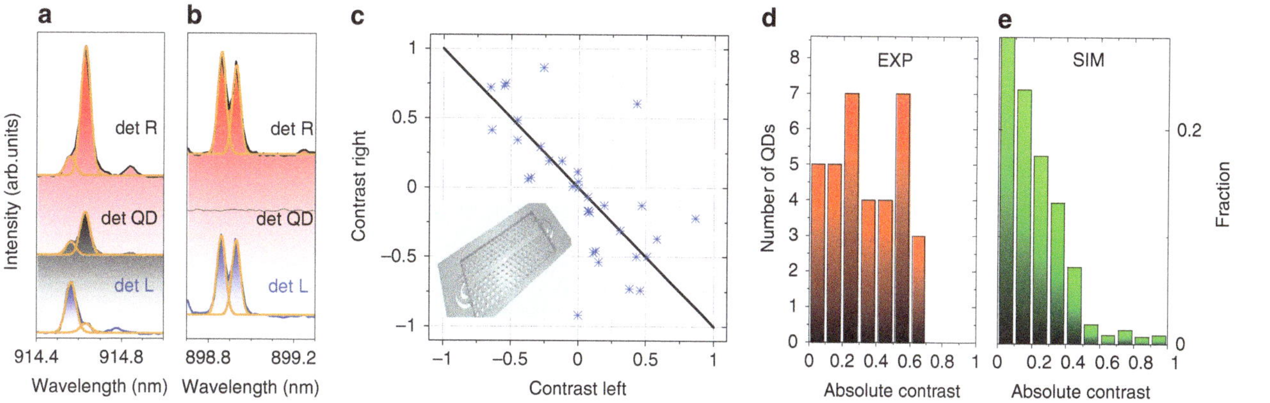

Figure 4 | Experimental demonstration of spin readout in photonic crystal waveguides. (**a**) Photoluminescence spectra of QDs with high contrast in PhC WGs. (**b**) Experimental photoluminescence spectra of QDs with low contrast in PhC WGs. Fitted peaks in **a** and **b** are shown with orange curves. (**c**) Readout contrast $C=(I_{\sigma+}-I_{\sigma-})/(I_{\sigma+}+I_{\sigma-})$ for left and right out-couplers for randomly distributed self-assembled QDs in PhC WGs (35 QDs). The points correspond to experimental data, and the black diagonal line shows the expected distribution for ideal circularly polarized dipoles. (**d**) Statistical distributions of the absolute average contrasts $C=(|C_{\mathrm{LEFT}}|+|C_{\mathrm{RIGHT}}|)/2$ of QD spin readout taken from the experimental data in **c** for PhC WGs. All experimental data are recorded at $B_Z=1\,\mathrm{T}$. (**e**) Simulated statistical distributions of absolute contrasts calculated from the coupling of circularly polarized dipoles distributed across the WGs in the *xy*-plane ($z=0$) taking into account the electric field distributions (Supplementary Note 4) due to the out-couplers for PhC WGs. The inset in **c** shows schematic of the PhC WGs.

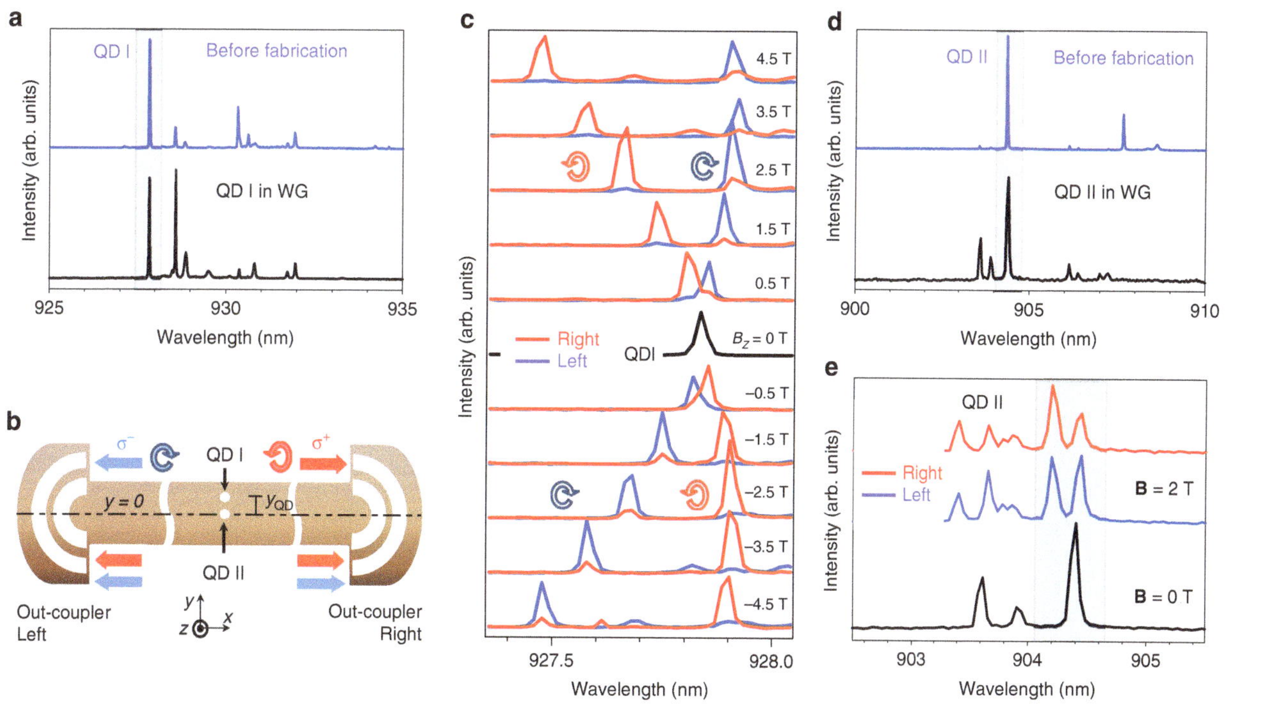

Figure 5 | Quantum dot registration and spin readout. (**a**) Spectra of QD I before and after NWG fabrication. (**b**) Schematic of spin readout for registered QD I displaced vertically from the centre by 32% of the waveguide width (the Sample details in the Methods) and the absence of spin-readout for registered QD II located at the centre of the waveguide. (**c**) Magnetic field dependence of emission from the registered QD I collected from the left out-coupler (blue) and right out-coupler (red). (**d**) Spectra of QD II before and after NWG fabrication. (**e**) Spectra of QD II in magnetic field $\mathbf{B}=2\,\mathrm{T}$ collected from left and right out-couplers and spectrum of QD II at $\mathbf{B}=0\,\mathrm{T}$ for reference.

(>90%) spin readout conversion by ~52%. Qualitative agreement between experiment (Fig. 3d) and simulation (Fig. 3e) is observed, with an increasing number of QDs to high contrast clearly seen, and levelling-off around a contrast of 80%. It is also notable that QDs with extremely high directionality exceeding 90% are observed experimentally (Fig. 3c,d). These results support the findings in the simulations which show chiral areas (Supplementary Fig. 6d) of 14% of the waveguide where the contrast exceeds 90% (and 34% exceeding 80%).

The circular dipole unidirectional coupling efficiency, defined as the fraction of power emitted into a waveguide mode propagating in one direction, is calculated to be 68% for a dipole positioned at a chiral point of the NWG (simulations in the Methods). This is an increase of more than ~2.8 compared with the coupling of a circular-dipole at the centre of the waveguide into one spatial direction. It is also ~1.4 times greater than for a linear-dipole co-polarized with the field at the centre of the NWG ($y=0$)[8]. Both comparisons show the favourable degrees of coupling efficiency obtained for spin readout at the chiral points.

In order to compare to PhC WGs, we perform similar experiments and simulations for them. QD spectra with high/low contrasts are shown in Fig. 4a,b. The contrasts for the 35 QDs studied are shown in Fig. 4c. The overall trends in the behaviour are similar to those seen in Fig. 3c for the NWGs, but with the very marked difference that no QDs exhibit high directionality values above 70% (Fig. 4c,d). This is expected as the chiral areas (Supplementary Fig. 9d) inside the PhC WG, where the contrast exceeds 90%, only constitute $\sim 0.8\%$ of the waveguide area ($\sim 1.5\%$ of area for $>80\%$ contrast) as shown in Supplementary Note 4. They are furthermore located in areas of low field intensity (Supplementary Fig. 9a), leading to poor dipole coupling. Both the high-fidelity chiral areas (~ 20 times smaller than for the NWGs) and field intensities for the PhCs indicate the advantageous characteristics of the NWGs for spin readout, with their near continuous translational symmetry, relative to PhC WGs. This situation can likely be improved for PhC WGs by the use of glide plane techniques, as shown in ref. 14.

QD registration and spin readout. Having identified the location of chiral points theoretically, and observed high contrast unidirectional emission for randomly distributed dots, we now demonstrate control of the emission direction, using QD registration techniques[17,18] (see the Methods for details). First, QD spectra were compared before and after the registration and NWG fabrication to confirm that the QD (we term this dot QD I) is successfully integrated inside the NWG (Fig. 5a). The waveguide is fabricated (see the Methods for details) such that QD I is displaced from the centre of the waveguide at the chiral point, with y_{QD} $\sim 30\%$ of the waveguide width (see Fig. 5b schematic for details). For this position, we expect the σ^+ dipole to couple to the right propagating mode and the σ^- dipole to the left (Fig. 5b). The spectra in magnetic field reveal high directionality ($|C_{LEFT}| = 92 \pm 3\%$, $|C_{RIGHT}| = 80 \pm 3\%$) of emission with absolute contrast independent of magnetic field to within 10%. Furthermore, the expected signs of the contrast correlate with the QD I position, with the σ^+ line appearing at the right out-coupler with high energy when the field is positive, whereas the σ^- line is collected from the left out-coupler. Moreover, when the direction of the magnetic field is changed, the order of the lines is reversed, but the dipole coupling direction remains the same (Fig. 5c). Thus, we not only achieve control of spin-directionality by registration but also the reversal of the emission direction at a given energy by magnetic field control[14]. We further demonstrate the effectiveness of the registration techniques by fabricating a control sample with a QD (QD II) positioned at the centre of the waveguide (see Fig. 5b schematic for details). Spectra before and after NWG fabrication are compared to prove the integration of the QD within the waveguide (Fig. 5d). Low spin-readout contrast of ($|C_{LEFT}|$ $\sim 3 \pm 6\%$, $|C_{RIGHT}| = 24 \pm 4\%$, Fig. 5e) is found, as expected for a dot at the centre of the waveguide. We note that the approach with dot registration is not limited to one QD and can be scaled up to create more complex circuits with deterministically coupled QDs to realize spin–photon and spin–spin entanglement on chip.

Discussion

As noted above, the experimental points in Figs 3c and 4c show scatter around the diagonal, compared with theoretical expectations. This behaviour is not fully understood. It may arise due to back reflections in the finite length waveguides, in combination with elliptically polarized QDs. Fine structure splitting[20] of the QDs can be excluded as we observe no magnetic field dependence for asymmetrically coupled QDs. The circular polarization for the QDs may also be affected in photonic structures by the nanostructure and surface proximity[21]. Moreover, the finite size of the dot, which is known to cause a breakdown in the point dipole approximation[22] used for our simulations in nanophotonic structures may also affect the contrast.

To conclude, we demonstrate experimentally and theoretically chiral effects in a simple nanophotonic system consisting of waveguides containing embedded quantum emitters. The unidirectional phenomena we report may be used for spin read out and to transfer spin information from localized emitters in waveguide geometries. Deterministic positioning of a QD at a chiral point of the waveguide is achieved, with accompanying spin readout, a possible route to scalability. Larger areas for chiral behaviour found in the NWGs because of their continuous translational symmetry together with simplified fabrication techniques and low loss may provide significant advantages over PhC WGs. The findings and techniques presented could contribute to the creation of spin-optical on-chip networks and processing devices based on nanophotonic waveguides.

Methods

Sample details. The sample was grown by molecular beam epitaxy on an undoped [100] GaAs substrate, and consisted of a 140-nm-thick GaAs membrane containing a single layer of self-assembled InGaAs QDs at the centre grown on a 1-µm sacrificial $Al_{0.6}Ga_{0.4}As$ layer. The photonic structures were patterned by electron beam lithography (EBL) and the pattern transferred to the membrane using inductively coupled plasma etching. To release the membranes, the sacrificial AlGaAs layer was removed by selective wet etch using a buffered hydrogen fluoride solution. The suspended NWGs were 15 µm long, 280 nm wide and 140 nm height. An scanning electron microscopy (SEM) image of the suspended NWG is shown in Supplementary Fig. 1a. The W1 PhC WGs were fabricated with lattice constant $a = 254$ nm and hole radius $r = 0.31$ a. An SEM image of the PhC WG is presented in Supplementary Fig. 1b. Both waveguides were terminated with semi-circular $\lambda/2n$ air-GaAs out-coupler gratings designed for optimum operation at $\lambda = 950$ nm at the centre of the QD ensemble PL emission[23].

Experimental set-up. All measurements were performed in a helium bath cryostat at $T = 4.2$ K within which a superconducting magnet provided magnetic field of up to 5 T normal to the sample plane. Ultra-stable positioning of the sample within the system is provided by X, Y, Z home-made piezo stages. The cryostat insert had optical access to the sample in a confocal scanning microscope arrangement[24]. The excitation and collection spots were below 1.5 µm in diameter and could be separately moved by more than 15 µm by scanning mirrors to obtain the exact geometry required for each experiment. Optical excitation was provided by an 808 nm diode laser coupled to an optical fibre. The collected PL signal was fibre coupled and spectral measurements were performed using a spectrometer with a LN2 cooled charged-coupled device (CCD). Using a second exit port on the spectrometer, spectrally filtered PL signals were sent to two avalanche photodiodes (APDs) for QD autocorrelation $g^{(2)}(\tau)$ measurements using a single-photon counting module.

QD registration method. Dot registration[17,18] is carried out in a scanning micro-PL set-up with two collection paths, as illustrated in Supplementary Fig. 2. The dot registration process involves three stages: pre-registration markers are fabricated, individual QDs are registered and photonic structures are deterministically fabricated around a QD. Registration markers are patterned using EBL and a positive resist. The sample is developed in xylene leaving the wafer surface exposed in the desired pattern, as shown in Supplementary Fig. 3a. The inner double markers are used to register the relative position of a QD, whereas the outer angular markers are used to re-align the EBL during the final fabrication stage. A layer of 5 nm of titanium followed by 20 nm of gold are then evaporated onto the wafer. The sample is then placed in a solvent bath to remove the remaining layer of resist and any unwanted titanium and gold.

A 1-mm diameter cubic zirconia (ZrO_2) Weierstrass solid immersion lens is placed directly above a registration grid, with an aberration free image size of ~ 50 µm in diameter. A low QD density of $\sim 1.6 \cdot 10^7$ cm^{-2} and a solid immersion lens-enhanced excitation spot size ~ 350 nm enables individual QDs to be measured. The closed-loop three axes scanning piezo stage, with a resolution of 1 nm, is used to scan the objective lens, which simultaneously moves the excitation and collection spots. One collection path is spectrally filtered through a monochromator to isolate a single exciton line before it is measured with an APD. The second collection path is used to measure the reflected laser signal from the gold markers to provide reference points in the scan from which the QD position is measured. No additional spectral filtering is used on the second collection path before it is detected with an APD, although it is heavily attenuated using neutral density filters. Horizontal and vertical line scans are used to determine the relative QD position. The signals from both APDs were recorded during the scans as a function of the stage position (Supplementary Fig. 3b).

The spacing between the gold markers is designed to be 15 μm. The relative position of the QD is then determined by repeating the line scans 100 times to provide a statistical average (Supplementary Fig. 3c). A Gaussian curve is fitted to the distribution to determine the centre. The minimum standard deviation observed is $\sigma < 5$ nm, which is exceeding the accuracy of the best position uncertainty previously achieved[17,18]. The position of the suspended NWGs to be fabricated is placed within the original EBL design file at a suitable location, such that the QD are in the centre or laterally displaced. To align the EBL writing field to the pattern previously fabricated, the electron beam is scanned over the outer (blue) registration markers. The reflected electron flux is recorded and used to determine the central position of each marker. The suspended nanobeam is then fabricated (SEM image in Supplementary Fig. 3d) as described above in the Sample details section.

Calculations of the fields in the waveguides. FDTD simulations were performed using freely available software package[25] and commercial-grade simulator[26]. The GaAs waveguide devices simulated had the dimensions described in the Sample details section above. Details of the simulations are available in Supplementary Notes 1–4.

References

1. Barz, S. *et al.* Demonstration of blind quantum computing. *Science* **335,** 303–308 (2012).
2. Kimble, H. J. The quantum Internet. *Nature* **453,** 1023–1030 (2008).
3. Kok, P. & Lovett, B. W. *Introduction to Optical Quantum Information Processing* (Cambridge Univ., 2010).
4. DiVincenzo, D. P. The physical implementation of quantum computation. *Fortschr. Phys.* **48,** 771–783 (2000).
5. Barrett, S. D. & Kok, P. Efficient high-fidelity quantum computation using matter qubits and linear optics. *Phys. Rev. A* **71,** 060310 (2005).
6. O'Brien, J. L. Optical quantum computing. *Science* **318,** 1567–1570 (2007).
7. Prtljaga, N. *et al.* Monolithic integration of a quantum emitter with a compact on-chip beam-splitter. *Appl. Phys. Lett.* **104,** 231107 (2014).
8. Makhonin, M. N. *et al.* Waveguide coupled resonance fluorescence from on-chip quantum emitter. *Nano Lett.* **14,** 6997–7002 (2014).
9. Rodriguez-Fortuño, F. J. *et al.* Near-field interference for the unidirectional excitation of electromagnetic guided modes. *Science* **340,** 328–330 (2013).
10. Junge, C., O'Shea, D., Volz, J. & Rauschenbeutel, A. Strong coupling between single atoms and nontransversal photons. *Phys. Rev. Lett.* **110,** 213604 (2013).
11. Mitsch, R., Sayrin, C., Albrecht, B., Schneeweiss, P. & Rauschenbeutel, A. Quantum state-controlled directional spontaneous emission of photons into a nanophotonic waveguide. *Nat. Commun.* **5,** 5713 (2014).
12. Petersen, J., Volz, J. & Rauschenbeutel, A. Chiral nanophotonic waveguide interface based on spin-orbit coupling of light. *Science* **346,** 67–71 (2014).
13. le Feber, B., Rotenberg, N. & Kuipers, L. Nanophotonic control of circular dipole emission. *Nat. Commun.* **6,** 6695 (2015).
14. Söllner, I. *et al.* Deterministic photon–emitter coupling in chiral photonic circuits. *Nat. Nano* **10,** 775–778 (2015).
15. Young, A. B. *et al.* Polarization engineering in photonic crystal waveguides for spin-photon entanglers. *Phys. Rev. Lett.* **115,** 153901 (2015).
16. Bliokh, K. Y., Rodríguez-Fortuño, F. J., Nori, F. & Zayats, A. V. Spin–orbit interactions of light. *Nat. Photon* **9,** 796–808 (2015).
17. Thon, S. M. *et al.* Strong coupling through optical positioning of a quantum dot in a photonic crystal cavity. *Appl. Phys. Lett.* **94,** 111115 (2009).
18. Dousse, A. *et al.* Controlled light-matter coupling for a single quantum dot embedded in a pillar microcavity using far-field optical lithography. *Phys. Rev. Lett.* **101,** 267404 (2008).
19. Luxmoore, I. J. *et al.* Optical control of the emission direction of a quantum dot. *Appl. Phys. Lett.* **103,** 241102 (2013).
20. Gammon, D. *et al.* Fine structure splitting in the optical spectra of single GaAs quantum dots. *Phys. Rev. Lett.* **76,** 3005–3008 (1996).
21. Stepanov, P. *et al.* Quantum dot spontaneous emission control in a ridge waveguide. *Appl. Phys. Lett.* **106,** 041112 (2015).
22. Tighineanu, P., Sørensen, A. S., Stobbe, S. & Lodahl, P. Unraveling the mesoscopic character of quantum dots in nanophotonics. *Phys. Rev. Lett.* **114,** 247401 (2015).
23. Faraon, A. *et al.* Dipole induced transparency in waveguide coupled photonic crystal cavities. *Opt. Express* **16,** 12154–12162 (2008).
24. Grazioso, F., Patton, B. R. & Smith, J. M. A high stability beam-scanning confocal optical microscope for low temperature operation. *Rev. Sci. Instrum* **81,** 093705 (2010).
25. Johnson, S. S. & Joannopoulos, J. J. Block-iterative frequency-domain methods for Maxwell's equations in a planewave basis. *Opt. Express* **8,** 173–190 (2001).
26. Lumerical FDTD Solutions. Lumerical Solutions, Inc. http://www.lumerical.com/tcad-products/fdtd/ (2014).

Acknowledgements

This work has been supported by the Engineering and Physical Sciences Research Council (Programme Grant EP/J007544/1). We thank D.M. Whittaker, D.N. Krizhanovksii, E. Cancellieri and F. Li for fruitful discussions.

Author contributions

D.M.P., J.E.D. and M.N.M. carried out the experiments, J.E.D. and B.R. performed the dot registration and sample fabrication, R.J.C. carried out the simulations, E.C. grew the samples. All authors analysed the data and contributed to the writing of the manuscript, with P.K. providing quantum technology context. A.M.F., M.S.S. and M.N.M. conceived the project and provided overall guidance and management.

Additional information

Competing financial interests: The authors declare no competing financial interests.

9

Approaching soft X-ray wavelengths in nanomagnet-based microwave technology

Haiming Yu[1,2], O. d' Allivy Kelly[3], V. Cros[3], R. Bernard[3], P. Bortolotti[3], A. Anane[3], F. Brandl[1], F. Heimbach[1]& D. Grundler[1,4]

Seven decades after the discovery of collective spin excitations in microwave-irradiated ferromagnets, there has been a rebirth of magnonics. However, magnetic nanodevices will enable smart GHz-to-THz devices at low power consumption only, if such spin waves (magnons) are generated and manipulated on the sub-100 nm scale. Here we show how magnons with a wavelength of a few 10 nm are exploited by combining the functionality of insulating yttrium iron garnet and nanodisks from different ferromagnets. We demonstrate magnonic devices at wavelengths of 88 nm written/read by conventional coplanar waveguides. Our microwave-to-magnon transducers are reconfigurable and thereby provide additional functionalities. The results pave the way for a multi-functional GHz technology with unprecedented miniaturization exploiting nanoscale wavelengths that are otherwise relevant for soft X-rays. Nanomagnonics integrated with broadband microwave circuitry offer applications that are wide ranging, from nanoscale microwave components to nonlinear data processing, image reconstruction and wave-based logic.

[1] Physik Department E10, Technische Universität München, James-Franck-Strasse 1, D-85748 Garching bei München, Germany. [2] Fert Beijing Institute, School of Electronic and Information Engineering, Beihang University, Xueyuan Road 37, Beijing 100191, China. [3] Unité Mixte de Physique, CNRS, Thales, Univ Paris-Sud, Université Paris-Saclay, 91767 Palaiseau, France. [4] Laboratory of Nanoscale Magnetic Materials and Magnonics, Institute of Materials, School of Engineering, École Polytechnique Fédérale de Lausanne, STI-IMX-LMGN, Station 17, CH-1015 Lausanne, Switzerland. Correspondence and requests for materials should be addressed to H.Y. (email: haiming.yu@buaa.edu.cn) or to D.G. (email: dirk.grundler@epfl.ch).

Magnonics is an emerging technology for low-power signal transmission and data processing based on spin waves (magnons) propagating in magnetic materials[1–5]. In contrast to, for example, plasmons in metals[6], magnons exist also in insulators, overcoming the damping issue. Thanks to the rapid advances of nanotechnology, magnonics is currently seeing an outburst. The wavelength λ of spin waves processed by magnonic crystals and magnonic waveguides have diminished from the millimetre to micrometre length scale at a fast pace[7–9]. Evidencing a wavelength of 900 nm, Demidov *et al.*[10] demonstrated a powerful wavelength-conversion technique, removing restrictions imposed by the size of microwave antennas. Still, for the real breakthrough, spin waves should be generated and manipulated down to the few 10-nm length scale. At this scale, exchange interaction dominates enabling large group velocities and fast signal transmission. From such spin waves, nanoscale microwave devices operating from GHz to THz can be conceived, for future mobile communications[11] and efficient computing based on coherent data processing[12]. However, sub-100 nm magnonics was argued to be far from realization[4]. Inelastic light scattering is widely used for studying propagating spin waves with $\lambda > 200$ nm[13,14]. Neutron scattering[15], electron loss spectroscopy[16] and scanning tunnelling microscopes[17] have addressed smaller λ. Unfortunately, none of these techniques allow direct integration on a chip. In the past, on-chip spin-wave emitters were realized by parametric pumping[18] and spin-transfer torque nanopillars[13,14]. However, the extremely large microwave signal and complex spin-wave emission, respectively, make the low-power consumption and coherent processing of information in interconnected magnonic networks unrealistic[12]. Au *et al.*[19] proposed a resonant microwave-to-magnon transducer exploiting spin-precessional motion in a ferromagnetic nanowire placed on top of a magnonic conduit. Following this intriguing concept, they evidenced spin-wave excitation at $\lambda = 6\,\mu$m. Our approach tackles the long-standing problem how to integrate sub-100 nm magnonics on-chip with microwave circuits[4].

In this paper, we report the inductive generation and detection of plane-wave spin waves in the low-power linear regime with a wavelength λ of 88 ± 2 nm. This record value of λ is achieved in thin yttrium iron garnet (YIG) by using conventional coplanar waveguides (CPWs) that cover small arrays of tailored magnetic nanodisks. The nanodisks reside between the CPWs and YIG, and are operated as microwave-to-magnon transducers[19]. For them, we use materials established in spintronics and magnonics, that is, polycrystalline $Ni_{80}Fe_{20}$ (Py) and amorphous $Co_{20}Fe_{60}B_{20}$ (CoFeB). Spin waves with wave vectors of up to $k = 2\pi/\lambda \approx 8\times10^5$ rad cm^{-1} show large signal-to-noise ratios and are exchange dominated exhibiting enlarged group velocities. At the same time, we discover our microwave-to-magnon transducers to be reconfigurable. They provide resonantly enhanced spin-wave amplitudes that are tuned ON and OFF by writing saturated and unsaturated states, respectively, on the disks. Our results pave the way for interconnecting conventional microwave electronics with exchange-dominated spin waves and substantially advancing magnonics-based technologies.

Results

Enhanced magnon excitation via resonantly driven nanodisks. Figure 1a,b illustrates sample and experiment. CPWs were integrated to ferrimagnetic YIG with a thickness $t_1 = 20$ nm. Underneath the CPWs, nanodisks of thickness $t_2 = 15$ nm were

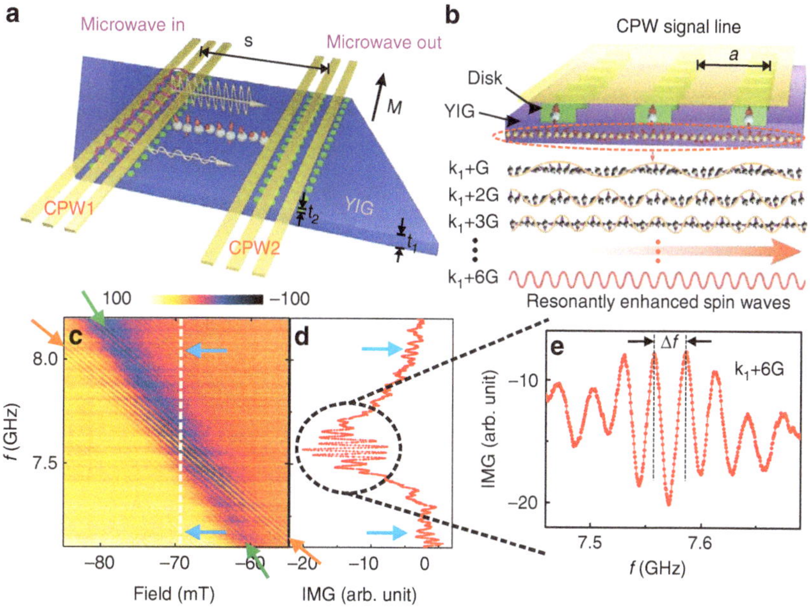

Figure 1 | Resonantly driven nanodisks for injection and detection of large-amplitude spin waves in thin YIG. (**a**) Sketch of the experiment. Large-magnetization ferromagnetic nanodisk arrays (green) were positioned between CPWs (yellow) and insulating YIG (violet). The magnetization M of YIG is parallel to the CPWs. (**b**) Sketched microwave-to-magnon transduction: the resonant spin-precessional motion in disks (large arrows) is exploited to excite short-wavelength spin waves (small arrows highlighted by the oval) in YIG. Depending on the frequency, different wave vectors $k_1 + nG$ ($n = 1, 2, 3, \ldots$) are induced. $G = 2\pi/a$ denotes a reciprocal lattice vector perpendicular to the CPW. Spin waves propagate to the detector CPW indicated by the horizontal arrow. (**c**) Colour-coded transmission signal S_{12} monitoring spin-wave propagation between CPW1 to CPW2. Green and orange arrows guide the eyes and highlight two branches that cross near the centre of the graph. Green (orange) arrows indicate the resonant excitation of the permalloy nanodisks (YIG at large wave vector k_{SW}). (**d**) Transmission spectrum S_{12} at -69 mT (broken line in **c**) displaying propagation of spin waves through YIG with enlarged amplitude when permalloy nanodisks resonate together with YIG. Blue arrows indicate non-resonant spin-wave excitation. (**e**) Enlarged oscillating transmission signal (imagninary (IMG) part) around 7.6 GHz attributed to $k_{SW} = k_1 + 6G = 48$ rad µm^{-1} $= 4.8 \times 10^5$ rad cm^{-1}. The corresponding wavelength amounts to 131 ± 3 nm.

fabricated from either evaporated Py or magnetron sputtered CoFeB using electron beam lithography and lift-off processing (Supplementary Fig. 1). The nanodisks with diameter of 350 ± 5 nm were arranged on a square lattice with a lattice constant $a = 800$ nm. The real and imaginary part of scattering parameters S_{pq} were measured in reflection and transmission configuration using a vector network analyser (VNA) connected to the CPWs ($p = 1,2$ and $q = 1, 2$ denote the different ports of the VNA and the CPWs). In this work we show the imaginary part of S_{12} and not the modulus to benefit from the phase resolution of the VNA and extract the group velocity v_g from oscillating transmission signals[20,21] (Methods). The microwave magnetic field h_{rf} generated by a CPW was inhomogeneous and provided excitation in the linear regime at a distinct set of wave vectors $\mathbf{k}_i$ with $i = 1, 2, 3, \ldots$ as explored in ref. 20. The propagation distance s between CPWs was 30 μm.

We found that the resonant spin-precessional motion, that is, ferromagnetic resonance (FMR), of saturated nanodisks occurred at much larger frequency f_0 compared with the uniform mode of the YIG film with wave vector $k_{SW} = 0$. Only for spin waves in YIG with large k_{SW}, the nanodisks were in resonance with spin precession in YIG. In Fig. 1c, we show colour-coded transmission data S_{12} obtained on the Py/YIG hybrid sample. We focus on a field regime where field-dependent FMR of Py nanodisks (branch highlighted by green arrows) becomes degenerate with spin waves in YIG (branch indicated by orange arrows). In Fig. 1d,e, the line spectrum S_{12} taken at the crossing point at $\mu_0 H = -69$ mT contains a large transmission signal detected around 7.6 GHz. The oscillating behaviour indicates the propagation of spin waves through the YIG film between the two CPWs[20,21]. Their corresponding wavelength λ amounts to only 131 ± 3 nm. At the crossing point (circle in Fig. 1d), the transmitted spin-wave signal is significantly enhanced by a factor of 6 compared with signals measured for non-degenerate eigenfrequencies of Py and YIG (blue arrows in Fig. 1d and Supplementary Fig. 2).

In Fig. 2a, we show absorption spectra S_{11} taken in reflection configuration (selected lineplots are depicted in Supplementary Fig. 3). For $\mu_0|H| > 32$ mT, all eigenfrequencies $f(H)$ (dark) are mirrored with respect to $\mu_0 H = 0$ mT. The two pairs of most prominent branches (black) are highlighted by blue and green arrows. In case of the low-frequency one (blue arrow), $f(H)$ follows roughly a square-root-like field dependence for $H > 0$. The branch is consistent with $f(H)$ of an unpatterned thin film of YIG when excited by a microstructured CPW[20]. It is noteworthy that the coercive field of the YIG film is only ~0.5 mT (ref. 20). Its hysteretic behaviour is thus not resolved in Fig. 2a.

The branches marked by green arrows are at much higher frequency values and follow a different $f(H)$ behaviour compared with YIG. These absorption features are in agreement with the FMR of the Py nanodisks that are magnetized along the in-plane magnetic field H. The nanodisks exhibit a large coercive field (switching field) of 32 mT and the hysteretic behaviour of the FMR frequencies is clearly observed. When $\mu_0 H$ is increased from -100 mT and reaches -6 mT, the upper prominent branch starts to vanish, indicating that the magnetization M of Py nanodisks turns away from the applied field orientation. The geometrical parameters of the nanodisks support the vortex state in the intermediate field regime[22], as sketched in the inset. When the field is further increased to $+33$ mT, the nanodisks become saturated, with M parallel to H. For $\mu_0|H| > 33$ mT, the upper prominent branch is hence present again. The black vertical lines in Fig. 2a highlight the three field regimes for different magnetic states of the nanodisks. The weak absorption lines (white arrow) at lower frequencies are attributed to channelling effects of spin

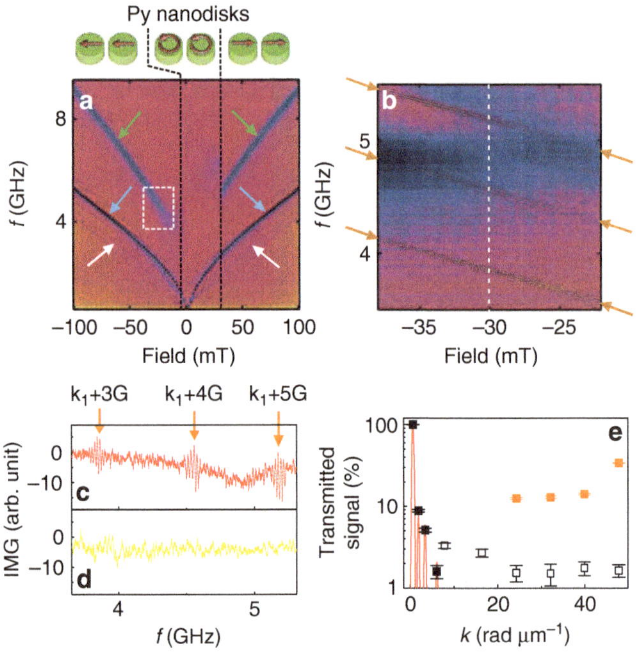

Figure 2 | Reconfigurable microwave-to-magnon transducer exciting exchange-dominated spin waves. (**a**) Colour-coded absorption spectra S_{11} measured on the permalloy (Py)/YIG hybrid sample in reflection configuration. The field is varied from -100 to $+100$ mT. The blue arrows highlight excitation of YIG at k_1. The green arrows highlight the FMR branch of Py nanodisks that show hysteretic behaviour. White arrows indicate spin-wave modes attributed to channelling effects due to the static stray field of the dots. (**b**) Colour-coded spectra S_{12} monitoring three sets of spin waves propagating through YIG (orange arrows). The parameter regime agrees with the white square in **a**. (**c**) Line spectrum S_{12} (imagninary (IMG) part) taken at -30 mT when nanodisks were saturated. Here, three resonantly enhanced modes are seen. (**d**) Line spectrum S_{12} taken at $+30$ mT when the nanodisks were in the unsaturated state. (**e**) Relative amplitudes of transmitted spin waves as a function of the wave vector k (experiment, symbols; expected values following ref. 20, red lines). Black, full (open) squares indicate modes k_1-k_4 (grating coupler modes without crossing of the nanodisk FMR branch) measured at -10 mT. Filled orange squares depict spin-wave modes in YIG with crossing of the nanodisk FMR branch. Error bars are a measure of the trace noise in S_{12} (as displayed in **d**) compared with the signal strength. All signals are normalized to the amplitude of the respective k_1 mode taken as 100%.

waves in the thin YIG induced by the inhomogeneous stray fields of the nanodisks[9]. They will not be discussed here.

Figure 2b shows further colour-coded spectra S_{12} taken in the field and frequency regime marked by the white frame of Fig. 2a. Strikingly, we resolve three branches (highlighted by orange arrows). We note that the field dependence $f(H)$ of each of the three branches agrees with $f(H)$ of the YIG thin film and not with the steeper dependence attributed to the Py nanodisks. In Fig. 2c, we show the lineplot of the spectrum S_{12} taken at -30 mT. The observed oscillating signals are consistent with spin-wave excitation in YIG via the magnonic grating coupler effect[9]. The square lattice of saturated nanodisks provides us with reciprocal lattice vectors $\pm\mathbf{G}$ that add to $\mathbf{k}_i$ of the CPW. The data of Fig. 2c are consistent with spin waves transmitted through YIG having wave vectors $k_{SW} = k_1 + 3G$, $k_1 + 4G$ and $k_1 + 5G$, where $G = 2\pi/a$. In the data of Fig. 2d taken at $+30$ mT we do not observe these transmission signals as will be discussed later.

We investigated the modes k_i and grating coupler modes $k_1 + nG$ ($n = 1, 2, 3 \ldots$) over a broad magnetic field range.

Following our analysis and comparison with calculated dispersion relations $f(k)$ (Methods), we resolved spin waves with wave vectors ranging from k_1 to k_4 and from k_1+1G to k_1+6G in the Py/YIG hybrid sample. As the grating coupler modes followed the weaker field dependence of the YIG film, crossings with the steeper FMR branch of the Py nanodisks took place for large k_{SW}. For each mode k_1+nG, the crossing occurred at a different field value H_n. In Fig. 2e, we summarize the signal strengths of k_i (filled black squares) and grating-coupler modes with (filled orange squares) and without (open squares) crossing the FMR branch. The red lines in Fig. 2e show the signal intensities as expected from the Fourier analysis of the field h_{rf} (compare with ref. 9). The first mode at $k_1 = 0.67\,\mathrm{rad\,\mu m^{-1}}$ ($\lambda = 9.4\,\mathrm{\mu m}$) is the most prominent one. The relative variation of the measured signal strengths taken at -10 mT matches quite well with the calculated variation in Fig. 2e. The open squares summarize the relative signal strengths of grating coupler modes k_1+1G to k_1+6G without provoking the crossing with the nanodisk FMR at -10 mT. For modes k_1+3G to k_1+6G, the crossing with the nanodisk FMR was possible. Strikingly, a considerably increased signal strength was found. The relevant fields $\mu_0 H_n$ ($n = 3\ldots6$) were -29, -42, -54 and -69 mT, respectively. For the wave vector of $48\,\mathrm{rad\,\mu m^{-1}}$ ($\lambda = 131$ nm), we obtained a large signal strength of 34 % of mode k_1. The measured group velocity of $\sim 0.86\,\mathrm{km\,s^{-1}}$ agrees rather well with the value calculated from the dispersion relation (Supplementary Fig. 4).

Resonant grating couplers and sub-100 nm magnons. In the following, we present results on nanodisk arrays prepared from CoFeB, offering a larger FMR frequency f_0 compared with Py. All other parameters are kept identical, such as the thickness, nanodisk diameter and lattice constant (Supplementary Table 1). Absorption spectra S_{11} of the CoFeB/YIG hybrid sample are colour coded in Fig. 3a. Owing to the large saturation magnetization M_S of CoFeB[21], the field-dependent FMR frequencies for the saturated CoFeB nanodisks were higher by ~ 3 GHz compared with Fig. 2a. This allowed us to induce crossings between the nanodisk FMR and spin waves in YIG at larger k_{SW} compared with Py. The observed spin-wave eigenfrequencies (symbols) and the calculated dispersion relation (line) are shown in Fig. 3b for $\mu_0|H| = 90$ mT. The experimental data follow the predicted dependence $f \propto k^2$ caused by exchange interaction[23]. We extract a large relative signal strength of 38 % at $k_{SW} = k_1 + 9G = 71\,\mathrm{rad\,\mu m^{-1}}$ (Fig. 3c and Supplementary Fig. 5). The corresponding wavelength λ amounts to 88 ± 2 nm. The propagation velocity is about $1.2\,\mathrm{km\,s^{-1}}$ for the k_1+9G mode. Even larger velocities would be achieved by further increasing k_{SW} as v_g grows with increasing k (ref. 4). To further explore the grating coupler effect, we rotated an applied field of 8 mT in the film plane similar to that in ref. 20. In a specific angular regime (Supplementary Fig. 6), we observed weak transmission signals S_{12} around 12 GHz from which we estimated the so far shortest wavelength λ of 68 ± 1 nm, that is, $k = 9.2 \times 10^5\,\mathrm{rad\,cm^{-1}}$. Here, the signal-to-noise ratio was close to 1 and too small to extract exact eigenfrequencies and group velocities. Still, these preliminary data substantiate the perspective of spin waves with λ of a few tens of nanometres excited by further optimized grating couplers.

Reconfigurable microwave-to-magnon transducer. In Fig. 2c,d we compare two spectra S_{12} taken at the same absolute field $\mu_0|H|$, but two different states of the Py nanodisks. Strikingly, in the spectrum measured at $+30$ mT (Fig. 2d), we do not resolve the grating coupler modes. By means of different magnetic histories, we thus control the magnonic grating coupler effect between ON and OFF.

Discussion

We now discuss the microscopic origin of the enhanced signal amplitudes. The CPW's microwave field $\mathbf{h}_{rf}$ coherently excites the FMR in the nanodisks at $f = f_0$, leading to a large dynamic magnetic permeability[23]. Considering ref. 19, the dynamic stray field h_{dip} of the precessing spins locally enhances the microwave magnetic field and enlarges the torque acting on the magnetic moments in the YIG film near each nanodisk. As the nanodisks are thin platelets, the precessional motion is elliptical[23] for which the in-plane dynamic magnetization components are larger compared with the out-of-plane components. This leads to surface charges mainly at their outer rim from where dynamic stray field lines originate. As nanodisks are positioned on the top surface of YIG, out-of-plane stray-field components are expected to be mainly relevant. Au *et al.*[19] presented an analytical formula for h_{dip} of an individual cylindrical transducer. For a fixed sample design, the enhancement of h_{rf} via h_{dip} depended on materials properties according to $(2\alpha(1+2H/M_S))^{-1}$ at f_0; α denoted the Gilbert damping parameter of the given transducer material. In contrast to ref. 19, our nanodisk arrays modulate spatially the enhancement effect. The spatial modulation leads to characteristic reciprocal lattice vectors and, in our case, to a resonantly enhanced grating coupler effect for the spin-wave excitation in YIG. The reciprocal lattice vectors add to k_i given by the bare CPW[9]. We attribute the slightly larger signal amplitude

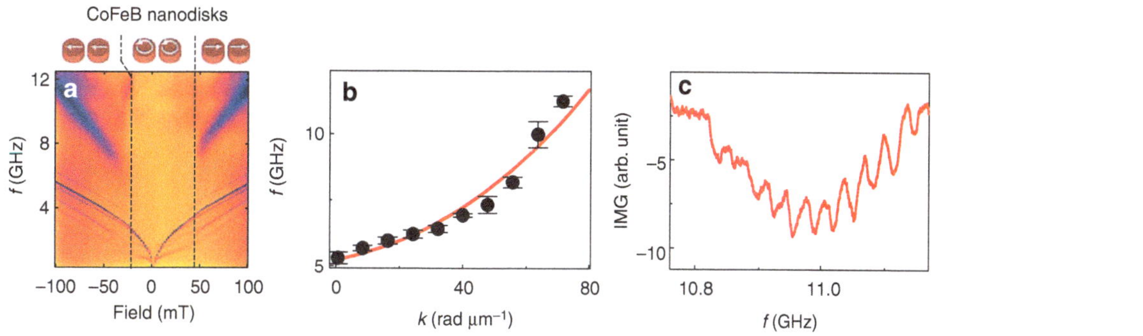

Figure 3 | Spin waves with $\lambda = 88 \pm 2$ nm induced by resonating CoFeB nanodisks. (**a**) Colour-coded absorption spectra S_{11} measured on a CoFeB/YIG hybrid sample. The field is swept from -100 to $+100$ mT. Field regions for saturated and unsaturated states are highlighted by vertical lines. (**b**) Dispersion relation of the YIG thin film calculated for 90 mT (line). The black circles show eigenfrequencies obtained on CoFeB/YIG at the same absolute field. Error bars reflect the uncertainty of the resonance frequency due to noise and are extracted by taking the s.e. from fitting a Lorentz curve to the respective resonant peak in the measured spectra. (**c**) Transmission signal S_{12} that we attribute to $k_{SW} = k_1 + 9G$.

observed for CoFeB compared with Py (Supplementary Fig. 5) to the larger saturation magnetization $\mu_0 M_S = 1.8\,T$ and the smaller damping parameter $\alpha = 0.007$ (ref. 21). Py shows $\mu_0 M_S = 1.0\,T$ and $\alpha = 0.0096$ (ref. 24). To decrease λ to the few 10-nm regime, the nanomagnets must provide an even larger resonance frequency than CoFeB. It is worth noting that a hexaferrite can exhibit an FMR eigenfrequency f_0 of 53 GHz in zero magnetic field[25]. Exploiting such a material as a transducer, a spin-wave wavelength of $\sim$60 nm would be realized in YIG even without applying a magnetic field. The frequency f_0 of hexaferrite can be further increased by $H > 0$. These features would allow one to observe branch crossings beyond $n = 9$. Furthermore, state-of-the-art nanolithography would allow to decrease the lattice constant a by about a factor of 10 compared with our prototype devices. The corresponding reciprocal lattice vectors G can thus be made one order of magnitude larger, suggesting wavelengths $\lambda = 2\pi/(k_1 + nG)$ on the 10-nm length scale for $n = 6$.

We assume the resonantly enhanced spin-wave amplitude to be even larger than given by the value of, for example, 38% stated for the CoFeB/YIG hybrid sample. This value is extracted from the inductive voltage picked up by the detector CPW and normalized to the amplitude at k_1. At k_1, the spin-precessional motion induces the maximum voltage signal in the CPW as the spin-precessional motion can be in phase across the full width of the signal line. This large signal is illustrated by the left-most peak (red line) in Fig. 2e. Spin waves with wave vectors $k > k_1$ induce much smaller voltage signals, as out-of-phase components underneath the signal line compensate each other at all times. For specific (commensurable) wavelengths, the induced signal is zero, such that a given CPW would not detect the corresponding spin wave at all. For the grating coupler mode with $n = 9$, the wavelength is extremely small compared with the CPW's signal line width and even smaller than the diameter of the nanodisks. The spin wave cannot uniformly excite a nanodisk, leading to a small inductive response in the CPW. For large n, we therefore expect the absolute spin-wave signal in YIG to be larger than evaluated from the width-averaged voltage induced in the CPW. To measure the absolute spin-precessional amplitude a dynamic technique with high spatial resolution would be needed, such as synchrotron-based X-ray magnetic circular dichroism. Following ref. 19, resonant microwave-to-magnon transducers should be as close as possible to magnonic devices as spacer layers degrade the excitation efficiency (compare Fig. 1e in ref. 19).

For the present study, we decided for a lattice constant of 800 nm and avoided to bring nanodisks too close together. In case of smaller separation, strongly coupled nanodisks are expected. An artificial band structure might be formed such that forbidden frequency gaps would open. Frequency gaps would counteract efficient spin-wave transmission over broad frequency and wave vector regimes. Still, studies on coupled dots would be very interesting in terms of excitation strength and damping.

Our CPW-based technique is versatile for both fundamental and applied research. A VNA provides a very high frequency resolution (about 1 Hz at 10 GHz), which is better compared with inelastic light scattering, neutron scattering, scanning tunnelling microscopy or electron loss spectroscopy at room temperature. We have focussed on the Damon–Eshbach configuration where k_{SW} is perpendicular to H. Other configurations work in a similar way. The excellent frequency resolution of the VNA combined with the resonant grating coupler effect demonstrated here thus opens a new horizon for exploration and functionalization of exchange-dominated spin waves. For example, one might excite coherent short-wavelength spin waves in different inputs of multi-terminal magnonic devices such as the magnon transistors presented in ref. 26. The relevant coherency can be induced by either a CPW common to the different inputs or a microwave cavity.

Importantly, the group velocity scales with wave vector $k = 2\pi/\lambda$ in the first third of the first Brillouin zone[4]. For YIG, that is, the key material in magnonics offering the smallest propagation losses, signal speeds of a few $km\,s^{-1}$ would be realistic. These velocities are comparable to surface acoustic waves (SAWs) that are currently used in microwave technology. SAW-based microwave electronics such as delay lines, filters and duplexers are building blocks of billions of cell phones. Exchange-dominated spin waves as discussed here offer a wavelength that is two orders of magnitude smaller compared with SAWs. Magnonics thus allows for unprecedented miniaturization of microwave components. This aspect is also interesting for the internet of things consisting of billions of smart devices communicating in a wireless manner[11].

The reconfigurable characteristics of Fig. 2c,d contribute to a topic of great current interest and provide additional functionality[11,27,28]. Reconfiguration may be achieved by an applied magnetic field as used here or by microwave-assisted switching. Microwave-assisted switching has already been demonstrated in mesoscopic magnets with and without vortex states[29,30]. Consequently, tailored microwave pulses would allow one to reconfigure nanodisks under a preselected antenna. This offers further possibilities in grating coupler-based microwave circuits. The constructive superposition of two microwave signals sent to the same CPW could be used to induce the switching of nanodisks. The superimposed signal could erase the grating coupler effect and the relevant microwave signal would not be transmitted to the detector CPW. This operation would mimic an Exclusive OR (XOR) logic gate functionality. Importantly, our experiments show that six periods of nanodisks between the ground lines of a CPW are already sufficient to induce a grating couler effect. This finding is important for the miniaturization and high-density integration of spin-wave transducers with microwave circuitry on a chip. In two dimensions, five Bravais lattices of different lattice symmetries are possible for microwave-to-spin-wave transduction at optimized magnonic properties. It is noteworthy that from small array to small array, different lateral or materials parameters could be used to implement different operational frequencies and wavelengths on the same chip if required. Beyond that, we expect grating couplers with selectively switched nanodisks to allow for the generation of complex wave patterns on a chip.

In ref. 23, it was discussed for a YIG sphere that magnon–magnon scattering processes were found to increase the linewidth $\Delta H = 4\pi\alpha f_0/\gamma$, measured at $k = 0$, with increasing wave vector k (γ is the gyromagnetic ratio). For our largest wave vector of $9.2 \times 10^5\,rad\,cm^{-1}$, four-magnon scattering is expected to contribute most[23]. Figure 11.6 of ref. 23 suggests the linewidth measured at $k = 0$ to increase by $\sim$0.9 Oe at $k = 9.2 \times 10^5\,rad\,cm^{-1}$ and 35 GHz. Assuming a spin wave with $9.2 \times 10^5\,rad\,cm^{-1}$ to be operated at 35 GHz for $\mu_0 H = 50\,mT$, the relevant relaxation time $\tau_0 = 1/(2\pi\alpha 35\,GHz)$ without four-magnon scattering amounts to $\approx$20 ns. From the calculated spin-wave dispersion relation $f(k)$, we extract a group velocity $v_g = 2\pi df/dk = 3.7\,km\,s^{-1}$ at $k = 9.2 \times 10^5\,rad\,cm^{-1}$. The decay length l_d given by $l_d = v_g \tau_0$ is then estimated to be around 73 μm, without additional four-magnon scattering. Considering the linewidth of $\sim$3 Oe at $k = 0$ for our thin-film YIG[31], the additional wave-vector-dependent damping of 0.9 Oe is expected to increase ΔH by a factor of 1.3. Assuming a linear relationship between ΔH and α at a given frequency, we expect the decay length to be reduced by a corresponding amount of $\sim$30%, leading to an estimated value l_d of 50 μm. This value outrivals the typical mean free path of electrons by three orders of magnitude. Hence, the key advantage of magnonics in information technology, that is, the coherent and parallel processing of information within large cellular networks[12], can now be realized.

In conclusion, we explored the spin-wave excitation using magnetic nanoresonators integrated to conventional microwave antenna. We realized reconfigurable microwave-to-spin-wave transducers at GHz frequencies that permitted us exciting and detecting exchange-dominated spin waves. In our non-optimized devices, the minimum wavelength for transmitted spin waves was clearly below 100 nm. Further reduction is straightforwardly possible. Thereby, wavelengths become accessible in solid-state devices that are known from soft X-rays. We are convinced that our results boost the integration of nanomagnonics with microwave circuits and allow to harvest the rich advantages of spin-wave-based applications.

Methods

Sample fabrication. We grew thin YIG films on Gadolinium Gallium Garnet (111) substrates using pulsed laser deposition. The growth temperature was set to 650 °C and oxygen pressure was 0.25 mbar. The YIG thickness was measured using X-ray reflectometry. Root-mean-square roughness was better than 0.5 nm. The Gilbert damping parameter α was evaluated to be 2.3×10^{-4}. We used electron beam lithography and Ar ion milling to form mesas with 45° edges to avoid reflection of the spin waves. The width of the mesa was 300 µm. Using electron beam lithography and lift-off processing, the nanodisks were prepared from either evaporated polycrystalline $Ni_{80}Fe_{20}$ or magnetron-sputtered amorphous CoFeB. The composition of the target was $Co_{20}Fe_{60}B_{20}$. We did not apply a cleaning process for the interface between YIG and the nanodisks. Using again electron beam lithography and lift-off processing, CPWs were integrated on the YIG mesa and nanodisks to perform broadband spin-wave spectroscopy. These CPWs were from Au and Cr, and were designed having 2.1-µm-wide signal lines and ground lines, and 1.4-µm-wide gaps between them. The distance s between two central conductors of adjacent CPWs was 30 µm. Supplementary Fig. 1 displays a microscopy image of one of the CPWs prepared on top of, both, a Py nanodisk array and the ferrimagnetic thin-film insulator YIG. We prepared the nanodisks on the top surface of the YIG film. Intentionally, we did not embed them, to allow for spin-wave propagation through the unpatterned YIG film. The arrays did not exist between neighbouring CPWs. We stress that the specific relation between precessional frequencies of nanodisks (FMR) and film (spin waves) was intentionally opposite to ref. 9. In ref. 9 the grating coupler effect was reported for all-metal-based magnetic devices. Owing to the reversed relation, we were able to force the field-dependent FMR of the nanodisks to be degenerate with short-wavelength spin-wave eigenfrequencies in YIG. Such a resonant grating coupler was not realized in ref. 9.

Propagating spin-wave spectroscopy. To measure spin-wave propagation between two CPWs via scattering parameter S_{12}, the CPWs were connected to a VNA that provided a microwave current at GHz frequencies f to one of the CPWs. Through $d\mathbf{M}/dt = -|\gamma|\mu_0\mathbf{M}\times(\mathbf{H}_{\text{eff}} + \mathbf{h}_{\text{rf}})$, the accompanying magnetic field $\mathbf{h}_{\text{rf}}$ excited the magnetic moments underneath the antenna[32,33]. The microwave power was set at −5 dBm (316 µW) or smaller. As the saturated nanodisks exhibited both different effective fields $\mathbf{H}_{\text{eff}}$ (due to shape anisotropy)[23] and larger magnetization values M compared with YIG, their FMR was at larger frequency f compared with the uniform precessional motion of the thin YIG film, that is, the spin excitation with wave vector $k_{\text{SW}} = 0$. The inhomogeneous magnetic field around a CPW excited the spins at a distinct set of wave vectors $\mathbf{k}_i$ with $i = 1, 2, 3, \ldots$ as was shown in ref. 20. The distribution of wave vectors k provided by a microwave antenna was calculated from Fourier transformation using the lateral dimensions, that is, signal line width, ground line width and their spacing[34]. Wave vectors $\mathbf{k}_i$ generated by the CPW were perpendicular to and in the plane of the CPW. The absolute amplitude S_{12} measured at k_1 depended on H and decreased with increasing H. The wavelengths of spin waves i excited directly be the CPW were calculated according to $\lambda_i = 2\pi/k_i$ with k_i extracted from the Fourier analysis of the microwave field. The wavelengths λ of grating-coupler-induced spin waves were determined via the calculated spin-wave dispersion relations $f(k) = f(2\pi/\lambda)$ (see main text) that allowed us to relate the given VNA frequency f to the relevant wave vector $k = 2\pi/\lambda$. The error bar for λ was taken from the linewidth δf of the envelope function and extracted from the slope of the dispersion relation $f(k)$ around the corresponding wave vector k. We attributed the observation of the FMR branch of the Py nanodisks in the S_{12} data of Fig. 1c (black) of the main text to the remaining electromagnetic cross-talk that existed between the two CPWs being separated by 30 µm. In Supplementary Fig. 2a, the eigenfrequencies of the two branches are summarized that were highlighted by orange and green arrows in Fig. 1c of the main text. Eigenfrequencies were extracted from transmission data displaying oscillating signals as shown in Fig. 1 of the main text in that we defined envelope functions connecting either the local maxima or minima. The eigenfrequency was taken at the frequency where the two envelope functions exhibited the largest difference. In Supplementary Fig. 2b we show the field dependence of the peak-to-peak amplitude in S_{12} when spin waves propagate through the YIG film (orange). Clearly, the amplitude is significantly enhanced at the crossing point of the two branches. Without the crossing, the spin-wave amplitude is always at a small signal level (Supplementary Fig. 2c,d) as was reported in ref. 9. Supplementary Table 1 summarizes wavelengths of spin waves observed in different samples with and without nanomagnet arrays for comparison. The grating coupler effect allowed us to increase the wave vector k_1 by adding multiples of the reciprocal lattice vector $G = 2\pi/a$ (ref. 9), where a amounted to 800 nm.

Group velocity calculation. The oscillation feature of propagating spin-wave signal[33] provided us with the group velocity v_g of the respective spin-wave mode. As studied in refs 21,35, the frequency separation Δf (defined in Fig. 1e of the main text) reflects a change in the phase by 2π, which is accumulated by spin waves propagating along the distance $s = 30$ µm. The group velocity v_g is calculated from Δf according to ref. 21

$$v_g = \frac{\partial\omega}{\partial k} \approx \frac{2\pi\Delta f}{2\pi/s} = \Delta f \cdot s. \tag{1}$$

where s is the distance between two CPW signal lines as indicated in Fig. 1a of the main text.

Absorption spectra. Supplementary Fig. 3a,b shows the reflection spin-wave spectra S_{11} (absorption) measured using one-and-the-same antenna for excitation and detection. In the absorption spectrum at −69 mT, when the crossing occurs, we observe two peaks separated by ~150 MHz (highlighted by arrows) and attribute them to the two grating coupler modes $k_1 \pm 6G$. In the transmission spectra, only one of them ($k_1 + 6G$) is prominent. The wave vector $k_1 + 6G$ was determined by comparison with the calculated spin-wave dispersion relation (see below).

Analysis of dispersion relations and group velocities. In our experiment the field **H** was applied perpendicular to the spin-wave vector k_{SW}. Thus, we addressed the so-called Damon–Eshbach mode configuration. Following refs 20,36, we calculated the relevant dispersion relation $f(k)$ in YIG (solid line in Supplementary Fig. 4a) using a thickness of 20 nm, the magnetization $\mu_0 M_S = 0.176$ T, an exchange constant 0.192×10^{-6} erg cm^{-1} and $\mu_0|H| = 69$ mT. In Supplementary Fig. 4a, we plot the extracted eigenfrequencies (symbols) as a function of the anticipated wave vector k. The experimental data taken on the Py/YIG hybrid sample follow quite well the calculated dispersion relation substantiating the mode allocation. In the literature, exchange constants up to $\sim 0.4 \times 10^{-6}$ erg cm^{-1} were reported for (thin) YIG. Using such a large value we remodelled well the spin-wave eigenfrequencies with the shortest wavelength in the CoFeB/YIG hybrid sample, but not the intermediate wavelengths and not the results obtained on the Py/YIG hybrid sample. Calculated spin-wave group velocities of all the different modes from k_1 up to $k_1 + 6G$ observed on the Py/YIG hybrid sample are shown in Supplementary Fig. 4b. It is noteworthy that v_g increases with k, consistent with the theoretical prediction (line). This behaviour is opposite to the long-wavelength regime reported for bare YIG in ref. 20 and underlines that we generated exchange-dominated spin waves in the thin-film YIG. We extracted 0.86 km s^{-1} for the spin wave with $\lambda = 131$ nm (mode $k_1 + 6G$). This value is larger by a factor of two compared with mode k_1 directly excited by the CPW. For the resonantly enhanced spin wave with $\lambda = 88$ nm in the CoFeB/YIG hybrid sample, the spin-wave velocity was measured to be 1.2 km s^{-1} and hence three times larger compared with the long-wavelength limit at the same applied magnetic field H.

Spin-wave amplitudes by resonant grating couplers. The microwave current in the CPW is in one direction in the signal line and the opposite direction in the two ground lines. This corresponding phase relation in the CPW's magnetic microwave field h_{rf} creates the finite wave vectors k_i ($i = 1, 2, 3,..$) for the spin waves and an excitation efficiency as displayed by the red curve in Fig. 2e of the main text. The red curve represents the outcome of the Fourier transformation of the current distribution described above. The measured signal strengths do follow the expected excitation efficiency for the spin waves with k_1 to k_4 directly excited by the bare CPW. For k being larger than k_4, the grating coupler effect comes into play and rules the excitation efficiency—the measured signal strengths do not follow the outcome of the Fourier transformation any longer.

In Supplementary Fig. 6 we show angular-dependent transmission signals from which we extracted the shortest spin-wave wavelength of 68 nm. At the small field of 8 mT that we rotated, the grating coupler effect was nonresonant. The corresponding spin-wave signal was weak and close to a signal-to-noise ratio of 1:1. A detailed inspection and high-resolution computer screens were used to identify angular-dependent branches. Circles highlight two of the regions where branches were suggested. The signal was too small to extract a group velocity from a contrast oscillation. From such weak angular-dependent branches compared with calculated dispersion relations, we concluded that our non-optimized nanomagnet array, prepared via conventional nanolithography, induced spin waves with a minimum wavelength of $\lambda = 68 \pm 1$ nm.

References

1. The next wave. *Nat. Phys.* **11,** 437 (2015).
2. Nikinov, D. E. & Young, I. A. Overview of beyond-CMOS devices and a uniform methodology for their benchmarking. *Proc. IEEE* **101,** 2498–2533 (2013).
3. Stamps, R. L. *et al.* The 2014 magnetism roadmap. *J. Phys. D Appl. Phys.* **47,** 333001 (2014).
4. Chumak, A. V., Vasyuchka, V. I., Serga, A. A. & Hillebrands, B. Magnon spintronics. *Nat. Phys.* **11,** 453–461 (2015).
5. Lenk, B., Ulrichs, H., Garbs, F. & Münzenberg, M. The building blocks of magnonics. *Phys. Rep.* **507,** 107–136 (2011).
6. High, A. A. *et al.* Visible-frequency hyperbolic metasurface. *Nature* **522,** 192–196 (2015).
7. Gulyaev, Y. *et al.* Ferromagnetic films with magnon bandgap periodic structures: magnon crystals. *JETP Lett.* **77,** 567–570 (2003).
8. Wang, Z. K. *et al.* Observation of frequency band gaps in a one-dimensional nanostructured magnonic crystal. *Appl. Phys. Lett.* **94,** 083112 (2009).
9. Yu, H. *et al.* Omnidirectional spin-wave nanograting coupler. *Nat. Commun.* **4,** 2702 (2013).
10. Demidov, V. E. *et al.* Excitation of short-wavelength spin waves in magnonic waveguides. *Appl. Phys. Lett.* **99,** 082507 (2011).
11. Grundler, D. Reconfigurable magnonics heats up. *Nat. Phys.* **11,** 438–441 (2015).
12. Khitun, A., Bao, M. & Wang, K. L. Magnetic cellular nonlinear network with spin wave bus for image processing. *Superlattices Microstruct.* **47,** 464–483 (2010).
13. Madami, M. *et al.* Direct observation of a propagating spin wave induced by spin-transfer torque. *Nat. Nanotechnol.* **6,** 635–638 (2011).
14. Urazhdin, S. *et al.* Nanomagnonic devices based on the spin-transfer torque. *Nat. Nanotechnol.* **9,** 509–513 (2014).
15. Hutchings, M. T. & Samuelsen, E. J. Measurement of spin-wave dispersion in NiO by inelastic neutron scattering and its relation to magnetic properties. *Phys. Rev. B* **6,** 3447–3461 (1972).
16. Vollmer, R., Etzkorn, M., Kumar, P. S. A., Ibach, H. & Kirschner, J. Spin-polarized electron energy loss spectroscopy of high energy, large wave vector spin waves in ultrathin fcc co films on cu(001). *Phys. Rev. Lett.* **91,** 147201 (2003).
17. Spinelli, A., Bryant, B., Delgado, F., Fernndez-Rossier, J. & Otte, A. F. Imaging of spin waves in atomically designed nanomagnets. *Nat. Mater.* **13,** 782–785 (2014).
18. Sandweg, C. W. *et al.* Spin pumping by parametrically excited exchange magnons. *Phys. Rev. Lett.* **106,** 216601 (2011).
19. Au, Y. *et al.* Resonant microwave-to-spin-wave transducer. *Appl. Phys. Lett.* **100,** 182404 (2012).
20. Yu, H. *et al.* Magnetic thin-film insulator with ultra-low spin wave damping for coherent nanomagnonics. *Sci. Rep.* **4,** 6848 (2014).
21. Yu, H. *et al.* High propagating velocity of spin waves and temperature dependent damping in a CoFeB thin film. *Appl. Phys. Lett.* **100,** 262412 (2012).
22. Cowburn, R. P., Koltsov, D. K., Adeyeye, A. O., Welland, M. E. & Tricker, D. M. Single-domain circular nanomagnets. *Phys. Rev. Lett.* **83,** 1042–1045 (1999).
23. Gurevich, A. & Melkov, G. *Magnetization Oscillations and Waves* (CRC Press, 1996).
24. Counil, G. *et al.* Temperature dependences of the resistivity and the ferromagnetic resonance linewidth in permalloy thin films. *IEEE Trans. Magn.* **42,** 3323–3325 (2006).
25. Appleton, S. G. Field dependence of FMR in ferrites and implications for microwave absorber design. *Proceedings of SPIE Smart Structures and Materials 1995: Smart Electronics* **2448,** 22–34 (1995).
26. Chumak, A. V., Serga, A. A. & Hillebrands, M. Magnon transistor for all-magnon data processing. *Nat. Commun.* **5,** 4700 (2014).
27. Heyderman, L. J. & Stamps, R. L. Artificial ferroic systems: novel functionality from structure, interactions and dynamics. *J. Phys. Condens. Matter* **25,** 363201 (2013).
28. Vogel, M. *et al.* Optically-reconfigurable magnetic materials. *Nat. Phys.* **11,** 487–491 (2015).
29. Thirion, C., Wernsdorfer, W. & Mailly, D. Switching of magnetization by nonlinear resonance studied in single nanoparticles. *Nat. Mater.* **2,** 524–527 (2003).
30. Podbielski, J., Heitmann, D. & Grundler, D. Microwave-assisted switching of microscopic rings: correlation between nonlinear spin dynamics and critical microwave fields. *Phys. Rev. Lett.* **99,** 207202 (2007).
31. d'Allivy Kelly, O. *et al.* Inverse spin Hall effect in nanometer-thick yttrium iron garnet/Pt system. *Appl. Phys. Lett.* **103,** 082408 (2013).
32. Silva, T. J., Lee, C. S., Crawford, T. M. & Rogers, C. T. Inductive measurement of ultrafast magnetization dynamics in thin-film permalloy. *J. Appl. Phys.* **85,** 7849–7862 (1999).
33. Bailleul, M., Olligs, D. & Fermon, C. Propagating spin wave spectroscopy in a permalloy film: a quantitative analysis. *Appl. Phys. Lett.* **83,** 972–974 (2003).
34. Vlaminck, V. & Bailleul, M. Spin-wave transduction at the submicrometer scale: experiment and modeling. *Phys. Rev. B* **81,** 014425 (2010).
35. Neusser, S. *et al.* Magnonic minibands in antidot lattices with large spin-wave propagation velocities. *Phys. Rev. B* **84,** 094454 (2011).
36. Kalinikos, B. A. & Slavin, A. N. Theory of dipole-exchange spin wave spectrum for ferromagnetic films with mixed exchange boundary conditions. *J. Phys. C Solid State Phys.* **19,** 7013–7033 (1986).

Acknowledgements

We thank Anna Fontcuberta i Morral for critical reading and comments on the manuscript. Financial support by the German Excellence Cluster Nanosystems Initiative Munich (NIM), the DFG via project GR1640/5-2 in the priority programme SPP1538, ANR-12-ASTR-0023 Trinidad and NSF China under grant number 11444005 are gratefully acknowledged.

Author contributions

H.Y. and D.G. planned the experiment, designed the samples and analysed the data. O.d'A., V.C., R.B., P.B., A.A., F.B., F.H. and H.Y. fabricated the samples. H.Y. conducted the experiments. F.B. and F.H. helped to conduct the experiment and analyse the data. H.Y. and D.G. wrote the manuscript. All authors commented on the manuscript.

Additional information

Competing financial interests: The authors declare no competing financial interests.

10

High-sensitivity acoustic sensors from nanofibre webs

Chenhong Lang[1,2], Jian Fang[2], Hao Shao[2], Xin Ding[1] & Tong Lin[2]

Considerable interest has been devoted to converting mechanical energy into electricity using polymer nanofibres. In particular, piezoelectric nanofibres produced by electrospinning have shown remarkable mechanical energy-to-electricity conversion ability. However, there is little data for the acoustic-to-electric conversion of electrospun nanofibres. Here we show that electrospun piezoelectric nanofibre webs have a strong acoustic-to-electric conversion ability. Using poly(vinylidene fluoride) as a model polymer and a sensor device that transfers sound directly to the nanofibre layer, we show that the sensor devices can detect low-frequency sound with a sensitivity as high as 266 mV Pa^{-1}. They can precisely distinguish sound waves in low to middle frequency region. These features make them especially suitable for noise detection. Our nanofibre device has more than five times higher sensitivity than a commercial piezoelectric poly(vinylidene fluoride) film device. Electrospun piezoelectric nanofibres may be useful for developing high-performance acoustic sensors.

[1]College of Textiles, Donghua University, Shanghai 201620, China. [2]Institute for Frontier Materials, Deakin University, Geelong, Victoria 3216, Australia. Correspondence and requests for materials should be addressed to T.L. (email: tong.lin@deakin.edu.au).

Detection of sound, an acoustic wave, has been applied in many fields including environment protection[1], industrial manufacture[2], healthcare[3], medical[4], scientific research[5] and defence[6]. Numerous acoustic sensors have been developed based on various transducing principles such as piezo-resistance[7], piezo-capacitance[8], piezo-optics[9] and piezoelectricity[10–12]. Piezoelectric acoustic sensors especially for those made of piezoelectric polymers are distinct from the others in that they have higher sensitivity and flexibility and can be processed into diverse geometries (for example, flat or non-flat film)[10–12]. However, most of the piezoelectric polymers need to be processed by a tedious multistep method involving mechanical stretching and high-voltage poling at an elevated temperature to attain the piezoelectricity before making transducers[13], which are also of energy consumption and high cost.

Recently, the piezoelectricity of nanofibres prepared by electrospinning have been reported[14–18]. Our group and other researchers' studies have indicated that nanofibres electrospun from piezoelectric polymers, such as poly(vinylidene fluoride) (PVDF), poly(vinylidene fluoride-*co*-hexafluoropropene) (PVDF-HFP) and poly(vinylidene fluoride-*co*-trifluoroethylene) (P(VDF-TrFE)), have strong piezoelectricity without stretching and poling treatments[16–24]. These piezoelectric nanofibres have shown great potential for making mechanical sensors and energy generators[16,18,19,22]. The voltage outputs of randomly orientated PVDF nanofibre webs under compressing or bending are even higher than those of commercial piezoelectric PVDF films. However, little has been reported on the acoustic-to-electric energy conversion of polymer nanofibres in research literature to date.

Here we show that piezoelectric nanofibres prepared by electrospinning have an incredible acoustic-to-electric conversion ability. Using PVDF as a model polymer and a device structure that allows sound to directly transfer to the nanofibre layer, we show that the nanofibre sensor devices are able to detect low-frequency sound with a sensitivity as high as 266 mV Pa^{-1}. In comparison with the sensor device made of a commercial piezoelectric PVDF dense film, our nanofibre devices have more than five times higher sensitivity. They can precisely distinguish sound waves in the low to middle frequency region. The devices show higher sensitivity to the sound of the pressure level above 100 dB, which is very suitable for detecting noise.

Results

Preparation of sensing material and device. PVDF nanofibres were prepared by a needle electrospinning technique. Figure 1a shows the morphology of the as-electrospun fibres. All fibres looked uniform without bead. The fibres had a diameter of 310 ± 60 nm (see the histogram of diameter distribution in Supplementary Fig. 1). The X-ray diffraction pattern and Fourier transform infrared spectrum indicated that the PVDF nanofibres mainly contained α and β crystal phases, with the β-phase content as high as 86% (see the detail calculation in Supplementary Fig. 2). These results are in good accordance with our previous findings[16,21], and the high β-phase content is beneficial to mechanical-to-electronic energy conversion[25].

The acoustic sensor was prepared by sandwiching a piece of PVDF nanofibre web with two transparent polyethylene terephthalate (PET) films (thickness 110 μm), which were gold sputter-coated on one side. The gold-coated surfaces were contacted with the nanofibre layer, and they functioned as electrodes to collect electrical signals. Figure 1b schematically illustrates the sensor structure and a photo of the actual device is shown in Fig. 1c. To allow the nanofibres to directly receive sound waves, a through hole was cut on each plastic film. Figure 1d shows the setup for measuring the sensing property. A speaker with a sound-collecting cone was used as sound source, and the narrow end of the cone faced the hole of the device. To avoid false signals generated from the vibration of the testing equipment and mechanical components other than the nanofibre web, the testing

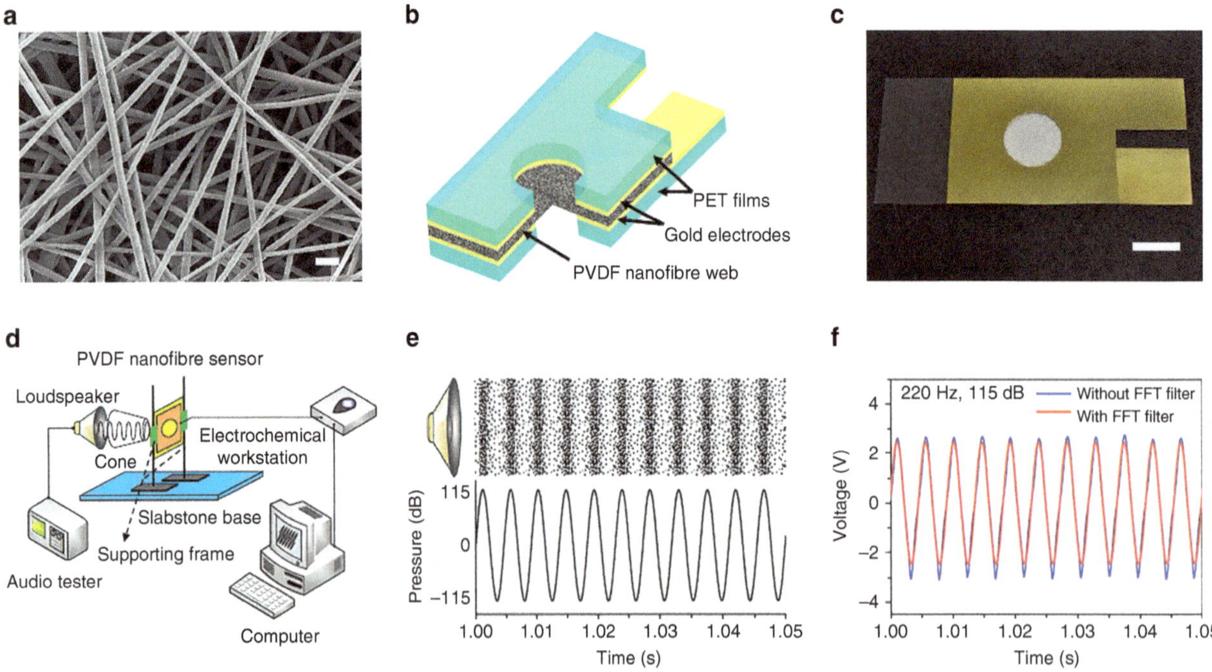

Figure 1 | Material and characterization. (**a**) SEM image of the PVDF nanofibres (scale bar, 1 μm), (**b**) schematic illustration of sensor structure, (**c**) digital photo of the device (scale bar, 1 cm), (**d**) schematic illustration of the setup for testing the sensor device, (**e**) illustration of sound wave (the black dots illustrate the motion of air molecules associated with sound), (**f**) voltage outputs of the device under sound with and without FFT treatment (hole diameter, 12.8 mm; web thickness, 40 μm; web area, 12 cm^2).

equipment and the sensor device were mounted on a heavy slabstone base. In this way, the mechanical vibrations from background, supporting base and frame had negligible contribution to the voltage output of the sensor device (see the test results in Supplementary Figs 3 and 4), even at low acoustic pressure levels (Supplementary Fig. 5).

Acoustic sensor behaviours. When sound waves (Fig. 1e) reached the nanofibre device, electric signals generated, as shown in Fig. 1f. At the sound wave of frequency 220 Hz (intensity level 115 dB), the device generated a periodic voltage output with peak value as high as 3.10 V (also see Supplementary Fig. 6). The overall output is of a typical alternating current. To eliminate the noise signals from the background sound in the testing environment, electronic instrument and connecting cables, we used a fast Fourier transform (FFT) technique to process the output signals[26] (see details in Supplementary Fig. 6). After FFT filtering using the sound frequency, the voltage output was very similar to the raw data (Fig. 1f). However, when the FFT filtering was conducted using the frequency beyond the sound frequencies, the output was very small (Supplementary Fig. 6). This suggests that the voltage response chiefly comes from the sound waves.

By maintaining the sound pressure at the same level while varying the sound frequency, the effect of sound wave frequency on voltage output was examined. As shown in Fig. 2a, the voltage output changes with sound frequency. The maximum voltage output was obtained at 220 Hz. When the frequency was above 220 Hz, a decrease in the voltage value resulted. A little fluctuation in the output was also observed in the frequency range of 400–1,500 Hz. The voltage output decreased to nearly zero when the sound frequency was over 2,000 Hz. It is known that sound is hearable typically in a frequency ranging between 20 and 20,000 Hz (refs 27,28). The noise pollution existing widely in road traffic, airports, industrial and many public places primarily comes from the sounds of low frequency (20–200 Hz). The nanofibre device could be useful for the detection of acoustic waves in those areas.

At three acoustic pressure levels, the device showed a similar voltage output profile, except for different voltage values. This indicates that the nanofibre acoustic sensor is sensitive to low to medium frequency sound waves, and the voltage output is affected by sound pressure level (SPL). Figure 2b shows the effect of SPL on the voltage output of the device. At the same sound frequency, the output increased from 1.2 to 245 mV when the SPL changed from 60 to 95 dB. When the SPL was higher than 100 dB, the voltage output increased nearly in linear manner with increasing the SPL. When the SPL was 115 dB, the output reached 2.29 V. It is known that the sound in the SPL range of 90–110, 110–120, 120–130 and the above is ranked as 'very noisy', 'extremely noisy', 'painful' and 'intolerable' noise, respectively[29]. The large voltage response to sound pressure higher than 100 dB suggests that the nanofibre device could be useful for monitoring noise.

Sensitivity is a key parameter to sensors. The sensitivity (S) for acoustic sensors can be calculated by the equation (1) (refs 10,30),

$$S = \frac{V}{P} = \frac{V}{P_o \cdot 10^{L_p/20}} \quad (1)$$

where P is the sound pressure, V is the voltage output of the device, P_o is the reference sound pressure of 0.00002 Pa and L_p is the SPL in decibel. Based on Fig. 2b, the sensitivity of the nanofibre device was estimated, being as high as 266 mV Pa^{-1}. For comparison, a commercial piezoelectric PVDF film (the same dimension and thickness) was also subjected to the same test. As also shown in Fig. 2b, at the same SPL, the outputs from the dense film device are all smaller than these from the nanofibre one. The maximum sensitivity achieved by the dense film based sensor was 42.5 mV Pa^{-1} (Supplementary Fig. 7). Slightly different to the nanofibre device, the PVDF film device had the highest voltage response to sound at 230 Hz though the response profile

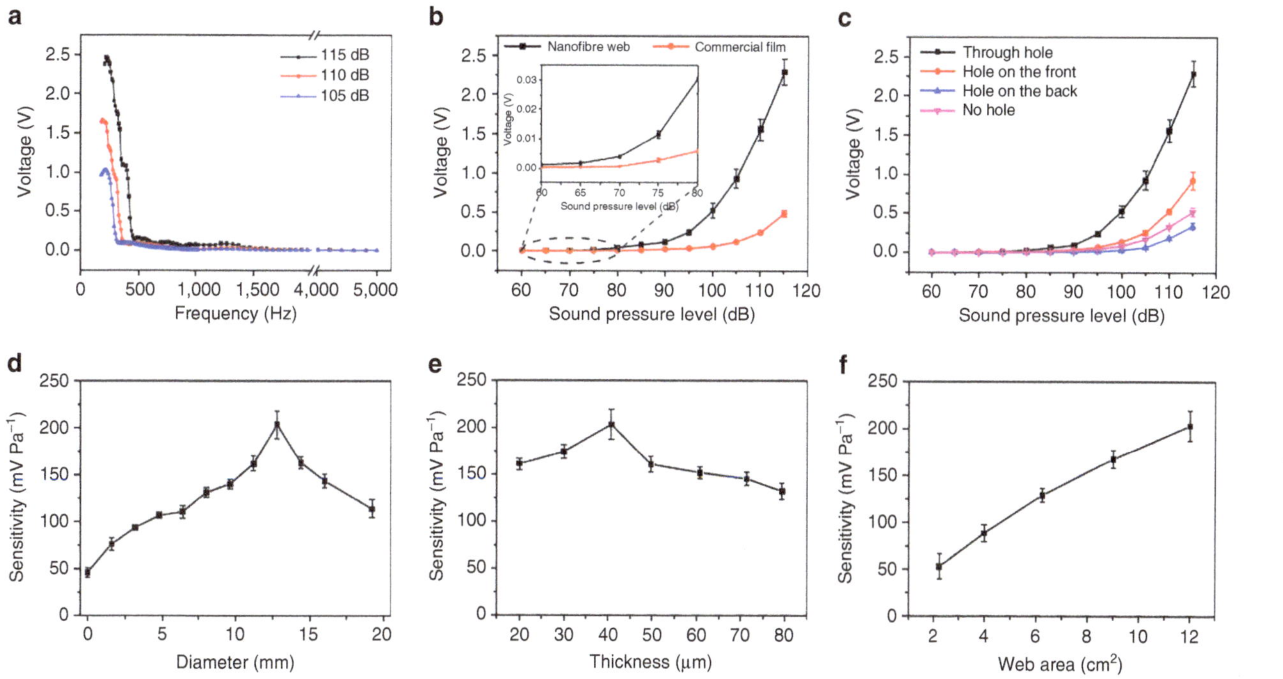

Figure 2 | Voltage outputs and sensitivity of the sensor devices. Effect of (**a**) sound wave frequency, (**b**) SPL (sound wave frequency, 220 Hz), (**c**) hole profile on device voltage outputs. Effect of (**d**) hole diameter (web thickness, 40 μm; sound wave frequency, 220 Hz; SPL, 115 dB), (**e**) thickness of the nanofibre web (hole diameter, 12.8 mm; sound wave frequency, 220 Hz; SPL, 115 dB), (**f**) nanofibre web area (hole diameter, 12.8 mm; web thickness, 40 μm; sound wave frequency, 220 Hz; SPL, 115 dB) on device sensitivity. The error bars represent the s.d. of the test results of at least three replicates.

was similar to our nanofibre device (Supplementary Fig. 8). Apparently, the device voltage output profile was affected by device structure and material property.

Effect of device structure. The hole in the plastic films was found to play an important part in deciding the voltage output value (Fig. 2c and Supplementary Fig. 9). Without hole, the device can generate only 20% of the voltage. When the hole was cut just on one of the plastic films, the device generated higher voltage output than the non-hole device but lower voltage output than the dual-hole one. The single-hole device with the hole faced the sound source generated higher voltage than that on the back. This can be explained by the effect of the hole on sound propagation. Direct exposure of nanofibres to sound waves facilitates to produce higher voltage output.

The hole size affected the voltage output and sensitivity. When the hole diameter was smaller than 6.4 mm, the sensitivity was lower than 110 mV Pa^{-1} (voltage output is 1.25 V). Larger hole increased the sensitivity until the diameter reached 12.8 mm. Further increasing the hole diameter resulted in decrease in the sensitivity (Fig. 2d).

Nanofibre web thickness also affected the voltage output. When the thickness increased from 20 to 40 μm, the sensitivity increased from 161 to 205 mV Pa^{-1} (output voltage from 1.81 to 2.29 V). Further increasing the fibre web thickness led to decrease in both sensitivity and voltage output (Fig. 2e).

The nanofibre web size showed an effect on the output voltage and sensitivity. When the hole size was maintained unchanged, the output voltage and sensitivity showed a linear increase with increasing the nanofibre web size (Fig. 2f), indicating that the nanofibres covered with the plastic films also contribute to converting sound waves into electricity. Therefore the plastic films should have an effect on sensor performance. A systematic understanding on the effect of plastic films on the acoustic-to-electric conversion of PVDF nanofibre web will be reported in our future publication.

The effect of PVDF nanofibres. The above results were based on randomly orientated PVDF nanofibre web with fibre diameter of 310 ± 60 nm. It was expected that the fibre diameter and orientation affected the acoustic-to-electric conversion. To examine the effect from fibre diameter, we prepared randomly orientated PVDF fibre webs with different average fibre diameters (from 220 to 770 nm, see fibre morphology and diameter distribution in Supplementary Fig. 10). The fibres with average diameter of 220 and 270 nm contain beads, while all the others are bead-free fibres. Indeed, fibre diameter influenced sensor behaviour. Although all the fibre webs showed acoustic-to-electric conversion ability, their piezo-potential varied. Under the sound of frequency below 500 Hz, the bead-free fibres with the smallest diameter (310 nm) had higher piezo-potential than the others (Supplementary Fig. 11). With the same fibre web thickness and device structure, the device made of the finest non-bead nanofibres showed the highest sensitivity. The sensitivity decreased with increasing the average fibre diameter. However, the devices made of the beaded fibres had lower sensitivity, though their average fibre diameter was smaller than that of the bead-free fibres.

We also measured the β-crystal phase content of the fibres, and noted that the β-phase content changed with fibre average diameter. The change of β-phase content with fibre average diameter followed a similar trend to the change of device sensitivity with average fibre diameter. This suggests that the piezo-potential is mainly controlled by the β-crystal phase of nanofibres. This result is in accordance with our previous report on the effect of PVDF fibre diameter on mechanical-energy-to-electricity conversion[25].

Apart from randomly orientated fibre web, aligned PVDF nanofibre mat was also prepared. With similar fibre diameter, the aligned nanofibre web showed even higher electric outputs than the randomly oriented ones (see fibre morphology and output result in Supplementary Fig. 12). It was also noted that the aligned PVDF nanofibres had higher β-phase content (90%) than the randomly orientated nanofibres (86%).

Detection of multiple sound waves. The ability to differentiate multiple frequencies of sound is critical for acoustic sensors. To prove this ability, two sound acoustic sources were used to generate bi-frequency sound waves (see the setup in Supplementary Fig. 13). Under two sound waves (190 and 260 Hz), the sensor device showed a voltage output with two amplitude peaks at 190 and 260 Hz after the FFT treatment (Fig. 3a). Similarly, the nanofibre device was also able to distinguish two sound waves with very closer frequencies (220.00 and 220.05 Hz), revealing the excellent detection resolution (Fig. 3b and Supplementary Fig. 14).

To prove the usability of the nanofibre sensors, we used a nanofibre device to record people's voice. During testing, the words 'one, two, three, four, five' were pronounced twice (louder in the second). Figure 3c shows the raw and the FFT-processed output data (also see the Supplementary Movie 1). Different words resulted in different output profiles. For comparison, the voice was also recorded by a laptop microphone and analysed by software, which showed the similar output profile (Fig. 3d). The nanofibre device has potential for sound recording.

Comparison with other nanofibre materials. Apart from PVDF, we also tested nanofibres made of other piezoelectric polymers such as polyacrylonitrile, PVDF-HFP and P(VDF-TrFE) and non-piezoelectric polymer such as poly(vinyl alcohol). With the same device structure, these piezoelectric nanofibre devices had even higher sensitivity than the PVDF nanofibre device (Supplementary Fig. 15). The highest sensitivity was 750 mV Pa^{-1}, for P(VDF-TrFE) at 100 dB. This suggests that the high acoustic-to-electric conversion property of electrospun piezoelectric nanofibres is general, suitable for various piezoelectric polymers. However, the device made of poly (vinyl alcohol) nanofibre web produced only negligible signals (Supplementary Fig. 16).

In addition, dense PET film, polyester film and cellulose paper were used separately to replace the PVDF nanofibre web in the sensor device. As expected, the replacement of piezoelectric nanofibres with those non-piezoelectric materials resulted in disappearance of the piezo-potential, and only negligible signal presumably from the vibration of the testing system was recorded (Supplementary Fig. 16).

Discussion

It is known that polymer nanofibres are flexible and they are easy to vibrate under acoustic waves[31]. However, fibre vibration caused by sound absorption is often localized, and fibre deformations in this case differ to those caused by compression or bending of a fibrous web because the mechanical deformation in the latter case spans the whole fibrous structure. Under sound waves, the plastic films in our sensor device also vibrate and this enhances the nanofibre vibration in the plastic film covered areas.

The vibration velocity of our PVDF nanofibre web and a commercial piezoelectric PVDF film (both 40 μm in thickness) in the central part of the hole was measured at different SPLs.

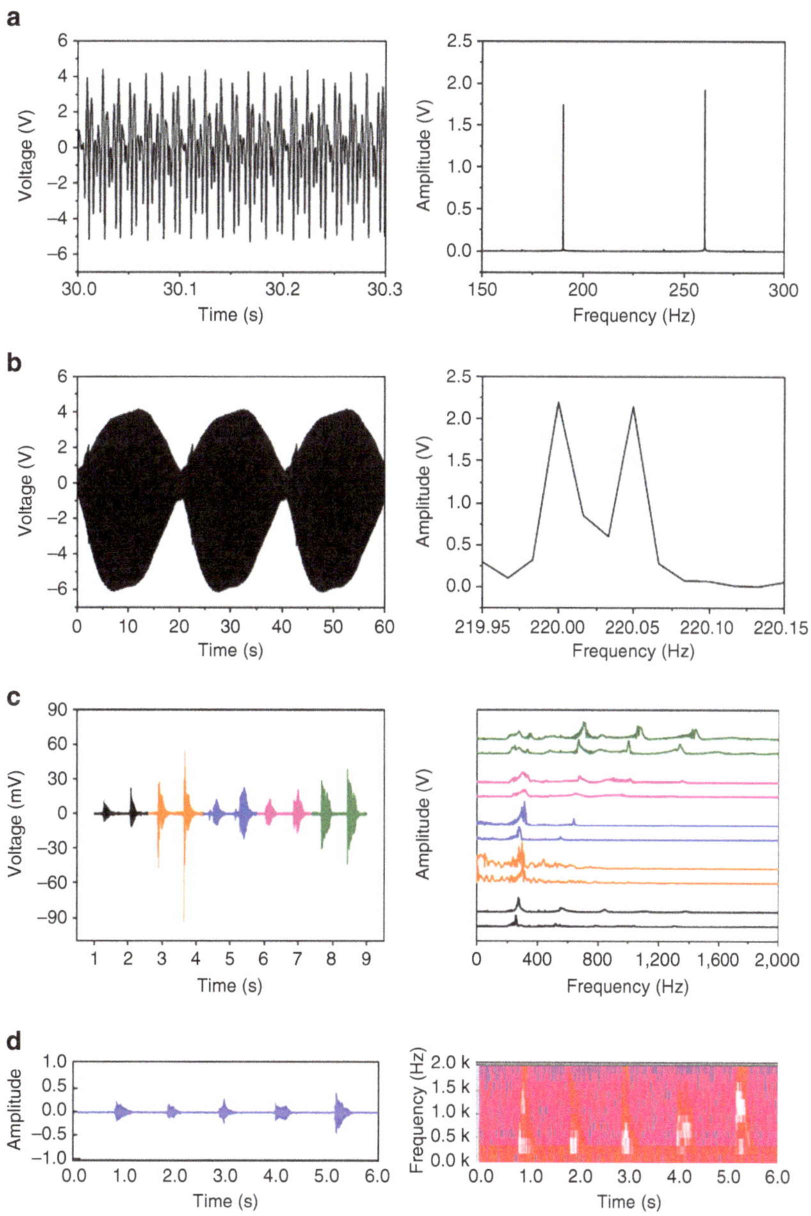

Figure 3 | Sensing resolution and sound detection. Voltage outputs of the nanofibre sensor device under bi-frequency sound waves and the FFT-processed frequency spectrum (**a**) 190 and 260 Hz, (**b**) 220.00 and 220.05 Hz (SPL 115 dB for both loudspeakers); (**c**) voltage outputs of people's voice 'one, two, three, four, five' and the FFT-processed frequency spectrum (SPL, 70–80 dB). The black, orange, blue, pink and green lines represent voltage outputs (left) and FFT profiles (right) of words 'one, two, three, four, five', respectively. (**d**) Sound waveform of the same voice recorded by a commercial microphone and the FFT-processed spectrogram (the colour from blue, pink to red and white in the spectrogram indicates sound intensity increases).

As shown in Fig. 4a, when the SPL increased from 60 to 90 dB, the nanofibre web vibration velocity increased. However, the vibration velocity was stabilized at a constant level when the SPL increased to the range of 90–115 dB, suggesting the saturated vibration state under sound. In comparison, the commercial PVDF film showed a completely different vibration feature under sound. The film showed almost no vibration when the SPL was below 85 dB. However, the vibration increased gradually to the same level of the nanofibre web, when the SPL increased to 115 dB. Figure 4b shows the effect of acoustic wave frequency on the vibration of the nanofibre web and the commercial PVDF film. At 85 dB, the nanofibre web had stronger vibration than the PVDF film when the sound wave frequency was below 800 Hz, and the vibration at higher sound wave frequencies was equally weak.

A finite element method (COMSOL Multiphysics 3.5a) was also used to analyse the vibration of nanofibre web and commercial film in the central part (Supplementary Fig. 17a,b). The modelling results indicated that the vibration velocity of the PVDF nanofibre web in the central part was much higher than that of the commercial PVDF film at the SPLs and sound frequencies studied. The calculated vibration velocity for the commercial PVDF film showed a similar trend to the experiment result (Supplementary Fig. 17c,d). For the PVDF nanofibre web, the modelling result on vibration matched well with the experiment measurement only when SPL was lower than 90 dB. At the SPL above 90 dB, the calculation value of the vibration velocity showed a continuous increase with increasing SPL, which differed from the experiment value. This discrepancy between the modelling result and the experimental data can be attributed to

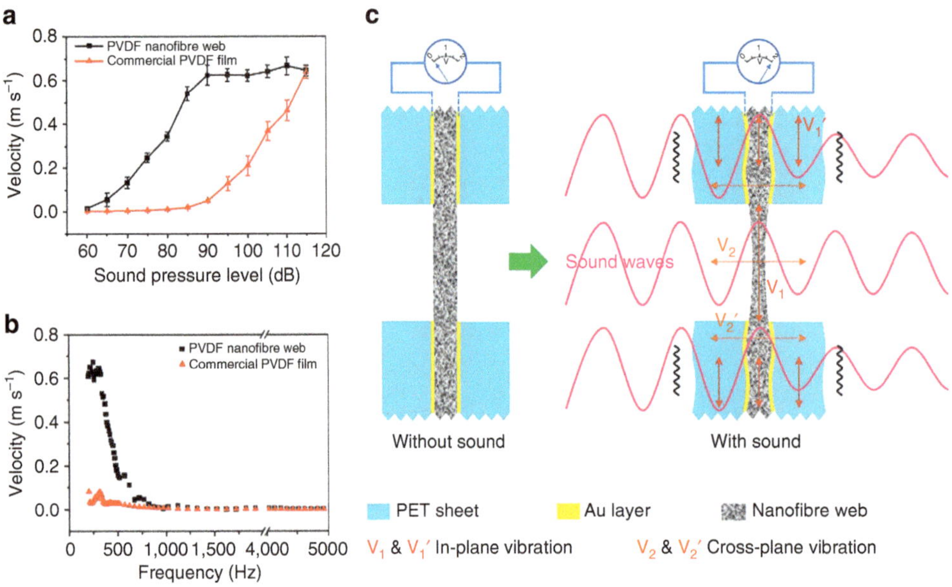

Figure 4 | Vibration velocity and proposed piezoelectric conversion mechanism. Vibration velocity of PVDF nanofibre web and commercial PVDF piezoelectric film in the central part at (**a**) different SPLs and (**b**) different sound frequencies (SPL, 85 dB), (**c**) proposed sound vibration modes and piezoelectric conversion mechanism for the nanofibre device (the error bars represent the s.d. obtained from the test results of at least three replicates).

the simplified modelling conditions and the complexity of nanofibre vibration.

Therefore, the acoustic-to-electric energy conversion of our nanofibre devices should come from the co-action of the piezoelectric electrospun nanofibres and the plastic electrodes. Figure 4c schematically illustrates the proposed sound sensing mechanism. When acoustic waves hit the piezoelectric nanofibres, vibration takes place on the fibres and plastic films because of the sound absorption. The through hole in the electrodes only allows a part of nanofibres to directly expose to sound waves. Under sound, the nanofibre web could vibrate in the in-plane (V_1) and cross-plane (V_2) directions. In the meanwhile, the nanofibres that are covered by the gold-coated plastic sheet also vibrate with the plastic electrodes both in-plane (V_1') and cross-plane (V_2') as well. The directly exposed nanofibres vibrate more intensively than those covered with the electrodes. Such asymmetric vibrations on the fibres and vibration propagation along the fibres enhance nanofibre deformation.

The piezoelectricity of a nanofibre web that is prepared by electrospinning of a piezoelectric polymer has been reported by our group and other researchers[16,17,21]. The high β crystal phase content and oriented macromolecular chain dipoles, in the direction across the web thickness as proposed[25], facilitate piezoelectric conversion under compressive impact. The cross-plane vibrations (V_2 and V_2') could lead to generation of piezoelectric power under acoustic waves. The in-plane vibrations assist vibration propagation throughout the entire nanofibre web.

In conclusion, we have proved the acoustic-to-electric energy conversion property of piezoelectric nanofibres and their interesting application as a transducing medium for making acoustic sensors. The sensor devices show much higher sensitivity than commercial piezoelectric PVDF film. They also have an incredible ability to distinguish sound waves in the low to middle frequency region. Since the nanofibre device shows higher sensitivity to the SPL above 100 dB, it is especially useful for detection of noise. Electrospun piezoelectric nanofibres may be useful for developing novel acoustic sensors for various applications.

Methods

Electrospinning. PVDF solution (20%) was prepared by dissolving 2 g PVDF pellets (M_W 275,000 g mol^{-1}, Sigma-Aldrich) in 10 ml solvent mixture of dimethylformamide (Sigma-Aldrich) and acetone (Sigma-Aldrich) (volume ratio of dimethylformamide and acetone, 4/6) and heated at 70 °C for 2 h. To examine the effect of fibre diameter on energy conversion, PVDF nanofibres with different diameters were prepared from PVDF solutions with the concentration ranging from 16 to 28%. The homogeneous PVDF solution was placed to a plastic syringe with a metal needle (21 gauge). A syringe pump (KD Scientific) was used to control the flow rate at 1 ml h^{-1} and 15 kV high voltage was applied to the syringe needle using a high-voltage power supply (Gamma High Voltage). A grounded rotating drum collector (100 r.p.m.) placed 15 cm away from the needle tip was used for collecting the PVDF nanofibres.

Sensor fabrication. A piece of electrospun PVDF nanofibre web was clamped by two gold-coated plastic films. A hole (diameter = 12.8 mm) was cut in the middle of each plastic film.

Characterizations. The morphology of nanofibres was observed under a scanning electron microscope (Zeiss Supra 55VP), and fibre diameter was measured based on the scanning electron microscope images using image processing software (ImageJ 1.45s). XRD was recorded on a diffractometer (Panalytical XRD) with Cu-Ka radiation operated at 40 kV and 30 mA. The sample was scanned in the 2θ range of 10° to 60°. FTIR spectra were measured on a FTIR spectrophotometer (Bruker Optics) in ATR mode. Vibration velocity was measured using a CLV-2534 compact laser vibrometer.

Test of sensor behaviour. The acoustic sensors were tested using a commercial PC speaker (Logitech, LS21) as sound source. The sound frequency was controlled by a computer and an Audio Tester software. A sound level metre (DIGITECH, QM-1589) was used to record the acoustic wave pressure. An electrochemistry working station (e-Corder 401) was used to record electric signals with a sampling frequency of 10,000 per second. To avoid false signals generated from the vibration of the testing equipment and mechanical components other than the nanofibre web, the testing equipment and the sensor device were mounted on a heavy slabstone base.

Finite element method (FEM) modelling. The vibration mode was calculated by a finite element method, using the Pressure acoustic and Solid mechanics interaction modulus in COMSOL Multiphysics 3.5. The FEM analysis used an integrated device structure model (Supplementary Fig. 17a) consisting of two PET films and a layer of PVDF as active layer. The geometries were set the same as the actual values. Due to the complexity of nanofibrous structure, the PVDF nanofibre web was simplified into a low-density film with the density of 450 kg m^{-3}. The commercial PVDF film and PET film have a density of 1,780 kg m^{-3} and

1,380 kg m^{-3}, respectively. The Young's modulus of the simplified nanofibre web, dense PVDF film and PET films were set at 1,450 and 500 MPa, and Poisson's ratio 0.42, 0.35 and 0.35, respectively.

References

1. Blumstein, D. T. *et al.* Acoustic monitoring in terrestrial environments using microphone arrays, applications, technological considerations and prospectus. *J. Appl. Ecol.* **48,** 758–767 (2011).
2. Liang, S. Y., Hecker, R. L. & Landers, R. G. Machining process monitoring and control, the state-of-the-art. *J. Manuf. Sci. Eng. Trans. ASME* **126,** 297–310 (2004).
3. Stansfeld, S., Haines, M. & Brown, B. Noise and health in the urban environment. *Rev. Environ. Health* **15,** 43–82 (2000).
4. Barnett, G. H., Kormos, D. W., Steiner, C. P. & Weisenberger, J. Intraoperative localization using an armless, frameless stereotactic wand. Technical note. *J. Neurosurg.* **78,** 510–514 (1993).
5. Suslick, K. S. Sonochemistry. *Science* **247,** 1439–1445 (1990).
6. Ghiurcau, M. V., Rusu, C., Bilcu, R. C. & Astola, J. Audio based solutions for detecting intruders in wild areas. *Signal Process* **92,** 829–840 (2012).
7. Scheeper, P. R., van der Donk, A. G. H., Olthuis, W. & Bergveld, P. A review of silicon microphones. *Sensor. Actuat. A-Phys.* **44,** 1–11 (1994).
8. Torkkeli, A. *et al.* Capacitive microphone with low-stress polysilicon membrane and high-stress polysilicon backplate. *Sensor. Actuat. A-Phys.* **85,** 116–123 (2000).
9. Consales, M. *et al.* Lab-on-fiber technology, toward multifunctional optical nanoprobes. *ACS Nano* **6,** 3163–3170 (2012).
10. Toda, M. & Thompson, M. L. Contact-type vibration sensors using curved clamped PVDF film. *IEEE Sens. J.* **6,** 1170–1177 (2006).
11. Xu, J., Dapino, M. J., Gallego-Perez, D. & Hansford, D. Microphone based on polyvinylidene fluoride (PVDF) micro-pillars and patterned electrodes. *Sensor. Actuat. A-Phys.* **153,** 24–32 (2009).
12. Kim, J. *Improvement of Piezoelectricity in Piezoelectric Paper Made with Cellulose*. Report No. AOARD-084035 (Inha University, 2009).
13. Lovinger, A. J. Ferroelectric polymers. *Science* **220,** 1115–1121 (1983).
14. Chang, C., Tran, V. H., Wang, J., Fuh, Y. K. & Lin, L. Direct-write piezoelectric polymeric nanogenerator with high energy conversion efficiency. *Nano Lett.* **10,** 726–731 (2010).
15. Hansen, B. J., Liu, Y., Yang, R. & Wang, Z. L. Hybrid nanogenerator for concurrently harvesting biomechanical and biochemical energy. *ACS Nano* **4,** 3647–3652 (2010).
16. Fang, J., Wang, X. & Lin, T. Electrical power generator from randomly oriented electrospun poly(vinylidene fluoride) nanofibre membranes. *J. Mater. Chem.* **21,** 11088 (2011).
17. Mandal, D., Yoon, S. & Kim, K. J. Origin of piezoelectricity in an electrospun poly(vinylidene fluoride-trifluoroethylene) nanofibre web-based nanogenerator and nano-pressure sensor. *Macromol. Rapid Commun.* **32,** 831 (2011).
18. Persano, L. *et al.* High performance piezoelectric devices based on aligned arrays of nanofibres of poly(vinylidenefluoride-co-trifluoroethylene). *Nat. Commun.* **4,** 1633 (2013).
19. Wang, Y. R., Zheng, J. M., Ren, G. Y., Zhang, P. H. & Xu, C. A flexible piezoelectric force sensor based on PVDF fabrics. *Smart. Mater. Struct.* **20,** 045009 (2011).
20. Ren, G., Cai, F., Li, B., Zheng, J. & Xu, C. Flexible pressure sensor based on a Poly(VDF-TrFE) nanofibre web. *Macromol. Mater. Eng.* **298,** 541–546 (2013).
21. Fang, J., Niu, H., Wang, H., Wang, X. & Lin, T. Enhanced mechanical energy harvesting using needleless electrospun poly(vinylidene fluoride) nanofibre webs. *Energ. Environ. Sci.* **6,** 2196–2202 (2013).
22. Li, B., Xu, C., Zheng, J. & Xu, C. Sensitivity of pressure sensors enhanced by doping silver nanowires. *Sensors* **14,** 9889–9899 (2014).
23. Mandal, D., Henkel, K. & Schmeisser, D. Improved performance of a polymer nanogenerator based on silver nanoparticles doped electrospun P(VDF-HFP) nanofibres. *Phys. Chem. Chem. Phys.* **16,** 10403–10407 (2014).
24. Adhikary, P., Garain, S. & Mandal, D. The co-operative performance of a hydrated salt assisted sponge like P(VDF-HFP) piezoelectric generator, an effective piezoelectric based energy harvester. *Phys. Chem. Chem. Phys.* **17,** 7275–7281 (2015).
25. Shao, H., Fang, J., Wang, H. & Lin, T. Effect of electrospinning parameters and polymer concentrations on mechanical-to-electrical energy conversion of randomly-oriented electrospun poly(vinylidene fluoride) nanofibre mats. *RSC Adv.* **5,** 14345–14350 (2015).
26. Duhamel, P. & Vetterli, M. Fast fourier transforms, a tutorial review and a state of the art. *Signal Process* **19,** 259–299 (1990).
27. Volandri, G., Di Puccio, F., Forte, P. & Carmignani, C. Biomechanics of the tympanic membrane. *J. Biomech.* **44,** 1219–1236 (2011).
28. Möser, M. & Muller, G. *Handbook of Engineering Acoustics* (Springer, 2013).
29. Anthrop, D. F. The noise crisis. *Univ. Toronto Law J.* **20,** 1–17 (1970).
30. Davis, D. & Patronis, E. *Sound System Engineering* 3rd edn (Focal Press, 2006).
31. Na, Y., Agnhage, T. & Cho, G. Sound absorption of multiple layers of nanofibres webs and the comparison of measuring methods for sound absorption coefficients. *Fiber. Polym.* **13,** 1348–1352 (2012).

Acknowledgements

Funding support from an ARC Future Fellow grant (ARC FT120100135) is acknowledged. C.H.L. acknowledges the scholarship from the Donghua University International Visiting Program for excellent PhD students and funding from the Nanofibre Group at Institute for Frontier Materials to support her research at Deakin University.

Author contributions

C.H.L. and J.F. performed the experiments and data analyses. H.S. conducted a part of measurement. X.D. supported the experiment. C.H.L., J.F. and T.L. wrote of the manuscript. T.L. conceived the overall project and finalized the manuscript.

Additional information

Competing financial interests: The authors declare no competing financial interests.

11

Tunable quasiparticle trapping in Meissner and vortex states of mesoscopic superconductors

M. Taupin[1], I.M. Khaymovich[1,2,†], M. Meschke[1], A.S. Mel'nikov[2,3] & J.P. Pekola[1]

Nowadays, superconductors serve in numerous applications, from high-field magnets to ultrasensitive detectors of radiation. Mesoscopic superconducting devices, referring to those with nanoscale dimensions, are in a special position as they are easily driven out of equilibrium under typical operating conditions. The out-of-equilibrium superconductors are characterized by non-equilibrium quasiparticles. These extra excitations can compromise the performance of mesoscopic devices by introducing, for example, leakage currents or decreased coherence time in quantum devices. By applying an external magnetic field, one can conveniently suppress or redistribute the population of excess quasiparticles. In this article, we present an experimental demonstration and a theoretical analysis of such effective control of quasiparticles, resulting in electron cooling both in the Meissner and vortex states of a mesoscopic superconductor. We introduce a theoretical model of quasiparticle dynamics, which is in quantitative agreement with the experimental data.

[1] Low Temperature Laboratory, Department of Applied Physics, Aalto University School of Science, P.O. Box 13500, FI-00076 Aalto, Finland. [2] Institute for Physics of Microstructures, Russian Academy of Sciences, GSP-105, 603950 Nizhni Novgorod, Russia. [3] Lobachevsky State University of Nizhni Novgorod, 23 Prospekt Gagarina, 603950 Nizhni Novgorod, Russia. † Present address: LPMMC, CNRS/Foundation Nanosciences under the aegis of Joseph Fourier University Foundation, BP 166, 38042 Grenoble, France. Correspondence and requests for materials should be addressed to M.T. (email: mathieu.taupin@aalto.fi).

The presence of excess quasiparticles (QPs) is often characterized by an effective electron temperature T that exceeds the temperature of the phonon bath T_0. The resulting overheating is known to be the origin of such effects as decoherence in qubit systems[1–3], decrease of the quality factor of superconducting resonators[4,5], the excess current in single-electron turnstiles[6], and low efficiency of electronic cooling in normal metal (N)–insulator (I)–superconductor (S) junctions[7,8]. In short, overheating is a major factor limiting the performance of S mesoscopic devices. More than the overall quasiparticle number N_{qp}, the critical parameter is the location of these excess quasiparticles. For instance, for tunnel junction circuits, it is crucial to avoid the quasiparticles in a superconductor nearby the junction, while the extra quasiparticles located further away are of less concern. To suppress overheating in a superconductor one aims at lowering the generation of extra quasiparticles in the whole superconductor using proper electromagnetic shielding of the device, and decreasing the quasiparticle density by introducing quasiparticle traps (see for example, refs 9,10), by optimizing the device geometry[6,11], or by cooling using the tunnel junction to another superconductor with a larger gap[12–14]. The second method allows one to move quasiparticles away from critical locations and relax them. Quasiparticle traps have become an important element in designing devices for mesoscopic physics and metrology.

The most common ones among different types of quasiparticle traps are normal metal sinks[15–17], Andreev bound states in weak links[18], special S gap engineering[19–23] and non-uniform superconducting states induced by an external magnetic field[24–27]. Here we focus on the magnetic field controlled trapping, a method which has a number of advantages. The regions with the reduced gap in this case are of the same material as the rest of the device and therefore match perfectly the S parts without barriers or interface potentials. Besides, magnetic field gives the possibility to make tunable traps allowing, for example, the modulation of a resonator quality factor, needed for giant pulse formation (or Q-switching) in pulse lasing (see, for example, a book[28]). The controllable use of such traps in various applications mentioned above assumes, certainly, understanding of their cooling capacities, which is necessary to optimize the designing of the particular trap configurations for different mesoscopic devices.

Our work aims to the solution of this ambitious and important problem focusing on both experimental and theoretical study of individual traps, which appear in the Meissner and vortex states. To build a quantitative model of these traps we choose to verify it by the experimental measurements of the characteristics of non-equilibrium quasiparticle distributions in a mesoscopic S island (Al) in a single-electron transistor set-up with normal metal (Cu) leads. This particular device appears to provide a very convenient way to tune both the trap pattern applying an external magnetic field to the S island and the number of non-equilibrium quasiparticles injected in the island in the Coulomb blockade conditions by operating it as a turnstile of single electrons[29]. The turnstile operation frequency f of the gate voltage modulation controls quasiparticle injection rate. This set-up allows one to probe single quasiparticle excitations in the superconducting dot by measuring the average turnstile current under pumping conditions[6] (ideally this current equals ef) and to independently control the vortex number in the superconductor[30]. The resulting trap model has perfectly proved its validity and efficiency in this set-up which can be used in future applications.

Results

Qualitative description. We illustrate the key idea of quasiparticle redistribution by Fig. 1 in an S island with a large central part and two narrow extensions, called Sample A. In the absence of magnetic field acting on the sample, $B=0$, the quasiparticle density n_{qp} is nearly uniform in the S island with constant gap $E_g(x)=\Delta_0$, provided the heat diffusion length $L_T \gg R$ is large compared to the size of the central part R (see Fig. 1a). A small perpendicular magnetic field, typically few millitesla (mT), which induces Meissner screening currents flowing along the superconductor edges reduces the gap $E_g(x)$ mostly in the wide central part of the island but not in the narrow extensions near the junctions[31]. Due to this non-uniform gap potential $E_g(x)$, quasiparticles illustrated by red circles are redistributed so that their density is small at the junctions (see Fig. 1b). However, the total quasiparticle number is larger than that at $B=0$ due to its exponential dependence $N_{qp} \propto e^{-E_{g,min}/k_B T}$ on the minimal gap $E_{g,min}=\min_x E_g(x)$ over the island, where k_B is the Boltzmann constant. A vortex in the island leads to further quasiparticle redistribution because it plays a role of a quasiparticle potential well containing a lot of quasiparticles as shown in Fig. 1c. Despite its simplicity, the theoretical model that we present below yields a quantitative fit to the experimental data on the magnetic field and frequency dependencies of the pumping current, and thus to the quasiparticle distribution, rendering the turnstile an efficient probe of quasiparticle dynamics and relaxation.

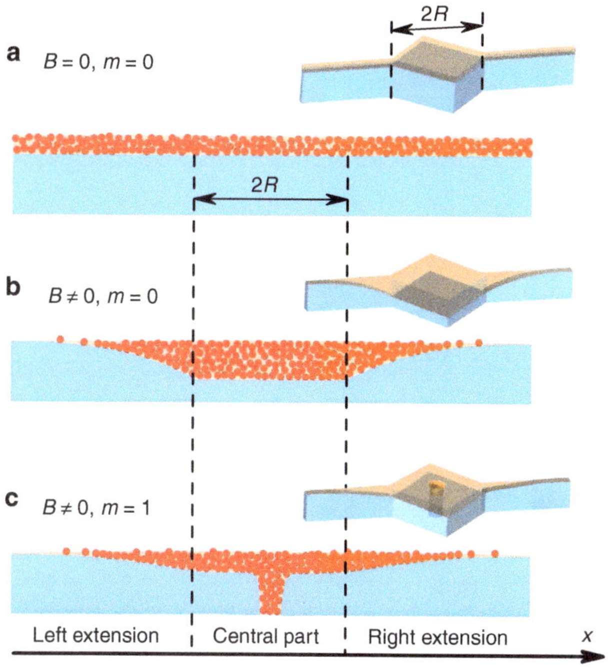

Figure 1 | QP density and gap distributions in a S disc. The S gap $E_g(B, x)$ in a S disc with narrow extensions (Sample A) is represented by the height of the blue volume, while the QP density $n_{qp}(x)$ is shown by red circles; m is the vorticity of the island, B is the magnetic field acting on the sample. The wide central part of a Sample A of size $2R$ is limited by vertical dashed lines, the narrow extensions are located on the sides. (**a**) Uniform zero magnetic field state; (**b**) Meissner state with reduced $E_g(B, x)$ in the central part at small fields in a vortex free state; (**c**) Single-vortex state with smaller gap reduction outside the vortex core than in **b**. The 3D schematics depict the corresponding $E_g(B, x)$ (in blue) and $n_{qp}(x)$ (in orange semitransparent).

DC measurements of the S gap. To probe the magnetic field induced changes in the gap of a S disc, we first measure a more basic structure, which we call Sample B (see Fig. 2b). It is formed of a S disc, mimicking the central part of Sample A (Fig. 2a), directly connected via tunnel junctions to normal leads at its edges. Measuring electron transport through the disc while

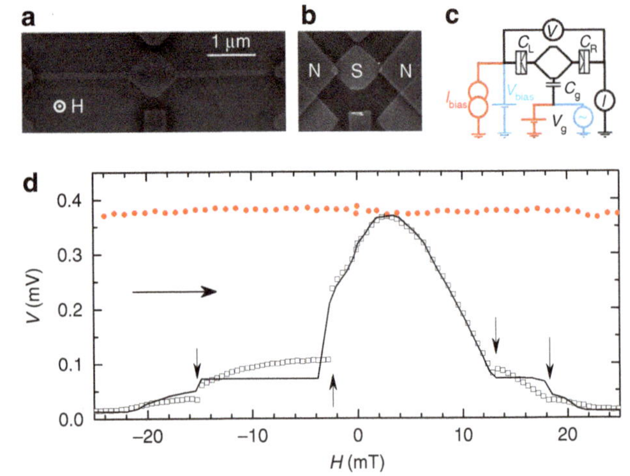

Figure 2 | Layouts of the samples and results of d.c. measurements. Electron micrographs of (**a**) Sample A and (**b**) Sample B. (**c**) Schematic picture of the device with its electrical connections where the red and black lines correspond to the d.c. measurements, while blue and black ones to the pump measurements. (**d**) Evolution of the voltage at a fixed bias current $I_{bias} = 10$ pA at the gate voltage $C_gV_g/e = n_g = 0.5$ suppressing the Coulomb energy with the magnetic field for Sample A (red filled circles) and Sample B (black open squares). The vertical arrows correspond to the applied field values at which the vorticity m increases step by step by one from -2 to 2 as the field is swept from -25 to 25 mT. The horizontal arrow shows the direction of the field sweep. The solid black line is the theoretical result for this measurement. The experimental uncertainty has been estimated as $\sim 3\,\mu$V from the noise of the amplifiers.

applying perpendicular magnetic field H allows us to access the field dependence of the gap value $E_g(H)$ at the edge of the disc and to control the vortex state. This way we can determine the critical fields for transitions between states with different vorticities m via simple d.c. transport measurements (similar approach as in ref. 30). We carried out current biased d.c. measurements at a gate voltage that suppresses the Coulomb energy (for the electrical configuration, see red and black lines in Fig. 2c). The experiments have been performed at a bath temperature of $T_0 \sim 60$ mK (well below the S gap Δ_0 at $B = 0$ and the Coulomb energy $E_C = e^2/(2C)$, where C is the total capacitance of the island). Note that B is the actual field seen by the sample, while the applied magnetic field H differs from later due to some screening by the sample holder used for shielding the sample from the environment (see Supplementary Note 1).

The d.c. drain–source voltage V measured versus the magnetic field H, swept from -25 to 25 mT, is shown for Sample A (filled circles) and Sample B (open squares) in Fig. 2d at a fixed current of $I_{bias} = 10$ pA through the device. In general larger voltage corresponds to larger gap $E_g(H)$ and vice versa. The sample parameters E_C, Δ_0, and a total normal state resistance across the two junctions R_T, have been extracted from IV measurements at zero magnetic field. For the Sample B, starting from -25 mT, the value of the voltage is small: the island is close to its normal state. The gap increases when decreasing the absolute value of the field till the maximum value reached at $+2.5$ mT with two intermediate step-like anomalies at $H^{(2)}_{out} \sim -15$ mT and $H^{(1)}_{out} \sim -2$ mT, corresponding to the exit of vortices, the first one from two-vortex state to one-vortex state, and the second one from one-vortex state to a vortex free state, respectively. Increasing H further to positive values from 2.5 to 25 mT leads to decrease of the gap again, with two knee-like anomalies at critical-field values $H^{(1)}_{in} \sim 14$ mT and $H^{(2)}_{in} \sim 18$ mT corresponding to the entry of the first and the second vortex, respectively. A minor distortion of the applied field (the offset in the applied field $\delta H \approx 2.5$ mT corresponding to the maximal $V(H)$ value and asymmetry of $V(H)$ in the Meissner state) is caused by the sample holder, and was corrected to theoretical curves only by applying the magnetization curve $B(H)$, with B the field acting on the sample, measured separately (Supplementary Note 1). Note that the magnetic field B acting on the sample itself, is zero at the maximal $V(B = 0)$ and corresponds to the symmetric $V(B) = V(-B)$ in the Meissner state. The central part of Sample A has nearly the same shape and size as Sample B; thus one can expect the critical fields of these samples to be close to each other. The anomalies are absent in Sample A, as the gap near the tunnel junctions is only weakly affected by H in the presented range.

Theoretical analysis of DC data. For the theoretical analysis of the above experimental data we simplify the standard Usadel model taking into account that the size of the central part R of the measured samples is small compared with the characteristic length scale of the Green's functions outside the vortex core regions (Supplementary Note 2 for details). Such approximation leads to the Usadel equation for the normal ($\cos\theta$) and anomalous ($-i\sin\theta$) Green functions

$$(iE - \Gamma\cos\theta)\sin\theta + \Delta\cos\theta = 0, \qquad (1)$$

with the effective depairing parameter $\Gamma = \frac{\hbar}{2D}\langle \mathbf{v}_S^2 \rangle$ expressed through the superfluid velocity $\mathbf{v}_S = D(\nabla\varphi - 2e\mathbf{A}/\hbar c)$ and averaged $\langle..\rangle$ over the sample volume (over the central part of Sample A) with the excluded vortex core regions. Here φ is the S order parameter phase, $\mathbf{A}$ is the vector potential determined by the magnetic field B acting on the sample, and D is the diffusion coefficient. The component of $\mathbf{v}_S$ perpendicular to the sample boundary and to the boundaries of vortex cores should be zero. Similarly to the previous works[19,20,24–26], the vortex cores are assumed to be normal metal cylinders of the radius r_v of the order coherence length $\xi = \sqrt{\hbar D/\Delta_0}$, that is, $\theta = 0$ inside the cores. The sample size $2R \sim 1\,\mu$m is also smaller than the effective screening length $\lambda_{eff} \sim \lambda^2/d_S \sim 2.6\,\mu$m, therefore we expect uniform field distribution in the island. Here $\lambda \sim 230$ nm (ref. 24) is a typical bulk penetration depth and $d_S \simeq 20$ nm is the thickness of the aluminium disc.

Solution of the Usadel equation gives us the standard expression for the hard gap E_g in the density of states and for the order parameter Δ as functions of Γ (see refs 32–35 or Supplementary Note 2). We made a fit of the field dependence of the voltage $V(B)$ at fixed currents I_{bias} using standard expressions for the current-voltage characteristic of a tunnel junction (see Supplementary Note 3 for details) and of the depairing parameter

$$\Gamma/\Delta_0 = \alpha_1(B/B_C)^2 - m\alpha_2 B/B_C + m^2\alpha_3, \qquad (2)$$

taking into account that the vector potential $\mathbf{A}$ in the superfluid velocity $\mathbf{v}_S = D(\nabla\varphi - 2e\mathbf{A}/\hbar c)$ is proportional to magnetic field B while the S phase distribution φ is determined by vortex sources. Here α_i are numerical fitting parameters, B_C denotes the field value of the first vortex entry and m is the total vorticity. The estimate $B_C \sim \Phi_0/\pi R\xi \sim 10$ mT based on $\xi \approx 100$ nm and $R \approx 0.5$ µm is rather close to the value $B(H^{(1)}_{in}) \simeq 14.4$ mT from our d.c. measurements. Here $\Phi_0 = h/2e$ is the flux quantum. More accurate estimates of B_C can be done numerically, for example, within the Ginzburg–Landau approach for a concrete sample geometry[36,37]. According to ref. 34, the parameter α_1 determines the critical value of Γ/Δ_0 for the first vortex to enter for the Usadel equation with homogeneous $\mathbf{v}_S$ in a narrow strip should be $\alpha_1^{homog} = 0.237$, while the parameters α_2 and α_3 depend on the vortex configuration in the sample. The best fits to the experimental data are obtained with $\alpha_1 = 0.38$, $\alpha_2 = 0.438$, and $\alpha_3 = 0.266$, where we take $B_C = B(H_{in}^{(1)}) \simeq 14.4$ mT from experimental data. Parameter α_1 for a rectangular sample is expected to be a bit larger than its value α_1^{homog} in a narrow strip[34]. In the fitting we assume that both jump-like and knee-like anomalies in the $V(H)$ are associated with the change of vorticity[30] and verify this applying the same parameters to $V(H)$ with different values of I_{bias} (Supplementary Note 3). The S gap in the narrow extensions of Sample A shown in Fig. 2d is close to its zero-field value Δ_0 up to ~ 30 mT with few % accuracy as the depairing parameter in this case $\Gamma/\Delta_0 = (\pi\xi wB/\Phi_0)^2/6$ is small[35]. Here $w \simeq 130$ nm is the width of the extensions.

Pumping measurements. As expected from the d.c. measurements, the magnetic field does not improve the electronic pumping on Sample B (Supplementary Note 4). We thus focus only on the pumping measurements on Sample A, which has the highly non-uniform distribution of the gap (Fig. 1) under magnetic field. To probe the magnetic field dependence of non-equilibrium quasiparticle states, we measure the current I in turnstile mode averaged over the period of the drive $\tau_0 = 1/f$ (ref. 29). We apply a fixed bias voltage $V_{bias} = 100\,\mu$V and sinusoidal gate voltage $C_g V_g/e = n_g(t) = n_g^0 + A_g \sin(2\pi f t)$ through the capacitor C_g with variable amplitude A_g.

The turnstile current is expected to assume values equal to integer multiples of ef in the absence of non-equilibrium effects and unwanted tunnelling events. The measurements are carried out around the gate offset point $n_g^0 = 0.5$, to maximize the expected plateau width, for several frequencies f. Overheating of the S island, in particular at $H = 2.4$ mT corresponding to $B(H) = 0$, leads to positive deviations of this current from $I = nef$ (n is an integer) by tens percents at the expected plateau positions which corresponds to thousands of quasiparticles per μm^3 near the junction (see Fig. 3a at $f = 5$ MHz). The magnetic field improves quasiparticle trapping: the deviation from ef (and the corresponding quasiparticle density) at large enough magnetic fields decreases by an order of magnitude in the whole-frequency range from 0.5 to 200 MHz (see Fig. 4 and Supplementary Note 4) and approaches a few-percent level related to an amplifier noise, even for large gate amplitudes when pumping up to $n = 11$ electrons per cycle. The zoom up of the first plateau shown in Fig. 3b demonstrates the magnetic field dependence of I. To separate the Meissner current from the vortex contribution, we present in Fig. 3c pumping current versus the field at a fixed gate amplitude value indicated by the vertical dashed line in Fig. 3a. The excess current $\delta I = I - ef$ increases when the field is swept from large negative values to low values with jumps at $H_{out}^{(1,2)}$ (see vertical dashed lines in Fig. 4). The following field increase to positive values leads to decreasing excess current without visible anomalies. This is related to the difference in the kth vortex entry (exit) fields $H_{in(out)}^{(k)}$. Indeed, for $k = 1$ at these fields, we have $E_g(H_{in}^{(1)}) < E_g(H_{out}^{(1)})$, which leads to the efficient redistribution of quasiparticle density even without any vortex (Fig. 1b). Despite the absence of anomalies at the vortex entries, it is possible to estimate the value $H_{in}^{(1)}$ by varying the value of the initial field: the discontinuous anomaly at $H_{out}^{(1)}$ is only visible for a field amplitude in a sweep exceeding $H_{in}^{(1)} = 13.5$ mT, which is close to the value found by d.c. measurements in Sample B. At even higher values of the field, $|H| \gtrsim 30$ mT, the current quantization is lost again due to the eventual suppression of the S gap near the junctions as well (Supplementary Note 4).

Theoretical analysis of pumping data. To model theoretically the excess current as a function of the field B and frequency f we calculate the electronic temperature T using a heat balance equation

$$\dot{Q}_{eph}(T) \simeq IV_{bias} = [ef + \delta I(T)]V_{bias}. \quad (3)$$

We keep in mind that T is nearly uniform and constant in time provided the heat diffusion length L_T is large compared with the size of the island R, and the heat relaxation time τ_{eph}, determined by electron–phonon coupling, is much larger than the operating period $\tau_0 = 1/f$, that is, $\tau_{eph} \gg \tau_0$ allowing us to average the heat

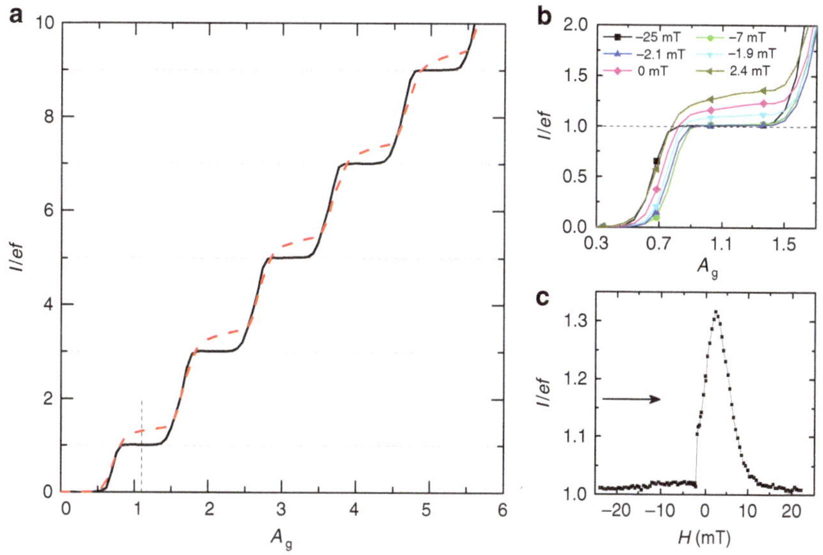

Figure 3 | Turnstile current driven by an a.c. gate. (**a**) The pumping current I normalized by ef at the operating frequency $f = 5$ MHz, the bias voltage $V_{bias} = 100\,\mu$V, and at the gate offset $n_g^0 = 0.5$ versus the normalized gate amplitude A_g at applied field $H = 2.4$ mT ($B = 0$) (red dashed curve) and at $H = -25$ mT (black solid curve). The horizontal dashed lines correspond to the expected current quantization $I = nef$; (**b**) the zoom up of the first plateau at several applied field values between −25 and 2.4 mT; (**c**) the evolution of the current along the vertical dashed line in **a** versus the field H varying from −25 to 25 mT; the sweep direction is shown by the horizontal arrow. In the further measurements we fixed the amplitude to the value shown by the vertical dashed line in **a**. The experimental uncertainty has been estimated from the noise of the amplifiers as ~10 fA, corresponding to 0.0125 ef at 5 MHz (not shown).

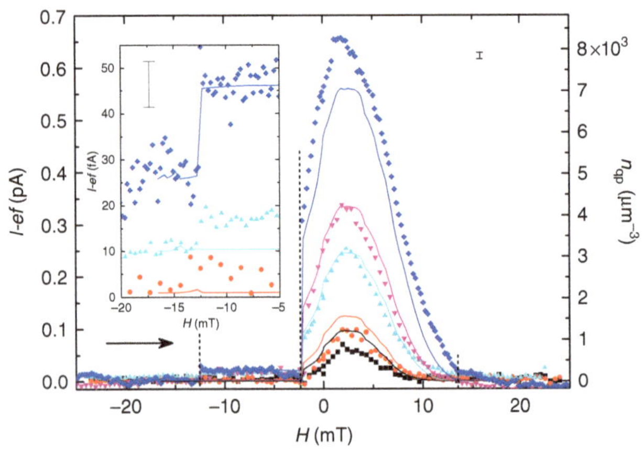

Figure 4 | Excess pumping current versus magnetic field. The excess current $I - ef$ at the first plateau for the driving frequencies 0.5 MHz (black), 1 MHz (red), 5 MHz (cyan), 10 MHz (violet), and 30 MHz (blue) with the maximum electronic temperature at $B(H) = 0$ of 274, 286, 320, 337 and 369 mK, respectively. Applied magnetic field H is swept from -25 to 25 mT (the sweep direction is shown by a horizontal arrow). The vertical dashed lines show the expected values corresponding to the entrance of an extra vortex into the island (for $H > 0$) and to the removal of one vortex ($H < 0$). (inset) A close-up of the current for a field close to the expulsion of the second vortex. For better visibility, the data sets for different frequencies have been shifted vertically. The measurements were performed at the bias voltage $V_{bias} = 100\,\mu V$, the gate offset $n_g^0 = 0.5$. In all panels symbols (solid lines) correspond to the experimental data (theoretical model). The scale on the right side shows the QP density n_{qp} near the junction. The experimental uncertainty is estimated from the noise of the amplifiers as ~ 10 fA, shown as an error bar on both panels.

diffusion equation across the sample volume and over the operating period. We assume further that most of the Joule dissipation occurs inside the S island and take into account the excess pumping current I averaged over the period from the first plateau on the right hand side (see Supplementary Note 5 for details). The value of $\dot{Q}_{eph} = \dot{Q}_{eph}^{(nv)} + \dot{Q}_{eph}^{(v)}$ describes the heat flow rate from the electronic subsystem to phonons for a non-zero depairing parameter Γ. We calculate its value $\dot{Q}_{eph}^{(nv)}(T)$ outside the vortex core regions for any Γ combining the procedure described, for example, in refs 38,39 and the solution of equation (1). For the experimental parameters $k_B T_0, \Gamma^{2/3} E_g^{1/3} \ll k_B T \ll E_g$ in low-temperature limit the electron–phonon heat flux

$$\dot{Q}_{eph}^{(nv)}(T) \simeq \frac{\Sigma(\mathcal{V} - \mathcal{V}_v)}{\zeta(5)} \left\{ \frac{64}{63} T^5 e^{-\frac{E_g}{k_B T}} + \frac{2\pi E_g^4}{3k_B^4} T e^{-\frac{2E_g}{k_B T}} \right\} \quad (4)$$

decomposes into recombination $\propto e^{-\frac{2E_g}{k_B T}}$ and scattering terms $\propto e^{-\frac{E_g}{k_B T}}$ (see, for example, ref. 40). Here Σ is the electron–phonon coupling constant, and $\mathcal{V}$ ($\mathcal{V}_v$) is the volume of the island (the vortex core regions). In the vortex cores in the same limit of negligible phonon temperature $T_0 \ll T$ the electron–phonon heat flow is modelled by the standard normal metal expression $\dot{Q}_{eph}^{(v)} = \Sigma \mathcal{V}_v T^5$ with the volume of m vortex cores assumed to be $\mathcal{V}_v = m r_v^2 d_S$. The recombination term in equation (4) becomes dominant at $k_B T \gtrsim 0.1 E_g$. Beyond the low-temperature limit, we use a numerically calculated expression for $\dot{Q}_{eph}^{(nv)}(T)$ instead of equation (4) (see Supplementary Note 5 for calculation details).

Eventually we obtain the magnetic field and frequency dependence of the measured excess current $\delta I(T)$ as

$$\delta I(T) = C \frac{\sqrt{2\pi \Delta_0 k_B T}}{e R_T} \exp[-\Delta_0 / k_B T]. \quad (5)$$

Note that the quasiparticle density near the junction is proportional to the excess current $n_{qp} = D(E_F) e R_T \delta I(T)/C$ and can be extracted from $\delta I(T)$ using the normal state density of states in the superconductor $D(E_F)$ (see the scale on the right side of Fig. 4 showing the quasiparticle density n_{qp}). Here $C \sim 1$ is a numerical coefficient determined by the wave form and the amplitude A_g of the gate drive, in particular the duration for one junction to be open for tunnelling in each cycle. A detailed derivation is given in Supplementary Note 6. Note that in equation (5) we neglected contributions of higher order processes in R_T^{-1} like Andreev tunnelling due to the small transparency of the junctions (see experimental results in ref. 41 and estimates in Supplementary Note 6). By solving (3) with the substituted expressions (2) and (5) we find the solution for T and $\delta I(T)$ (solid lines in Fig. 4). We used the constant $C = 1$ for a fixed drive amplitude A_g. The main uncertainty in the fitting procedure originates from the parameter $\Sigma\mathcal{V}$. The volume of the S sample can be estimated based on the electron micrograph (Fig. 2b) as $\mathcal{V} \simeq 3 \times 10^{-20}$ m^3, but usually this value is overestimated due to additional uncontrolled oxidation of Al. On the other hand, the typical range of the measured values of the electron–phonon relaxation constant Σ in the bulk aluminium[40,42,43] is within 2×10^8 to 5×10^8 W K^{-5} m^{-3}. Our fitting gives results agreeing reasonably well with the experimental data within the range of $\Sigma\mathcal{V}$ from 4×10^{-12} to 9×10^{-12} W K^{-5}. In Fig. 4, we present a fit for a certain middle value $\Sigma\mathcal{V} = 6 \times 10^{-12}$ W K^{-5}, which is in the best agreement with the experiment at moderate frequencies. Assuming $\mathcal{V} \simeq 3 \times 10^{-20}$ m^3 we get $\Sigma = 2 \times 10^8$ W K^{-5} m^{-3}, which is towards the low end due to the overestimated $\mathcal{V}$ but within the range given above. We have extracted the optimal value of the vortex core radius within the range $r_v = 2.5$–2.7ξ both from the d.c. measurements (Supplementary Note 3) and from the pumping data, which is in perfect agreement with the previous theoretical results[19,20].

Discussion

According to the theoretical model, equation (5), the maximal electronic temperature at $A_g = 1.1$ and $f = 30$ MHz is $T \simeq 370$ mK. It corresponds to a number of non-equilibrium quasiparticles $N_{qp} = n_{qp}\mathcal{V} \simeq 250$ in the uniform state (Fig. 1a). In the field, increasing from $B(H) = 0$ the Meissner supercurrents sufficiently improve the electron–phonon relaxation by reducing the gap $E_g(\Gamma)$ in the central part of the island even before the first vortex enters the island. This leads to at least 10–20 times reduction of the quasiparticle density near the junctions when the excess current approaches the amplifier noise level. The vortex contribution is clearly seen in the decreasing field regime due to the hysteresis caused by vortices. Indeed, the vortices that entered the island at a certain value of the field stay there till smaller fields (where the effect of Meissner current is smaller) and improve the relaxation of hot quasiparticles most effectively. Such hysteresis allows us to see the vortex contribution alone (see the larger step in Fig. 4 at $H \sim -2$ mT) and the improvement of relaxation in the two-vortex state with respect to the one-vortex state (the smaller step at $H \sim -13$ mT). We estimate the recombination rate in the vortex state $\Gamma_{rec} \simeq f / N_{qp,vort}$ as the quasiparticle injection rate f divided by the quasiparticle number $N_{qp,vort} \simeq 2D(E_F)\mathcal{V}_v k_B T \ln 2$ in the vortex core volume $\mathcal{V}_v$ (see Supplementary Note 5 for details). At $f = 30$ MHz it gives $N_{qp,vort} \sim 100$, $\Gamma_{rec} \simeq 0.3$ MHz of the recombination rate, that is, 20 times higher than $\Gamma_{rec}^0 = 16$ kHz estimated in ref. 40 at $B = 0$.

In conclusion, we demonstrate effective control of the number of excess quasiparticles and their spatial distribution in a mesoscopic superconducting disc by applying a small magnetic field on it. We find that both the Meissner supercurrents and vortices entering the disc one by one each give important observable contributions to the trapping of non-equilibrium quasiparticles. We demonstrate that a single-vortex contribution is sufficient to keep the superconducting disc near equilibrium up to 30 MHz injection frequency with $n_{qp} \simeq 400\,\mu m^{-3}$ quasiparticle density near the junctions and recombination rate of order of $\Gamma_{rec} \simeq 0.3$ MHz. Our d.c. and pumping measurements confirm the assumption[19,20] that a vortex can be considered as a normal metal cylinder with the effective radius $r_v = 2.5$–2.7ξ both in charge and heat transport problems. Our theoretical analysis of the quasiparticle trapping has proven its validity and efficiency in the set-up being in quantitative agreement with the experimental data.

Methods

Device fabrication. The hybrid devices with aluminium as the superconductor, copper as the normal metal, and aluminium oxide as the tunnel barrier in between, have been fabricated by standard electron-beam lithography and two-angle shadow evaporation technique. The aluminium island is $d_S = 20$ nm thick and it is oxidized with O_2 for 2 min at 2 mbar. The copper leads, 25 nm thick, are placed on the oxidized Al forming tunnel junctions.

Sample geometries and parameters. Two different island geometries have been employed in the measurements: Sample B has a nearly square-shaped island, as shown Fig. 2b, and Sample A with the same central part as geometry B has two additional long narrow aluminium extensions from each side towards the junctions (Fig. 2a). The diagonals of the island are $2R \sim 1\,\mu m$ both in A and B, and the narrow extensions of the island in A are 2 μm long and $w \sim 0.13\,\mu m$ wide. The sum of the tunnel resistances of the two junctions is $R_T \simeq 577\,k\Omega$ for Sample A and $R_T \simeq 714\,k\Omega$ for Sample B. We measured the *IV* characteristics of single-electron transistors at various values of the d.c. gate voltage at the base temperature to determine the zero-field S gap value $\Delta_0 \simeq 190$ (207) μeV and the charging energy $E_C \simeq 173$ (133) μeV for Sample A (B).

Reproducibility and noise. All the results presented here are reproducible between different runs and between samples of similar geometry, in particular, as concerns the values of the critical fields of vortex entry (exit). The results depend only on whether the absolute field value increases or decreases, provided by the hysteresis in vortex entry/exit events, but they do not depend on the sign of the field as such. The samples are cooled down through the superconducting transition with a zero-field cooled magnet. The uncertainties of current and voltage are estimated to be 10 fA and 3 μV, respectively. They are taken as the s.d. of the signal from the amplifiers.

References

1. Martinis, J. M., Ansmann, M. & Aumentado, J. Energy decay in superconducting Josephson-junction qubits from nonequilibrium quasiparticle excitations. *Phys. Rev. Lett.* **103,** 097002 (2009).
2. Paik, H. *et al.* Observation of high coherence in Josephson junction qubits measured in a three-dimensional circuit qed architecture. *Phys. Rev. Lett.* **107,** 240501 (2011).
3. Còrcoles, A. D. *et al.* Protecting superconducting qubits from radiation. *Appl. Phys. Lett.* **99,** 181906 (2011).
4. Wang, H. *et al.* Improving the coherence time of superconducting coplanar resonators. *Appl. Phys. Lett.* **95,** 233508 (2009).
5. Barends, R. *et al.* Minimizing quasiparticle generation from stray infrared light in superconducting quantum circuits. *Appl. Phys. Lett.* **99,** 113507 (2011).
6. Knowles, H. S., Maisi, V. F. & Pekola, J. P. Probing quasiparticle excitations in a hybrid single electron transistor. *Appl. Phys. Lett.* **100,** 262601 (2012).
7. Pekola, J. P. *et al.* Trapping of quasiparticles of a nonequilibrium superconductor. *Appl. Phys. Lett.* **76,** 2782 (2000).
8. Rajauria, S., Courtois, H. & Pannetier, B. Quasiparticle-diffusion-based heating in superconductor tunneling microcoolers. *Phys. Rev. B* **80,** 214521 (2009).
9. Nguyen, H. Q. *et al.* Trapping hot quasi-particles in a high-power superconducting electronic cooler. *New J. Phys.* **15,** 085013 (2013).
10. Ullom, J. N., Fisher, P. A. & Nahum, M. Magnetic field dependence of quasiparticle losses in a superconductor. *Appl. Phys. Lett.* **73,** 2494–2496 (1998).
11. Yamamoto, T., Nakamura, Y., Pashkin, Y. A., Astafiev, O. & Tsai, J. S. Parity effect in superconducting aluminum single electron transistors with spatial gap profile controlled by film thickness. *Appl. Phys. Lett.* **88,** 212509 (2006).
12. Chi, C. C. & Clarke, J. Enhancement of the energy gap in superconducting aluminum by tunneling extraction of quasiparticles. *Phys. Rev. B* **20,** 4465–4473 (1979).
13. Blamire, M. G., Kirk, E. C. G., Evetts, J. E. & Klapwijk, T. M. Extreme critical-temperature enhancement of Al by tunneling in Nb/AlOx/Al/AlOx/Nb tunnel junctions. *Phys. Rev. Lett.* **66,** 220–223 (1991).
14. Heslinga, D. R. & Klapwijk, T. M. Enhancement of superconductivity far above the critical temperature in double-barrier tunnel junctions. *Phys. Rev. B* **47,** 5157–5164 (1993).
15. Goldie, D. J., Booth, N. E., Patel, C. & Salmon, G. L. Quasiparticle trapping from a single-crystal superconductor into a normal-metal film via the proximity effect. *Phys. Rev. Lett.* **64,** 954–957 (1990).
16. Ullom, J. N., Fisher, P. A. & Nahum, M. Measurements of quasi-particle thermalization in a normal metal. *Phys. Rev. B* **61,** 14839–14843 (2000).
17. Rajauria, S. *et al.* Efficiency of quasiparticle evacuation in superconducting devices. *Phys. Rev. B* **85,** 020505 (R) (2012).
18. Levenson-Falk, E. M., Kos, F., Vijay, R., Glazman, L. & Siddiqi, I. Single-quasiparticle trapping in aluminum nanobridge Josephson junctions. *Phys. Rev. Lett.* **112,** 047002 (2014).
19. Golubov, A. A. & Houwman, E. P. Quasiparticle relaxation rates in a spatially inhomogeneous superconductor. *Phys. C* **205,** 147–153 (1993).
20. Golubov, A. A. *et al.* Quasiparticle lifetimes and tunneling times in a superconductor-insulator-superconductor tunnel junction with spatially inhomogeneous electrodes. *Phys. Rev. B* **49,** 12953–12968 (1994).
21. Friedrich, S. *et al.* Experimental quasiparticle dynamics in a super-conducting, imaging X-ray spectrometer. *Appl. Phys. Lett.* **71,** 3901–3903 (1997).
22. Aumentado, J., Keller, M. W., Martinis, J. M. & Devoret, M. H. Nonequilibrium quasiparticles and 2e periodicity in single-cooper-pair transistors. *Phys. Rev. Lett.* **92,** 066802 (2004).
23. Court, N. A., Ferguson, A. J., Lutchyn, R. & Clark, R. G. Quantitative study of quasiparticle traps using the single-Cooper-pair transistor. *Phys. Rev. B* **77,** 100501 (R) (2008).
24. Peltonen, J. T., Muhonen, J. T., Meschke, M., Kopnin, N. B. & Pekola, J. P. Magnetic-field-induced stabilization of nonequilibrium superconductivity in a normal-metal/insulator/superconductor junction. *Phys. Rev. B* **84,** 220502 (R) (2011).
25. Nsanzineza, I. & Plourde, B. L. T. Trapping a single vortex and reducing quasiparticles in a superconducting resonator. *Phys. Rev. Lett.* **113,** 117002 (2014).
26. Wang, C. *et al.* Measurement and control of quasiparticle dynamics in a superconducting qubit. *Nat. Commun.* **5,** 5836 (2014).
27. van Woerkom, D. J., Geresdi, A. & Kouwenhoven, L. P. One minute parity lifetime of a NbTiN Cooper-pair transistor. *Nat. Phys.* **11,** 547–550 (2015).
28. Früngel, F. B. A. *Optical Pulses—Lasers—Measuring Techniques* (Academic Press, 2014).
29. Pekola, J. P. *et al.* Hybrid single-electron transistor as a source of quantized electric current. *Nat. Phys.* **4,** 120–124 (2008).
30. Kanda, A., Baelus, B. J., Peeters, F. M., Kadowaki, K. & Ootuka, Y. 2004Experimental evidence for giant vortex states in a mesoscopic superconducting disk. *Phys. Rev. Lett.* **93,** 257002 (2004).
31. Stan, G., Field, S. B. & Martinis, J. M. Critical field for complete vortex expulsion from narrow superconducting strips. *Phys. Rev. Lett.* **92,** 097003 (2004).
32. Skalski, S., Betbeder-Matibet, O. & Weiss, P. R. Properties of superconducting alloys containing paramagnetic impurities. *Phys. Rev* **136,** A1500–A1518 (1964).
33. Maki, K. & Fulde, P. Equivalence of different pair-breaking mechanisms in superconductors. *Phys. Rev* **140,** A1586–A1592 (1965).
34. Fulde, P. Tunneling density of states for a superconductor carrying a current. *Phys. Rev* **137,** A783–A787 (1965).
35. Anthore, A., Pothier, H. & Esteve, D. Density of states in a superconductor carrying a supercurrent. *Phys. Rev. Lett.* **90,** 127001 (2003).
36. Schweigert, V. A., Peeters, F. M. & Singha Deo, P. Vortex phase diagram for mesoscopic superconducting disks. *Phys. Rev. Lett.* **81,** 2783–2786 (1998).
37. Palacios, J. J. Vortex matter in superconducting mesoscopic disks: Structure, magnetization, and phase transitions. *Phys. Rev. B* **58,** R5948–R5951 (R) (1998).
38. Kopnin, N. B. *Theory of Nonequilibrium Superconductivity* (Oxford Univ. Press, 2001).
39. Timofeev, A. V. *et al.* Recombination-limited energy relaxation in a Bardeen-Cooper-Schrieffer superconductor. *Phys. Rev. Lett.* **102,** 017003 (2009).
40. Maisi, V. F. *et al.* Excitation of single quasiparticles in a small superconducting Al island connected to normal-metal leads by tunnel junctions. *Phys. Rev. Lett.* **111,** 147001 (2013).
41. Aref, T., Maisi, V. F., Gustafsson, M. V., Delsing, P. & Pekola, J. P. Andreev tunneling in charge pumping with SINIS turnstiles. *Europhys. Lett.* **96,** 37008 (2011).
42. Kautz, R. L., Zimmerli, G. & Martinis, J. M. Self-heating in the Coulomb-blockade electrometer. *J. Appl. Phys.* **73,** 2386–2396 (1993).

43. Giazotto, F., Heikkilä, T. T., Luukanen, A., Savin, A. M. & Pekola, J. P. Opportunities for mesoscopics in thermometry and refrigeration: Physics and applications. *Rev. Mod. Phys.* **78,** 217–274 (2006).

Acknowledgements

We thank V. Maisi for valuable discussions, and T. Faivre for technical help and useful comments. This work has been supported in part by Academy of Finland (Project Nos. 284594, 272218), by the European Union Seventh Framework Programme INFERNOS (FP7/2007–2013) under Grant Agreement No. 308850, by Microsoft Project Q, by the EMRP (project no. SIB01-REG2), by the Russian Foundation for Basic Research, and the grant of the Russian Science Foundation (No. 15–12–10020). We acknowledge the availability of the facilities and technical support by Otaniemi research infrastructure for Micro and Nanotechnologies (OtaNano).

Author contributions

M.T., M.M. and J.P.P. conceived and designed the experiments; M.T. performed the experiments; M.T., I.M.K. and A.S.M. analysed the data. M.T., I.M.K., M.M., A.S.M., and J.P.P. contributed with materials/analysis tools; M.T., I.M.K., M.M., A.S.M. and J.P.P. wrote the paper.

Additional information

Competing financial interests: The authors declare no competing financial interests.

12

Viscoelastic lithography for fabricating self-organizing soft micro-honeycomb structures with ultra-high aspect ratios

Gi Seok Jeong[1,2], Da Yoon No[1,3], JaeSeo Lee[4], Junghyo Yoon[5], Seok Chung[5] & Sang-Hoon Lee[1,4]

High-aspect ratio micro- and nano-structures have been used for the production of a variety of applications. In this paper, we describe a simple and cost-effective approach to fabricate an arrayed microarchitecture with an ultra-high aspect ratio using soft materials. The shapes and sizes of the honeycomb structure can be easily modulated by changing the dimensions and position of the base mould pattern and the pressure. The honeycomb structure is used to prepare a drug delivery patch and a microwell array to form cell spheroids without cell loss. The honeycomb structures prepared using natural ECM (collagen-Matrigel) materials are successfully fabricated. The hepatocytes and endothelial cells are seeded and co-cultured in the ECM-based micro-honeycomb to prepare a 3D liver model successfully mimicking an ultrastructure of liver and providing enhanced liver function.

[1] Department of Biomedical Engineering, College of Health Science, Korea University, Anam-ro 145, Seongbuk-gu, Seoul 02841, Korea. [2] Biomedical Engineering Research Center, Asan Institute for Life Sciences, Asan Medical Center, 88 olympic-Ro, Songpa-gu, Seoul 05505, Korea. [3] Department of Bioengineering, School of Engineering and Medicine, Stanford University, 443 Via Ortega, Stanford, California 94305, USA. [4] KU-KIST Graduate School of Converging Science and Technology, Korea University, Seoul 02841, Korea. [5] School of Mechanical Engineering, Korea University, Anam-ro 145, Seongbuk-gu, Seoul 02841, Korea. Correspondence and requests for materials should be addressed to S.-H.L. (email: dbiomed@korea.ac.kr).

Recent progress in the field of micro and nanotechnologies has led to the production of a variety of complex micro- or nano-structures for use in sensors, actuators and battery applications[1,2]. Advanced micro- and nano-scale technologies have enabled broad applications of these structures, even in the biomedical fields of drug delivery, tissue engineering and medical devices[3–5]. Effective drug or cell delivery applications and tissue engineering techniques require structures that can pack a high number of cells[6] or drugs as compactly as possible[7]. The construction of a structure with minimal dead space has received significant attention from scientists and engineers. One of the nature-inspired structures considered optimal for this purpose is the honeycomb structure prepared with a high aspect ratio (HAR) wall. For practical applications in the biomedical field, honeycomb structures must be formed from a soft material or a natural extracellular matrix (ECM). The fabrication of such honeycomb architectures using soft materials has presented a significant challenge due to the mechanical weakness inherent in HAR walls fabricated from soft materials. Conventional methods for fabricating HAR structures include deep reactive ion etching (D-RIE)[8], lithographie–galvanoformung–abformung processes[9] and UV lithography[10,11]. Although these approaches have provided ingenious solutions to fabricating diverse and challenging structures, they are low-throughput approaches, and the fabrication costs increase rapidly as the aspect ratio is increased. Some of these methods are limited to hard materials, and the construction of such structures using soft materials, such as poly(dimethylsiloxane) (PDMS) or ECM materials has been challenging[12,13]. The mechanical instabilities of natural ECM materials make it extremely difficult to build micro-honeycomb structures with ultra-thin walls[14].

In this study, we propose a novel method of constructing self-organizing micro-honeycomb structure arrays consisting of ultra-high aspect ratio walls (with a maximum aspect ratio exceeding 500). The self-organization of the micro-honeycomb structures relies on the surface tension of the viscoelastic materials and the vacuum pressure. The shapes and dimensions could be readily controlled by changing the distance between the base mould patterns or by altering the vacuum pressure. Soft micro-honeycomb structures prepared from PDMS or natural ECM materials (collagen–Matrigel) were successfully fabricated. The PDMS structures displayed an excellent capacity for drug or cell loading. Hepatocyte and endothelial cells were seeded and co-cultured in the ECM hydrogel micro-honeycomb structures to fabricate a three-dimensional (3D) liver model consisting of compact cell spheroids and vessel-like structures that provided enhanced liver functions.

Results

Principles of viscoelastic lithography. The principles underlying viscoelastic lithography are illustrated in Fig. 1a and Supplementary Fig. 1. A 10-mm-thick layer of a highly viscous PDMS (HV-PDMS) solution was poured onto a base mould with a patterned array of holes. The HV-PDMS solution did not permeate the holes in the array due to its viscosity. The air trapped in the array of holes increased in volume on application low pressures between − 20 and − 70 kPa, forming spherical bubbles in the HV-PDMS (isotropic expansion, Supplementary Fig. 1b). As the bubbles increase in volume, their shapes deformed in the presence of the growing neighbouring bubbles (anisotropic expansion, Supplementary Fig. 1c). Bubble growth in the viscoelastic materials depended mainly on three key parameters: the pressure, the size of each hole (P_d), and the distance between the holes (P_S). The formation of single-static bubbles under reduced pressure could be described using the Laplace equation[15,16].

The Laplace equation for bubble formation is presented in equations (1–3),

$$P_V = (F_\partial + F_S)/\pi D_B^2, \tag{1}$$

where

$$F_\partial = \sigma(\pi D_B), \tag{2}$$

$$F_S = \sigma(\pi P_d)\sin\theta, \tag{3}$$

where P_V is the vacuum pressure, σ is the surface tension of the liquid, and D_B is the diameter of the bubble.

As shown in equations (1–3) and Supplementary Fig. 1a, the PDMS prepolymer required a high surface tension to balance the forces within the bubbles. However, conventional PDMS prepolymers have a low surface tension and an insufficient viscosity for constructing micro-honeycomb structures under a negative pressure. The PDMS prepolymer (using a 10:1 mixture of the PDMS prepolymer and the curing agent) viscosity was increased by thermally curing at 40 °C for 2 h, which raised the viscosity of the PDMS solution to ~120 Pa · s. Shape-tunable micro-honeycomb structures were constructed using a two-phase expansion process, as shown in Fig. 1b and the Supplementary Fig. 1b,c. The bubbles in the HV-PDMS solution expanded isotropically as the pressure decreased, and the isotropic expansion was disrupted by interactions among neighbouring bubbles as the bubble size approached P_S. The walls between growing bubbles became very thin and tall, forming tightly packed honeycomb structures. The schematic diagrams and optical microscopy images shown in Fig. 1b clearly illustrated this effect. Seen from the top, the bubble array formed a polygonal network as the bubbles expanded anisotropically, and the wall thicknesses between the bubbles decreased as the bubble depth increased. Top and side optical images acquired during the bubble growth process are shown in the Supplementary Fig. 1b. The bubble growth parameters could be controlled to produce micro-honeycomb structures with diverse shapes. A scanning electron microscopy (SEM) image of a micro-honeycomb structure is shown in Fig. 1c. The minimum wall thickness was 2 μm, as shown in the inset in Fig. 1c, and the structure had a maximum aspect ratio of ~500. The proposed method is much cheaper and requires much less time than other methods[8,17]. During the fabrication process, PDMS underwent three status transitions, from a liquid (low viscoelastic property), to a viscoelastic liquid (highly viscoelastic), to a solid. The bubble growth process was performed during the viscoelastic stage. Diverse bubble shapes were produced using different combinations of three parameters: (1) the pattern parameters (P_S and P_d), (2) the material parameters (the viscoelastic properties of the PDMS), and (3) the force parameter (pressure). P_S, P_d and the width of a bubble are defined in Supplementary Fig. 2. During isotropic expansion, the polygonal factor (D_{min}/D_{max}, where D_{min} is the horizontal diameter of a polygonized bubble, D_{max} is the diagonal diameter of a polygonized bubble) is defined as illustrated in Supplementary Fig. 2b. A polygonal factor of '1' produces bubble growth during isotropic expansion. Supplementary Fig. 3 shows the bubble shape change as a function of P_S and the pressure. During expansion, the bubbles underwent isotropic and anisotropic expansion, finally reaching equilibrium. The effects of the pattern and force parameters on the polygonal factor were measured and are illustrated in Fig. 1e,f. These graphs reveal that the shape of a bubble becomes polygonal at lower P_S values and higher negative pressures. Figure 1g shows the height of a bubble as a function of the viscosity over a range of pressures (30–60 kPa below atmospheric pressure). The height of the bubble gradually increased on application of increasingly negative pressures or

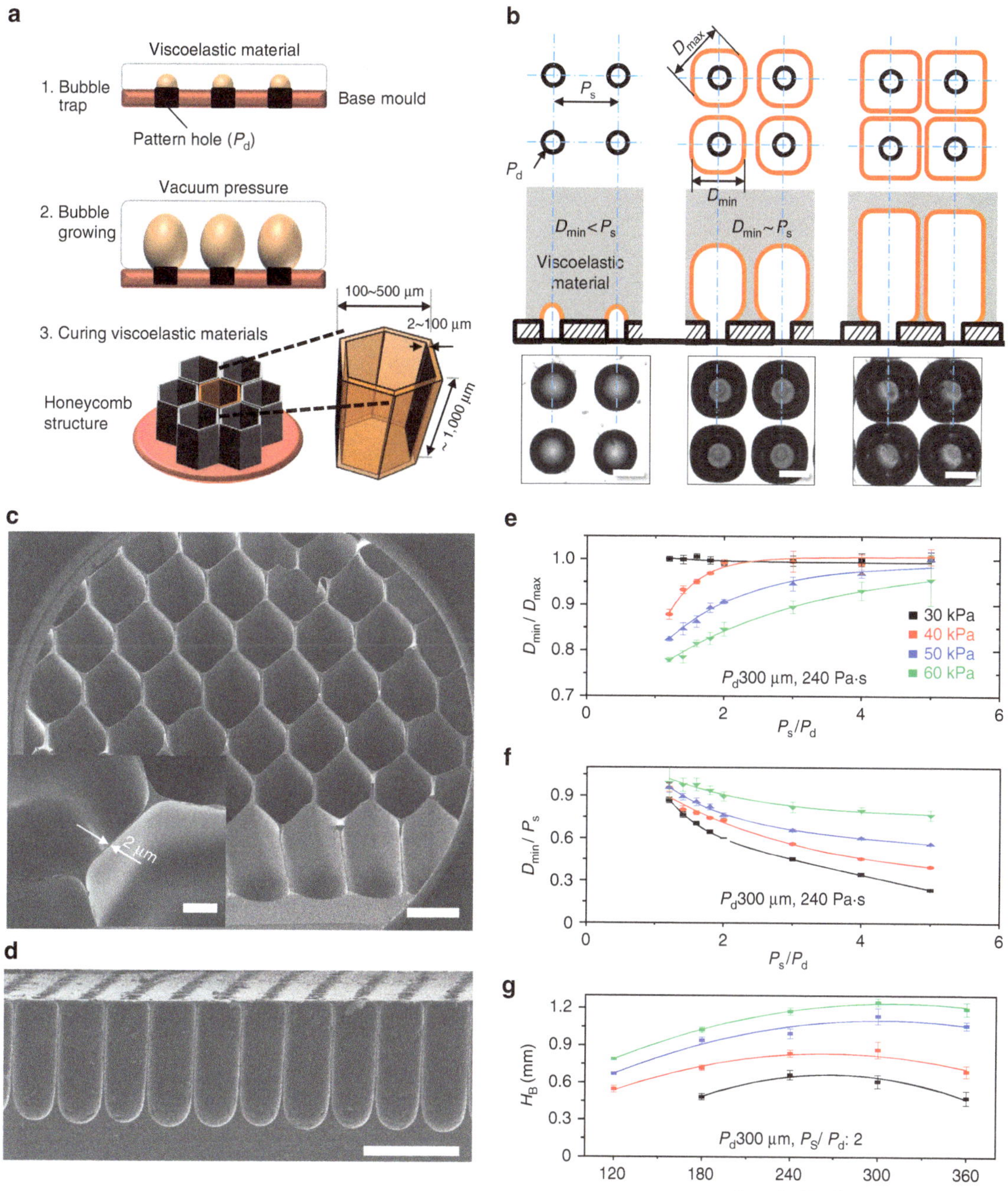

Figure 1 | The principles underlying viscoelastic lithography. (**a**) The fabrication of tunable honeycomb microstructures using viscoelastic lithography. (**b**) Growth and interactions among bubbles in the HV-PDMS under a negative pressure (20 to 70 kPa below atmospheric pressure). The trapped air slowly increased in volume to form spherical bubbles in the HV-PDMS matrix (isotropic expansion phase). Polygonal bubbles were formed as the bubbles interacted with neighbouring bubbles (anisotropic expansion phase). The scale bar, 300 μm. (**c**) SEM image of the micro-honeycomb structures fabricated using viscoelastic lithography. The minimum wall thickness was 2 μm (inset, scale bar, 100 μm), and the aspect ratio was ~500 (height to wall thickness). The scale bar, 500 μm. (**d**) Cross-sectional image of the micro-honeycomb structure. The scale bar, 500 μm. (**e**,**f**) The relationship between the polygonal factor D_{min}/D_{max} and the pattern parameters over a range of pressures. Small separation distances between bubbles gave D_{min}/D_{max} values approaching 1. (**g**) The height of the bubble as a function of the HV-PDMS properties over a range of pressure (30 – 60 kPa below atmospheric pressure). The bubble height in the HV-PDMS gradually increased, although it decreased under higher viscosity conditions (360 Pa · s). All error bars indicate the s.d.'s.

increasing viscosity, whereas the bubble height decreased at a viscosity of 360 Pa · s.

Modulation of the micro-honeycomb structure shape. Figure 2 shows the diverse micro-honeycomb structure shapes that resulted from various parameter combinations. The base mould was held constant, and the growing bubble patterns were changed by controlling the vacuum pressure (Fig. 2a,b). Under low vacuum conditions (30 kPa), the growing bubbles formed spherical shapes (Fig. 2a). Intermediate vacuum conditions (50 kPa) could be used to tune the growing hemispherical bubbles to form a jar-like structure (Fig. 2b). High vacuum conditions (>60 kPa) created a densely packed honeycomb structure. The locations and

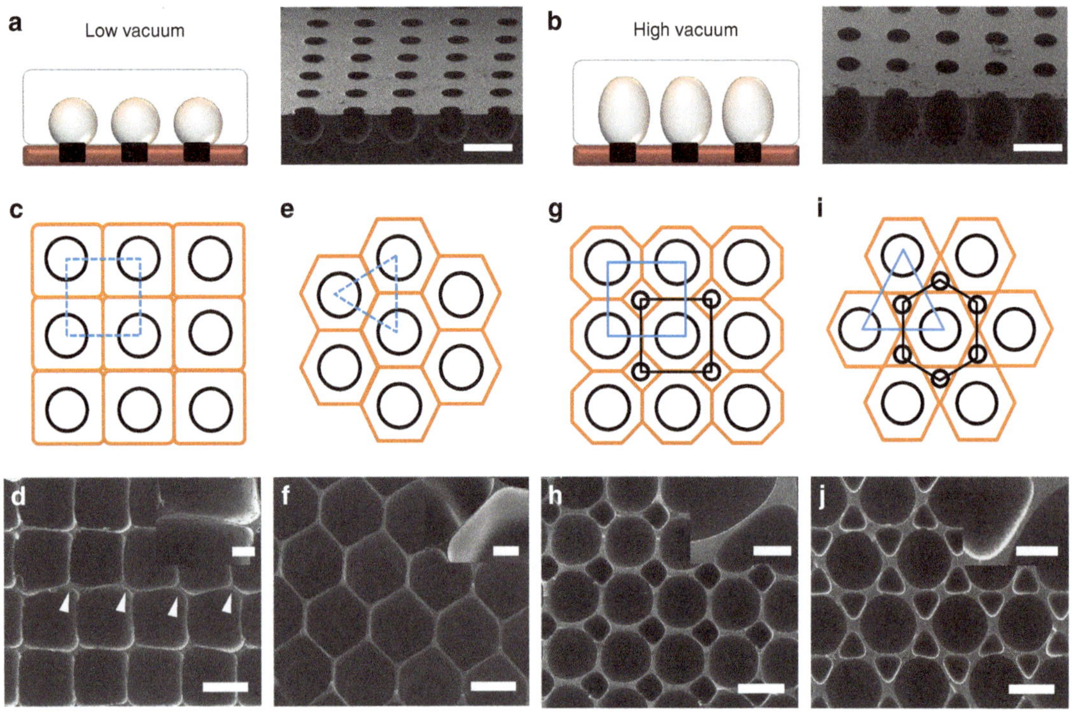

Figure 2 | Modulation of micro-honeycomb structures. (**a**,**b**) The same base mould was used to tune the growing bubble patterns by controlling the vacuum pressure. (**a**) During the initial stage, the growing bubbles were spherical in shape. (**b**) Under high vacuum conditions, the growing hemispherical bubbles formed a jar-like structure. (**c**) A gridiron-like (blue dotted line and black circles) hole arrangement produced cuboidal full-sized bubbles (orange rectangles). (**d**) SEM image of the cuboidal structure formed. The minimum wall thickness between cuboids was <2 µm (scale bar in the inset: 20 µm). Some misalignment events (white arrowheads) were present, indicating that unstable forces were generated as the bubbles grew. (**e**,**f**) A triangular (blue dotted lines) hole-pattern (black circles) generated a well-organized uniform hexagonal structure with a HAR. (**f**) SEM image of a hexagonal bubble pattern. The inset shows that the wall thickness was <2 µm (scale bar in the inset: 20 µm). (**g**,**h**) A pattern formed from holes of two sizes: 300 and 100 µm. (**g**) The 300 and 100 µm holes were arranged in a rectangular pattern (blue and black dotted lines). A large octagon surrounded by four small rhombus shapes was created (orange line). The small rhombus-shaped bubbles exerted supportive forces on the large octagonal bubbles. (**h**) SEM image of the large octagonal and small rhombus-shaped structures. The walls were ~5-µm-thick, as shown in the inset, thicker than the walls of the rectangular and hexagonal structures (scale bar in the inset: 50 µm). (**i**) The 300 and 100 µm holes were arranged in triangular and hexagonal patterns, respectively (indicated by the blue and black dotted lines). A large hexagonal bubble surrounded by six small triangular bubbles was created as the bubbles expanded. (**j**) SEM image of the large hexagonal structures surrounded by six small triangular structures (scale bar in the inset: 50 µm). Scale bars in **a**,**b**,**d**,**f**,**h** and **j** indicate 400 µm.

sizes of the hole-pattern could be adjusted to modulate the shape and size of the honeycomb structures. Densely packed polygonal structures with different spacings and arrangements could be produced using a single hole-pattern, as shown in Fig. 2c–j. A gridiron-like pattern of holes (Fig. 2c, black circles) produced cuboid structures with square cells when viewed from the top (Fig. 2c, orange line). An SEM image of the cuboidal structure is shown in Fig. 2d. The minimum wall thickness between cuboid cells was <2 µm (Fig. 2d, inset), and the structure was ~1 mm high. A few instances of misalignment among the cuboid wells were observed (indicated by white arrowheads in Fig. 2d), possibly caused by imbalances among the forces operating at the position at which the four bubbles met. By contrast, a triangular pattern (indicated by the blue dotted line in Fig. 2e) created a well-organized and uniform hexagonal structure with a HAR (Fig. 2e, orange line). The SEM image of the hexagonal bubble pattern formed readily reveals the honeycomb structure, as shown in Fig. 2f. Unlike the cuboidal structures, the hexagonal structures were completely aligned and no deformations were found. These results suggested that the forces were balanced and well regulated at the three-bubble intersections in the hexagonal structures. Although the structures were 1 mm high, the wall thickness was <2 µm (Fig. 2f, inset) due to the stable balance of forces between the growing bubbles. Hole-patterns composed of holes with different sizes (300 and 100 µm) were used to fabricate structures consisting of a heterogeneous honeycomb shape by appropriately positioning the holes. Positioning the 300 and 100 µm holes in a rectangular array (illustrated by the blue and black dotted lines in Fig. 2e) led to the formation of large octagonal bubbles, each of which was surrounded by four small rhombus-shaped bubbles (Fig. 2h, orange line). The small rhombus-shaped bubbles exerted supportive forces on the large octagonal bubbles. An SEM image of the resulting large octagonal and small rhombus-shaped bubble is shown in Fig. 2h. Unlike the rectangular bubbles (Fig. 2c,d) described above, larger stable bubbles could be grown using the heterogeneous structure because of the support provided by the small rhombus-shaped bubbles. In other words, the smaller bubbles present between the larger bubbles stabilized the forces as the bubbles grew. The wall of this structure was ~5-µm-thick (thicker than the walls formed in the cuboid and hexagonal structures), as shown in the inset of Fig. 2h. Arranging the 300 and 100 µm holes in triangular and hexagonal patterns (illustrated with blue and black dotted lines in Fig. 2i), respectively, permitted the formation of large hexagonal bubbles, each of which was surrounded by six small triangular bubbles, as the bubbles expanded (Fig. 2j, orange line). An SEM image of the large hexagonal bubbles surrounded by six small triangular bubbles is shown in Fig. 2j. These heterogeneous micro-honeycomb structures were uniform in shape, and the walls were ~10-μm-thick, as shown in the inset of Fig. 2j. These results indicated that heterogeneous and densely packed micro-honeycomb structures could be produced simply by altering the positions and sizes of the hole-pattern. Additional useful structures could potentially be produced using appropriate patterns and designs to optimize the bubble expansion process.

Applications of the micro-honeycomb structures. Figure 3 illustrates several applications of the micro-honeycomb structures. Cell spheroid formation plays an important role in tissue engineering[18,19], drug screening[20] and stem cell research[21,22]. During cell spheroid formation in a microwell array, lossless cell seeding and harvesting of intact cell spheroids is important[22]. The use of thin microwell array walls enables the lossless seeding. Cell losses in various concave PDMS microwell arrays with different inter-wall distances were measured, and the results are presented in Supplementary Fig. 4. Comparing to theses microwell arrays, the rectangular microstructure allowed the lossless cell seeding and the washing steps to remove undocked cells were not needed (Fig. 3a,b). Spheroid formation was achieved using mouse embryonic stem cells. Three days after seeding, the mouse embryonic stem cells self-aggregated and formed spheroids, as shown in Fig. 3c. A fluorescent image of the live/dead assay (green: live, red: dead) revealed that most cells were alive in the microwell culture array (Fig. 3c). These spheroids could be used as embryoid bodies to induce stem cell differentiation.

In another application, we fabricated jar-like micro-honeycomb structures for use as drug-releasing patches (Fig. 3d,e). A patch comprising the jar-like structures loaded with an anticancer drug (10 μg ml^{-1} doxorubicin hydrochloride) was placed on a cell culture dish. The anticancer drug (doxorubicin) diffused out from the drug patch, and 1 day after drug patch application, most MCF-7 cells in the dish detached from the surface (Fig. 3f). A variety of viscoelastic materials could be used for the viscoelastic lithography. We fabricated 10% (w/v) polycaprolactone micro-honeycomb structures with a wall thickness on the submicron scale (Fig. 3g,h). The inset shown in Fig. 3h reveals the submicron wall thickness. A mixture of fibrinogen and thrombin were also used to fabricate ECM micro-honeycomb structures (Fig. 3i).

3D liver model in the micro-honeycomb structure. A proof-of-concept tissue engineering application of the scaffold was demonstrated by seeding hepatocytes and Endothelial cells in the natural ECM micro-honeycomb structure to fabricate a 3D liver-like tissue (Fig. 4a). We used an ECM (a 7:3 mixture of collagen and Matrigel) honeycomb structure as a scaffold to pack as many cells as possible into a given volume (Fig. 4b). An SEM image of the natural ECM honeycomb structure (with an aspect ratio of *ca.* 400) is shown in Fig. 4c. The sidewall was ~2-μm-thick (Fig. 4c, inset). Before hepatocyte seeding, endothelial cells were seeded (Fig. 4d) and permitted to grown uniformly over the scaffold (Fig. 4e, day 0) across the bottom and sidewalls of each well (as indicated by the white arrowheads in Fig. 4f,g, day 1). Primary hepatocytes harvested

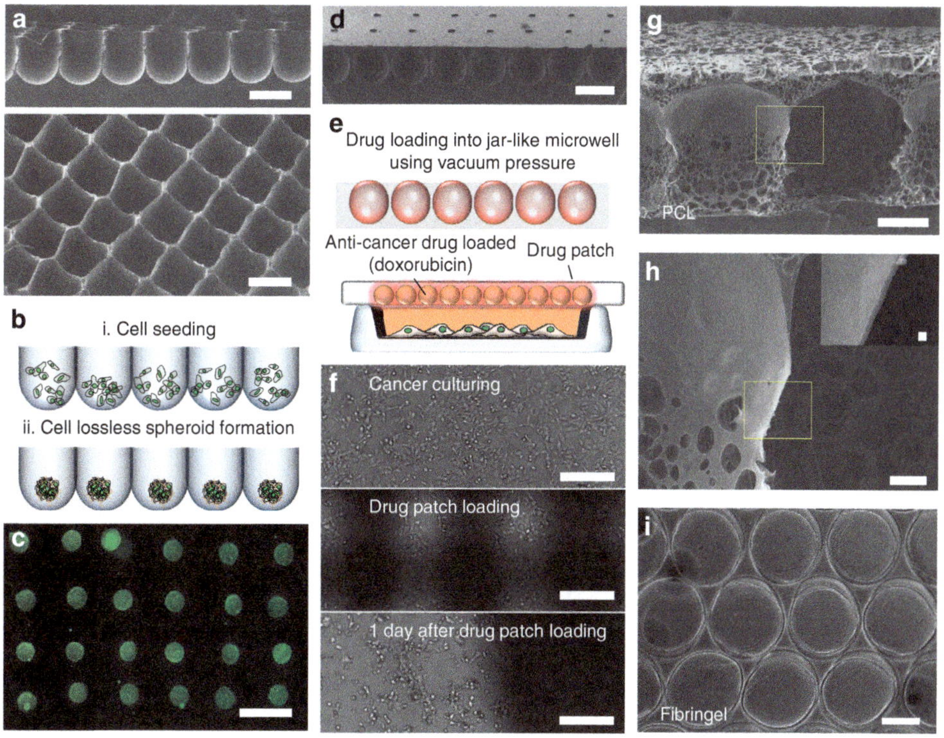

Figure 3 | Diverse applications of the tunable HAR honeycomb microstructures. (**a**) SEM images of the rectangular microstructures for lossless cell seeding. (**b**) Schematic diagram of the lossless cell seeding process. (**c**) Mouse embryonic stem cells (mESCs) were seeded into the HAR microstructure without the need for additional washing processes. Three days after seeding, the mESCs self-aggregated and formed spheroids in the HAR microwell. The spheroids are shown after staining using a live/dead assay. (**d**) Drug delivery patch using the jar-like microstructure array. (**e**) Schematic diagram of drug loading into the jar-like HAR honeycomb microstructure. (**f**) Breast cancer cells (MCF-7) were cultured on the 35-mm culture dish. The drug-loaded (1% doxorubicin hydrochloride, Sigma-Aldrich, MO, USA) patch was applied to cover the dish. One day after drug patch loading, most MCF-7 cells in the dish were dead. (**g,h**) polycaprolactone (PCL)-based HAR honeycomb microstructure with a wall thickness on the submicron scale (inset, scale bar, 1 μm). (**i**) Fabrication of the fibrin gel-based HAR honeycomb microstructure. (The scale bar indicates 300 μm.) The scale bars in **a**,**c** and **d** indicate 500 μm, **f** and **g** indicate 200 μm and **h** indicates 30 μm.

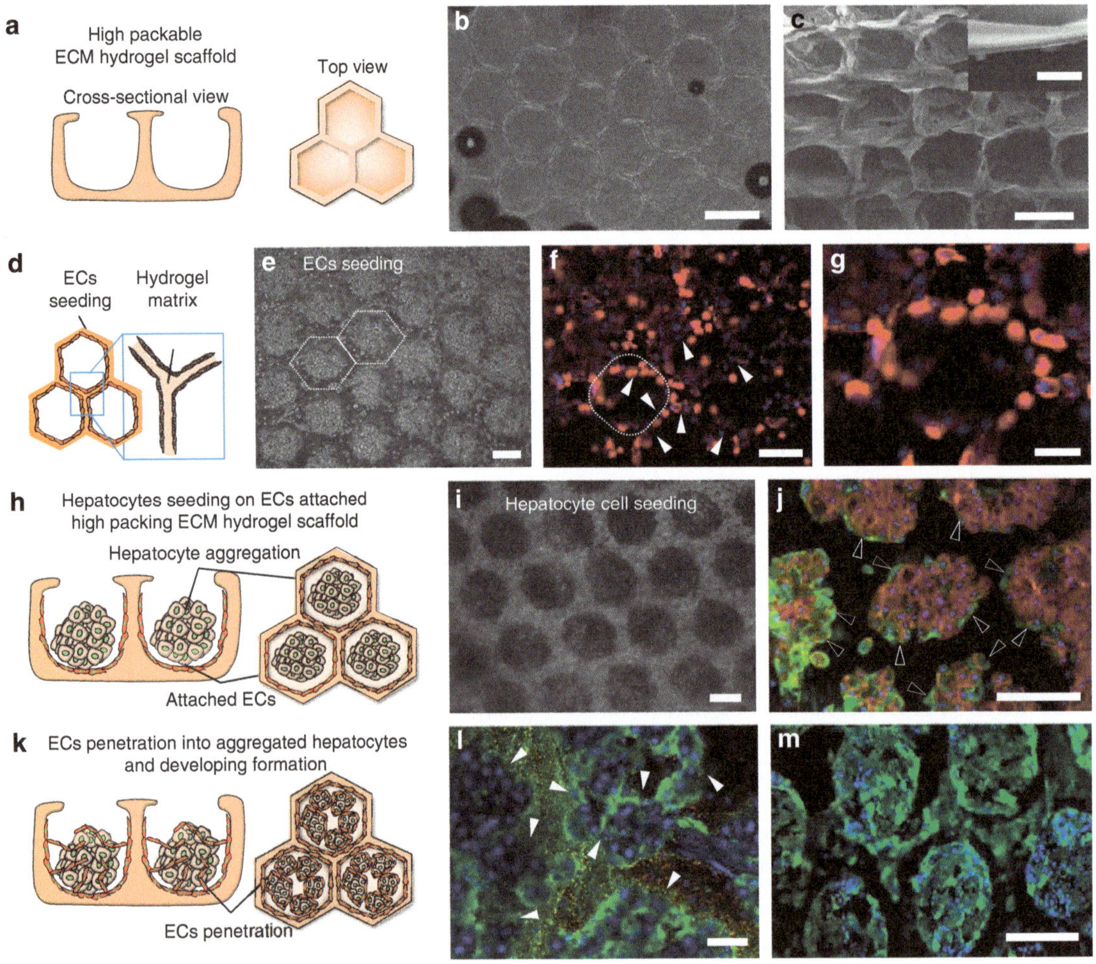

Figure 4 | 3D-tunable micro-honeycomb structure of an ECM matrix for use as a co-culturing system. (**a**) Schematic diagram of the ECM matrix used as a co-culturing system. (**b**) Optical microscopy image of the patterned collagen-Matrigel (ratio of 7:3) matrix. (**c**) SEM image of the patterned Matrigel micro-honeycomb structure (aspect ratio of 400). The inset shows the sidewall thickness (2 μm) produced at a pressure 70 kPa below atmospheric pressure (scale bar, 10 μm). (**d**) Schematic diagram of the endothelial cells attached to the patterned hydrogel matrix. (**e**) Optical microscopy image collected after endothelial cell seeding. (**f**) Endothelial cells were attached to both sides of the cuboidal patterned hydrogel matrix sidewalls. Red and blue indicate CD31 and DAPI, respectively. (**g**) Magnified view of **f**. (**h**) Schematic diagram of the hepatocyte seeding process and cell aggregation on the endothelial cell-patterned hydrogel. (**i**) Optical microscopy image after hepatocyte seeding. (**j**) Immunostaining of the serum albumin (green, black arrowheads) and actin (red), revealing that albumin was secreted by the patterned hepatocytes. (**k**) Schematic diagram of the endothelial cells that penetrated the hepatocyte aggregates and void structure. (**l**) Immunostaining of the serum albumin (green) revealed that albumin was secreted into the void structures of the hepatocytes (white arrowheads; scale bar, 10 μm). (**m**) The void structures of the hepatocytes displayed high levels of CYP450 activity (green). The scale bars in **b**,**c**,**e**,**f**,**g**,**i**,**j** and **m** indicate 500 μm.

from rat liver were seeded and cultured (Fig. 4h,i) and inoculated onto the hydrogel scaffolds coated with endothelial cells. The hepatocytes aggregated very quickly and displayed excellent albumin expression, as indicated by the black arrowheads in Fig. 4i,j. Figure 4k shows a schematic diagram of the EC penetration into the hepatocyte aggregates. Supplementary Fig. 5g presents a confocal image of a spheroid cross-section, revealing void structures within the hepatocyte aggregates. Some endothelial cells were present in the void structures and penetrated into the hepatic aggregates of the scaffold. The liver functions of this engineered tissue were characterized by analysing the serum albumin (Fig. 4j,l) expression and CYP450 (Fig. 4m) activity using immunofluorescence staining. As shown in Fig. 4j, the hepatic structures expressed serum albumin actively (green and black arrowheads). Figure 5a,b show that the hepatocyte-EC co-culture models in the hydrogel microwell displayed higher levels of albumin and urea secretion than do conventional spheroid culturing models over a period of 1 week. Albumin secretion in the hydrogel microwell group was 2.55 times higher than the value obtained from spheroid co-culturing models in PDMS microwells (Fig. 5b). As shown in Fig. 5c, the activity of CYP3A4 the hydrogel microwell group over 1 week was 131% of the value obtained from the spheroids prepared using PDMS microwells. Confocal images of fluorescein diacetate accumulation suggested that bile duct formation occurred in the spheroids prepared in the hydrogel microwells (Fig. 5d, left). The black arrows indicate the proliferating endothelial cells that formed vessel-like structures (Fig. 5d, right). Prevascularization of the engineered tissues with a controlled 3D structure is advantageous for implantation and engraftment. The hydrogel microwell played a huge role in inducing the formation of these structures and endothelial cells are more viable and able to preserve their specific functions in a natural hydrogel. The enhanced hepatic function and endothelial characteristics were evaluated by conducting reverse transcription PCR. As shown in Fig. 5e, the expression of ALB and CYP2E1, a member of the cytochrome P450 family, were significantly higher

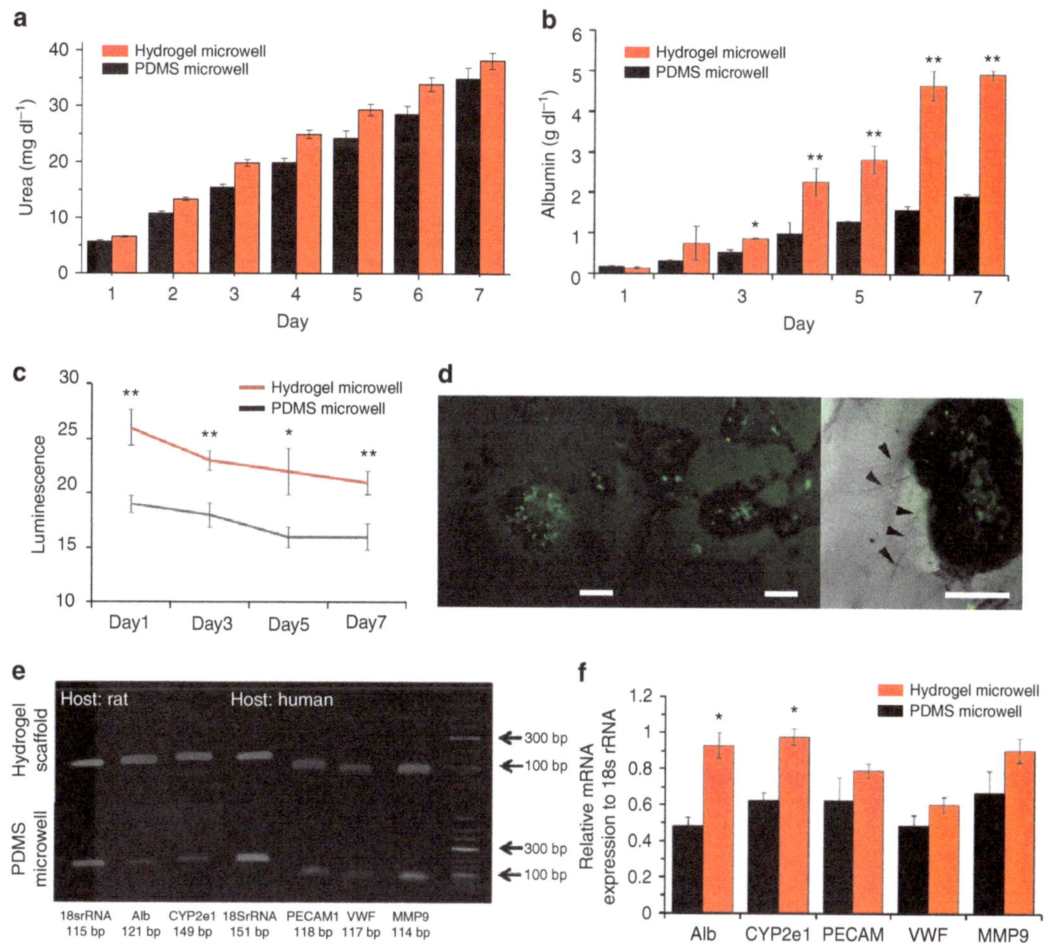

Figure 5 | Functional assays of the co-cultured spheroids in a 3D hydrogel scaffold and PDMS microwells. (**a**) Urea secretion test. No significant differences were observed between the PDMS microwells and the hydrogel microwells. (**b**) Albumin secretion test. The hydrogel microwell group displayed significantly higher levels of albumin secretion after day 3. (**c**) CYP3A4 enzymatic activity test. The *y* axis corresponds to the luminescence level. (**d**) Confocal images of fluorescein diacetate staining. The accumulated green fluorescence demonstrates bile duct formation. The black arrows indicate proliferated endothelial cells that formed vessel-like structures (scale bars, 200 μm). (**e**) Gene expression of the co-cultured spheroids and quantified results of mRNA gene PCR based on 18s rRNA expressions. (**f**) Alb and CYP2E1 displayed significantly higher expression levels in the 3D matrix group (*$P<0.05$, **$P<0.01$). All error bars indicate the s.d.

in the hydrogel microwell group than in the control group. We analysed the genes related to endothelial and proangiogenic characteristics, such as PECAM1, vWF and MMP9; however, no significant differences between the two culture methods were observed (Fig. 5e). Although EC-related gene expression also showed no significant differences between the two culture methods, the morphological features of the endothelial cells were quite different. Co-cultured spheroids in the PDMS microwells did not reveal proliferating endothelial cells on the PDMS surface (Supplementary Figs 6 and 9), whereas the endothelial cells in the hydrogel microwell actively proliferated. Seeding the hepatocytes in the EC pre-seeded hydrogel microwell resulted in the formation of hepatocyte aggregates. A 3D liver model with vessel-like structures around the spheroids and on the hydrogel matrix appeared on day 5 (Fig. 5d). Similar amounts of albumin were secreted by both models between days 1 and 3; however, the co-culture model expressed more albumin from day 4 onwards compared with the mono-culture model. The differences between the amounts of albumin secreted by the two models increased as time passed (Supplementary Fig. 5h,i). The albumin and urea secretion rate per cell are described in the Supplementary Note 1. These results were further supported by results obtained in previous studies, in which co-cultured endothelial cells were found to assist hepatocytes in maintaining their features *in vitro* over an extended period of time[23–26].

Discussion

Recent rapid progress in material and microfabrication technologies enables the development of a variety of soft structures, including 3D scaffolds for tissue engineering applications. Despite such progress, the construction of 3D soft structures with HAR thin wall has presented a significant challenge. The novel self-organizing method proposed here facilitates the laboratory-based construction of a variety of soft micro-honeycomb structure arrays with ultra-thin walls without the use of complicated or expensive processes and facilities. The sizes and shapes of the HAR structures could be modulated simply by changing the base mould pattern and applied vacuum pressure. The fabricated structures were used to demonstrate a variety of applications, including uniform cell spheroid formation and drug delivery patches. With the proposed method, we demonstrated the preparation of a 3D HAR scaffold composed of a very soft hydrogel, including collagen. Such scaffolds are beneficial in providing an *in vivo*-like environment. Although recent progress in technology has enabled the development of biological substitutes that can restore, maintain or improve the functions of tissues or whole organs, the scale-up and creation of physiologically relevant tissues with well-distributed vascular systems continues to pose a challenge. The development of highly organized functional tissue-level organoids requires spatially controlled co-cultures of heterogeneous cell types. Various patterned hydrogel scaffolds have been developed using photo-patterning[27,28], microscale stamping[29], lift-off[2] techniques and 3D printing; however, it is difficult to construct patterned hydrogel scaffolds that can support the development of complex native ultrastructures while maintaining organ-specific function. The proposed method allows the construction of HAR honeycomb structures using a soft hydrogel without the need for demoulding processes, followed by the successful culture of heterogeneous cells. As shown in Supplementary Fig. 5a,b, residual stress at the solid–solid interface between the base mould and the natural ECM hydrogel material was not observed, indicating that large-area densely packed micro-honeycomb structures could be easily constructed (Supplementary Fig. 5c–e). We successfully cultured a 3D liver structure consisting of hepatocytes and endothelial cells. The co-cultured hepatocytes and endothelial cells displayed high levels of albumin and urea secretion over the course of 1 week, in addition to excellent cell–cell interactions. These results indicated that the hydrogel microwell provided a suitable environment for engineering liver tissues consisting of compactly packed cells ($>10^{10}$ cells) and vascular structures.

In summary, we successfully produced shape-tunable densely packed soft micro-honeycomb structures using only the viscoelastic properties of a material. The PDMS and natural ECM structures described here are, to the best of our knowledge, the first honeycomb structures with ultra-HAR walls reported to date. The shapes and sizes of the structures could be adjusted by controlling the applied pressure and the distance between bubbles, as characterized experimentally. Natural ECM hydrogel micro-honeycomb structures will be extensively applied towards the fabrication of engineered 3D tissues for failed organ function regeneration. The tunable pattern size, shape and the possibility of mass-producing the structures are critical advantages for the construction of a range of structures outside of the biomedical field that require a high-packing density, for examples, devices that require high-void densities, such as batteries.

Methods

Preparation of HV-materials. The HV-PDMS was prepared by pre-heating the PDMS prepolymer (a mixture of PDMS and the curing agent in a ratio of 10:1) for 120 min at 40 °C. The viscosity of the PDMS prepolymer was critical and was monitored using a rotational viscometer (HBDV-II; Brookfield Engineering, Middleboro, MA, USA, Supplementary Fig. 7). A 5-mm-thick layer of the HV-PDMS solution was poured onto a base mould with an appropriate pattern of holes, and the HV-PDMS did not penetrate the holes. The polycaprolactone honeycomb microstructure was fabricated by preparing polycaprolactone (10% w/v, Sigma, St Louis, MO, USA, dissolved in dimethylformamide/tetrahydrofuran). The pressure was decreased to a value between 20 and 70 kPa below atmospheric pressure, and the air trapped in the array of small holes increased in volume to form spherical bubbles in the HV-PDMS (Supplementary Fig. 8).

Natural ECM hydrogel patterning. The collagen–Matrigel matrix was fabricated from Matrigel and type 1 collagen purchased from BD Bioscience (San Jose, CA, USA). The gel stiffness could be varied by varying the pH of the pre-polymerized collagen solution over the range pH 7.4–11 (ref. 30). In this study, a collagen solution with an *in vivo*-like pH of 7.4 was used to prepare the collagen–Matrigel matrix polymerization process. The collagen was diluted in sterile water, and the pH was adjusted by adding a 10 × phosphate-buffered saline (PBS) solution containing phenol red and 0.5 N NaOH. Once the pH-regulated collagen had been prepared, the Matrigel and collagen were mixed in a collagen:Matrigel ratio of 7:3. A desiccator was pre-heated to 40 °C, the collagen–Matrigel mixture was gently poured onto a patterned PDMS plate in the desiccator, and a vacuum was applied to the mixture for 30 min to allow the required patterns to form in the collagen–Matrigel mixture (Supplementary Fig. 5). The fibrin honeycomb microstructure was fabricated using a 1:1 fibrin:thrombin mixture (Sigma-Aldrich, St Louis, MO, USA) poured onto the base mould within 1 min of mixing.

Preparation of cells and culture. The human umbilical vein endothelial cells (HUVECs) were purchased from Lonza (Basel, Switzerland) and were cultured in an endothelial cell growth medium (ScienCell Research Laboratories, Carlsbad, CA, USA) containing 5% foetal bovine serum, 20 ng ml^{-1} vascular endothelial cell growth factor and 1% penicillin/streptomycin. The HUVECs were cultured for no more than seven passages. Primary hepatocytes were isolated from 8-week-old male Sprague–Dawley rats (DBL, Seoul, South Korea) using a two-step collagenase perfusion procedure. The cell viability was determined using the trypan blue exclusion method. The isolated cells were >90% viable and were used in subsequent experiments. The hepatocytes were cultured in high-glucose Dulbecco's modified Eagle's medium supplemented with 20 mM HEPES (Sigma-Aldrich, St Louis, MO, USA), 25 mM $NaHCO_3$, 30 mg ml 21 L-proline, 10% foetal bovine serum, 25 U ml^{-1} penicillin, 25 μg ml^{-1} streptomycin, 10 μg ml^{-1} gentamicin, 10 ng ml^{-1} epidermal growth factor, 50 ng ml^{-1} insulin, 10^{-4} M dexamethasone, 10 mM nicotinamide and 100 mM L-ascorbic acid.

A 100-μl aliquot of a HUVEC suspension (5×10^4 cells ml^{-1}) was seeded onto the patterned hydrogel, and the device was placed in an incubator at 37 °C for 4 h to allow the cells to become attached to the hydrogel matrix. A 500-μl aliquot of the primary hepatocyte suspension (2×10^5 cells ml^{-1}) was then seeded onto the HUVEC—hydrogel matrix. The matrix was then kept at 37 °C in an incubator (with a 5% CO_2 atmosphere) to culture the cells. Cell responses were monitored daily using phase contrast (AMG; Westover Scientific, Bothell, WA, USA) and confocal microscopy (Olympus, Tokyo, Japan). The HUVECs mixture and the hepatocyte medium were replaced every other day during the cell culture process. All animal procedures were conducted according to the Korean University IRB guidelines and after approval by the appropriate institutional review committees.

Albumin and urea analyses. Albumin and urea secretion were analysed by measuring the albumin and urea concentrations in a medium conditioned by the cultured models (Supplementary Note 1). Patterned cell sheets were cultured in 48 wells over 7 days. A 500-μl aliquot of the medium was removed at each of the chosen time points for the assay, and this volume was replaced with 500 μl of fresh medium.

CYP450 luminescence assay and fluorescein diacetate staining. Cytochrome P450 3A4 (CYP3A4) enzymatic activity was measured using the luminescent P450-Glo CYP3A4 Cell-based Assay (Luciferin-PFBE, Promega), as described by the manufacturer. Briefly, cells were incubated with a luminogenic substrate (Luciferin-PFBE) in a culture medium for 4 h at 37 °C. The medium was subsequently transferred to a 96-well opaque white plate, and an equal volume of Luciferin Detection Reagent was added to initiate a luminescent reaction. The plate was incubated at room temperature for 20 min and luminescence was read using a Multimode Plate Reader (PerkinElmer). The spheroids were stained with fluorescein diacetate to reveal bile duct formation. Spheroids were incubated in a culture medium containing 2.5 μg ml^{-1} fluorescein diacetate (Sigma) for 40 min in an incubator. After 40 min, the medium was replaced, and the spheroids were incubated in the absence of fluorescein diacetate for 40 min. The fluorescein distribution over the spheroids was observed using confocal microscopy (Olympus, Tokyo, Japan).

Immunostaining. Endothelial and primary hepatocyte cells were fixed using 4% PFA by incubating them at 4 °C for 30 min in 0.1% Triton X-100, then incubating them in PBS for 20 min at room temperature. The cells were rinsed with 0.1% BSA in PBS, incubated with Block Ace (Dainippon Pharma, Tokyo, Japan) at 4 °C for 30 min, and then incubated with CD31 antibody (Abcam, Cambridge, UK), cytochrome P450 reductase antibodies (Abcam, Cambridge, UK) and serum albumin (Santa Cruz Biotechnology Inc., Santa Cruz, CA) overnight at 4 °C. The cells were then rinsed with 0.1% BSA in PBS and incubated with Alexa Fluor 488-conjugated anti-rabbit IgG secondary antibodies at 4 °C for 90 min. The cells were incubated with 4,6-diamidino-2-phenylindole for 5 min at room temperature.

Gene analysis. After 9 days of culturing, co-cultured spheroids composed of hepatocytes and Endothelial cells were retrieved from the PDMS microwells and from the 3D ECM scaffold after collagenase treatment for 12 h (Collagenase type 1, 1 mg ml^{-1}). RNA was purified using the RNeasy kit (Qiagen, CA, USA) and cDNA was generated using reverse transcriptase (TAKARA, Japan) according to the manufacturer's instructions. The primer sequences are listed in the Supplementary Table 1.

References

1. Zhao, Q., Gan, Z. & Zhuang, Q. Electrochemical sensors based on carbon nanotubes. *Electroanalysis* **14**, 1609–1613 (2002).
2. Macak, J. M., Tsuchiya, H. & Schmuki, P. High-aspect-ratio TiO_2 nanotubes by anodization of titanium. *Angew. Chem. Int. Ed.* **44**, 2100–2102 (2005).

3. Gibson, M. C., Patel, A. B., Nagpal, R. & Perrimon, N. The emergence of geometric order in proliferating metazoan epithelia. *Nature* **442,** 1038–1041 (2006).
4. Tsang, V. L. *et al.* Fabrication of 3D hepatic tissues by additive photopatterning of cellular hydrogels. *FASEB J.* **21,** 790–801 (2007).
5. Zimmermann, W.-H. Tissue engineering: polymers flex their muscles. *Nat. Mater.* **7,** 932–933 (2008).
6. Engelmayr, G. C. *et al.* Accordion-like honeycombs for tissue engineering of cardiac anisotropy. *Nat. Mater.* **7,** 1003–1010 (2008).
7. Yang, R., Chen, T., Chen, H. & Wang, W. Microfabrication of biodegradable (PLGA) honeycomb-structures and potential applications in implantable drug delivery. *Sens. Actuators B Chem.* **106,** 506–511 (2005).
8. Li, X., Abe, T. & Esashi, M. Deep reactive ion etching of Pyrex glass using SF6 plasma. *Sens. Actuators A Phys.* **87,** 139–145 (2001).
9. Ruprecht, R. *et al.* Various replication techniques for manufacturing three-dimensional metal microstructures. *Microsyst. Technol.* **4,** 28–31 (1997).
10. Xia, Y. & Whitesides, G. M. Soft lithography. *Annu. Rev. Mater. Sci.* **28,** 153–184 (1998).
11. Whitesides, G. M., Ostuni, E., Takayama, S., Jiang, X. & Ingber, D. E. Soft lithography in biology and biochemistry. *Annu. Rev. Biomed. Eng.* **3,** 335–373 (2001).
12. Mapili, G., Lu, Y., Chen, S. & Roy, K. Laser-layered microfabrication of spatially patterned functionalized tissue-engineering scaffolds. *J. Biomed. Mater. Res. B Appl. Biomater.* **75,** 414–424 (2005).
13. Chandra, D., Taylor, J. A. & Yang, S. Replica molding of high-aspect-ratio (sub-) micron hydrogel pillar arrays and their stability in air and solvents. *Soft Matter* **4,** 979–984 (2008).
14. Khetan, S. & Burdick, J. A. Patterning hydrogels in three dimensions towards controlling cellular interactions. *Soft Matter* **7,** 830–838 (2011).
15. Fainerman, V. B., Miller, R. & Joos, P. The measurement of dynamic surface tension by the maximum bubble pressure method. *Colloid Polym. Sci.* **272,** 731–739 (1994).
16. Mysels, K. J. The maximum bubble pressure method of measuring surface tension, revisited. *Colloids Surf.* **43,** 241–262 (1990).
17. Becker, E., Ehrfeld, W., Hagmann, P., Maner, A. & Munchmeyer, D. Fabrication of microstructures with high aspect ratios and great structural heights by synchrotron radiation lithography, galvanoforming, and plastic moulding (LIGA process). *Microelectron. Eng.* **4,** 35–56 (1986).
18. No, D. Y., Jeong, G. S. & Lee, S.-H. Immune-protected xenogeneic bioartificial livers with liver-specific microarchitecture and hydrogel-encapsulated cells. *Biomaterials* **35,** 8983–8991 (2014).
19. Jeong, G. S. Networked neuro-spheres formed by topological attractants for engineering of 3-dimensional nervous system. *Tissue Eng. Regen. Med.* **11,** 297–303 (2014).
20. Friedrich, J., Seidel, C., Ebner, R. & Kunz-Schughart, L. A. Spheroid-based drug screen: considerations and practical approach. *Nat. Protoc.* **4,** 309–324 (2009).
21. Jeong, G. S. *et al.* Surface tension-mediated, concave-microwell arrays for l arge-scale, simultaneous production of homogeneously sized embryoid bodies. *Adv. Healthc. Mater.* **2,** 119–125 (2013).
22. Jeong, G. S., Jun, Y., Song, J. H., Shin, S. H. & Lee, S.-H. Meniscus induced self organization of multiple deep concave wells in a microchannel for embryoid bodies generation. *Lab Chip* **12,** 159–166 (2012).
23. Yamada, M. *et al.* Controlled formation of heterotypic hepatic micro-organoids in anisotropic hydrogel microfibers for long-term preservation of liver-specific functions. *Biomaterials* **33,** 8304–8315 (2012).
24. Morin, O. & Normand, C. Long-term maintenance of hepatocyte functional activity in co-culture: requirements for sinusoidal endothelial cells and dexamethasone. *J. Cell. Physiol.* **129,** 103–110 (1986).
25. Takayama, G., Taniguchi, A. & Okano, T. Identification of differentially expressed genes in hepatocyte/endothelial cell co-culture system. *Tissue Eng.* **13,** 159–166 (2007).
26. Hwa, A. J. *et al.* Rat liver sinusoidal endothelial cells survive without exogenous VEGF in 3D perfused co-cultures with hepatocytes. *FASEB J.* **21,** 2564–2579 (2007).
27. Qi, H. *et al.* Patterned differentiation of individual embryoid bodies in spatially organized 3D hybrid microgels. *Adv. Mater.* **22,** 5276–5281 (2010).
28. Wylie, R. G. *et al.* Spatially controlled simultaneous patterning of multiple growth factors in three-dimensional hydrogels. *Nat. Mater.* **10,** 799–806 (2011).
29. Majd, S. & Mayer, M. Hydrogel stamping of arrays of supported lipid bilayers with various lipid compositions for the screening of drug—membrane and protein—membrane interactions. *Angew. Chem.* **117,** 6855–6858 (2005).
30. Chung, S. *et al.* Cell migration into scaffolds under co-culture conditions in a microfluidic platform. *Lab Chip* **9,** 269–275 (2009).

Acknowledgements

This work was supported by National Research Foundation of Korea (NRF-2015R1A2A1A09004998), Republic of Korea.

Author contributions

G.S.J. designed and carried out the experiments, prepared most of the data and wrote the paper; D.Y.N. carried out the cell studies and assisted with the experiments; JS.L. carried out cell studies and assisted with the experiments; J.Y. carried out and assisted with the experiments; S.C. consulted on the manuscript and contributed to writing the paper; S.-H.L. proposed the idea, managed the research process and wrote the paper.

Additional information

Competing financial interests: The authors declare no competing financial interests.

13

Towards controlled polymer brushes via a self-assembly-assisted-grafting-to approach

Tian Zhou[1], Hao Qi[1], Lin Han[2], Dmitri Barbash[1] & Christopher Y. Li[1]

Precise synthesis of polymer brushes to modify the surface of nanoparticles and nanodevices for targeted applications has been one of the major focuses in the community for decades. Here we report a self-assembly-assisted-grafting-to approach to synthesize polymer brushes on flat substrates. In this method, polymers are pre-assembled into two-dimensional polymer single crystals (PSCs) with functional groups on the surface. Chemically coupling the PSCs onto solid substrates leads to the formation of polymer brushes. Exquisite control of the chain folding in PSCs allows us to obtain polymer brushes with well-defined grafting density, tethering points and brush conformation. Extremely high grafting density (2.12 chains per nm^2) has been achieved in the synthesized single-tethered polymer brushes. Moreover, polymer loop brushes have been successfully obtained using oddly folded PSCs from telechelic chains. Our approach combines some of the important advantages of conventional 'grafting-to' and 'grafting-from' methods, and is promising for tailored synthesis of polymer brushes.

[1] Department of Materials Science and Engineering, Drexel University, Philadelphia, Pennsylvania 19104, USA. [2] School of Biomedical Engineering, Science and Health Systems, Drexel University, Philadelphia, Pennsylvania 19104, USA. Correspondence and requests for materials should be addressed to C.Y.L. (email: chrisli@drexel.edu).

Long-chain polymer molecules can be attached at their chain end(s) to a surface or interface by either physical or chemical bonds. Depending on the surface grafting density of these polymer chains, their morphology can be described as brushes, mushrooms or pancakes[1,2]. In polymer brushes, because the distance between the neighbouring tethering points is small, there is significant overlay among adjacent polymer chains, and pronounced chain stretching takes place orthogonal to the substrate. The past few decades witnessed exciting progresses in studies on polymer brushes, and they show great promises in various fields, including coating, biomedical, sensing, catalysis and so on[3–9]. On the synthesis side, two major synthetic approaches, including grafting-to and grafting-from, have been developed, with clear advantages and disadvantages associated with each of the method[10–13]. For example, a grafting-from approach produces dense polymer brushes with large thickness, but the chemical reactions required are usually complex and the as-prepared polymer brushes need to be detached from the substrate for characterization purposes[14–18]. Using grafting-to approaches, the chemical structures and dispersity ($Đ$) of polymers can be better controlled/characterized. However, the resulting grafting density is strongly dependent on the molar mass of the polymer chains; polymer brushes with high grafting density are challenging to achieve using this method. In this paper, we disclose an approach, which is fundamentally different from the previously reported methods, to synthesize polymer brushes with controlled structure and high grafting density on flat surfaces. In this approach, we first pre-assemble end-functionalized polymers into two-dimensional (2D) polymer single crystals (PSCs) using controlled crystallization. These 2D single-crystal sheets can then be covalently coupled onto a flat substrate. On removing unbound polymers, well-defined polymer brushes can be obtained. We therefore name the approach 'self-assembly-assisted-grafting-to' method.

PSCs have been extensively investigated during the past 60 years. They are typically used as the model system to determine polymer crystalline structures and related polymer physics[19–22]. Recently, we demonstrated that surface-functionalized 2D PSCs can be considered as free-standing 'self-assembled monolayers'—they can collect nanoparticles, and the resultant hybrid materials can be used in applications such as nanoparticle asymmetric functionalization, nanomotors, surface-enhanced Raman spectroscopy and catalysis supports[23–28]. These surface-functionalized 2D single crystals also provide a unique opportunity for the synthesis of well-defined polymer brushes. The key step in our method is pre-assembling polymers into PSCs before coupling them onto the substrate. More importantly, these PSCs should adapt a so-called integral folding conformation, that is, the number of folds for each polymer chain should be an integral number (n)[21]. This unique conformation ensures the end-functional groups reside on the surface of the crystal for subsequent chemical coupling. As the first demonstration of this method, alkoxysilane-terminated poly (ε-caprolactone) (PCL) was used as the model system to grow the PSC templates whose thickness was tuned by the crystallization temperature (T_c). PCL single-crystal templates with different thicknesses were covalently bound to a glass substrate, yielding polymer brushes with well-controlled grafting density and tethering points. Exceptionally high grafting density ($\sim$2.12 chains per nm^2) has been achieved by using extended chain single crystals as the templates. Furthermore, using bifunctional telechelic PCL to form oddly folded PSCs, we were able to synthesize polymer loop brushes (PLBs) with precisely controlled tethering points and high grafting density (0.53 chain per nm^2 and 1.06 tethering points per nm^2). Compared with previously reported conventional brush synthesis approaches, our strategy combines the advantages of both grafting-to and grafting-from methods, leading to polymer brushes with well-controlled structures and high grafting density.

Results

Polymer brushes with precisely tuned grafting density. The synthetic route of polymer brushes via self-assembly-assisted-grafting-to method is shown in Fig. 1. In this study, alkoxysilane group was used to functionalize the chain ends to form chemically tethered polymer brushes on glass substrates. Hydroxyl-terminated PCL (PCL-OH) was first synthesized via ring-opening polymerization using tetraethylene glycol monomethyl ether as the initiator followed by chain end functionalization using 3-(triethoxysilyl)propyl isocyanate. Supplementary Fig. 2 summarizes the proton NMR (nuclear magnetic resonance) and gel permeation chromatography (GPC) results of the functionalized PCL. According to the chain end analysis using NMR, the degree of polymerization of the polymer is $\sim$30 (denoted as TEG-PCL30-SiOR), while GPC shows the $Đ$ is 1.22. Solution crystallization via self-seeding was then used to prepare PSCs following the temperature programme as shown in Supplementary Fig. 1 and Supplementary Table 1. The seeding temperature (T_s) is critical to forming uniform single crystals, and in this work T_s was fixed at 42 °C, while three different T_c were chosen to prepare PSCs with different lamellar thicknesses. Figure 1 shows the atomic force microscopy (AFM) images of the resultant TEG-PCL30-SiOR single crystals at $T_c = 5$, 15 and 35 °C, respectively. In all of the three cases, hexagonal-shaped single crystals have been obtained. The shape slightly varies and the aspect ratio of the crystal changes from 2/1 to 3/1 and 5/1 as the T_c increases from 5 to 15 and 35 °C. The increased aspect ratio is due to the change of the relative growth rates of each crystal planes as the T_c increases[19–21]. More interestingly, height analysis of the AFM images shows the lamellar thickness of these PSCs is 5.5, 9 and 12 nm, respectively, which is due to the different undercooling during the crystallization process[29]. PCL has an orthorhombic unit cell, with $a = 0.747$ nm, $b = 0.498$ nm and $c = 1.705$ nm, space group of $P2_12_12_1$ and a

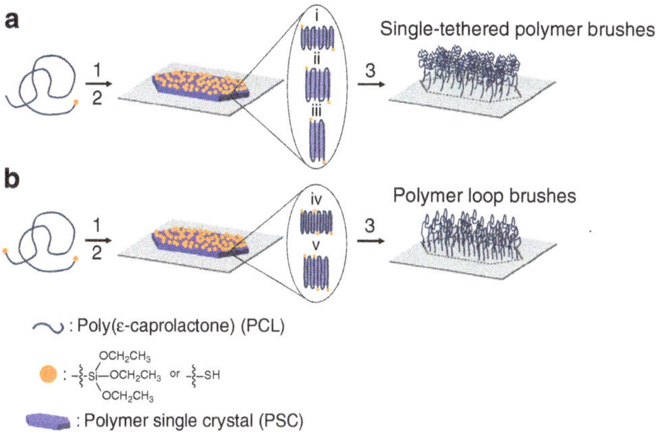

Figure 1 | Synthetic route for polymer brushes via the self-assemble-assisted-grafting-to method. (**a**) Synthesis of STPBs using mono-functionalized PCL; (**b**) preparation of PLBs using α, ω-functionalized PCL. Steps 1, 2 and 3 are solution crystallization via a self-seeding method, chemical immobilization of PSCs on a solid substrate and the removal of unbound polymer chains, respectively. The integral number folding structure of polymer chains is illustrated in (i–v), and $n = 4$, 2, 1, 5 and 3 for i–v, respectively. In the process of **b**, oddly folded chain structure is required for synthesizing PLBs, as shown in (iv) and (v). The dotted line on the substrates represents the outline of polymer brush-grafted regime templated from the hexagonal-shaped PSC, indicating chemical tethering of polymer brushes only takes place at the interface between PSCs and the solid substrate.

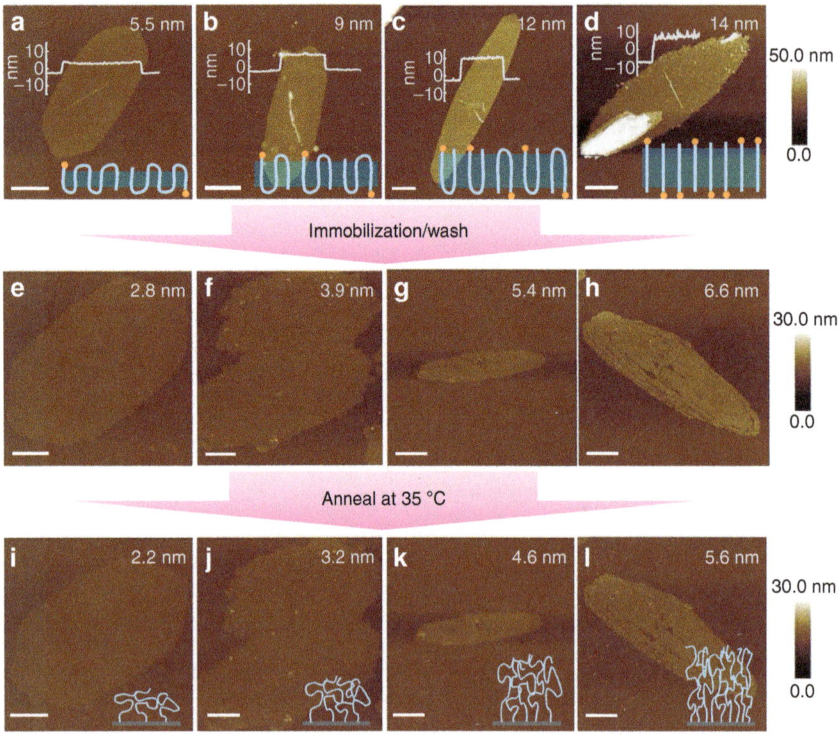

Figure 2 | Precise tuning of brush-grafting densities. Tapping-mode-AFM height images of TEG-PCL30-SiOR single crystals grown in dilute 1-butanol solution at 5 (**a**), 15 (**b**) and 35 °C (**c**), and TEG-PCL16-SiOR single crystals grown at 30 °C (**d**). The inset curves show the height profile of the corresponding crystal lamellae, and the schematics illustrate the chain-folding behaviours. Folding numbers are 4, 2, 1 and 0 for **a**–**d**, respectively. (**e**–**g**,**i**–**k**) The corresponding PCL brushes prepared from TEG-PCL30-SiOR single-crystal templates before (**e**–**g**) and after (**i**–**k**) thermal annealing. (**h**,**l**) PCL brushes prepared from TEG-PCL16-SiOR single-crystal templates before (**h**) and after (**l**) annealing. Scale bars, 1 μm. The inset numbers on the top right corner of each image are the measured thicknesses using AFM.

molecular helix of 7^*2_1 (ref. 30). Calculation based on this unit cell structure of PCL and the degree of polymerization suggests that the polymer chains fold 4, 2 and 1 times for the three crystals, respectively (schematics at the bottom of Fig. 2a–c). Since all the polymer chains adopt integral folding number, their alkoxysilane chain ends should be exclusively exposed on the lamellar surfaces, and can then serve as reaction sites for the following immobilization reaction. On the basis of the chain-folding structure of each crystal lamella, the areal density of alkoxysilane on one lamellar surface can be calculated to be 0.53, 0.89 and 1.33 -SiOR per nm^2, respectively. If we assume all the alkoxysilane groups on one side of the lamellae can be chemically coupled onto a glass substrate, these numbers are then the theoretical grafting density of the as-prepared polymer brushes.

The above-mentioned PCL single crystals were spin-coated onto a glass substrate. Ammonia was then applied to catalyse the solid-state reaction between silanol groups on the glass surface and the exposed alkoxysilane groups on the crystal lamellae. After 20 min of reaction, polymer-coated slides were thoroughly washed with acetone to remove unbound polymer chains. The as-prepared polymer brushes were then characterized using AFM as shown in Fig. 2e–g. It can be noticed that the polymer brush-covered domain retains the original hexagonal shape of the PSCs. Height analysis of the as-prepared polymer brush domains indicates the thickness significantly decreases compared with the corresponding single-crystal lamellae. This is because during solution crystallization, polymer chain ends are randomly exposed on the two surfaces of the 2D PSC (schematics in Fig. 2a–c). If all the alkoxysilane chain ends on one side of the lamella were successfully coupled with the substrate, we then anticipate a decrease of the coating thickness to 50% of the original value. The experimental results show a slightly greater decrease (Fig. 2), which can be attributed to the unreacted chain ends during immobilization—when PSCs were deposited onto a substrate, because of the surface roughness of the PSCs, not all the alkoxysilane chain ends have intimate contact with the substrate to ensure a successful coupling.

Grafting density of the PCL brush samples prepared using different PSC templates can be calculated using the following equation:

$$\sigma = \frac{\rho \times h \times N_A}{MW} \quad (1)$$

where ρ is the polymer density, σ is the grafting density, h is the polymer brush layer height, MW is the molar mass of polymer brushes and N_A is the Avogadro's number. Because the thickness measured using as-prepared PCL brushes can be inaccurate due to the possible 'constrained dewetting', which could result in inhomogeneous coating during the evaporation of acetone after the washing process[31,32], thermal annealing of the as-prepared polymer brush samples was performed overnight at 35 °C. This is the onset temperature of TEG-PCL30-SiOR crystallization according to the differential scanning calorimetry (DSC) thermogram (Supplementary Fig. 2c), annealing at this temperature leads to the crystallization of the PCL brushes, which eventually removes excessive free volume in the polymer brush coating, hence more precise brush density calculation. Height images obtained from AFM of these annealed samples are summarized in Fig. 2i–k. Compared with as-prepared polymer brushes, morphology of the polymer brush retained, while the thickness of annealed polymer brushes slightly decreased with a smoother surface according to image analysis results summarized in Table 1. Calculation of

Table 1 | Summary of the polymer brush synthesis using mono-functionalized PCL single-crystal templates prepared at different crystallization temperatures (T_c).

Polymer	T_c (°C)	$h_{crystal}$* (nm)	h_{brush}† (nm)	R_q‡ (nm)	h_{brush}§ (nm)	R_q‖ (nm)	σ_{theory}¶ (nm^{-2})	σ_{real}# (nm^{-2})
TEG-PCL30-SiOR	5	5.5	2.8	0.36	2.2	0.18	0.53	0.44
	15	9.0	3.9	0.38	3.2	0.29	0.89	0.65
	35	12.0	5.4	0.34	4.6	0.23	1.33	0.94
TEG-PCL16-SiOR	30	14.0	6.6	0.42	5.6	0.28	2.67	2.12

*Thickness of single-crystal lamellae measured using AFM.
†As-prepared polymer brush thickness measured using AFM.
‡Root mean squared (r.m.s.) average of height deviations of polymer brush-covered regime obtained from AFM images of as-prepared polymer brushes.
§Polymer brush thickness after thermal annealing obtained from AFM.
‖R.m.s. average of height deviations of polymer brushes covered regime obtained from AFM images of polymer brushes after thermal annealing.
¶Theoretical grafting density of polymer brushes assuming all alkoxysilane chain ends on one side of the crystal are coupled with the glass surface.
#Calculated grafting density of polymer brushes using equation (1) based on experimental results.

polymer brush grafting density was then performed based on the annealed sample thickness, which yields the grafting densities of 0.44, 0.65 and 0.94 chain per nm^2. Figure 3 shows the linear correlation of the grafting density with the chain-folding number n in the PSCs. The results therefore demonstrate the capability of precise control over the grafting density of the polymer brushes formed by a given polymer using our method. Note that the polymer chain grafting density calculated from AFM characterizations is slightly smaller than the grafting density of -SiOR groups on the single crystals, which is likely due to the reaction efficiency issue during solid-state coupling between single-crystal lamellae and glass substrates. We can define the coupling efficiency (f) as the ratio between the experimental and theoretical grafting densities. For PSCs with three thicknesses, f can be found to be 0.71, 0.73 and 0.78, respectively.

Polymer brushes with a grafting density >1 chain per nm^2 represent extremely dense brushes, and have been rarely reported even using the grafting-from approach[33–36]. The highest grafting density achieved in the above-mentioned system is 0.94 chain per nm^2. Figure 3 suggest that if extended-chain single crystals ($n=0$) prepared from mono-functionalized PCL are used as templates, the grafting density can be further increased. To test this hypothesis, single crystals from PCL with degree of polymerization of 16 (TEG-PCL16-SiOR, see Supplementary Fig. 2 for characterizations of its chemical structure) was prepared via the self-seeding method at $T_c = 30$ °C. As shown in Fig. 2d, the obtained lamellar thickness is ~14 nm, which confirms that extended-chain structure has been achieved. In this case, the surface areal density of alkoxysilane groups on one side of the lamellar surface is 2.67 –SiOR per nm^2, which in theory is the highest for symmetric mono-functionalized PCL crystals. Using the same ammonia-catalysed solid-state coupling reaction, the polymer brushes were successfully prepared and studied using AFM. As seen in Fig. 2h, the as-prepared brush domain has a thickness of ~6.6 nm, which decreased to 5.6 nm after thermal annealing at 35 °C (Fig. 2l). The grafting density of the polymer brushes can therefore be estimated to be ~2.12 chains per nm^2 using equation (1). Compared with other studies on the synthesis of polymer brushes on planar surfaces using either grafting-to or grafting-from methods, the grafting density of polymer brushes prepared using our method can reach a considerably higher value.

PSCs can be conveniently deposited onto a glass substrate to prepare a uniform, large-scale coated surface. As an example, TEG-PCL30-SiOR PSCs grown at 15 °C were drop casted onto a 5 mm × 5 mm glass slides. After the residual solvent was evaporated, similar solid-state coupling reaction was conducted to form chemically bounded polymer brushes on the glass slide. Over 10 random areas from the substrate were subject to phase-contrast optical microscopy imaging, and a typical graph is shown

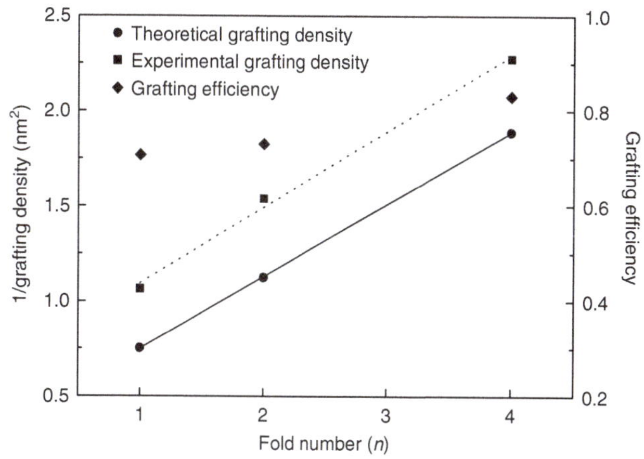

Figure 3 | Chain-folding determines grafting density. Correlation between the grafting density of the polymer brushes and the folding number of the polymer single-crystal templates.

in Fig. 4a. Nearly 100% of the surface is covered with polymer brushes within this large area (scale bar, 50 μm). Furthermore, AFM characterizations were also applied to study the nanostructures of polymer brushes prepared herein as shown in Fig. 4b. The polymer brush-covered area is homogeneous with a surface roughness comparable to those prepared using isolated PSCs as templates summarized in Table 1. Phase-contrast microscopy and AFM experiments demonstrate that our method can be used to functionalize a macroscale surface with a uniform coverage. Of interest is that no obvious 'grain boundaries' between PSCs were observed in AFM images. When casting PSCs onto a solid substrate, one expects that due to the defined shape/sizes, regions between adjacent PSCs might not be coated with brushes as shown in Fig. 4c. The observed uniform coating of the PSCs can be explained by the combination of elasto-capillary bending and chain sliding within PSCs[19–21,37,38]. For overlapped PSCs as shown in Fig. 4c, on solvent evaporation, the elasto-capillary bending would cause the top PSC deform towards the substrate, and the resultant interface width can be estimated as ~190 nm (ref. 39). However, detailed AFM analysis in Supplementary Fig. 4 suggests that the collapsed interface has an average width of ~33 nm (average of 10 measurements). The further sharpening of the interface can be explained by the chain sliding in the PSC caused by the capillary force. Because of the weak van der Waals force between adjacent chains in the PSC, it is relatively easy for chains to slide among each other[40]. As shown in Fig. 4d,e, as stacking occurs, the polymer chains of the top lamella slide down and fill in the bottom areas.

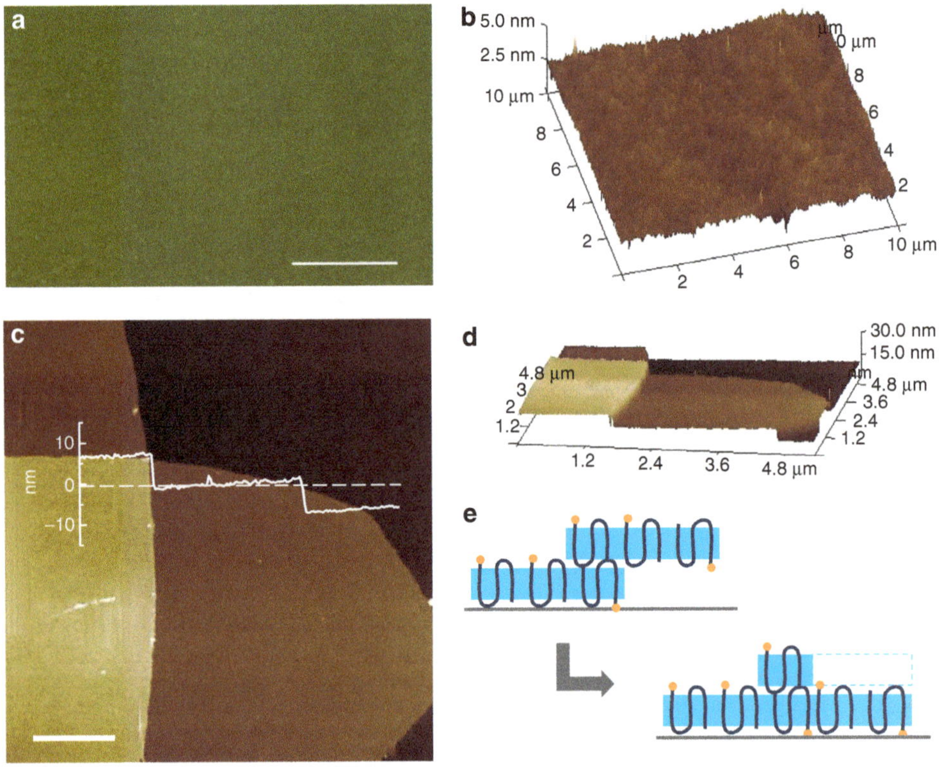

Figure 4 | Uniform coating of the polymer brushes at a macroscopic length scale. (**a**) Phase-contrast optical microscopy image of polymer brush-coated glass surface prepared using TEG-PCL30-SiOR single crystal grown at 15 °C as templates. Large surface grafting can be achieved with small amount of unmodified area showing relatively lighter colour compared with brush-covered regime; (**b**) three-dimensional (3D) height images of a 10 × 10-μm AFM scan randomly picked from image **a**; (**c**) AFM height image shows the stacking of two TEG-PCL30-SiOR PSCs prepared at 15 °C. The inset of **c** shows the height profile of the crystal lamellae, and the sudden change (instead of gradual decrease) in height at the edge of overlapped area indicates chain sliding. (**d**) 3D height image of **c**; (**e**) schematic illustration of the chain sliding when two PSC lamellae stack with each other when deposited onto a solid substrate. Scale bars, 50 μm (**a**); 1 μm (**c**).

This chain sliding apparently accounts for the filling of the otherwise 'empty' surface due to lamella stacking; it leads to the formation of uniform and dense polymer brushes.

Synthesis of dense PLBs. Another important polymer conformation on substrate is PLBs, where the polymer chain is tethered via both chain ends. The chain conformation, surface dynamics, friction and mechanical response of PLBs are found quite different compared with their single-tethered polymer brush (STPB) counterparts[41–48]. Detailed experimental studies on these materials however have been quite limited due to the lack of efficient approach towards uniform PLBs coating with precisely controlled ring size, molar mass and high grafting density. Up till now, only solution-based physisorption of triblock terpolymer[49–52] and grafting-to of telechelic polymers[53–56] have been successfully applied for the synthesis of PLBs on flat surfaces. Yet it is challenging to precisely control final brush structures and grafting density. Reversible addition-fragmentation chain transfer polymerization has been used to synthesize poly(methyl acrylate) loops on silica nanoparticle; chain branching of the brushes was noticed due to the close proximity of growing radicals[57]. Our method can also be used to synthesize PLBs with well-controlled anchoring points and high grafting density. To this end, α, ω-alkoxysilane-terminated telechelic PCL (diPCL-2SiOR) was used for PSC preparation. The chemical structures of these polymers were systematically studied using NMR and GPC as shown in Supplementary Fig. 3, which confirmed the average degree of polymerization of these PCL was 56, while the Đ was ~1.3. When a telechelic polymer chain folds odd-number times in the crystal lamellae, both chain ends are exposed onto the same side of lamellar surfaces as illustrated in Fig. 1b. After depositing such PSCs onto a solid substrate, coupling chain ends to the solid substrate would lead to PLBs. In our experiment, the lamellar thickness of the single crystals was adjusted by changing the T_c as shown in Fig. 5a,b. When PCL was crystallized at 5 °C, the obtained lamellae thickness was 8.5 nm, which increased to 12 nm when crystallized at 35 °C. These two lamellar thicknesses correspond to five and three times chain folding, respectively. Oddly folded chain structures were then successfully achieved. Using the similar solid-state grafting-to method, polymer brushes were successfully obtained. Figure 5 summarizes the AFM height images of the as-prepared polymer brushes templated from PSCs grown at 5 and 35 °C. Similar with the STPBs templated from PCL-SiOR PSCs, polymer brush domain herein also retained the same shape of crystal templates, and the as-prepared polymer brushes templated from diPCL-2SiOR PSCs prepared at 5 °C is about 4.8 nm in thickness, while the ones using PSCs grown at 35 °C show a dry thickness of around 5.6 nm.

To further confirm polymer brushes prepared herein were indeed loops, a control experiment was designed using PSCs of dithiol functionalized PCL (diPCL-2SH, degree of polymerization ~56 as calculated from NMR, see Supplementary Fig. 3) grown at 35 °C as the polymer brush templates. As seen in Fig. 6a, the thickness of crystal lamellae is the same as diPCL-2SiOR PSCs prepared at 35 °C, which confirms the identical chain-folding morphology within these two crystal lamellae (three-fold). In this case, polymer brushes prepared from these templates should have exactly the same morphology with those prepared from diPCL-

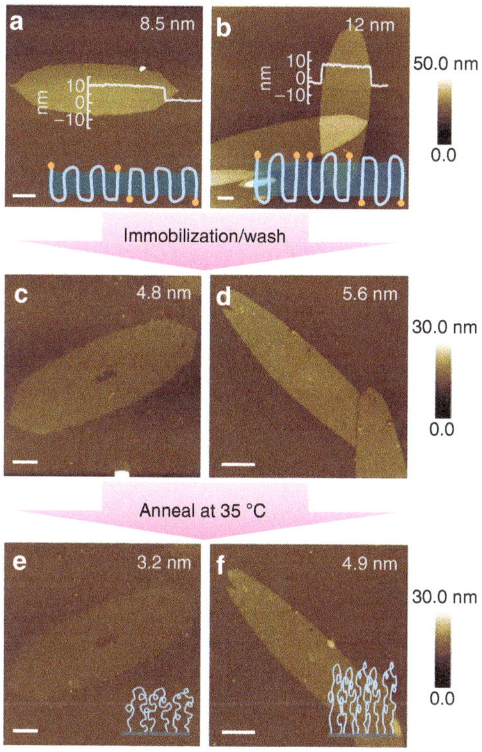

Figure 5 | Controlled grafting of dense polymer loops. Tapping-mode-AFM height images of polymer single crystals prepared from diPCL-2SiOR at 5 (**a**) and 35 °C (**b**). The inset curves show the corresponding height profile of crystal lamellae, and the schematics illustrate the oddly folded chain conformations ($n=3$ and 5). (**c-f**) The corresponding PLBs before (**c**,**d**) and after (**e**,**f**) annealing. The schematics illustrate the polymer loops with different grafting density. Scale bars, 1 µm. The inset numbers on the top right corner of each image are the measured thicknesses using AFM.

2SiOR. Polymer brushes were then grafted onto a gold substrate using diPCL-2SH PSCs as templates via Au–S bonds, which formed readily under ambient condition followed by the removal of unbound polymer chains. The successful immobilization of PCL brushes was confirmed using high-resolution (HR) C1*s* X-ray photoelectron spectroscopy (XPS) spectrum as shown in Fig. 6b. The deconvolution of the signal clearly shows three different components with binding energies at about 284.6, 286.2 and 288.7 eV, attributed to the C–H, C–O and O=C–O species, respectively[58]. The area ratio of [C–H]:[C–O]:[O=C–O] is around 73:14:13, which is in good agreement with the theoretical value of 5:1:1 for the PCL structure[59].

HR S2*p* XPS spectra were then employed to investigate the formation of chemical bonds between polymer chain ends and Au substrates to confirm whether these polymer brushes were singly or doubly tethered[54]. The chemical structures of the tethering sites of as-prepared polymer brushes were first examined as shown in the first spectrum of Fig. 6c with clear peaks of the 162/163.1 eV S2*p* spin–orbital pair (S2*p*, 3/2, 1/2), which are attributed to the Au–S bond formation, demonstrating successful chemical tethering of polymer brushes. Moreover, no free thiol-binding energy signals at 163.5/164.8 eV can be observed, which suggests all thiol groups at the polymer chain ends are chemically bounded, confirming the loop-structured brushes. This can be further supported by oxidizing the polymer brushes under ambient condition, and the spectrum after oxidation shown in the second spectrum of Fig. 6c does not exhibit any noticeable signals from oxidized thiol. A control sample prepared by skipping the washing process after immobilization of diPCL-2SH PSCs on Au substrate was subject to the same oxidation procedure, and the S2*p* HR-XPS spectrum of this sample is shown in the bottom spectrum of Fig. 6c. Other than similar thiolated sulfur signals, strong sulfonic acid peaks due to oxidized unbound thiol groups with binding energy of 169.1/170.3 eV (S2*p*, 3/2, 1/2) can be clearly seen in the curve. On the basis of these results, the PCL brushes prepared herein had all the chain ends being chemically tethered, hence forming loop conformation (PLBs).

The grafting density of PLBs templated from diPCL-2SiOR PSCs was also calculated based on the dry thickness obtained from AFM imaging using equation (1). Thermal annealing at 35 °C was conducted to remove free volume within brush layers, and the series of AFM images is summarized in Fig. 5. Similar effects on the thickness of polymer brushes and the surface roughness can be observed compared with STPBs described in the previous discussion. On the basis of the annealed layer thickness, the grafting density σ can be calculated as 0.35 and 0.53 loops per nm^2 for PLBs templated from PSCs grown at 5 and 35 °C, respectively. Meanwhile, since two tethering points are required to make a single-loop brush, the grafting density of effective tethering points is 0.7 and 1.06 tethering points per nm^2, respectively. Compared with other preparation methods reported in the literatures[49,60], the grafting density of PLBs obtained herein has been greatly improved.

The reported 'self-assembly-assisted-grafting-to' method is a fundamentally new approach to synthesize polymer brushes. The key of this new strategy is pre-assembling polymer chains into a 2D sheet. This process leads to the guided assembly of the end-functional groups of the polymer chain into a 2D pattern with a controlled areal density. Subsequent coupling the pre-assembled sheet therefore leads to well-defined polymer brushes. Advantages of this method include the following: (1) grafting densities of the polymer brushes can be easily controlled by PSC chain folding; (2) unprecedented high grafting densities of 2.12 chains per nm^2 can be achieved; (3) PLBs can be easily synthesized and the PLBs also have well-defined structure and controlled grafting densities; and (4) the tethering points of the polymer brushes are controlled by chain folding in the PSCs, and therefore their locations on the substrates are also precisely controlled. For example, in the case of PLB, if we assume the chain-folding direction in PCL is typically (200) or (110), three-fold conformation dictates the chain end-to-end distance to be 1.12 and 1.24 nm. Similarly, five-fold conformation leads to the chain end-to-end distance of 1.87 and 2.07 nm. The tethering point of the PLB can therefore be precisely controlled. Note that since the pre-assembly process is required for the polymer brush synthesis in this study, it is limited to crystalline polymers. Nevertheless, we envisage that block copolymers containing one crystalline segment can be used for single-crystal growth to synthesize noncrystalline polymer brushes. More detailed work will be reported in the near future.

Discussion

In this work, we report a fundamentally new 'self-assembly-assisted-grafting-to' method to synthesize polymer brushes. End-functionalized PCL has been pre-assembled into 2D single-crystal sheets, chemical coupling of which with solid substrates led to polymer brushes. For a given molar mass, by changing the crystallization temperature, the chain-folding structures can be precisely controlled, which eventually can be used for tuning grafting density of the subsequent polymer brushes. Polymer brushes with a grafting density as high as 2.12 chains per nm^2 have been achieved. Moreover, we

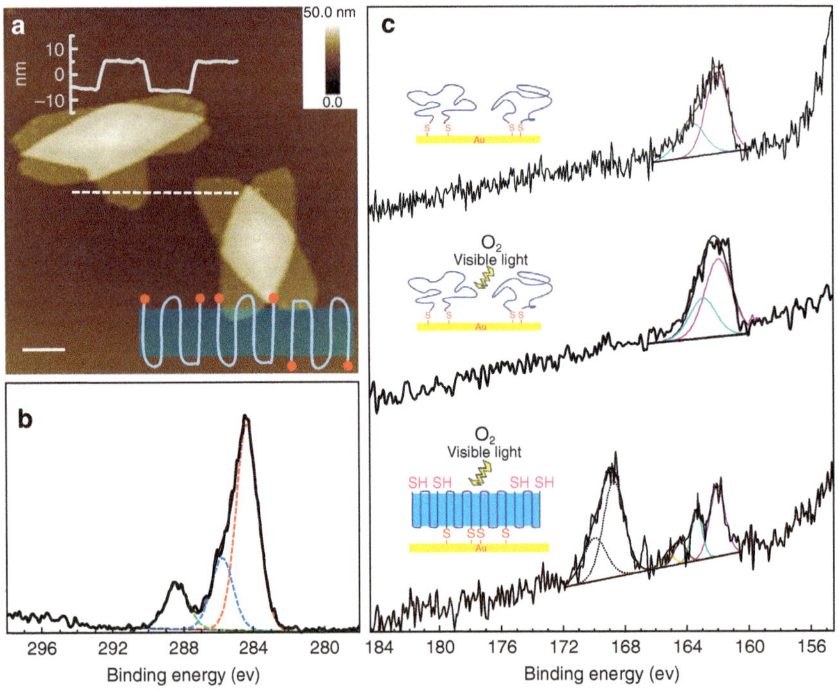

Figure 6 | Confirmation of the polymer loop conformation. (**a**) AFM height image of single crystal grown from diPCL-2SH at 35 °C. The inset curve shows the height profile of crystal lamellae along the dotted line, and the cartoon illustrates the three-time chain-folding morphology of PCL molecule perpendicular to lamellae surface. Red solid spheres represent exposed thiol chain ends. Scale bar, 1 μm. (**b**) C1s HR-XPS spectrum and its deconvolution results of PCL brushes on Au substrate templated from diPCL-2SH PSCs crystallized at 35 °C; (**c**) S2*p* HR-XPS spectra and corresponding deconvolution results. From top to bottom: as-prepared PCL PLBs; PCL loop brushes after oxidation; and diPCL-2SH PSCs on Au substrate after oxidation. The insets illustrate the chemical structures of different sulfur species existed in polymer brushes and PSCs.

have demonstrated that by using oddly folded PSCs with bi-functionalized chain end, PLBs with controlled tethering points, and grafting densities can be synthesized. Our method therefore provides a new pathway to synthesizing polymer brushes, particularly super-dense polymer brushes, with controlled architecture and grafting density.

Methods

Materials. (3-isocyanatopropyl)triethoxysilane (95%), dibutyltin dilaurate (95%), tin(II) 2-ethylhexanoate (92.5–100.0%) and dichloromethane (anhydrous, 99.8%) were purchased from Aldrich and used as received. Tetraethylene glycol mono-methyl ether (98%) was bought from Alfa Aesar and used as received. ε-Caprolactone (97%) was purchased from Aldrich and distilled under reduced pressure before use. 1-butanol (≥99.5%) was purchased from Aldrich and distilled before use. All other reagents were purchased from Aldrich and used as received.

Synthesis of model polymers, procedure of polymer single-crystal growth and brush synthesis can be found in Supplementary Methods.

Characterizations. Proton nuclear magnetic resonance spectra of synthesized PCL were recorded on a Varian 500 MHz spectrometer using deuterated chloroform ($CDCl_3$) as the solvent and tetramethylsilane as the internal standard. GPC characterizations were conducted using Waters GPC system with a 1525 binary HPLC pump, and a Waters 2414 refractive index detector was used. All measurements were performed using tetrahydrofuran as the carrier solvent with a flow rate of 1.0 ml min^{-1} at 30 °C. Standard monodispersed PS (Shodex standard, Kawasaki, Japan) were used for calibration. DSC experiments were carried out using a Perkin-Elmer DSC 7. The samples with an average weight of 2 mg were heated from 0 to 100 °C at a scanning rate of 10 °C min^{-1} under a nitrogen atmosphere and were cooled and reheated using the same rate. Phase-contrast optical microscopy images were collected using an Olympus BX-51 equipped with an Insight digital camera. Tapping-mode-AFM experiments were conducted on a Bruker Multimode 8 AFM and a Bruker Dimension Icon AFM (Bruker Nano, Santa Barbara, CA). Specifically, TESPA silicon probes (Bruker, Camarillo, CA) with spring constant $k \sim 42$ N m^{-1} and resonance frequency ~320 kHz were used, and the images were acquired with 512 × 512 points at a scan rate of ~1.0 Hz per line. Nanoscope Analysis software (Version 1.40, Bruker, Camarillo, CA) was used for image analysis to obtain polymer brush thickness and surface roughness data.

References

1. Milner, S. T. Polymer brushes. *Science* **251,** 905–914 (1991).
2. Rubinstein, M. & Colby, R. H. *Polymer Physics* (Oxford Univ. Press, 2003).
3. Hucknall, A., Rangarajan, S. & Chilkoti, A. In pursuit of zero: polymer brushes that resist the adsorption of proteins. *Adv. Mater.* **21,** 2441–2446 (2009).
4. Chen, T., Ferris, R., Zhang, J., Ducker, R. & Zauscher, S. Stimulus-responsive polymer brushes on surfaces: transduction mechanisms and applications. *Prog. Polym. Sci.* **35,** 94–112 (2010).
5. Stuart, M. A. C. *et al.* Emerging applications of stimuli-responsive polymer materials. *Nat. Mater.* **9,** 101–113 (2010).
6. Ma, X. *et al.* Polymer brush electrets. *Adv. Funct. Mater.* **23,** 3239–3246 (2013).
7. Hess, L. H. *et al.* Graphene transistors with multifunctional polymer brushes for biosensing applications. *ACS Appl. Mater. Interfaces* **6,** 9705–9710 (2014).
8. Krishnamoorthy, M., Hakobyan, S., Ramstedt, M. & Gautrot, J. E. Surface-initiated polymer brushes in the biomedical field: applications in membrane science, biosensing, cell culture, regenerative medicine and antibacterial coatings. *Chem. Rev.* **114,** 10976–11026 (2014).
9. Wolski, K., Szuwarzynski, M. & Zapotoczny, S. A facile route to electronically conductive polyelectrolyte brushes as platforms of molecular wires. *Chem. Sci.* **6,** 1754–1760 (2015).
10. Zhao, B. & Brittain, W. J. Polymer brushes: surface-immobilized macromolecules. *Prog. Polym. Sci.* **25,** 677–710 (2000).
11. Edmondson, S., Osborne, V. L. & Huck, W. T. S. Polymer brushes via surface-initiated polymerizations. *Chem. Soc. Rev.* **33,** 14–22 (2004).
12. Barbey, R. *et al.* Polymer brushes via surface-initiated controlled radical polymerization: synthesis, characterization, properties, and applications. *Chem. Rev.* **109,** 5437–5527 (2009).
13. Zdyrko, B. & Luzinov, I. Polymer brushes by the 'grafting to' method. *Macromol. Rapid Commun.* **32,** 859–869 (2011).
14. Wu, T., Efimenko, K. & Genzer, J. Combinatorial study of the mushroom-to-brush crossover in surface anchored polyacrylamide. *J. Am. Chem. Soc.* **124,** 9394–9395 (2002).
15. Kobayashi, M., Terada, M., Terayama, Y., Kikuchi, M. & Takahara, A. Direct synthesis of well-defined poly {2-(methacryloyloxy)ethyl}trimethylammonium chloride brush via surface-initiated atom transfer radical polymerization in fluoroalcohol. *Macromolecules* **43,** 8409–8415 (2010).
16. Schuh, C., Lomadze, N., Ruehe, J., Kopyshev, A. & Santer, S. Photomechanical degrafting of azo-functionalized poly(methacrylic acid) (PMAA) brushes. *J. Phys. Chem. B* **115,** 10431–10438 (2011).

17. Jia, X., Zhang, G., Li, W., Sheng, W. & Li, C. Tuning the low critical solution temperature of polymer brushes grafted on single-walled carbon nanotubes and temperature dependent loading and release properties. *J. Polym. Sci. Polym. Chem.* **52**, 1807–1814 (2014).
18. Kang, C., Crockett, R. M. & Spencer, N. D. Molecular-weight determination of polymer brushes generated by SI-ATRP on flat surfaces. *Macromolecules* **47**, 269–275 (2014).
19. Geil, P. *Polymer Single Crystals* (Krieger, 1973).
20. Wunderlich, B. *Macromolecular Physics* (Academic, 1976).
21. Cheng, S. Z. D. *Phase Transitions in Polymers* (Elsevier, 2008).
22. Chen, W. Y. *et al.* Onset of tethered chain overcrowding. *Phys. Rev. Lett.* **93**, 028301–028304 (2004).
23. Li, B. & Li, C. Y. Immobilizing Au nanoparticles with polymer single crystals, patterning and asymmetric functionalization. *J. Am. Chem. Soc.* **129**, 12–13 (2007).
24. Wang, B. B., Li, B., Dong, B., Zhao, B. & Li, C. Y. Homo- and hetero-particle clusters formed by janus nanoparticles with bicompartment polymer brushes. *Macromolecules* **43**, 9234–9238 (2010).
25. Dong, B., Wang, W., Miller, D. L. & Li, C. Y. Polymer-single-crystal@nanoparticle nanosandwich for surface enhanced Raman spectroscopy. *J. Mater. Chem.* **22**, 15526–15529 (2012).
26. Dong, B., Zhou, T., Zhang, H. & Li, C. Y. Directed self-assembly of nanoparticles for nanomotors. *ACS Nano* **7**, 5192–5198 (2013).
27. Zhou, T., Dong, B., Qi, H., Lau, H. K. & Li, C. Y. One-step formation of responsive 'dumbbell' nanoparticle dimers via quasi-two-dimensional polymer single crystals. *Nanoscale* **6**, 4551–4554 (2014).
28. Zhou, T., Dong, B., Qi, H., Mei, S. & Li, C. Y. Janus hybrid hairy nanoparticles. *J. Polym. Sci. B Polym. Phys.* **52**, 1620–1640 (2014).
29. Lauritzen, J. I. & Hoffman, J. D. Theory of formation of polymer crystals with folded chains in dilute solution. *J. Res. Natl Bur. Stand. A* **64A**, 73–102 (1960).
30. Hu, H. & Dorset, D. L. Crystal structure of poly(ε-caprolactone). *Macromolecules* **23**, 4604–4607 (1990).
31. Lee, T., Hendy, S. C. & Neto, C. Control of nanoparticle formation using the constrained dewetting of polymer brushes. *Nanoscale* **7**, 2894–2899 (2015).
32. Lee, T., Hendy, S. C. & Neto, C. Tunable nanopatterns via the constrained dewetting of polymer brushes. *Macromolecules* **46**, 6326–6335 (2013).
33. Zdyrko, B., Klep, V. & Luzinov, I. Synthesis and surface morphology of high-density poly(ethylene glycol) grafted layers. *Langmuir* **19**, 10179–10187 (2003).
34. Liu, Y., Klep, V., Zdyrko, B. & Luzinov, I. Polymer grafting via ATRP initiated from macroinitiator synthesized on surface. *Langmuir* **20**, 6710–6718 (2004).
35. Yoshikawa, C. *et al.* Fabrication of high-density polymer brush on polymer substrate by surface-initiated living radical polymerization. *Macromolecules* **38**, 4604–4610 (2005).
36. Patil, R. R., Turgman-Cohen, S., Srogl, J., Kiserow, D. & Genzer, J. On-demand degrafting and the study of molecular weight and grafting density of poly(methyl methacrylate) brushes on flat silica substrates. *Langmuir* **31**, 2372–2381 (2015).
37. King, H., Schroll, R. D., Davidovitch, B. & Menon, N. Elastic sheet on a liquid drop reveals wrinkling and crumpling as distinct symmetry-breaking instabilities. *Proc. Natl Acad. Sci. USA* **109**, 9716–9720 (2012).
38. Chang, C.-C., Wang, Z., Sheng, Y.-J. & Tsao, H.-K. Nanostructure collapse by elasto-capillary instability. *Soft Matter* **10**, 8542–8547 (2014).
39. Roman, B. & Bico, J. Elasto-capillarity: deforming an elastic structure with a liquid droplet. *J. Phys. Condens. Matter* **22**, 493101 (2010).
40. Liu, K. *et al.* Extracting a single polyethylene oxide chain from a single crystal by a combination of atomic force microscopy imaging and single-molecule force spectroscopy: toward the investigation of molecular interactions in their condensed states. *J. Am. Chem. Soc.* **133**, 3226–3229 (2011).
41. Shull, K. R. Wetting autophobicity of polymer melts. *Faraday Discuss.* **98**, 203–217 (1994).
42. Jones, R. L. & Spontak, R. J. Effect of chain-length and surface density on looped polymers grafted to an impenetrable surface. *J. Chem. Phys.* **103**, 5137–5143 (1995).
43. Gulati, H. S., Hall, C. K., Jones, R. L. & Spontak, R. J. Equilibrium conformations and dynamic relaxation of double-tethered chain molecules at an impenetrable interface. *J. Chem. Phys.* **105**, 7712–7722 (1996).
44. Honeker, C. C. & Thomas, E. L. Impact of morphological orientation in determining mechanical properties in triblock copolymer systems. *Chem. Mater.* **8**, 1702–1714 (1996).
45. Irvine, D. J., Mayes, A. M. & GriffithCima, L. Self-consistent field analysis of grafted star polymers. *Macromolecules* **29**, 6037–6043 (1996).
46. Maeda, N., Chen, N. H., Tirrell, M. & Israelachvili, J. N. Adhesion and friction mechanisms of polymer-on-polymer surfaces. *Science* **297**, 379–382 (2002).
47. Meng, X. X. & Russel, W. B. Telechelic associative polymers: Interactions between strongly stretched planar adsorbed layers. *Macromolecules* **36**, 10112–10119 (2003).
48. Makke, A., Lame, O., Perez, M. & Barrat, J.-L. Influence of Tie and loop molecules on the mechanical properties of lamellar block copolymers. *Macromolecules* **45**, 8445–8452 (2012).
49. Alonzo, J. *et al.* Looped polymer brushes formed by self-assembly of poly(2-vinylpyridine)-polystyrene-poly(2-vinylpyridine) triblock copolymers at the solid-fluid interface. Kinetics of preferential adsorption. *Macromolecules* **39**, 8434–8439 (2006).
50. Huang, Z. *et al.* Impact of solvent quality on the density profiles of looped triblock copolymer brushes by neutron reflectivity measurements. *Macromolecules* **41**, 1745–1752 (2008).
51. Alonzo, J., Mays, J. W. & Kilbey, II S. M. Forces of interaction between surfaces bearing looped polymer brushes in good solvent. *Soft Matter* **5**, 1897–1904 (2009).
52. Banquy, X., Burdynska, J., Lee, D. W., Matyjaszewski, K. & Israelachvili, J. Bioinspired bottle-brush polymer exhibits low friction and amontons-like behavior. *J. Am. Chem. Soc.* **136**, 6199–6202 (2014).
53. Haung, Z. *et al.* Polymer loop formation on a functionalized hard surface: quantitative insight by comparison of experimental and monte carlo simulation results. *Langmuir* **26**, 202–209 (2010).
54. Patton, D., Knoll, W. & Advincula, R. C. Polymer loops vs. brushes on surfaces: adsorption, kinetics, and viscoelastic behavior of alpha,omega-thiol telechelics on gold. *Macromol. Chem. Phys.* **212**, 485–497 (2011).
55. Ehlert, S. *et al.* Polymer ligand exchange to control stabilization and compatibilization of nanocrystals. *ACS Nano* **8**, 6114–6122 (2014).
56. Sakurai, S., Watanabe, H. & Takahara, A. Preparation and characterization of looped polydimethylsiloxane brushes. *Polym. J.* **46**, 117–122 (2014).
57. Rotzoll, R. & Vana, P. Synthesis of poly(methyl acrylate) loops grafted onto silica nanoparticles via reversible addition-fragmentation chain transfer polymerization. *J. Polym. Sci. Polym. Chem.* **46**, 7656–7666 (2008).
58. Cai, T., Li, M., Neoh, K.-G. & Kang, E.-T. Preparation of stimuli responsive polycaprolactone membranes of controllable porous morphology via combined atom transfer radical polymerization, ring-opening polymerization and thiol-yne click chemistry. *J. Mater. Chem.* **22**, 16248–16258 (2012).
59. Hu, Y., Zhao, N. N., Li, J. S., Yang, W. T. & Xu, F. J. Temperature-responsive porous polycaprolactone-based films via surface-initiated ATRP for protein delivery. *J. Mater. Chem.* **22**, 21257–21264 (2012).
60. Seo, J.-H. *et al.* Modulation of friction dynamics in water by changing the combination of the loop- and graft-type poly(ethylene glycol) surfaces. *Soft Matter* **11**, 936–942 (2015).

Acknowledgements
This work was supported by the National Science Foundation Grant DMR-1308958. AFM was purchased under grant #W911NF-14-1-0450.

Author contributions
C.Y.L. conceived and directed the project; T.Z., H.Q. and D.B. designed and performed the experiments; L.H. helped with the AFM experiments; T.Z., H.Q. and C.Y.L. discussed the results and wrote the manuscript.

Additional information

Competing financial interests: The authors declare no competing financial interests.

14

Thermal selectivity of intermolecular versus intramolecular reactions on surfaces

Borja Cirera[1], Nelson Giménez-Agulló[2], Jonas Björk[3], Francisco Martínez-Peña[1], Alberto Martin-Jimenez[1], Jonathan Rodriguez-Fernandez[4], Ana M. Pizarro[1], Roberto Otero[1,4], José M. Gallego[5], Pablo Ballester[2,6], José R. Galan-Mascaros[2,6] & David Ecija[1]

On-surface synthesis is a promising strategy for engineering heteroatomic covalent nanoarchitectures with prospects in electronics, optoelectronics and photovoltaics. Here we report the thermal tunability of reaction pathways of a molecular precursor in order to select intramolecular versus intermolecular reactions, yielding monomeric or polymeric phthalocyanine derivatives, respectively. Deposition of tetra-aza-porphyrin species bearing ethyl termini on Au(111) held at room temperature results in a close-packed assembly. Upon annealing from room temperature to 275 °C, the molecular precursors undergo a series of covalent reactions via their ethyl termini, giving rise to phthalocyanine tapes. However, deposition of the tetra-aza-porphyrin derivatives on Au(111) held at 300 °C results in the formation and self-assembly of monomeric phthalocyanines. A systematic scanning tunnelling microscopy study of reaction intermediates, combined with density functional calculations, suggests a [2 + 2] cycloaddition as responsible for the initial linkage between molecular precursors, whereas the monomeric reaction is rationalized as an electrocyclic ring closure.

[1] IMDEA Nanoscience, c/Faraday 9, Cantoblanco, 28049 Madrid, Spain. [2] Institute of Chemical Research of Catalonia, Barcelona Institute of Science and Technology, Avinguda Països Catalans 16, Tarragona 43007, Spain. [3] Department of Physics, Chemistry and Biology, IFM, Linköping University, 58183 Linköping, Sweden. [4] Departamento de Física de la Materia Condensada, Universidad Autónoma de Madrid, c/Francisco Tomás y Valiente 7, Cantoblanco, 28049 Madrid, Spain. [5] Instituto de Ciencia de Materiales de Madrid, c/ Sor Juana Inés de la Cruz 3, Cantoblanco, 28049 Madrid, Spain. [6] Catalan Institution for Research and Advanced Studies, Passeig Lluis Companys 23, Barcelona 08010, Spain. Correspondence and requests for materials should be addressed to P.B. (email: pballester@iciq.es) or to J.R.G.-M. (email: jrgalan@iciq.es) or to D.E. (email: david.ecija@imdea.org).

Surface-mediated synthesis of low-dimensional polymers from simple molecular precursors (monomers) under ultra-clean conditions is a rapidly emerging field with relevance for molecular electronics, optoelectronic devices, magnetism, molecular recognition and sensing, catalysis, filtration and membranes[1–5]. This promising bottom-up strategy relies on a careful selection of building blocks equipped with functional groups and surfaces, whereby one-dimensional and two-dimensional nanostructures have been engineered and a plethora of new materials is envisioned[6–21].

Regarding the molecular precursors, porphyrin and phthalocyanine derivatives are the subject of increasing attention, inspired by scientific curiosity, biological relevance and potential technological impact[22].

Several chemical reactions have been exploited to produce surface-assisted lateral covalent designs inspected with ultimate spatial resolution[23]. These include cyclocondensation of boronic acids[8], tetramerization of 1,2,4,5-tetracyanobenzene with iron[16], Schiff-base reaction of aldehydes and amines to form imine[24], acylation reactions (yielding polyamides[25], polyimides[7] or polyesters[26]), Bergmann cyclization[27], azide–alkyne cycloadditions[28,29] and carbon–carbon couplings exemplified by alkane polymerizations[30], the Ullmann[6,17], the Glaser[18] and the Sonogashira reactions[21] or dehalogenation reactions followed by radical combination[13,14,31]. Despite their importance in bulk chemistry and their growing relevance in 3D polymer science, [2+2] cycloaddition reactions have not yet been exploited to produce surface-confined polymeric nanostructures on surfaces.

Importantly, the selectivity of reaction pathways under distinct thermal stimuli, although being a promising strategy to increase synthetic versatility, has remained mostly elusive on surfaces[17]. Here, we introduce surface-confined thermally tunable reaction pathways as a route to select intramolecular versus intermolecular covalent reactions yielding either monomeric phthalocyanines or low-dimensional phthalocyanine polymers, respectively. To this end, we deposit 2,3,7,8,12,13,17,18-octaethyl-5,10,15,20-tetraaza-porphyrin (OETAP) under ultra-high vacuum on a pristine Au(111) crystal held at room temperature, forming close-packed supramolecular assemblies. This precursor phase is gently annealed to 300 °C giving rise to quasi-one-dimensional polymers that can be rationalized as phthalocyanine derivatives. Further insights of the polymerization process with high-resolution low-temperature scanning tunnelling microscopy (STM) and state-of-the-art density functional (DFT) calculations reveal that the linkage between two OETAPs is likely to be initiated by a [2+2] cycloaddition, involving the chemical transformations of the peripheral ethyl substituents of two adjacent molecular species. Importantly, by raising the temperature of the substrate prior deposition of the molecular precursors, the intermolecular covalent coupling can be precluded. For substrate temperatures of 300 °C or higher, it is observed that the mechanisms for polymeric growth are blocked and the OETAP species are transformed into individual phthalocyanines, via an electrocyclic ring closure (ERC) reaction, and then self-assembled into islands. DFT simulations support our experimental findings indicating a lower energetic barrier for dimerization as compared with the intramolecular reaction of the monomer. We envision that our results will pave the way for the development of low-dimensional materials exploiting the propensity of molecular precursors equipped with convenient peripheries to undergo tunable covalent reactions under thermal or light stimuli.

Results

Self-assembly of OETAP precursors on Au(111). Figure 1a shows a high-resolution STM image of an OETAP array on Au(111) self-assembled after room temperature deposition. Each molecule is visualized as a dim centre, attributed to the macrocyclic core, surrounded by eight brighter protrusions, which are assigned to the ethyl moieties, in excellent agreement with simulated STM images (cf. Fig. 1b and Supplementary Fig. 1) and consistent with surface-confined assemblies of identically substituted porphyrin compounds[32]. The dense-packed architecture displays two distinctly oriented molecular species, related by a ~16° rotation, which features a rhombic lattice ($a = 14.1 \pm 1$ Å, $b = 14.1 \pm 1$ Å, $\theta = 60°$), and is stabilized by lateral non-covalent interactions between the ethyl substituents.

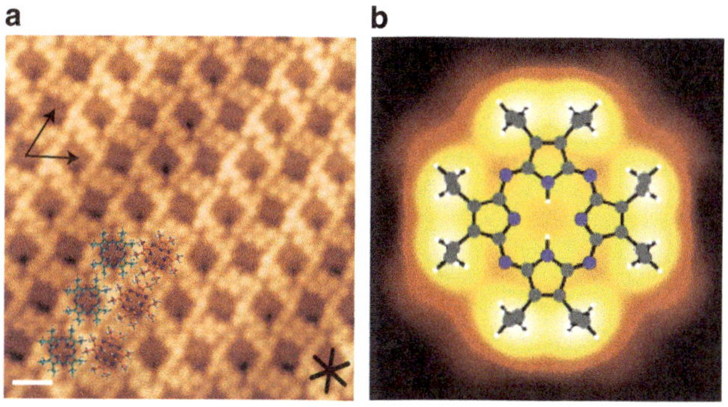

Figure 1 | Self-assembly of OETAP on Au(111) after deposition at room temperature. (**a**) High-resolution STM images revealing submolecular features. Black arrows depict the lattice vectors of the rhombic unit cell. Black star shows the close-packed directions of the surface. Superimposed coloured atomistic models address the two distinctly oriented molecular species (green and red, respectively). Tunnelling parameters: $V_b = 0.5$ V, $I = 0.1$ nA. Scale bar, 1 nm. (**b**) STM simulated image of an individual OETAP on Au(111) at 0.5 V, with a superimposed ball-and-stick model of the molecular species. Hydrogen, carbon and nitrogen atoms are depicted in white, grey and violet, respectively.

Polymerization of OETAP species. Next we explore the polymerization of OETAP precursors on Au(111). After submonolayer deposition at room temperature, gentle annealing of the substrate to 75, 100 and 150 °C results in no appreciable changes. At 225 °C, the polymerization of OETAP species is initiated. High-resolution images of the assemblies after holding the substrate at 275 °C reveal the formation of molecular chains, most of them confined on the fcc regions of Au(111) (cf. Fig. 2a). A careful inspection of these chains allows us to discern submolecular features and, thus, clarify the molecular organization. The tetrapyrrolic macrocyclic cores are now visualized as dim crosses, whereas the ethyl substituents have reacted in two distinct ways: (i) by covalently linking adjacent species through the intermediacy of multiple covalent reactions (cf. below) and (ii) by undergoing an intramolecular ERC reaction. The final result is the formation of polymeric chains, in which the repeating unit (monomer) derives from OETAP. A statistical analysis shows that the average polymeric size is ~7 monomers, that the most frequent polymer comprises 6 macrocycles and that extended polymers (>20 monomers) can be formed. Importantly, the majority of the monomeric entities are joined together via two covalent structural motifs, denoted L and V (cf. Fig. 2c,d). Both structural motifs are the product of multiple covalent reactions between peripheral ethyl groups from adjacent OETAP precursors, where two molecules face each other. When the two ethyl groups that initiate the intermolecular reaction lie in opposite sides of the OETAP precursors the product is an L-type motif (cf. Fig. 2c and Supplementary Figs 1 and 2 for agreement with

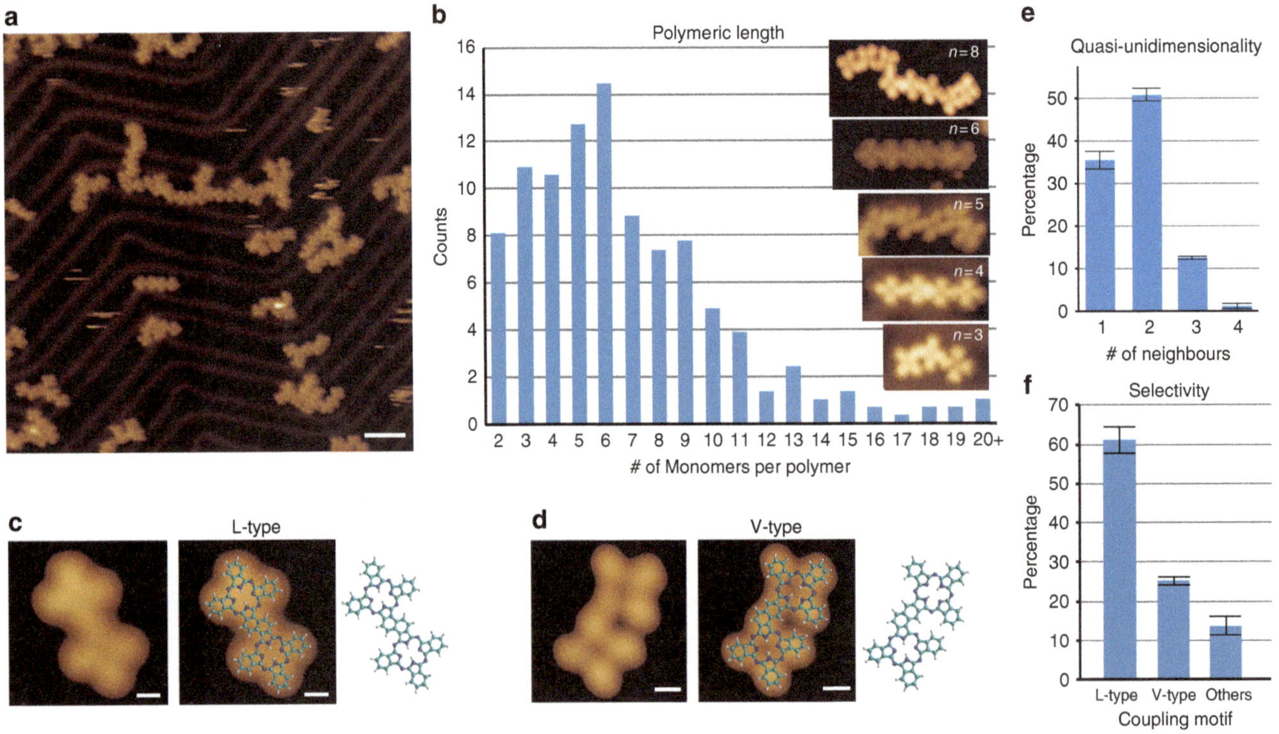

Figure 2 | Surface-confined synthesis of phthalocyanine tapes. The deposition of 0.3 ML of OETAP species (precursors) on Au(111) held at room temperature and subsequent gently annealing at 275 °C affords the formation of phthalocyanine polymers. (**a**) Long-range STM image displaying the formation of quasi-unidimensional polymers mainly confined to the fcc regions. (**b**) High-resolution STM images of polymers of distinct size (3, 4, 5, 6 and 8) and histogram of the polymeric length. (**c,d**) High-resolution STM image and models of the majority of products between coupled monomers (**c**, L-type and **d**, V-type). (**e**) Histogram of the number of neighbours per monomer, highlighting the quasi-unidimensional feature of the assemblies. (**f**) Histogram of different types of coupling motifs exhibiting selectivity towards L-type-binding motif. Errors bars in **e,f** are the standard deviation of the mean. Tunnelling parameters: **a**, $V_b = -1.5$ V; **b**, $V_b = -0.5$ V for $n = 3$, $V_b = -1.5$ V for $n = 4$, $V_b = -1$ V for $n = 5$, $V_b = -1$ V for $n = 6$, $V_b = -1.2$ V for $n = 8$; **c**, $V_b = -1.5$ V and **d**, $V_b = -0.7$ V. Scale bars, **a**, 5 nm; **c,d**, 0.5 nm.

DFT simulated image), whereas the V-type motif is the product when the ethyl groups lie in the same side of opposite molecules. In both cases, the resulting polymer can be interpreted as a phthalocyanine tape with locally straight (L) or slightly curved (V) appearance. Notably, the thermally induced polymerization of OETAP precursors into phthalocyanine tapes seems to present a preference for the L-type coupling motif versus the V-type motif (ca 2:1), as seen in Fig. 2f. In addition, polymers are mostly quasi-unidimensional, with over 85% of the monomers connected to only one or two neighbours. The remaining 15% of monomers are connected to three or four neighbours mainly through cross motifs based on L or V coupling schemes.

Regarding the proposed covalent binding motifs, we are certain that the observed molecular chains are polymers and not supramolecular assemblies due to the following reasons: (i) the centre-to-centre distance between macrocycles is too short for a non-covalent interaction and our DFT study of the reaction pathway (see below for details) is fully consistent with the experimental observations; and (ii) using perturbative scanning conditions, the polymeric chains can be displaced as entire units for long distances (over 10 nm) preserving their size and shape, thus revealing a strong covalent interaction between the monomers (cf. Supplementary Fig. 3 and Supplementary Movie 1).

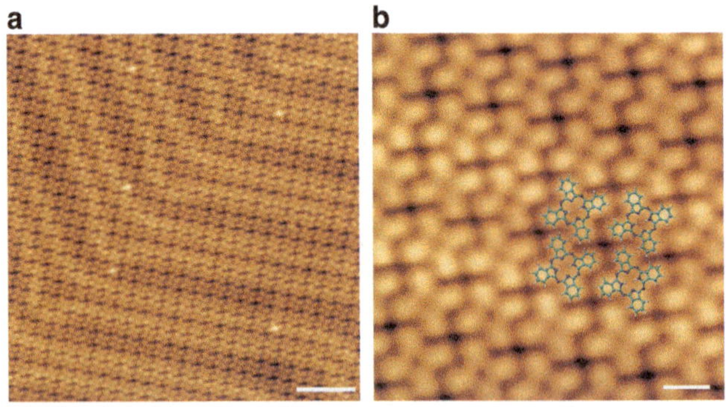

Figure 3 | Surface-assisted synthesis and self-assembly of phthalocyanines. The deposition of OETAP species on Au(111) held at 300 °C affords the on-surface chemical transformation of precursors into phthalocyanine species. (**a**) Long-range STM image, revealing the close-packed supramolecular assembly and highlighting the presence of the herringbone reconstruction of Au(111). (**b**) High-resolution STM image and superimposed modelling of the molecular constituents, which matches the reported assembly of 2H-Pc on Au(111). **a**,**b**, $V_b = -1$ V. Scale bars: **a**, 5 nm; **b**, 1 nm.

Transformation of OETAP precursors into phthalocyanines. To test the feasibility of thermal control of the covalent reaction pathways, OETAP precursors were deposited on pristine Au(111) held at 300 °C. As depicted in Fig. 3, a new scenario is manifested. Instead of polymeric chains, we observe the formation of individual entities, which self-assemble into close-packed arrays based on a square unit cell. High-resolution STM imaging allows us to discern submolecular features of the assemblies, whereby each molecular species presents four bright protrusions and a central void forming a cross-like shape. Both the molecular appearance and the self-assembled architecture are identical to those found after deposition of 2H-Pc (free base phthalocyanine) on Au(111)[33]. Thus, based on our results and a comparison with the literature, we suggest that the deposition of OETAP precursors on Au(111) held at 300 °C gives rise to the formation of 2H-Pc species, thanks to the dehydrogenation and ring closure of their ethyl peripheries, a phenomenon observed for porphyrin derivatives[32].

Reaction intermediates during polymerization. To address the reaction mechanisms that could plausibly explain the formation of the covalent phthalocyanine polymers and phthalocyanine monomers, a detailed analysis of the STM images collected during the annealing process was carried out.

Figure 4 illustrates reaction intermediates observed after depositing OETAP species holding the substrate at room temperature, subsequently followed by gentle annealing to 225 °C. Long-range STM images show the formation of small islands mainly distributed over the fcc regions of the Au(111) surface (cf. Fig. 4a,e), thus revealing the dissolution of the former OETAPs islands and subsequent self-assembly into smaller patches. High-resolution topographs (cf. Fig. 4b–d) allow us to discern submolecular features and elucidate the molecular packing. For clarity, Fig. 4f–h display a rationalization of the distinct molecular species by superimposing a coloured model. Here, dots and rods of the same colour represent intact ethyl moieties and ERC-reacted ethyl termini of the same molecule, respectively. In addition, bi-coloured rods indicate links between tetrapyrrole molecular units. A minority of unreacted OETAP species, characterized by the eight-dotted appearance described above, are still observed either isolated on the surface or forming part of the supramolecular islands. However, most of the molecular precursors have experienced covalent reactions, giving rise to the reaction intermediates of the polymeric chains. The majority of OETAPs have undergone one or two ring closure reactions (intramolecular electrocyclizations) that transformed their diethyl-pyrrole units into isoindole components, which are imaged as rods. At this stage of the reaction, some of the OETAPs are also covalently linked to one another, affording distinct covalent bonding motifs, assignable to intermediates of the two main final structures (L and V). **1-L** and **2-L** represent reaction intermediates of the L-type coupling motif (cf. Fig. 4i,j), whereas **1-V** and **2-V** are the corresponding analogous reaction intermediates of the V-type-binding motif. We observe that the intermolecular bond in intermediate **1-L** comprises the reaction between two ethyl moieties in opposite sides of the OETAP precursors (cf. Fig. 4i and Supplementary Fig. 4), keeping the other two ethyl peripheries of the pyrroles intact, which are visualized as bright lobes with the same height as in individual unaltered OETAPs.

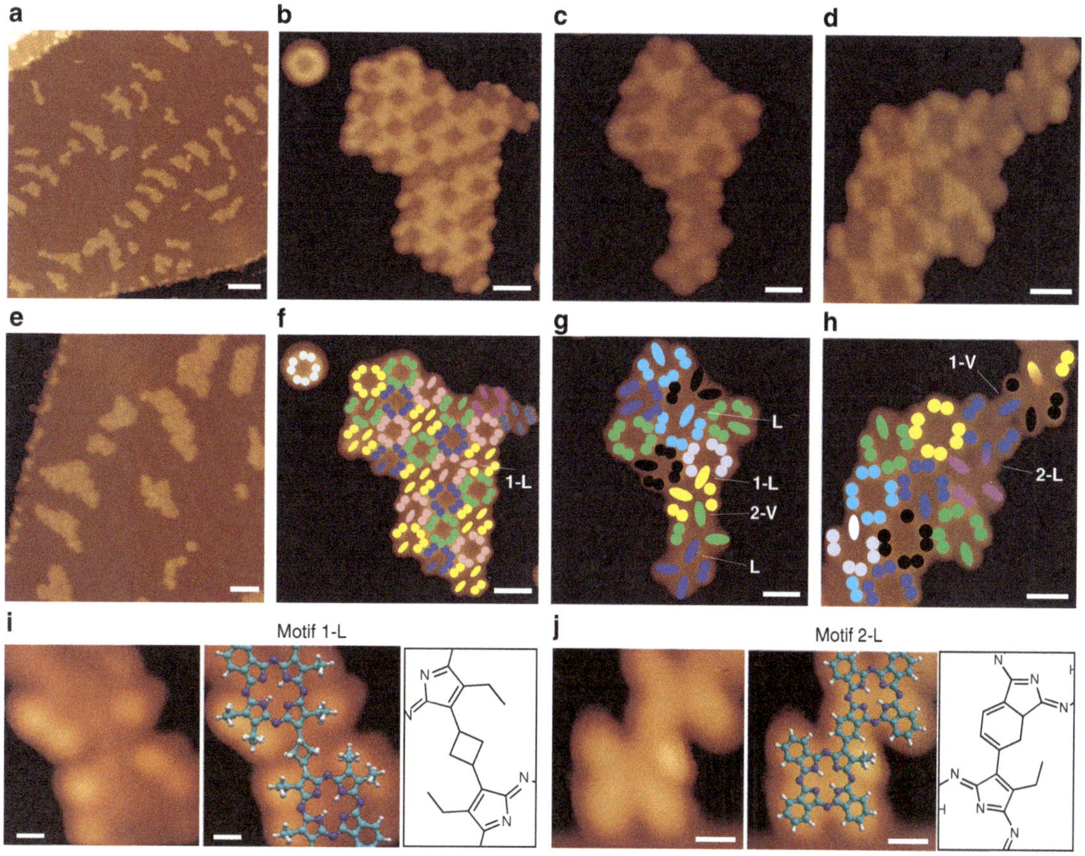

Figure 4 | Surface-assisted formation of phthalocyanine derivatives after deposition of OETAP precursors on Au(111). (**a**,**e**) Large-scale STM images showing the formation of patches of initially reacted molecules coexisting with unreacted species. (**b**–**d**) High-resolution STM images and (**f**–**h**) corresponding schematic colouring of the initial steps of reactions giving motifs **1-L**, **1-V**, **2-L**, **2-V** and **L**. (**i**,**j**) Zoom-in of motif **1-L** and **2-L**, and superimposed modelling, together with chemical drawing insets. Coloured filled dots represent ethyl moieties, bi-coloured ellipses show covalent connections between adjacent species, and mono-coloured ellipses represent full electrocyclic ring closures. Tunnelling parameters: **a**, $V_b = -0.8$ V; **b**, $V_b = -0.7$ V; **c**, $V_b = -0.7$ V; **d**, $V_b = -0.8$ V; **e**, $V_b = -1.2$ V; **i**, $V_b = 0.5$ V and **j**, $V_b = -1.5$ V. Scale bars, **a**, 10 nm; **b**, 1.5 nm; **c**, 1.0 nm; **d**, 1.0 nm; **e**, 5 nm; **f**, 1.5 nm; **g**, 1.0 nm; **h**, 1.0 nm; **i**, 0.3 nm and **j**, 0.5 nm.

In **2-L**, the reaction appears to have evolved, involving another ethyl group (cf. Fig. 4j), thus keeping only one of the four initial ethyl groups unreacted. Finally, in **L**, the four ethyl groups have reacted to produce the seed of a polymeric chain.

Discussion

Taking into account the initial precursors, the identified intermediates and the final products, we tentatively propose multi-step reaction mechanisms for both pathways, the polymerization of OETAPs into phthalocyanine quasi-unidimensional tapes (cf. Fig. 5a and Supplementary Figs 5 and 6), and the transformation of OETAP species into unsubstituted monomeric phthalocyanines (cf. Fig. 5b).

At the first stage of the polymeric reaction, two OETAPs face each other, positioning four ethyl moieties opposite each other (shown in green in Fig. 5a), which will undergo a series of complex reactions activated by the ramp annealing of the substrate. We propose that the ethyl substituents undergo an initial dehydrogenation reaction to produce an ethenyl residue per molecule, either in s-*cis* or s-*trans* conformation. Two opposite ethenyl groups adequately oriented are susceptible to react via a [2 + 2] cycloaddition affording intermediate **1-L** (or **-V**), which is repeatedly observed by STM imaging. Further work is necessary to corroborate this mechanism. The L-to-V ratio in the final coupling motifs (ca. 2:1) corresponds to a preferential ethyl-to-ethenyl dehydrogenation favouring the conformation where the ethenyl group lies s-*trans* to the bond opposite the pyrrolic nitrogen, as shown in Fig. 5a. Intermediate **1-L/V** can be considered as a dimer of two OETAP macrocycles joined together through a four-membered ring, fitting the appropriate angles and bond distances. This type of reaction, thermally forbidden by the Woodward–Hoffmann rules, is unprecedented on metallic surfaces and has only been explored on semiconductor surfaces to covalently functionalize the substrate support, whereby dangling bonds play a major role[34]. Encouragingly, DFT simulations support the tentative [2 + 2] cycloaddition (vide infra).

To explain intermediate **2-L/V**, also observed by STM imaging, a series of chemical reactions, including the loss of two methylene groups (as ethene), is tentatively suggested (cf. Supplementary Figs 5 and 6, and Supplementary Note 1). An additional dehydrogenation reaction of the remaining ethyl substituent and subsequent 6-π electrocyclization reaction followed by an aromatization process results in the formation of STM-characterized dimer **3**, observed in both L and V conformations, with a naphthalene spacer bridging the two tetrapyrrolic macrocycles.

During the polymerization process, we observe that the ethyl groups not involved in the construction of the covalent bridging spacer undergo a series of dehydrogenation and electrocyclic ring closures, giving rise to isoindole units, and thus providing the final phthalocyanine aspect to the polymers. In short, the dehydrogenation of the β-ethyl groups followed by the intramolecular electrocyclization of the resulting ethenyl resides transformed the 3,4-diethylpyrrole units into isoindole analogues (cf. Fig. 5b). We provide theoretical results (see below for details) supporting our suggestion that an analogous mechanism takes place in the transformation of OETAPs into 2H-Pcs, during the deposition process of the tetraazaporphyrins on Au(111) held at 300 °C, as previously observed for β-ethyl-substituted porphyrins[32].

To gain additional insight into reaction mechanisms, we performed DFT calculations for a model system on the Au(111) surface. We studied and compared the monomer electro-cyclization and the initial dimerization process. The complete dimerization process involves a complex multi-step sequence of events and a more in-depth analysis of all reaction steps will be accounted for elsewhere. Both the monomer electro-cyclization and the dimerization reaction of monomers require four preliminary dehydrogenation reactions that converted two ethyl substituents into ethenyl residues, with a largest energy barrier of 1.60 eV (cf. Supplementary Fig. 7 for details).

From this starting point, the monomer electro-cyclization is followed by the actual ring-closing reaction, with an effective potential energy barrier of 1.82 eV as seen in Fig. 6a–c (the ring closing reaction is associated with several barriers, which are

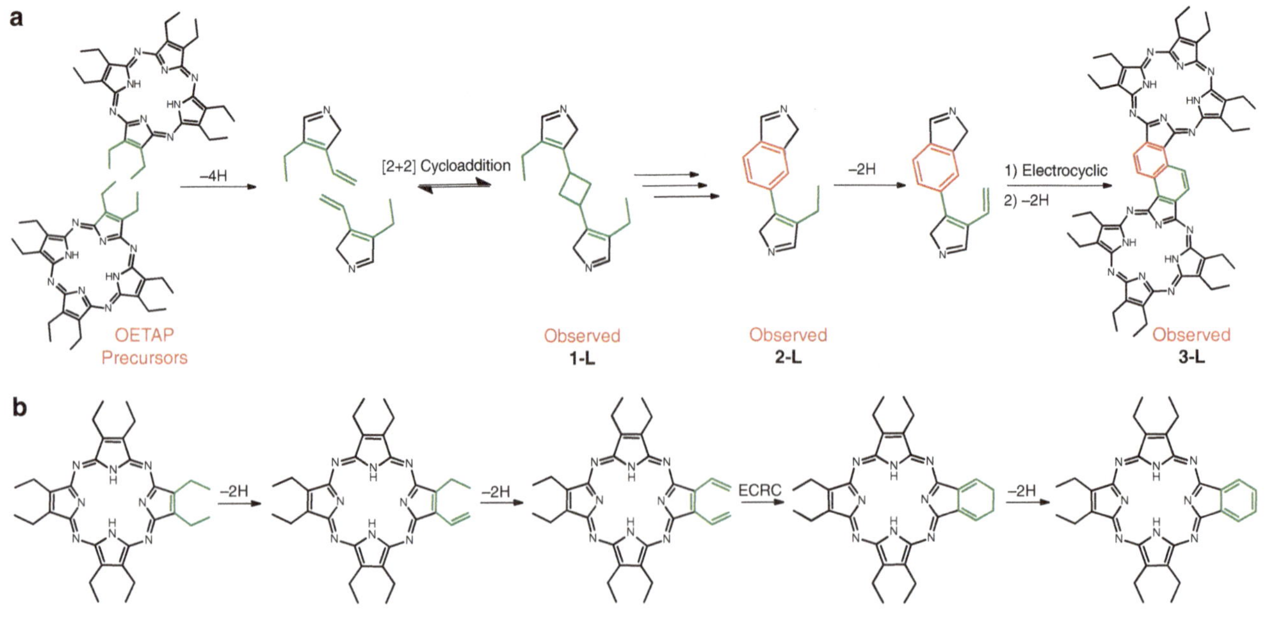

Figure 5 | Proposed reaction mechanisms describing the surface-assisted formation of phthalocyanine polymers and monomers on Au(111). (**a**) Reaction steps involving STM-imaged intermediates **1-L** and **2-L** in the phthalocyanine dimer formation (**3-L** reaction product) via L-type linking motif. For a full step-wise reaction mechanism and for the analogous reaction of the V-type motif, see Supplementary Figs 5 and 6. (**b**) Diethyl-pyrrole to isoindole moiety conversion.

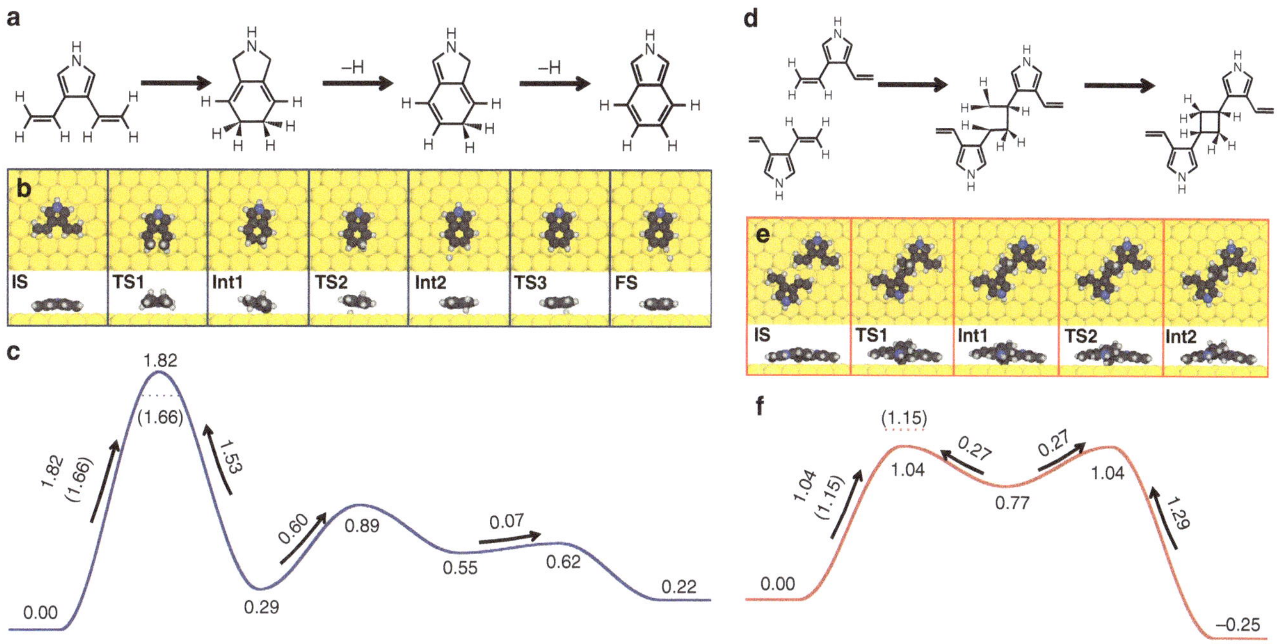

Figure 6 | Reaction pathway considerations for a model pyrrole molecule on Au(111). (**a-c**) Monomer cyclization. (**d-f**) Initial steps of the dimerization process. For both reactions, it is assumed that the ethyl legs have been transformed into ethenyl groups through dehydrogenation reactions, the details of which are shown in the Supplementary Information. (**a,d**) Chemical stick models and (**b,e**) top and side view ball models for reactants (**IS**), transition states (**TS**), reaction intermediates (**Int**) and the monomer reaction final state (**FS**). (**c,f**) Energy profiles for the monomer cyclization and initial steps of the dimerization process, respectively. Free energy barriers are shown in parentheses and indicated by the dotted lines for **TS1** of each reaction, calculated by including vibrational enthalpy and entropy at 275 °C. Energies are given in units of eV.

shown in Supplementary Fig. 8). Two dehydrogenation steps with relatively small energy barriers are required for the final aromatization process providing the isoindole unit. These results suggest the possibility to tune the reaction temperature such that the ethyl-to-ethenyl transformation is triggered, without activating the ring closuring reaction owing to its significantly higher energy barrier.

Next, we investigate the dimerization reaction, starting from two monomers that have experienced a complete dehydrogenation of the ethyl residues. The reaction is initiated by a [2 + 2] cycloaddition with an effective potential energy barrier of 1.04 eV (cf. Fig. 6d–f, **IS** to **Int2**). Notably, the [2 + 2] cycloaddition proceeds in a two-step mechanism, via an intermediate state (**Int1**) in which a carbon atom is chemically bonded to the surface. Importantly, the cycloaddition is just slightly exothermic with a reaction energy of − 0.25 eV, providing a Boltzmann factor of ∼200 between **Int2** and **IS** at 275 °C, thereby substantiating the reversibility of the coupling reaction. This supports the suggested pathway in Fig. 5a, and further detailed in Supplementary Fig. 5, in which **IS** and **Int2** are in thermal equilibrium before the reaction proceeds in a Diels–Alder step. Future studies will unravel the information of the final steps of the dimerization.

In addition to potential energy barriers, we also considered the effect of zero-point and thermal contributions on the decisive steps of the monomer cyclization and dimerization by including vibrational enthalpy and entropy. Free energy barriers were evaluated at a temperature of 275 °C and are indicated for the initial monomer and dimerization steps in Fig. 4c,f, respectively. The monomer ring-closing reaction is slightly lowered by 0.16 eV, whereas the free energy barrier for the dimerization is 0.11 eV larger than the corresponding potential energy barrier. Supplementary Table 1 shows the individual contributions from vibrational enthalpy and entropy on the two barriers.

Although zero-point and thermal vibrational contributions have a small quantitative effect on barriers, both the potential energy and the free energy landscape unambiguously demonstrate that the decisive step for the dimerization has a significantly lower barrier than the monomer cyclization, with potential energy and free energy barrier differences of 0.72 and 0.51 eV, respectively. As a result, the monomer ring-closure will occur only if the dimerization is kinetically hindered, such as for extremely low coverage or in the absence of nearby non-reacted molecules. Hereby, the balance between diffusion and reaction barriers plays a crucial role, as the molecules need to meet to undergo the polymerization. These results explain why the monomer ring closure is observed for the deposition at the already heated substrate, in which situation the molecules experience the electrocyclic ring closure before meeting other species, while the intermolecular connections are formed when depositing the molecules at a room temperature substrate followed by thermal annealing allowing the precursors sufficient time to diffuse.

In summary, we have successfully introduced thermally tunable covalent reactions on surfaces to engineer quasi-unidimensional phthalocyanine tapes or self-assembled phthalocyanines by depositing molecular precursors on a well-defined Au(111) surface and regulating the substrate temperature. Our results open avenues to thermally control reaction pathways on surfaces, selecting intermolecular versus intramolecular reactions, and thus allowing to induce the growth of unprecedented polymeric heteroatomic nanoarchitectures or to produce monomeric reactions. DFT simulations corroborate the experimentally observed behaviour, demonstrating that the decisive step determining the reaction product has a lower free energy barrier for the dimerization compared with monomer intramolecular cyclization reaction. Our study discloses strategies to grow uni- and two-dimensional polymeric nano-architectures embedding heteroatomic monomers. Such systems bear prospects for molecular electronics, optoelectronics and photovoltaics.

Methods

Experiments. The experiments were performed in a custom-designed ultra-high vacuum system that hosts a low-temperature Omicron scanning tunnelling microscope, where the base pressure was below 5×10^{-10} mbar. All STM images were taken in constant-current mode with electrochemically etched tungsten tips, applying a bias (V_b) to the sample and at a temperature of ~ 80 K. The Au(111) substrate was prepared by standard cycles of Ar+ sputtering (800 eV) and subsequent annealing to 723 K for 10 min. OETAP molecules were synthesized according to the procedure described by Fitzgerad and co-workers[35,36] and deposited by organic molecular-beam epitaxy from a quartz crucible held at 450 K onto a clean Au(111) at room temperature, if not stated otherwise. If necessary, in a subsequent step, the samples were annealed with a thermal gradient of 1 °C s^{-1} and kept at the desired temperature for 30 min. Next, they were cooled down to room temperature with a thermal gradient of -1 °C s^{-1} and finally transferred to the STM stage held at 77 K.

Theory. Periodic DFT calculations were performed with the VASP code[37], using the projector-augmented wave method[38]. Exchange-correlation effects were described by the version of the van der Waals density functional[39] (vdWDF) introduced by Hamada[40] denoted as rev-vdWDF2. Transition states were calculated using the climbing image nudged elastic band[41] and the Dimer[42] methods. STM simulations were carried out with the Tersoff–Hamann approximation[43] using the implementation by Lorente and Persson[44]. Detailed information about DFT calculations are provided in Supplementary Methods.

References

1. Gourdon, A. On-surface covalent coupling in ultrahigh vacuum. *Angew. Chem. Int. Ed.* **47,** 6950–6953 (2008).
2. Franc, G. & Gourdon, A. Covalent networks through on-surface chemistry in ultra-high vacuum: state-of-the-art and recent developments. *Phys. Chem. Chem. Phys.* **13,** 14283–14292 (2011).
3. Zhang, X., Zeng, Q. & Wang, C. On-surface single molecule synthesis chemistry: a promising bottom-up approach towards functional surfaces. *Nanoscale* **5,** 8269–8287 (2013).
4. Liu, X.-H., Guan, C.-Z., Wang, D. & Wan, L.-J. Graphene-like single-layered covalent organic frameworks: synthesis strategies and application prospects. *Adv. Mater.* **26,** 6912–6920 (2014).
5. Zhuang, X. *et al.* Two-dimensional soft nanomaterials: a fascinating world of materials. *Adv. Mater.* **27,** 403–427 (2015).
6. Grill, L. *et al.* Nano-architectures by covalent assembly of molecular building blocks. *Nat. Nanotechnol.* **2,** 687–691 (2007).
7. Treier, M., Richardson, N. V. & Fasel, R. Fabrication of surface-supported low-dimensional polyimide networks. *J. Am. Chem. Soc.* **130,** 14054–14055 (2008).
8. Zwaneveld, N. A. A. *et al.* Organized formation of 2D extended covalent organic frameworks at surfaces. *J. Am. Chem. Soc.* **130,** 6678–6679 (2008).
9. Matena, M. *et al.* Transforming surface coordination polymers into covalent surface polymers: Linked polycondensed aromatics through oligomerization of N-heterocyclic carbene intermediates. *Angew. Chem. Int. Ed.* **47,** 2414 (2008).
10. Lafferentz, L. *et al.* Conductance of a single conjugated polymer as a continuous function of its length. *Science* **323,** 1193–1197 (2009).
11. Jensen, S., Früchtl, H. & Baddeley, C. J. Coupling of triamines with diisocyanates on Au(111) leads to the formation of polyurea networks. *J. Am. Chem. Soc.* **131,** 16706–16713 (2009).
12. Bombis, C. *et al.* Single molecular wires connecting metallic and insulating surface areas. *Angew. Chem. Int. Ed.* **48,** 9966–9970 (2009).
13. Cai, J. *et al.* Atomically precise bottom-up fabrication of graphene nanoribbons. *Nature* **466,** 470–473 (2010).
14. Bieri, M. *et al.* Two-dimensional polymer formation on surfaces: insight into the roles of precursor mobility and reactivity. *J. Am. Chem. Soc.* **132,** 16669–16676 (2010).
15. Lipton-Duffin, J. A. *et al.* Step-by-step growth of epitaxially aligned polythiophene by surface-confined reaction. *Proc. Natl Acad. Sci. USA* **107,** 11200–11204 (2010).
16. Abel, M. *et al.* Single layer of polymeric Fe-Phthalocyanine: An organometallic sheet on metal and thin insulating film. *J. Am. Chem. Soc.* **133,** 1203–1205 (2011).
17. Lafferentz, L. *et al.* Controlling on-surface polymerization by hierarchical and substrate-directed growth. *Nat. Chem.* **4,** 215–220 (2012).
18. Zhang, Y.-Q. *et al.* Homo-coupling of terminal alkynes on a noble metal surface. *Nat. Commun.* **3,** 1286 (2012).
19. Wiengarten, A. *et al.* Surface-assisted dehydrogenative homocoupling of porphine molecules. *J. Am. Chem. Soc.* **136,** 9346–9354 (2014).
20. Gao, H.-Y. *et al.* Decarboxylative polymerization of 2,6-naphthalenedicarboxylic acid at surfaces. *J. Am. Chem. Soc.* **136,** 9658–9663 (2014).
21. Sanchez-Sanchez, C. *et al.* Sonogashira cross-coupling and homocoupling on a silver surface: chlorobenzene and phenylacetylene on Ag(100). *J. Am. Chem. Soc.* **137,** 940–947 (2015).
22. Auwärter, W., Ecija, D., Klappenberger, F. & Barth, J. V. Porphyrins at interfaces. *Nat. Chem.* **7,** 105–120 (2015).
23. Lackinger, M. & Heckl, W. M. A STM perspective on covalent intermolecular coupling reactions on surfaces. *J. Phys. D Appl. Phys.* **44,** 464011 (2011).
24. Weigelt, S. *et al.* Covalent interlinking of an aldehyde and an amine on a Au(111) surface in ultrahigh vacuum. *Angew. Chem. Int. Ed.* **46,** 9227–9230 (2007).
25. Schmitz, C. H., Ikonomov, J. & Sokolowski, M. Two-dimensional ordering of poly(p-phenylene-terephthalamide) on the Ag(111) surface investigated by scanning tunneling microscopy. *J. Phys. Chem. C* **113,** 11984–11987 (2009).
26. Marele, A. C. *et al.* Formation of a surface covalent organic framework based on polyester condensation. *Chem. Commun.* **48,** 6779–6781 (2012).
27. Sun, Q. *et al.* On-surface formation of one-dimensional polyphenylene through Bergman cyclization. *J. Am. Chem. Soc.* **135,** 8448–8451 (2013).
28. Bebensee, F. *et al.* On-surface azide–alkyne cycloaddition on Cu(111): does It "Click" in ultrahigh vacuum? *J. Am. Chem. Soc.* **135,** 2136–2139 (2013).
29. Díaz Arado, O. *et al.* On-surface azide–alkyne cycloaddition on Au(111). *ACS Nano* **7,** 8509–8515 (2013).
30. Zhong, D. *et al.* Linear alkane polymerization on a gold surface. *Science* **334,** 213–216 (2011).
31. Liu, J. *et al.* Toward Cove-edged low band gap graphene nanoribbons. *J. Am. Chem. Soc.* **137,** 6097–6103 (2015).
32. Heinrich, B. W. *et al.* Change of the magnetic coupling of a metal–organic complex with the substrate by a stepwise ligand reaction. *Nano Lett.* **13,** 4840–4843 (2013).
33. Komeda, T., Isshiki, H. & Liu, J. Metal-free phthalocyanine (H_2Pc) molecule adsorbed on the Au(111) surface: formation of a wide domain along a single lattice direction. *Sci. Technol. Adv. Mater.* **11,** 054602 (2010).
34. Wong, K. T. & Bent, S. F. in *Functionalization of Semiconductor Surfaces* 51–88 (John Wiley & Sons, Inc. Hoboken, New Jersey, 2012).
35. Fitzgerald, J., Taylor, W. & Owen, H. Facile synthesis of substituted fumaronitriles and maleonitriles: precursors to soluble tetraazaporphyrins. *Synthesis* **9,** 686–688 (1991).
36. Giménez-Agulló, N. *et al.* Single-molecule-magnet behavior in the family of [Ln(OETAP)2] double-decker complexes (Ln = Lanthanide, OETAP = Octa(ethyl)tetraazaporphyrin). *Chem. Eur. J* **20,** 12817–12825 (2014).
37. Kresse, G. & Furthmüller, J. Efficient iterative schemes for ab initio total-energy calculations using a plane-wave basis set. *J. Phys. Rev. B* **54,** 11169–11186 (1996).
38. Blöchl, P. E. Projector augmented wave method. *Phys. Rev. B* **59,** 17953–17979 (1994).
39. Dion, M. *et al.* Van der Waals density functional for general geometries. *Phys. Rev. Lett.* **92,** 246401 (2004).
40. Hamada, I. van der Waals density functional made accurate. *Phys. Rev. B* **89,** 121103(R)–121107(R) (2014).
41. Henkelman, G., Uberuaga, B. P. & Jónsson, H. A climbing image nudged elastic band method for finding saddle points and minimum energy paths. *J. Chem. Phys.* **113,** 9901–9904 (2000).
42. Kästner, J. & Sherwood, P. Superlinearly conveging dimer method for transition state search. *J. Chem. Phys.* **128,** 014106 (2008).
43. Tersoff, J. & Hamann, D. R. Theory and application for the scanning tunneling microscope. *Phys. Rev. Lett.* **50,** 1998–2001 (1983).
44. Lorente, N. & Persson, M. Theoretical aspects of tunneling-current-induced bond excitation and breaking at surfaces. *Farad. Discuss* **117,** 277–290 (2000).

Acknowledgements

We thank the financial support of the EU (projects ERC Starting Grant CHEMCOMP no 279313, EC FP7-PEOPLE-2011-COFUND AMAROUT II program, EC FP7-PEOPLE-2013-CIG (no. 631396) and FEDER CTQ2014-56295-R), the Spanish Ministerio de Economía y Competitividad (MINECO) (projects RYC-2012-11133, RYC-2012-11231, CTQ2014-56295-R, CTQ2014-52974-REDC and Severo Ochoa Excellence Accreditation 2014-2018 SEV-2013-0319), the Comunidad de Madrid (project MAD2D), the Generalitat de Catalunya (2014-SGR-797), the ICIQ Foundation and the IMDEA Foundation. We acknowledge Professor R. Miranda for fruitful discussions.

Author contributions

D.E. and B.C. designed the study. N.G.-A., J.R.G.-M. and P.B. synthesized the molecular precursor. B.C., J.R.-F., A.M.-J. and D.E. performed the experiments. J.B. realized the DFT calculations. B.C., F.M.-P., A.M.P. and D.E. rationalized the tentative reaction mechanisms. B.C., J.B., J.M.G., R.O. and D.E. analysed the data. All authors co-wrote the paper. D.E. supervised the whole project.

Additional information

Competing financial interests: The authors declare no competing financial interests.

15

In situ microscopy of the self-assembly of branched nanocrystals in solution

Eli Sutter[1], Peter Sutter[2], Alexei V. Tkachenko[3], Roman Krahne[4], Joost de Graaf[5], Milena Arciniegas[4] & Liberato Manna[4]

Solution-phase self-assembly of nanocrystals into mesoscale structures is a promising strategy for constructing functional materials from nanoscale components. Liquid environments are key to self-assembly since they allow suspended nanocrystals to diffuse and interact freely, but they also complicate experiments. Real-time observations with single-particle resolution could have transformative impact on our understanding of nanocrystal self-assembly. Here we use real-time *in situ* imaging by liquid-cell electron microscopy to elucidate the nucleation and growth mechanism and properties of linear chains of octapod-shaped nanocrystals in their native solution environment. Statistical mechanics modelling based on these observations and using the measured chain-length distribution clarifies the relative importance of dipolar and entropic forces in the assembly process and gives direct access to the interparticle interaction. Our results suggest that monomer-resolved *in situ* imaging combined with modelling can provide unprecedented quantitative insight into the microscopic processes and interactions that govern nanocrystal self-assembly in solution.

[1]Department of Mechanical and Materials Engineering, University of Nebraska-Lincoln, Lincoln, Nebraska 68588, USA. [2]Department of Electrical and Computer Engineering, University of Nebraska-Lincoln, Lincoln, Nebraska 68588, USA. [3]Center for Functional Nanomaterials, Brookhaven National Laboratory, Upton, New York 11973, USA. [4]Department of Nanochemistry, Istituto Italiano di Tecnologia (IIT), via Morego 30, Genova IT-16163, Italy. [5]Faculty 8: Mathematics and Physics, Institute for Computational Physics (ICP), University of Stuttgart, Allmandring 3, Stuttgart 70569, Germany. Correspondence and requests for materials should be addressed to E.S. (email: esutter@unl.edu) or to L.M. (email: liberato.manna@iit.it).

The self-assembly of nanocrystals into ordered superstructures promises access to photonic[1,2], plasmonic[3], electronic[4,5], magnetic[6] and catalytic[7] functionalities that are tunable and may even be reconfigurable in real time[8]. Besides technological interest, research on crystallization of colloids is driven by the recognition that they are not simply large atom equivalents but represent a form of matter with complex collective behaviour that can be controlled via the interplay of entropic, van der Waals, steric and dipolar interactions[9–11], by attaching molecular linkers such as coordinated ligands or DNA[12–14], or by using shaped particles[15]. Understanding the structure, formation and transformation of colloidal matter requires probing particle configurations from monomers to extended periodic or aperiodic assemblies. Computer simulations provide this capability[16,17], but experiments are complicated by the need to interrogate processes in liquids that allow suspended particles to diffuse and interact freely. Larger micro-colloids can be imaged with single-particle resolution by optical microscopy[18,19], but offer limited diversity of materials and shapes[20]. For nanocrystals, *in situ* measurements have largely been restricted to reciprocal space techniques[14,21], which yield little information on the assembly process or the underlying forces. Liquid-cell electron microscopy (LCEM) has recently been developed into a tool capable of imaging processes in liquids with nanometre resolution[22–24]. LCEM has been used previously to study single-layer assembly of compact nanocrystals driven by convective transport during solvent drying[25]. A large class of self-assembly processes, for instance the entropy-driven assembly of polyhedral nanocrystals[15] or the DNA-guided crystallization[12–14,26] and reconfiguration[27] of spherical nucleic acid conjugates occurs in bulk solutions with transport by Brownian motion.

Using the example of the one-dimensional (1D) self-assembly of CdSe/CdS octapods, we demonstrate the ability of LCEM to dynamically probe such solution-phase self-assembly processes in real space with nanometre resolution. Octapods, branched nanocrystals with a complex but still highly symmetric shape[28–30], have been found to self-assemble into a variety of superstructures: interlocked 1D chains in bulk liquid, two-dimensional square lattices on substrates[31] or micron-sized three-dimensional (3D) structures of aligned 1D chains[21]. Strongly anisotropic interactions are thought to be key to their rich set of accessible superstructures. By real-time imaging with monomer resolution, the *in situ* LCEM observations reported here establish several key aspects of the self-assembly of ordered linear chains of octapods in solution. The trimer, that is, a three-octapod chain, is the shortest stable chain of interlocked octapods. Two-octapod chains (dimers) are suppressed, which implies that they are unstable. Dynamic fluctuations and the incorporation of trimers and of longer oligomers are mostly responsible for the growth of the interlocked chains, which in this way can reach up to ~20 octapods in length. Besides tracking nucleation and growth, the need for *in situ* observations is illustrated by measurements of important properties, such as the interparticle spacing in the chains. In their native solution environment, the octapod chains are much less densely packed than observed in conventional *ex situ* analysis after drying or vitrification of the solvent. On the basis of the characteristics obtained from *in situ* microscopy, a statistical mechanics model is built that explains the chain-length distribution (notably the absence of stable dimers), establishes the relative importance of van der Waals and entropic forces in the self-assembly, and determines the strength of the particle–particle interaction. Our findings illustrate the ability of *in situ* microscopy combined with modelling to obtain detailed quantitative insight into nanocrystal self-assembly processes in solution.

Results

***In situ* liquid-cell electron microscopy.** *In situ* microscopy was performed by scanning transmission electron microscopy (STEM) using solutions of CdSe/CdS octapods (pod length $L \sim (47 \pm 3)$ nm, $L/D_p \sim 4.2$; see Fig. 1) in toluene with a nominal particle concentration of 10^{-8} M, prepared according to reported methods[21].

The solution was loaded in the liquid cell and observed within a viewing area where it was held between two electron transparent SiN_x membranes (see Methods). Immediately after starting the observations, there was a higher concentration of octapods near the corners of the cell (for example, Supplementary Figs 1 and 6) and along the edges of the window, whereas the centre of the cell remained less populated.

Figure 2 shows a representative sequence of liquid-cell STEM images of octapods in toluene (extracted from Supplementary Movie 1) that follows the time evolution near the boundary between high and low concentration. Initially (observation time $t < 30$ s) the solution within these areas appears homogeneous, with diffuse contrast and streaking suggesting dispersed individual octapods and possibly chains submerged in the bulk solvent. As the self-assembly progresses, octapods and linear chains are clearly observed ($t \geq 35$ s). Chains of interlocked octapods form exclusively in the region with higher concentration near the edges and corners of the viewing window (Supplementary Fig. 7). The first clear contrast is due to two short-chain segments, each containing three interlocked octapods ($t = 35$ s). In the next ~20 s, additional short chains become visible and the first longer chains appear; surrounding diffuse contrast indicates continuing rapid motion in the solvent. During the further evolution, the existing chains grow longer so that the field of view becomes progressively crowded with interlocked octapod chains of different length. An overall downward directionality might be interpreted as a sign of progressive drying of the solvent and concomitant increase of the octapod density in the remaining

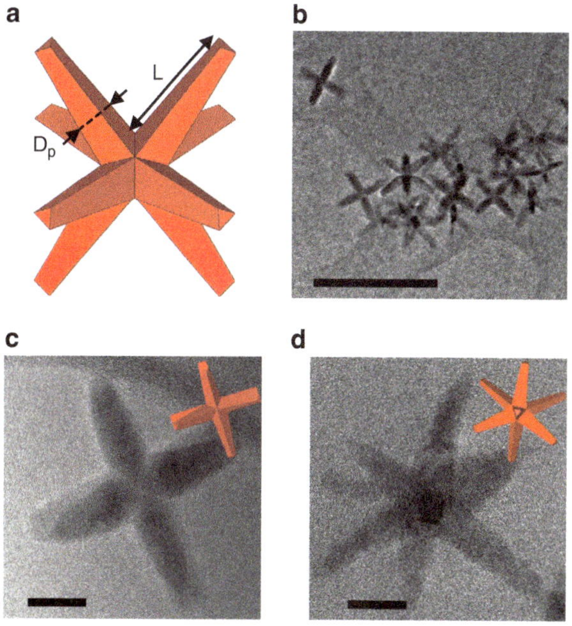

Figure 1 | CdSe/CdS octapods. (**a**) Schematic representation of geometry of a single octapod-shaped CdSe/CdS nanocrystal. (**b**) TEM image of a group of CdSe/CdS octapods drop-cast onto amorphous carbon film. Scale bar, 200 nm. (**c**,**d**) High-magnification TEM images of octapods with different orientations. Scale bar, 20 nm. TEM, transmission electron microscopy.

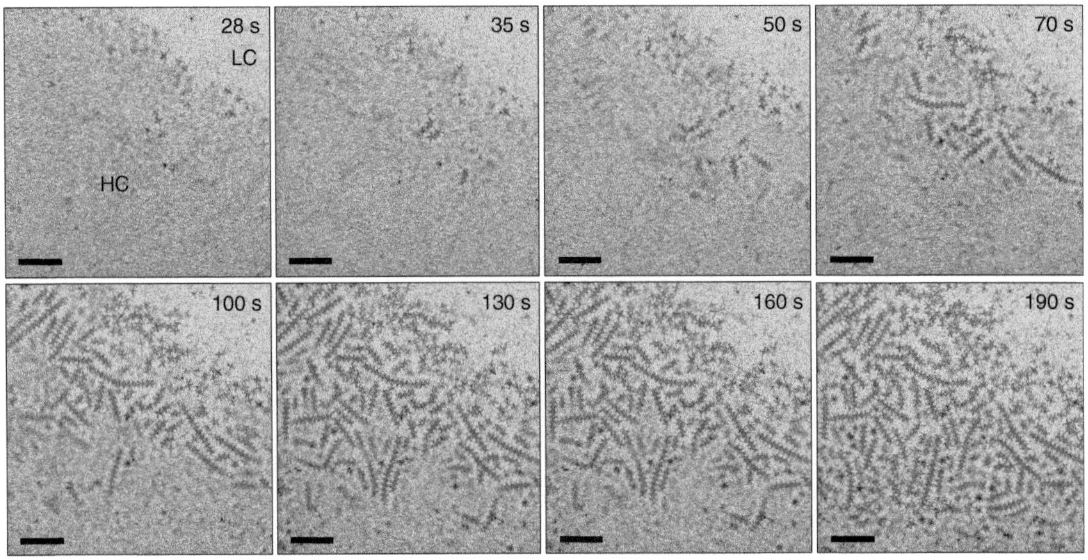

Figure 2 | *In situ* LCEM of the assembly of octapods into linear chains. Selected inverted dark-field STEM images from Supplementary Movie 1 showing the formation of ordered chains of octapods in toluene. Scale bar, 500 nm. The images show the evolution of the system starting from single octapods dispersed in the entire liquid volume (for $t<30$ s, 'HC'/'LC': high/low concentration of octapods) to interlocked chain-like structures that aggregate near the top membrane of the liquid cell.

solution, such as in evaporative self-assembly processes[32,33]. Further examination shows, however, that the formation of octapod chains occurs in a 3D solution environment of constant thickness, and is not significantly influenced by evaporation of the toluene solvent (see Supplementary Note 1 and Supplementary Figs 3–6,8 and 9).

Analysis of real-time image sequences allows us to develop a detailed picture of the self-assembly of 1D octapod chains in solution. Important elements, such as image contrast, predominant octapod orientations and a limited 3D imaging capability in STEM are discussed in the Methods and Supplementary Fig. 2. Figure 2 and additional data (Supplementary Movies 2–6; extracted images in Fig. 3 and Supplementary Figs 10–14) document the assembly mechanism. The process is initiated by the emergence of short segments of tightly interlocked octapod chains. The majority of chains are built from interlocked units, similar to those analysed *ex situ* in vitrified solvent[21], but observations of pairs or larger numbers of monomers interacting tip-to-tip and diffusing as a non-interlocked cluster (Supplementary Movies 2 and 3) suggest a richer set of interactions between monomers than the pod-to-pod van der Waals forces invoked so far.

Short interlocked octapod chains typically emerge from the bulk solution, which indicates that they are formed there. Observations of the fluid with focal plane near the top SiN_x membrane show the attachment and detachment of such short-chain segments as the predominant process of forming long 1D chains (Fig. 2). Chain growth is clearly a dynamic process, subject to fluctuations that add and remove segments of varying length (Fig. 3a,b). The smallest interlocked oligomer unit observed in our experiments is the trimer, that is, consists of three octapods. A survey of a large number of chains *ex situ* after opening the liquid cell (Fig. 3c) and *in situ* at different stages of the assembly (Fig. 4a) shows only isolated instances of interlocked dimers, whereas trimers and longer chains are much more abundant. The addition of trimers and longer oligomers is also the predominant elementary step in chain growth (Fig. 3b), while monomer incorporation is quite rare and is often followed by detachment of the same unit, suggesting incomplete locking (for example, Fig. 3b). Two-octapod increments are occasionally seen (Fig. 3b) but the absence of stable dimers in the solution suggests that these instances may actually represent a sequential addition of monomers or of non-interlocked dimers.

The observed solution-phase self-assembly deviates from a classic nucleation and growth scenario, which presumes the formation of a stable nucleus followed by attachment of monomer units[34]. While there appears to be a stable nucleus, the trimer, the further growth in most cases does not involve the incorporation of individual octapods, even though monomers are abundant and show extensive Brownian motion (Supplementary Note 2 and Supplementary Fig. 16). Instead, the stable nucleus grows primarily by attachment of other supercritical units, all of which are mobile in the solution. Attachment is facilitated by a considerable bending flexibility of the chains, which, for instance, allows long chains near the membrane to reach into the bulk liquid to capture a diffusing segment (Fig. 3a, Supplementary Fig. 10 and Supplementary Movie 4).

Statistical mechanical model. We constructed a model to probe the physical origin of the striking suppression of dimers and the prevalence of trimers as the shortest chain of interlocked octapods in solution. Using the measured chain-length distribution and other characteristics identified uniquely by *in situ* microscopy, the model provides insight into the interactions underlying the self-assembly process. In equilibrium, the probability of finding a chain of length n can be determined from the grand canonical distribution function:

$$C_n \propto \int d\Gamma e^{-\beta[H-\mu n]} \tag{1}$$

Here the chemical potential of octapods is determined by the monomer concentration, C_1, as $\mu = kT \ln C_1$; H is the interaction Hamiltonian of the system, $\beta = 1/kT$ and $d\Gamma$ represents an integration over internal degrees of freedom of the chain, that is, positions and orientations of $n-1$ octapods. If we neglect physical binding between the octapods (that is, assume $H=0$, but exclude their mutual overlap), the confinement of a single octapod within a cage made by two of its neighbours can be represented as a factor ν. This factor has the meaning of an

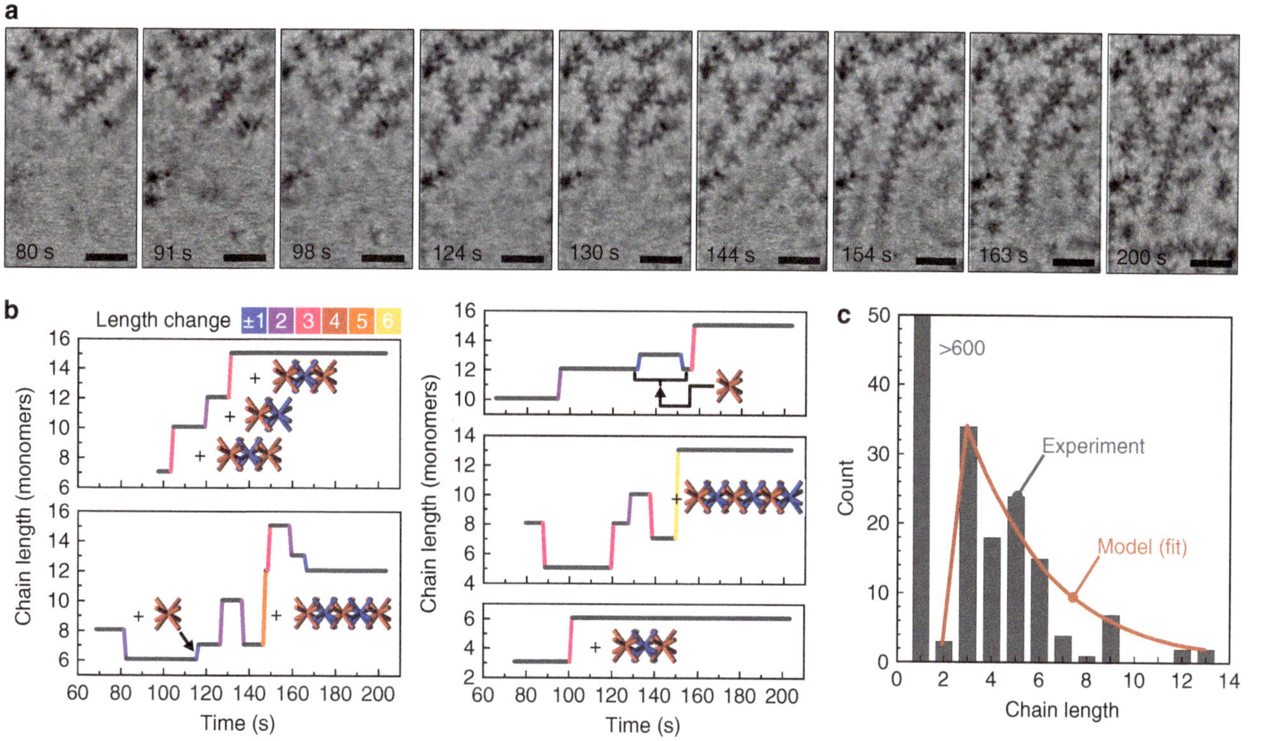

Figure 3 | Imaging octapod chains and tracking their growth with monomer resolution; modelling of the assembly process. (**a**) Representative time-lapse series of inverted dark-field STEM images showing the growth of a long chain of interlocked octapods. Scale bar, 200 nm. (**b**) Representative traces of the chain length as a function of time, showing length changes occurring only in exceptional cases by addition or removal of monomers, but instead mostly via incorporation and detachment of longer oligomers. Length changes are colour coded as shown. Longer horizontal lines imply that the length stays unchanged for extended time periods. (**c**) Length distribution of interlocked octapod chains, determined by *ex situ* STEM after opening the liquid cell (grey bars; total sample size: 752 objects, of which 642 were individual octapods). Note the absence of interlocked dimers. Red line: best fit of the length distribution within the statistical mechanics model described in the text.

effective localization volume and it accounts for the loss of translational and orientational entropies of the confined octapod. If Δ is a typical range of translational fluctuations, the corresponding amplitude of the angular degrees of freedom is Δ/R, where R is a characteristic radius of the octapod. Therefore an integration over three translational and three orientational degrees of freedom gives an effective localization volume $v \sim \Delta^6/R^3$.

We can now take into account mutual binding of the octapods. In our model, two octapods in contact create reversible bonds, each having the same characteristic binding energy $-\epsilon$ and localization range δ ($\sim$2.5 nm, set by the length of the ligands). We represent these bonds as a collection of two-state systems that can be in a bound or an unbound state independently of each other. This leads to the following distribution function:

$$C_n \sim (C_1 v)^{n-1}\left(1 + \frac{\delta}{\Delta} e^{\epsilon/kT}\right)^{Z_n} \quad (2)$$

An alternative, more detailed derivation of this result is presented in Supplementary Note 3. Z_n is the total number of contacts between the octapods that can be independently formed within the chain. This parameter is of crucial importance for understanding the observed chain statistics, in particular the substantial suppression of dimers compared with longer chains. We assume the maximum number of contacts that can be made between nearest neighbours to be 4. This implies that a tighter interlocking of two octapods with eight mutual contacts is not sterically allowed, presumably because of the geometry of their octahedral cores. This is consistent with the experimental observations (see below). In addition to the interactions with its nearest neighbours, it is assumed that fluctuations in the chain allow an octapod to create up to four contacts with any of its next-nearest neighbours. This consideration sets an upper bound to the number of contacts in a chain: $Z_n \leq 4(n-1) + 4(n-2) = 8n - 12$. However, this is not the only limitation. Each octapod has six degrees of freedom: three translational and three rotational. This means that for trimers ($n = 3$) the number of possible contacts (12) is exactly equal to the number of internal degrees of freedom of the chain, $6(n-1)$. For dimers ($n = 2$), the number of contacts is less than the number of degrees of freedom (4 and 6, respectively), while for chains longer than ($n \geq 3$), the situation is opposite. The maximally connected structure with $Z_n = 8n - 12$ would be over-constrained, and can only exist if the shapes are identical and symmetric to a very high precision. Any imperfection would lead to a loss of some of the contacts. We therefore postulate that the number of internal degrees of freedom sets another bound on the value of Z_n, that is,

$$Z_n = \min(8n - 12, 6(n-1)) \quad (3)$$

This model can successfully describe the observed chain statistics, including the suppression of dimers. In fact, its major predictions are very robust with respect to the choice of the model parameters. For $n \geq 3$, $Z_n = 6(n-1)$, that is, the overall distribution function is proportional to an exponential, $C_n \propto p^{n-1}$, where $p = (C_1 v)\left(1 + \delta e^{\frac{\epsilon}{kT}}/\Delta\right)^6$. For $n = 2$, $Z_n = 4$, that is, dimers are suppressed by a factor $(C_1 v)^{1/3}$.

As far as the specific parameters are concerned, the characteristic radius R of the octapod is set to $\sim$30 nm, half the pod length; $\Delta \approx 15$ nm was estimated as the difference between the apparent interparticle distance in the chain

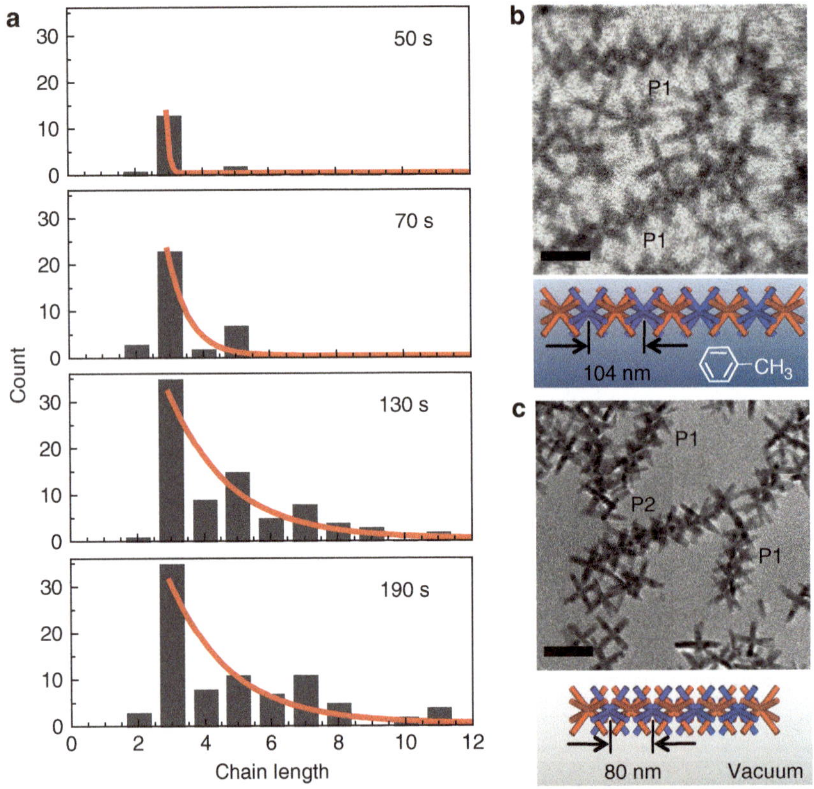

Figure 4 | Characteristics of octapod chains determined *in situ*. (**a**) Length distribution at different times during the self-assembly. Red lines are exponential fits for chain lengths ≥3. (**b**) Octapod chains in toluene, imaged *in situ* by liquid-cell STEM. Two longer chains are imaged in projection **P1** (see Methods and Supplementary Fig. 2). (**c**) Octapod chains in projections **P1** and **P2** imaged by TEM in vacuum after opening the liquid cell, shown at the same scale as **b**. Scale bars, 100 nm. Note the substantially larger interparticle spacing observed in toluene (52 ± 2 nm) compared with the dried assemblies (40 ± 2 nm), which suggests that capillary forces upon drying (and during freezing for cryo-TEM)[21] cause a significant tightening of the chains. TEM, transmission electron microscopy.

(50 nm) and its minimal value allowed geometrically (35 nm). The estimated average concentration of free octapods in the area in which self-assembly takes place is $C_1 \sim 10^{-6}\,\mathrm{M} \sim 10^{-6}\,\mathrm{nm}^{-3}$, so that $C_1 v \sim C_1 \Delta^6/R^3$ is of order 10^{-3}.

Discussion

Once the geometrical parameters are fixed, the only free parameter of the model is the energy per contact, ϵ. The value of the energy that gives the best fit to the experimental data is $\epsilon = 2.7kT$ (Fig. 3c), approximately twice the energy expected for a pure van der Waals interaction between two crossed pods[21]. The difference may be due to effects of inter-digitation of the ligands. The robustness of our model can be illustrated by the fact that changing $C_1 v$ by a factor of 2 leads to a correction to the binding strength ϵ of $0.1kT$, and to a relative change of the dimer fraction of only 25%. Additional molecular dynamics simulations also show that a pure van der Waals interaction is not sufficient to induce assembly (see Supplementary Note 4, Supplementary Methods and Supplementary Figs 17–19). Adding a short-range attraction does lead to chain formation.

In our model the assembly is driven not only by the energy gained because of binding. The fact that each contact fluctuates between bound and unbound states results in an additional entropic-free energy contribution $-6kT\ln\left(1+\frac{\Delta}{\delta}e^{-\epsilon/kT}\right) \approx -2kT$ per particle that stabilizes the chain. A final piece of evidence that supports the proposed mechanism stems from measurements of the octapod spacing in chains, determined *in situ* to avoid uncertainties due to capillary forces that may affect *ex situ* measurements of the lattice constant of dried or vitrified assemblies. The observed next-nearest neighbour octapod distance in solution (≈100 nm, Fig. 4b) is in fact greater than the one needed for physical contact between four tips of the two octapods (≈85 nm, close to the value found *ex situ* in vacuum (Fig. 4c) or vitrified solvent[21]). This distance is however sufficient to make up to two simultaneous contacts upon some rotation. In this way, the system can achieve the required value of Z_n (6 per particle) without paying an excessive entropic penalty for the extra confinement. The loose packing in solution also rationalizes characteristics important to chain growth, such as the facile addition and removal of segments and the bending of chains in tight radii, involved in capturing chain segments at later growth stages.

By imaging the self-assembly of branched nanocrystals with monomer resolution in solution, our experiments have pointed out a novel avenue for studying key aspects of self-assembly, such as the mechanism of formation and growth of ordered nanocrystal arrays and the interparticle spacing in solution. On the basis of characteristics obtained from *in situ* microscopy, a statistical mechanics model could be built that established the relative importance of van der Waals and entropic forces and determined the strength of the particle–particle interaction. A similar approach can be applied to other systems, such as spherical or polyhedral particles that arrange in two-dimensional or 3D assemblies, binary or more complex nanocrystal mixtures, or the folding and assembly of suitably tagged proteins. Once cells are developed that allow for controlled mixing of different liquids, external stimuli (for example, changes in PH, addition of co-solvents or depletants and so on.) may be used to actively change the assembly process while observing the outcome in real time.

Methods

Materials preparation. CdSe/CdS octapod-shaped nanocrystals were synthesized following a procedure based on the growth of CdS pods on cuboctahedral CdSe seeds that are prepared after a cation exchange process from $Cu_{2-x}Se$ nanocrystals. Briefly, a reaction flask was loaded with 120 mg of CdO and various surfactants (6 g of trioctylphosphine oxide, 160 mg of hexylphosphonic acid and 580 mg of octadecylphosphonic acid). The flask was kept under vacuum at 150 °C for 120 min. The temperature was increased to 350 °C and 1.6 g of trioctylphosphine was injected. After the recovery of the temperature, an aliquot of previously prepared $Cu_{2-x}Se$ nanocrystals with a concentration of 0.3×10^{-9} M was quickly injected in the flask after the addition of 16:0.5 mg S/trioctylphosphine solution. The reaction was run for 10 min and cooled down to the room temperature. After several washing steps with an anti-solvent, the final product was dispersed in 5 ml of toluene.

***In situ* liquid-cell STEM experiments.** The liquid-cell experiments were carried out in a dedicated specimen holder (Hummingbird Scientific) using micro-fabricated liquid cells, consisting of two 30-nm-thick SiN_x membrane windows with 50 μm × 50 μm window area supported by Si frames. The spacing between the windows was controlled using 200 and 300 nm polystyrene beads or 100 and 200 nm SiO_2 spacers. STEM imaging was performed in a FEI Titan 80–300 environmental Cs image corrected microscope operated at 300 kV. Low-loss electron energy loss spectra (EELS) were acquired using a Gatan Enfina spectrometer with an entrance aperture semi-angle of 5 mrad. The weighted average of the ratio of the thickness and the inelastic mean free path for the toluene and the two SiN membranes (each 30 nm thick) was determined from low-loss EELS, and the toluene thickness was calculated from it according to ref. 35 (Supplementary Figs 3,5 and 6). Typically, ~1 μl of the octapod-containing solution with an initial concentration of particles of 10^{-8} M was introduced in the liquid cell. The holder was inserted in the microscope immediately after loading of the cell. The time delay between loading of the solution and initiating the observations was ~10–20 min. STEM imaging was performed with ~2 Å probe size and ~0.2–0.3 nA beam current, measured in vacuum before introduction of the liquid cell. Time-lapse series of STEM images were acquired at several positions along the cell window, and at all positions the assembly of octapods was observed in real time. The results of the observations at two different areas are shown in Supplementary Movies 1 and 8. For Supplementary Movies 1–7 time-lapse STEM images of the assembly of the octapods were recorded with frame interval of 1 s and 3.96 μs dwell time/pixel. The conditions for the acquisition of STEM images were: electron dose rate of $1.35\,e^-\,Å^{-2}\,s^{-1}$, electron dose per image of $1.35\,e^-\,Å^{-2}$, total electron dose for the movie (204 frames) of $275.4\,e^-\,Å^{-2}$. The Supplementary Movies 1–7 are played at 20 frames per second. The original STEM images are shown in Supplementary Fig. 1. The STEM images in which the octapods appear dark on a light background have been inverted (in the program ImageJ) to facilitate visual perception and analysis. Supplementary Movie 8 is a series of time-lapse images recorded at a frame interval of 9 s.

Image contrast in *in situ* liquid STEM of octapod assemblies. The use of a focused probe in STEM, whose focal plane coincides with one of the membranes (for example, the top membrane) of the liquid cell, implies a limited 3D imaging capability. As an octapod chain transits from the bulk solution towards the membrane (Supplementary Fig. 2a, top), it initially appears with soft contrast that progressively sharpens (Supplementary Fig. 2a, bottom). Electron beam exposure generally induces octapods and chains to move from the interior of the solution towards the SiN_x membrane, possibly as a result of a beam-induced charging of the membrane and interaction with the tips of the polar (wurtzite) CdS pods via Coulomb or higher order multipole forces. While this behaviour facilitates high-resolution imaging by concentrating material near the focal plane, which allows to clearly distinguish individual octapods and their orientation in solution and within 1D chains, the attraction to the membrane is sufficiently weak that free octapod monomers can move via Brownian motion in the solution (see Supplementary Note 2 and Supplementary Fig. 16) with a substantially larger diffusion coefficient ($D_t \sim 10^{-15}\,m^2\,s^{-1}$) than found previously for particles in ultrathin liquid layers[36]. Octapod chains also show significant freedom of motion, including in-plane and out-of plane bending (Supplementary Fig. 10) and diffusion (Supplementary Fig. 11).

The interaction with the SiN_x membrane also acts to reduce the phase space of possible orientations of octapod chains (Supplementary Figs 2d,e and 15), which aids in the analysis of the chain length and allows counting the number of octapods in interlocked 1D chains with monomer resolution. Surveying a large number of chains, a large majority appears in projection **P1**, in which the pods of neighbouring nanoparticles project symmetrically towards the membrane. A minority of chains shows projection **P2**, in which they are rotated by 22.5° around their axis relative to **P1**. Occasionally, octapod chains with mixed appearance (that is, partial **P1** and **P2**) are observed. While still interlocked, these chains contain a stacking defect in which one section (a four-octapod segment in orientation **P2**, in Supplementary Fig. 2e) is rotated by 22.5° relative to the remainder of the chain (a five-octapod segment in projection **P1**, Supplementary Fig. 2e).

References

1. Konstantatos, G., Clifford, J., Levina, L. & Sargent, E. H. Sensitive solution-processed visible-wavelength photodetectors. *Nat. Photon.* **1,** 531–534 (2007).
2. Sun, L. *et al.* Bright infrared quantum-dot light-emitting diodes through inter-dot spacing control. *Nat. Nanotechnol.* **7,** 369–373 (2012).
3. Tao, A., Sinsermsuksakul, P. & Yang, P. Tunable plasmonic lattices of silver nanocrystals. *Nat. Nanotechnol.* **2,** 435–440 (2007).
4. Dong, A., Jiao, Y. & Milliron, D. J. Electronically coupled nanocrystal superlattice films by *in situ* ligand exchange at the liquid-air interface. *ACS Nano* **7,** 10978–10984 (2013).
5. Choi, J.-H. *et al.* Bandlike transport in strongly coupled and doped quantum dot solids: a route to high-performance thin-film electronics. *Nano Lett.* **12,** 2631–2638 (2012).
6. Cheon, J. *et al.* Magnetic superlattices and their nanoscale phase transition effects. *Proc. Natl Acad. Sci. USA* **103,** 3023–3027 (2006).
7. Yamada, Y. *et al.* Nanocrystal bilayer for tandem catalysis. *Nat. Chem.* **3,** 372–376 (2011).
8. Kim, Y., Shah, A. A. & Solomon, M. J. Spatially and temporally reconfigurable assembly of colloidal crystals. *Nat. Commun.* **5,** 3676 (2014).
9. Collier, C. P., Vossmeyer, T. & Heath, J. R. Nanocrystal superlattices. *Annu. Rev. Phys. Chem.* **49,** 371–404 (1998).
10. Shevchenko, E. V., Talapin, D. V., Kotov, N. A., O'Brien, S. & Murray, C. B. Structural diversity in binary nanoparticle superlattices. *Nature* **439,** 55–59 (2006).
11. Frenkel, D. Entropy-driven phase transitions. *Phys. A Stat. Mech. Appl.* **263,** 26–38 (1999).
12. Mirkin, C. A., Letsinger, R. L., Mucic, R. C. & Storhoff, J. J. A DNA-based method for rationally assembling nanoparticles into macroscopic materials. *Nature* **382,** 607–609 (1996).
13. Park, S. Y. *et al.* DNA-programmable nanoparticle crystallization. *Nature* **451,** 553–556 (2008).
14. Nykypanchuk, D., Maye, M. M., van der Lelie, D. & Gang, O. DNA-guided crystallization of colloidal nanoparticles. *Nature* **451,** 549–552 (2008).
15. Damasceno, P. F., Engel, M. & Glotzer, S. C. Predictive self-assembly of polyhedra into complex structures. *Science* **337,** 453–457 (2012).
16. Auer, S. & Frenkel, D. Prediction of absolute crystal-nucleation rate in hard-sphere colloids. *Nature* **409,** 1020–1023 (2001).
17. Cacciuto, A., Auer, S. & Frenkel, D. Onset of heterogeneous crystal nucleation in colloidal suspensions. *Nature* **428,** 404–406 (2004).
18. Gasser, U., Weeks, E. R., Schofield, A., Pusey, P. N. & Weitz, D. A. Real-space imaging of nucleation and growth in colloidal crystallization. *Science* **292,** 258–262 (2001).
19. Yethiraj, A. & van Blaaderen, A. A colloidal model system with an interaction tunable from hard sphere to soft and dipolar. *Nature* **421,** 513–517 (2003).
20. Vutukuri, H. R., Imhof, A. & van Blaaderen, A. Fabrication of polyhedral particles from spherical colloids and their self-assembly into rotator phases. *Angew Chem. Int. Ed. Engl.* **53,** 13830–13834 (2014).
21. Miszta, K. *et al.* Hierarchical self-assembly of suspended branched colloidal nanocrystals into superlattice structures. *Nat. Mater.* **10,** 872–876 (2011).
22. Williamson, M. J., Tromp, R. M., Vereecken, P. M., Hull, R. & Ross, F. M. Dynamic microscopy of nanoscale cluster growth at the solid-liquid interface. *Nat. Mater.* **2,** 532–536 (2003).
23. Zheng, H. M. *et al.* Observation of single colloidal platinum nanocrystal growth trajectories. *Science* **324,** 1309–1312 (2009).
24. Sutter, E. *et al.* *In situ* liquid cell electron microscopy of Ag-Pd galvanic replacement reactions on Ag nanoparticles. *Nat. Commun.* **5,** 4946 (2014).
25. Park, J. *et al.* Direct observation of nanoparticle superlattice formation by using liquid cell transmission electron microscopy. *ACS Nano* **6,** 2078–2085 (2012).
26. Rogers, W. B. & Manoharan, V. N. Programming colloidal phase transitions with DNA strand displacement. *Science* **347,** 639–642 (2015).
27. Maye, M. M. *et al.* Switching binary states of nanoparticle superlattices and dimer clusters by DNA strands. *Nat. Nanotechnol.* **5,** 116–120 (2010).
28. Li, H., Kanaras, A. G. & Manna, L. Colloidal branched semiconductor nanocrystals: state of the art and perspectives. *Acc. Chem. Res.* **46,** 1387–1396 (2013).
29. Glotzer, S. C. & Solomon, M. J. Anisotropy of building blocks and their assembly into complex structures. *Nat. Mater.* **6,** 557–562 (2007).
30. Conca, E. *et al.* Charge separation in Pt-decorated CdSe@CdS octapod nanocrystals. *Nanoscale* **6,** 2238–2243 (2014).
31. Qi, W. *et al.* Ordered two-dimensional superstructures of colloidal octapod-shaped nanocrystals on flat substrates. *Nano Lett.* **12,** 5299–5303 (2012).
32. Brinker, C. J., Lu, Y., Sellinger, A. & Fan, H. Evaporation-induced self-assembly: nanostructures made easy. *Adv. Mater.* **11,** 579–585 (1999).

33. Rabani, E., Reichman, D. R., Geissler, P. L. & Brus, L. E. Drying-mediated self-assembly of nanoparticles. *Nature* **426,** 271–274 (2003).
34. De Yoreo, J. J. & Vekilov, P. in *Biomineralization* (eds Dove, P.M., De Yoreo, J.J. & Weiner, S.) 57–93 (Mineralogical Society of America, 2003).
35. Egerton, R. F. *Electron Energy Loss Spectroscopy in the Electron Microscope* (Plenum Press, 1996).
36. Lu, J. Y., Aabdin, Z., Loh, N. D., Bhattacharya, D. & Mirsaidov, U. Nanoparticle dynamics in a nanodroplet. *Nano Lett.* **14,** 2111–2115 (2014).

Acknowledgements

This research has been carried out in part at the Center for Functional Nanomaterials, Brookhaven National Laboratory, which is supported by the US Department of Energy, Office of Basic Energy Sciences, under Contract No. DE-SC0012704. L.M. and M.A. acknowledge financial support from the European Union through the FP7 starting ERC grant NANO-ARCH (contract number 240111). We thank K. Jungjohann for participating in some of the experiments and measuring EELS spectra, and E. Krings for analysing the chain-length distributions. J.d.G. acknowledges financial support from the NWO Rubicon grant (#680501210) and Prof. A. Arnold and Dr O. Hickey for useful discussions.

Author contributions

E.S. conceived the study, performed *in situ* microscopy experiments, analysed the data and co-wrote the paper. P.S. analysed the data and co-wrote the paper. R.K. conceived the experiment and contributed to data analysis, discussion, interpretation of data and paper writing. J.d.G. performed numerical simulations, contributed to the discussion of the results and paper writing. A.T. conceived the statistical mechanical model explaining the octapod chain-length distribution. A.T. contributed to discussion, interpretation of the data, and writing of the paper. M.A. contributed to the data analysis, discussion, interpretation of the data and paper writing. L.M. contributed to the data analysis, discussion, interpretation of the data and paper writing. All authors commented on the manuscript.

Additional information

Competing financial interests: The authors declare no competing financial interests.

16

Exciton-exciton annihilation and biexciton stimulated emission in graphene nanoribbons

Giancarlo Soavi[1,†], Stefano Dal Conte[2], Cristian Manzoni[2], Daniele Viola[1], Akimitsu Narita[3], Yunbin Hu[3], Xinliang Feng[3], Ulrich Hohenester[4], Elisa Molinari[5,6], Deborah Prezzi[6], Klaus Müllen[3] & Giulio Cerullo[1,2]

Graphene nanoribbons display extraordinary optical properties due to one-dimensional quantum-confinement, such as width-dependent bandgap and strong electron-hole interactions, responsible for the formation of excitons with extremely high binding energies. Here we use femtosecond transient absorption spectroscopy to explore the ultrafast optical properties of ultranarrow, structurally well-defined graphene nanoribbons as a function of the excitation fluence, and the impact of enhanced Coulomb interaction on their excited states dynamics. We show that in the high-excitation regime biexcitons are formed by nonlinear exciton-exciton annihilation, and that they radiatively recombine via stimulated emission. We obtain a biexciton binding energy of ≈ 250 meV, in very good agreement with theoretical results from quantum Monte Carlo simulations. These observations pave the way for the application of graphene nanoribbons in photonics and optoelectronics.

[1] Dipartimento di Fisica, Politecnico di Milano, Piazza Leonardo Da Vinci 32, Milano 20133, Italy. [2] Istituto di Fotonica e Nanotecnologie, CNR, Piazza Leonardo Da Vinci 32, Milano 20133, Italy. [3] Max Planck Institute for Polymer Research, Ackermannweg 10, Mainz 55128, Germany. [4] Institute of Physics, University of Graz, Universitätsplatz 5, Graz 8010, Austria. [5] Dipartimento di Scienze Fisiche, Informatiche e Matematiche, Università di Modena e Reggio Emilia, Modena 41125, Italy. [6] Istituto Nanoscienze, CNR, via G. Campi 213/a, Modena 41125, Italy. † Present address: Cambridge Graphene Centre, University of Cambridge, Cambridge CB3 0FA, UK. Correspondence and requests for materials should be addressed to G.S. (email: gs544@cam.ac.uk) or to D.P. (email: deborah.prezzi@nano.cnr.it) or to G.C. (email: giulio.cerullo@polimi.it).

Slicing graphene into nanoribbons (GNRs) allows to open a bandgap in the graphene electronic structure, owing to the quasi-one-dimensional confinement[1]. This has important implications for electronic devices, such as GNR transistors[2], and forms the basis of the emerging field of graphene nanoplasmonics[3,4]. Especially appealing is also the possibility to additionally tailor specific properties through edge-structure engineering, such as magnetic ordering in zigzag terminated GNRs[5,6]. Further improvements along these lines are envisaged on the basis of recent advances in fabrication[7–10] and processing routes[11,12]. In particular, the bottom-up synthesis[7,8] of GNRs based on molecular precursors designed on purpose has proven capable of reaching nanometric widths with atomically precise edges, a regime where the GNR properties are widely tunable[1]. While single-walled carbon nanotubes (SWNTs) cannot be prepared with single chirality and require further processing with surfactants for the sorting[13,14], the bottom-up synthesis directly affords GNRs with a uniform chemical structure with 100% selectivity. Thus, prepared GNRs show well-defined electronic and optical properties, which are fully determined by their specific structure and can be further tuned by modulation of their width and edge configuration[10,15,16] as well as by atomically controlled doping[12,17]. All of this holds promise for application in next-generation optoelectronic and photonic devices, as recently suggested by the realization of all-GNR-based heterojunctions[12,18] and by other proposals for photovoltaic applications[16,19,20].

In spite of this interest, the field of GNRs is still in its infancy and little is known about their photophysical properties, especially in the non-equilibrium regime. Extraordinary optical properties were predicted[21–23], such as width-dependent bandgap and the formation of excitons with extremely high binding energies, which have been only recently demonstrated in bottom-up GNRs[15,24–26]. In particular, the pronounced excitonic effects[26] are accompanied by a significant increase of the optical absorbance, as compared with graphene, for light that is linearly polarized along the ribbon axis. At this stage, the understanding of the excited-state relaxation dynamics of GNRs would offer not only a deeper insight into the fundamental physics of these ideal one-dimensional systems but also a benchmark for their integration in advanced optoelectronic devices[27].

In the following, we apply resonant ultrafast pump-probe spectroscopy to nanometre-wide atomically precise GNRs obtained by a bottom-up solution synthesis[8] to study the kinetics of excitons and their interactions in the saturation (that is, nonlinear) excitation regime. For high-excitation fluences, we observe bimolecular exciton annihilation and the concomitant ultrafast ($\approx$1 ps) buildup of a stimulated emission (SE) signal from an excited biexciton state that is populated via a nonlinear process, a result that is extremely promising in view of applications of GNRs as tuneable active materials in lasers and light-emitting diodes.

Results

Linear absorption of GNRs. The GNRs studied here, characterized by cove-shaped edge morphology and hereafter labelled 4CNR (following the notation in ref. 16), were chemically synthesized as described in ref. 8. Their aromatic core structure, displayed in Fig. 1a (inset), features a modulated width of 0.7–1.1 nm, and is functionalized with long and branched alkyl chains (2-decyltetradecyl) at the outer benzene rings to guarantee dispersibility in organic solvents. The linear absorption spectrum of the GNR sample in tetrahydrofuran (THF) dispersion is shown in Fig. 1a (black curve), and compared with the simulated gas-phase spectrum obtained from *ab initio* *GW* plus Bethe–Salpeter (*GW*-BS) calculations (blue arrows), performed on H-passivated 4CNR (details concerning the calculations are reported in the Methods section). THF was chosen as a solvent to minimize aggregation of GNRs, which would significantly alter their optical properties. The effect of the solvent on the spectrum is instead expected to be minor (see, for example, ref. 28). The experimental spectrum is dominated by an optical transition of excitonic origin located at $\approx$570 nm, in good agreement with simulations. According to our *GW*-BS results, the first two excitons both arise from the linear combination of transitions among the two highest valence and two lowest conduction bands around the Γ point (mainly E_{12} and E_{21} transitions for the first and second exciton, respectively, as indicated in the band structure of Fig. 1b). In vacuum, excitons are tightly bound, with a giant binding energy of $\sim$1.5 eV (defined as the difference between the quasi-particle gap and the energy of the excitonic states), which is expected to diminish in presence of a dielectric environment.

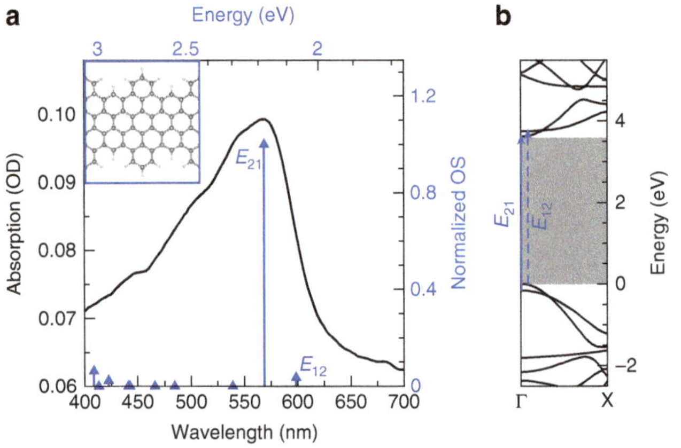

Figure 1 | Linear absorption and excitons in GNRs. (**a**) Linear absorption spectrum of the 4CNR sample in THF solution (black curve). A ball-and-stick model of the GNR without alkyl chains at the edges is shown in the inset. The experimental spectrum is compared with the result of *GW*-BS calculations, with excitonic transitions indicated by blue arrows. (**b**) *GW* quasi-particle band structure. The lines indicate the transitions that are mainly contributing to the first and second exciton. The 1.5 eV difference between the *GW* gap (**b**; grey area) and the excitonic transition reported in **a** defines the exciton binding energy in vacuum.

Ultrafast pump-probe spectroscopy of GNRs. We performed broadband pump-probe spectroscopy of the 4CNR samples using resonant excitation at 570 nm and white-light probing covering the 500–700 nm range, with an overall temporal resolution of $\approx$100 fs (see Methods for details of the experimental setup). Figure 2 shows the differential transmission ($\Delta T/T$) spectra for different pump-probe delays (Fig. 2a), and the $\Delta T/T$ dynamics at 600 nm probe wavelength (Fig. 2b), when the sample is excited with a low fluence of $\approx$100 μJ cm^{-2}. In the $\Delta T/T$ spectra we clearly distinguish two bands: (i) an intense photo-bleaching (PB) of the excitonic transition peaked at $\approx$570 nm, which we assign to ground-state depletion and/or phase space filling of the excited state; (ii) a relatively weak red-shifted photo-induced absorption (PA) band starting from $\approx$620 nm. Since these two bands have the same decay kinetics (Fig. 2b), we attribute the PA signal to excited-state absorption from the exciton to higher energy states or to the e-h continuum. The comparison of the $\Delta T/T$ spectra (Fig. 2a) for different pump-probe delays (from 500 fs to 50 ps) also highlights that the signal decays without any significant

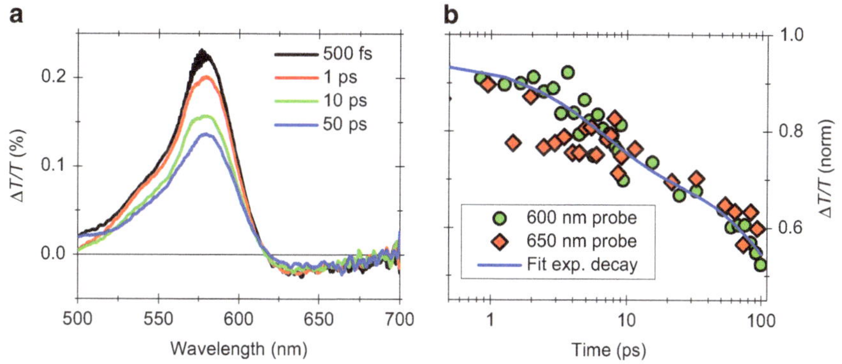

Figure 2 | Transient absorption spectra and dynamics at low fluence. (**a**) $\Delta T/T$ spectra of 4CNRs at different pump-probe delays and (**b**) decay dynamics at 600- (green circles) and 650- (red diamonds) nm probe wavelengths for an excitation fluence of $\approx$100 μJ cm^{-2}. The fit (blue line) in **b** correspond to a bi-exponential function with time constants $\tau_1 \approx 6$ ps and $\tau_2 \approx 330$ ps.

spectral evolution at this excitation fluence, thus indicating a simple relaxation mechanism of the excited state (either radiative via emission of photons and/or non-radiative via interaction with phonons).

To understand the carrier relaxation process, we can take as a reference the large amount of experimental results on SWNTs, since we expect them to display similar recombination dynamics. SWNTs also show complex, multi-component decay kinetics. In the low-excitation fluence regime, the long-lived (>5 ps) decay components in SWNTs have been reproduced with a bi-exponential model, which includes both the radiative and non-radiative lifetime[29–32]. Other studies describe these decay dynamics by a model for a diffusion-limited regime[33,34] or geminate e–h recombination in one dimension[35], thus leading to a power law ($\approx t^{-0.5}$) kinetics. Since a detailed study of the dynamics of these long-lived photoexcitations for GNRs is beyond the scope of this paper, we fit the kinetics of the PB signal at 600 nm probe wavelength with a simple bi-exponential decay model (Fig. 2b), which gives us the timescale of the relaxation processes. From the fit we obtain $\tau_1 \approx 6$ ps and $\tau_2 \approx 330$ ps, in good agreement with the results obtained for SWNTs[30].

Instead, we concentrate on the ultrafast (<5 ps) decay dynamics, and in particular on their dependence on the excitation fluence, as we show in Fig. 3. The normalized $\Delta T/T$ spectra at different excitation fluences for a 1 ps pump-probe delay (Fig. 3a) display very similar, fluence-independent PB and PA spectral features, while for a 5 ps pump-probe delay (Fig. 3b) we unambiguously observe the buildup, with increasing fluence, of a positive and red-shifted $\Delta T/T$ peak, at $\approx$650 nm. From a general point of view, a positive $\Delta T/T$ signal in pump–probe experiments describes either a PB or a SE process. We here assign the peak at $\approx$650 nm to a SE process, since (i) it does not correspond to any resonant feature in the linear absorption spectrum (Fig. 1a), as also confirmed by simulations, being instead red-shifted with respect to the main excitonic transition (at $\approx$570 nm); and (ii) it appears with a $\approx$ps delay with respect to the pump pulse and only at high-excitation fluences. We note that photons produced by SE are identical (phase, energy and momentum) to the probe photons and thus can be detected in pump–probe experiments, at variance with those produced by spontaneous emission.

Exciton–exciton annihilation and biexciton formation in GNRs. To clarify the origin of this SE signal, we concentrate on the fluence-dependent dynamics at the probe wavelengths of 600 nm, that is, the PB signal, and 650 nm, that is, the SE signal (Fig. 3c and d, respectively). We immediately observe that, for increasing fluence, the PB signal (Fig. 3c) displays a faster decay, while, correspondingly, the signal at 650 nm (Fig. 3d) undergoes a clear change in sign (from negative to positive) that corresponds to the delayed formation of the SE signal. First, let us analyse the fast PB decay at 600-nm probe wavelength (Fig. 3c). In semiconducting SWNTs the appearance of an ultrafast fluence-dependent decay component is explained by exciton–exciton annihilation, a two-exciton interaction process, in which one exciton recombines to the ground state and the other either dissociates into a free e–h pair or is promoted into a higher energy level[36–38]. Such process is also commonly observed in other one-dimensional semiconductors, such as conjugated polymers[39,40], and it has been recently observed also in monolayer MoS_2 (ref. 41). Theoretical calculations[42] also predict that an Auger-like mechanism occurs in semiconducting armchair GNRs because of effectively enhanced Coulomb interaction. In their work, Konabe *et al.*[42] find an exciton–exciton annihilation time in the order of few ps for 1.2–2.5 nm wide GNRs. After the initial ultrafast nonlinear decay process, the dynamics at all pump fluences are instead the same. This can be noticed by comparing the kinetics of the PB signal at high and low fluences over the full temporal range (inset in Fig. 3c). Being a two-body interaction process, exciton–exciton annihilation is expected to display a nonlinear dependence on the exciton density (see Methods for details) and/or the excitation fluence[37], in agreement with our experimental results (inset in Fig. 3a).

Second, we need to understand the origin of the delayed formation of the SE signal at 650 nm (Fig. 3d). As we have already discussed, following exciton–exciton annihilation both free e–h pairs and/or higher energy-excited states can be formed. The first scenario has been observed in SWNTs, where the creation of charges in the high-excitation regime leads to the formation of trions[43,44], which are detected as a negative (PA) and red-shifted $\Delta T/T$ signal. Clearly, this scenario is in contrast with our experimental results, which present a positive (SE) signal. Instead, the formation of a delayed and red-shifted SE signal in the high-excitation regime was observed in semiconducting quantum-dots (QDs)[45–48], another prototype of quantum-confined systems. In the case of QDs, the SE signal was explained in terms of emission from biexcitons and the energy distance between the main exciton PB signal and the biexciton SE signal gives the biexciton binding energy, which is typically of the order of few tens of meV. For our GNRs, the SE peak corresponds to a much larger value for the biexciton binding energy, that is, $E_b \approx 250$ meV. This value is in accordance with the comparably larger exciton binding energies in GNRs, that are at least one order of magnitude larger than in QDs due to the reduced screening as well as the extreme two-dimensional and transversal confinements. The following

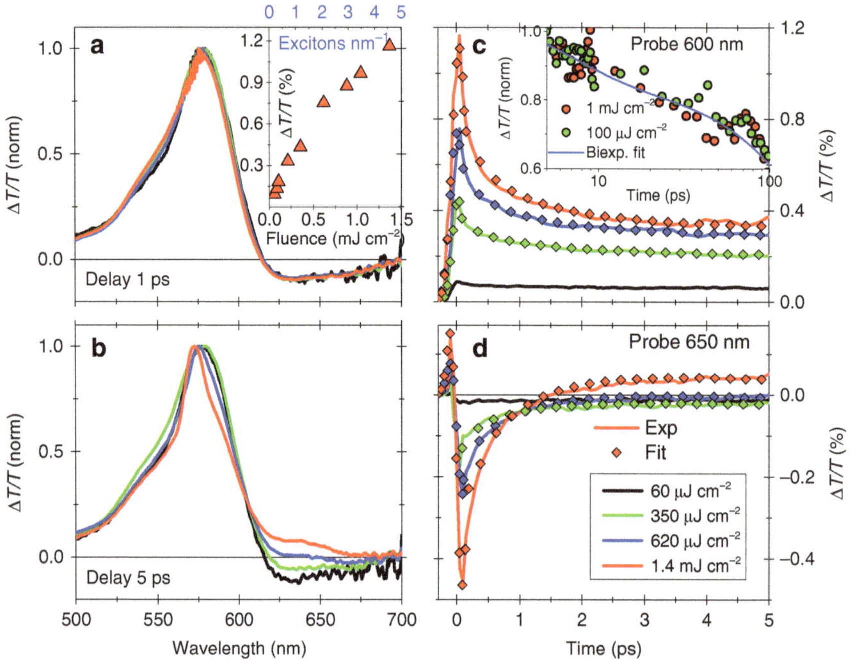

Figure 3 | Exciton-exciton annihilation and biexciton formation. Normalized $\Delta T/T$ spectra of 4CNRs for different excitation fluences at a fixed pump-probe delay of (**a**) 1 ps and (**b**) 5 ps. The inset in **a** reports the peak amplitude of the signal at 600 nm probe wavelength as a function of the excitation fluence (bottom *x* axis) and the exciton linear density (top *x* axis). Excitation-fluence-dependent dynamics at (**c**) 600 nm probe wavelength and (**d**) 650 nm probe wavelength. The fit (diamonds) is obtained from the coupled rate-equations (described in the text) based on exciton-exciton annihilation in one dimension. Inset in **c** represents the dynamics on a 100 ps timescale for 100 μJ cm^{-2} (low) and 1 mJ cm^{-2} (high) fluences, together with the bi-exponential fit used in Fig. 2b.

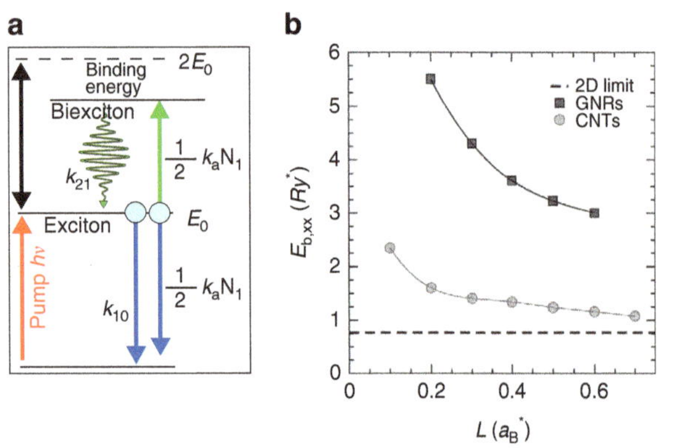

Figure 4 | Photoexcitation scenario in GNRs and biexciton binding energy. (**a**) Sketch of the energetic levels and the kinetic model described in the text. (**b**) Biexciton (XX) binding energy ($E_{b,XX}$) as a function of the lateral dimension (L), that is, width for GNRs (black squares) and diameter for SWNTs (grey circles). The data, obtained by guide-function QMC simulations, are shown in dimensionless exciton units, as detailed in the text. In these units, the binding energy of the 4CNR is $\sim 3.4\,Ry^*$. The dashed line indicates the two-dimensional value of $E_{b,XX} = 0.77\,Ry^*$.

photoexcitation scenario in GNRs thus emerges: at high-excitation fluences, exciton–exciton annihilation leads to the population of a radiative biexciton state, which then undergoes SE to the one-exciton level upon interaction with the probe pulse (Fig. 4a).

Discussion

To support our assignment, we first compute the biexciton binding energy in GNRs by means of guide-function quantum Monte Carlo (QMC) simulations, using an envelope function approach with effective masses, as previously reported in ref. 49 for SWNTs. Full description of the method is reported in the Methods section. In Fig. 4b, we show the biexciton binding energy dependence on the lateral confinement in dimensionless exciton units, that is, distances are measured in units of the effective Bohr radius $a^*_B = a_B\varepsilon/\mu$ and energies in units of the effective Rydberg $Ry^* = \frac{e^2}{2\varepsilon a^*_B}$, where μ is the reduced e–h mass, ε is the dielectric constant, a_B is the Bohr radius and e is the electron charge. By considering an average reduced mass as obtained from *ab initio* calculations (computed as the weighted average relative to the E_{21} and E_{12} transitions, that is, $\mu = 0.22$), the average width of the 4CNR ($w = 0.84$ nm), and the dielectric constant of the solvent ($\varepsilon = 7.5$), we obtain a biexciton binding energy of 180 meV, in good agreement with the experimental result of 250 meV. The discrepancy with respect to experiments is reasonable in view of the simplified description scheme used here, which is not expected to capture all microscopic details.

In Fig. 4b, we also report the biexciton binding energy for SWNTs, calculated by the same approach[49]: for a SWNT of similar lateral dimension, we find a binding energy that is less than one-half of that of the 4CNR in dimensionless units. This can be understood by considering the different biexciton confinement, since in the case of SWNTs the biexciton wavefunction is delocalized over the whole circumference, whereas in GNRs it becomes strongly confined in the transversal direction. The value obtained for GNRs is quite large also in comparison with the biexciton binding energy $E_b \approx 50$ meV of different monolayer transition metal dichalcogenides obtained by recent transient absorption[50] and photoluminescence experiments[51].

To further confirm our interpretation, we finally reproduce the temporal evolution of the exciton (that is, decay of the PB signal) and the biexciton (that is, formation of the SE signal) populations with the following coupled rate-equations[52–54]:

$$\frac{dn_E}{dt} = -k_{10}n_E - k_{e-e}\frac{n_E^2}{\sqrt{t}} + k_{21}n_B \quad (1)$$

$$\frac{dn_B}{dt} = +\frac{1}{2}k_{e-e}\frac{n_E^2}{\sqrt{t}} - k_{21}n_B \quad (2)$$

where n_E and n_B are the exciton and biexciton population, respectively, k_{10} is the decay rate from the exciton to the ground state, k_{21} is the radiative decay rate from the biexciton to the exciton producing the SE signal and k_{e-e} is the exciton annihilation rate constant (see Fig. 4a for the adopted model). The $t^{-0.5}$ dependence of the exciton annihilation rate arises from the one-dimensional diffusion mechanism of excitons, and its divergence for $t \to 0$ is cured by truncating it for times shorter than the width of the instrumental response function (IRF) (≈ 100 fs in our case). The quality of the fit (Fig. 3c,d) further confirms the validity of our model. From the fit we find that $k_{10} \approx 0$ ps and $k_{21} \approx 0.15\,ps^{-1}$, meaning that both mechanisms occur on a timescale that is longer with respect to the temporal window (5 ps) that we use for this analysis.

It is worth noting that we can fit our experimental data considering that all the excitons that undergo exciton–exciton annihilation form biexcitons. In particular, we can exclude that the observed SE signal is because of the dissociation of excitons into free e–h carriers, since SE from electrons in the continuum is not expected. Another possibility is that free e–h pairs get trapped into low-energy states; for example, due to defects. Nevertheless, since the selection rules for emission and absorption of photons are the same, the presence of such bright low-energy states should be detectable also in the linear absorption spectrum, and thus this scenario can be also excluded by looking at the absorption spectrum and theoretical simulations in Fig. 1. Thus, although we cannot exclude that free e–h carriers are formed, we can strongly assert that the observed SE arises from biexcitons. Since our model is able to correctly reproduce also the amplitude of the pump–probe signal without additional loss channels, we conclude that biexciton formation upon exciton–exciton annihilation is extremely efficient. Finally, for a fixed pump–probe delay we can also evaluate the annihilation rate $k_a = \frac{k_{e-e}}{\sqrt{t}}$, which corresponds to an initial $k_a \approx 2\,ps^{-1}$ for a pump–probe delay of 180 fs, close to our temporal resolution, and thus to an annihilation lifetime of ≈ 0.5 ps, in excellent agreement with experimental results on SWNTs[54] and theoretical calculations on GNRs[42].

In conclusion, we studied the transient photophysical properties of ultranarrow structurally well-defined GNRs by means of ultrafast pump–probe spectroscopy. We show that a nonlinear decay channel for the main excitonic transition sets in at high excitation densities, and, correspondingly, we unambiguously observe a red-shifted SE signal. Our experiments demonstrate that exciton–exciton annihilation populates a radiative biexciton state, with an extremely high binding energy ≈ 250 meV, in agreement with estimates from QMC simulations. The high efficiency we find for both exciton–exciton annihilation and biexciton formation is of great importance, not only for gaining fundamental understanding on strongly enhanced quantum effects in low-dimensional materials but also for its implications in GNR-based optoelectronic devices. Indeed, the clear observation of a strong SE is extremely promising in view of using GNRs as active light-amplifying materials in tuneable lasers and light-emitting diodes. Moreover, our results suggest that also multiple-exciton generation[55–57] can be extremely efficient in GNRs since it is governed by the same exciton–exciton annihilation rate. A detailed understanding of the photophysics of biexcitons and the mechanism of multiple-exciton generation will help to improve the efficiency of photovoltaic devices, with GNRs acting as light absorbers.

Methods

Pump-probe experimental setup. The experimental setup used for pump–probe measurements has been described in detail elsewhere[58]. In brief, the setup is based on a regeneratively amplified Ti:sapphire laser (Coherent, Libra) producing 100 fs, 4 mJ pulses at 800 nm wavelength and 1 kHz repetition rate. The probe pulse is obtained by focusing a fraction of the laser beam in a 2-mm-thick sapphire plate to generate a broadband single-filament white-light continuum. The pump pulse, generated by an optical parametric amplifier, is centred at 570 nm (at resonance with the main excitonic transition) with a bandwidth of ≈ 10 nm, corresponding to ≈ 70 fs duration. The probe light transmitted by the sample is dispersed on an optical multichannel analyser equipped with fast electronics, allowing single-shot recording of the probe spectrum at the full 1 kHz repetition rate. By changing the pump–probe delay, we record two-dimensional maps of the differential transmission ($\Delta T/T$) signal as a function of probe wavelength and delay. The temporal resolution (taken as full-width at half-maximum of pump–probe cross-correlation) is ≈ 100 fs over the entire probe spectrum.

Exciton linear density estimation. To estimate the exciton linear density reported in the inset of Fig. 3a, we proceed as follows: (i) we calculate the number of absorbed photons cm^{-2} from the measured pump fluence, the measured absorbance, and the pump photon energy (2.18 eV); (ii) we calculate the number of excitons cm^{-2} by considering an exciton photogeneration efficiency (Quantum Yield) of 96% based on optical pump—THz probe experiments on similar GNRs[59]; and (iii) we multiply this value by the concentration ($0.0021\,g\,l^{-1}$) of the dispersion and divide by the mass of a unit length of the GNRs ($0.83 \times 10^{-21}\,g\,nm^{-1}$). This estimate gives a density of $\sim$0.2–0.3 excitons nm^{-1} for pump fluences below $100\,\mu J\,cm^{-2}$, corresponding to an average distance between excitons of $\sim$4–5 nm. Such a value appears to be sufficient to prevent exciton–exciton interactions.

Coherent artefacts in pump-probe dynamics. The coupled-rate equations used to model the evolution of the exciton and biexciton populations are solved by taking into account both the IRF and possible coherent artefacts present in pump–probe measurements. For our experimental setup the IRF is the cross-correlation of the pump and probe pulses, and can be quite accurately modelled by a Gaussian function with ≈ 100 fs full-width at half-maximum. Possible coherent artefacts in pump–probe experiments are stimulated Raman amplification and cross-phase modulation (XPM)[60]. We fitted the initial 100 fs of the PB signal at 600 nm (Fig. 3c) including the IRF to reproduce the initial buildup and a Gaussian function (the same used for the IRF) to reproduce the initial ultrafast stimulated Raman amplification coherent artefact. For the SE dynamics at 650 nm (Fig. 3d), instead, we include also XPM, which we fit as the first derivative of a Gaussian function (the same used for the cross-correlation). Although extremely simple, this model correctly reproduces not only the evolution but also the initial steps of the pump–probe signal.

Thermal effects and sample heating. Time-resolved experiments were carried out at room temperature, assuming negligible temperature effects on the spectra on the basis of previous results on SWNTs (see, for example, ref. 61). Regarding the sample heating during experiments, we also expect negligible temperature changes upon photoexcitation. In fact, we can estimate a maximum increase in temperature of $\sim$0.1 K, based on a comparison with the work of Abdelsayed *et al.*[62], and by considering the following parameters: volume of the sample (0.3 ml), THF heat capacity ($123\,J\,mol^{-1}\,K^{-1}$), density ($889\,kg\,m^{-3}$), molar mass ($72\,g\,mol^{-1}$) and laser total energy ($\sim$100 nJ per pulse at 500 Hz for 10 min of irradiation, corresponding to 30 mJ). This indicates that we are working in a perturbative regime for what concerns thermal effects.

***GW*-BS calculations for the static absorption.** The ground-state atomic structure of the 4CNR was optimized by using the PWscf code of the Quantum ESPRESSO package (ref. 63), which is based on a plane-wave pseudopotential implementation of density functional theory. Calculations were performed by employing local density approximation exchange correlation (xc) functional and norm-conserving pseudopotentials, with a cutoff energy for the wavefunctions of 60 *Ry*. The atomic positions within the cell were fully relaxed until forces were $<5 \times 10^{-4}\,Ry\,bohr^{-1}$, while the lattice constant along the periodic direction was optimized separately. The Brillouin zone was sampled with 16 **k**-points along the periodic direction. The Kohn-Sham band structure obtained for the optimized geometry was improved by introducing many-body corrections within the G_0W_0 approximation for the self-energy operator. Here, the dynamic dielectric function was obtained within the plasmon-pole approximation, by employing a box-shaped truncation of the Coulomb potential[64] to remove the long-range interaction between periodic images. The optical absorption spectrum was then computed as the imaginary part of the macroscopic dielectric function starting from the solution of the BS equation, which allows for the inclusion of e–h interaction effects.

The static screening in the direct term was calculated within the random-phase approximation with inclusion of local field effects; the Tamm–Dancoff approximation for the BS Hamiltonian was employed. The aforementioned *GW*-BS calculations, performed by using the YAMBO code (ref. 65), were carried out for the fully H-passivated 4CNR, that is, by removing the alkyl side chains, in order to make them computationally affordable. We have checked that the different passivation does not affect the band structure properties at the density functional theory and local density approximation level in the energy window of interest for the determination of the optical absorption. Similar results were reported in ref. 66.

The guide function quantum Monte Carlo approach. Atomistic simulations as the ones described above are presently unfeasible for the investigation of biexcitons in realistic systems. Here we resort to an effective model based on guide function QMC simulations, as previously reported in ref. 49 for SWNTs. For the (unnormalized) guide function we use

$$\Psi_{\mathrm{T}}(r_1, r_2, r_{\mathrm{a}}, r_{\mathrm{b}}) = \exp[-(r_{1\mathrm{a}} + r_{1\mathrm{b}} + r_{2\mathrm{a}} + r_{2\mathrm{b}})] \tag{3}$$

where $r_{1,2}$ ($r_{\mathrm{a,b}}$) are the dimensionless positions of the two electrons (holes), and $r_{\mathrm{i,j}} = \left[(x_{\mathrm{i}} - x_{\mathrm{j}})^2 + (y_{\mathrm{i}} - y_{\mathrm{j}})^2\right]^{1/2}$ is the distance between particles confined to the two-dimensional nanoribbon. The Monte Carlo simulation approach is the same as in ref. 49 with the only exception that we add a box-like confinement potential

$$V(y) = V_0 / \left[e^{\beta\left(0.5 - \frac{|y|}{w}\right)} + 1\right] \tag{4}$$

along the transversal direction, with $V_0 = 1{,}000$, $\beta = 20$ and w the GNR width. In the simulations we use 20,000 walkers[67], a time step of $\Delta t = 0.25 \times 10^{-4}$, an equilibration interval of 20,000 and a measurement interval of 30,000 time steps. The smoothed confinement potential and the time step were chosen such that for 'typical' paths the inequality $|V(y_t) - V(y_{t+1})| \ll |(y_t - y_{t+1})/(2\Delta t^2)|$ holds[68]. We checked that the biexciton binding energy did not change substantially upon modifying Δt or other simulation parameters. The QMC approach has been able to predict quite accurately the biexciton binding energy of SWNTs. In fact, Colombier *et al.*[69] detected the presence of biexcitons in SWNTs embedded in a gelatine matrix by means of nonlinear optical spectroscopy, reporting a binding of 106 meV energy for the (9, 7) tube. The QMC model predicts a value of the binding energy of $\approx$100 meV assuming a dielectric constant of $\varepsilon = 3$ (instead of 2.3 as appropriate for a gelatin matrix). Such an overestimation in the case of small values of ε is known for phenomenological models, whereas for larger values of ε (like the one considered in this work, i.e., $\varepsilon = 7.5$) the QMC model is expected to give results similar to more refined (though not yet atomistic) approaches, as discussed in ref. 70.

References

1. Castro Neto, A. H., Guinea, F., Peres, N. M. R., Novoselov, K. S. & Geim, A. K. The electronic properties of graphene. *Rev. Mod. Phys.* **81,** 109–162 (2009).
2. Schwierz, F. Graphene transistors. *Nat. Nanotechnol.* **5,** 487–496 (2010).
3. Chen, J. *et al.* Optical nano-imaging of gate-tunable graphene plasmons. *Nature* **487,** 77–81 (2012).
4. Rodrigo, D. *et al.* Mid-infrared plasmonic biosensing with graphene. *Science* **349,** 165–168 (2015).
5. Son, Y.-W., Cohen, M. L. & Louie, S. G. Half-metallic graphene nanoribbons. *Nature* **444,** 347–349 (2006).
6. Magda, G. Z. *et al.* Room-temperature magnetic order on zigzag edges of narrow graphene nanoribbons. *Nature* **514,** 608–611 (2014).
7. Cai, J. *et al.* Atomically precise bottom-up fabrication of graphene nanoribbons. *Nature* **466,** 470–473 (2010).
8. Narita, A. *et al.* Synthesis of structurally well-defined and liquid-phase-processable graphene nanoribbons. *Nat. Chem.* **6,** 126–132 (2014).
9. Jacobberger, R. M. *et al.* Direct oriented growth of armchair graphene nanoribbons on germanium. *Nat. Commun.* **6,** 8006 (2015).
10. Narita, A., Wang, X.-Y., Feng, X. & Müllen, K. New advances in nanographene chemistry. *Chem. Soc. Rev.* **44,** 6616–6643 (2015).
11. Bennett, P. B. *et al.* Bottom-up graphene nanoribbon field-effect transistors. *Appl. Phys. Lett.* **103,** 253114 (2013).
12. Cai, J. *et al.* Graphene nanoribbon heterojunctions. *Nat. Nanotechnol.* **9,** 896–900 (2014).
13. Yang, F. *et al.* Chirality-specific growth of single-walled carbon nanotubes on solid alloy catalysts. *Nature* **510,** 522–524 (2014).
14. Omachi, H., Nakayama, T., Takahashi, E., Segawa, Y. & Itami, K. Initiation of carbon nanotube growth by well-defined carbon nanorings. *Nat. Chem.* **5,** 572–576 (2013).
15. Narita, A. *et al.* Bottom-up synthesis of liquid-phase-processable graphene nanoribbons with near-infrared absorption. *ACS Nano* **8,** 11622–11630 (2014).
16. Osella, S. *et al.* Graphene nanoribbons as low band gap donor materials for organic photovoltaics: quantum chemical aided design. *ACS Nano* **6,** 5539–5548 (2012).
17. Kawai, S. *et al.* Atomically controlled substitutional boron-doping of graphene nanoribbons. *Nat. Commun.* **6,** 8098 (2015).
18. Chen, Y.-C. *et al.* Molecular bandgap engineering of bottom-up synthesized graphene nanoribbon heterojunctions. *Nat. Nanotechnol.* **10,** 156–160 (2015).
19. Yan, X., Cui, X., Li, B. & Li, L. Large, solution-processable graphene quantum dots as light absorbers for photovoltaics. *Nano Lett.* **10,** 1869–1873 (2010).
20. Lee, S. L. *et al.* An organic photovoltaic featuring graphene nanoribbons. *Chem. Commun. (Camb.)* **51,** 9185–9188 (2015).
21. Prezzi, D., Varsano, D., Ruini, A., Marini, A. & Molinari, E. Optical properties of graphene nanoribbons: the role of many-body effects. *Phys. Rev. B* **77,** 041404 (2008).
22. Yang, L., Cohen, M. L. & Louie, S. G. Excitonic effects in the optical spectra of graphene nanoribbons. *Nano Lett.* **7,** 3112–3115 (2007).
23. Prezzi, D., Varsano, D., Ruini, A. & Molinari, E. Quantum dot states and optical excitations of edge-modulated graphene nanoribbons. *Phys. Rev. B* **84,** 041401 (2011).
24. Ruffieux, P. *et al.* Electronic structure of atomically precise graphene nanoribbons. *ACS Nano* **6,** 6930–6935 (2012).
25. Linden, S. *et al.* Electronic structure of spatially aligned graphene nanoribbons on Au(788). *Phys. Rev. Lett.* **108,** 216801 (2012).
26. Denk, R. *et al.* Exciton-dominated optical response of ultra-narrow graphene nanoribbons. *Nat. Commun.* **5,** 4253 (2014).
27. Bonaccorso, F., Sun, Z., Hasan, T. & Ferrari, A. C. Graphene photonics and optoelectronics. *Nat. Photon.* **4,** 611–622 (2010).
28. Ohno, Y. *et al.* Excitonic transition energies in single-walled carbon nanotubes: dependence on environmental dielectric constant. *Phys. Status Solidi B Basic Solid State Phys.* **244,** 4002–4005 (2007).
29. Styers-Barnett, D. J., Ellison, S. P., Park, C., Wise, K. E. & Papanikolas, J. M. Ultrafast dynamics of single-walled carbon nanotubes dispersed in polymer films. *J. Phys. Chem. A* **109,** 289–292 (2005).
30. Chou, S. *et al.* Phonon-assisted exciton relaxation dynamics for a (6,5)-enriched DNA-wrapped single-walled carbon nanotube sample. *Phys. Rev. B* **72,** 195415 (2005).
31. Dyatlova, O. A. *et al.* Ultrafast relaxation dynamics via acoustic phonons in carbon nanotubes. *Nano Lett.* **12,** 2249–2253 (2012).
32. Huang, L., Pedrosa, H. N. & Krauss, T. D. Ultrafast ground-state recovery of single-walled carbon nanotubes. *Phys. Rev. Lett.* **93,** 017403 (2004).
33. Russo, R. M. *et al.* One-dimensional diffusion-limited relaxation of photoexcitations in suspensions of single-walled carbon nanotubes. *Phys. Rev. B* **74,** 041405 (2006).
34. Allam, J. *et al.* Measurement of a reaction-diffusion crossover in exciton-exciton recombination inside carbon nanotubes using femtosecond optical absorption. *Phys. Rev. Lett.* **111,** 197401 (2013).
35. Soavi, G. *et al.* High energetic excitons in carbon nanotubes directly probe charge-carriers. *Sci. Rep.* **5,** 9681 (2015).
36. Ma, Y.-Z., Valkunas, L., Dexheimer, S. L., Bachilo, S. M. & Fleming, G. R. Femtosecond spectroscopy of optical excitations in single-walled carbon nanotubes: evidence for exciton-exciton annihilation. *Phys. Rev. Lett.* **94,** 157402 (2005).
37. Valkunas, L., Ma, Y.-Z. & Fleming, G. R. Exciton-exciton annihilation in single-walled carbon nanotubes. *Phys. Rev. B* **73,** 115432 (2006).
38. Huang, L. & Krauss, T. D. Quantized bimolecular auger recombination of excitons in single-walled carbon nanotubes. *Phys. Rev. Lett.* **96,** 057407 (2006).
39. Maniloff, E. S., Klimov, V. I. & McBranch, D. W. Intensity-dependent relaxation dynamics and the nature of the excited-state species in solid-state conducting polymers. *Phys. Rev. B* **56,** 1876 (1997).
40. Stevens, M. A., Silva, C., Russel, D. M. & Friend, R. H. Exciton dissociation mechanisms in the polymeric semiconductors poly(9,9-dioctylfluorene) and poly(9,9-dioctylfluorene-co-benzothiadiazole). *Phys. Rev. B* **63,** 165213 (2001).
41. Sun, D. *et al.* Observation of rapid exciton–exciton annihilation in monolayer molybdenum disulfide. *Nano Lett.* **14,** 5625–5629 (2014).
42. Konabe, S., Onoda, N. & Watanabe, K. Auger ionization in armchair-edge graphene nanoribbons. *Phys. Rev. B* **82,** 073402 (2010).
43. Santos, S. M. *et al.* All-optical trion generation in single-walled carbon nanotubes. *Phys. Rev. Lett.* **107,** 187401 (2011).
44. Matsunaga, R., Matsuda, K. & Kanemitsu, Y. Observation of charged excitons in hole-doped carbon nanotubes using photoluminescence and absorption spectroscopy. *Phys. Rev. Lett.* **106,** 037404 (2011).
45. Levy, R., Hönerlage, B. & Grun, J. B. Time-resolved exciton-biexciton transitions in CuCl. *Phys. Rev. B* **19,** 2326 (1979).
46. Klimov, V. I. *et al.* Optical gain and stimulated emission in nanocrystal quantum dots. *Science* **290,** 314–317 (2000).
47. Kreller, F., Lowisch, M., Puls, J. & Hennenberger, F. Role of biexcitons in the stimulated emission of wide-gap II-VI quantum wells. *Phys. Rev. Lett.* **75,** 2420 (1995).
48. Klimov, I. V. Spectral and dynamical properties of multiexcitons in semiconductor nanocrystals. *Annu. Rev. Phys. Chem.* **58,** 635–673 (2007).
49. Kammerlander, D., Prezzi, D., Goldoni, G., Molinari, E. & Hohenester, U. Biexciton stability in carbon nanotubes. *Phys. Rev. Lett.* **99,** 126806 (2006).

50. Sie, E. J., Lee, Y.-H., Frenzel, A. J., Kong, J. & Gedik, N. Intervalley biexcitons and many-body effects in monolayer MoS_2. *Phys. Rev. B* **92,** 125417 (2015).
51. You, Y. *et al.* Observation of biexcitons in monolayer WSe2. *Nat. Phys.* **11,** 477–481 (2015).
52. Lüer, L. *et al.* Size and mobility of excitons in (6, 5) carbon nanotubes. *Nat. Phys.* **5,** 54–58 (2009).
53. Martini, I. B., Smith, A. D. & Schwartz, B. J. Exciton-exciton annihilation and the production of interchain species in conjugated polymer films: comparing the ultrafast stimulated emission and photoluminescence dynamics of MEH-PPV. *Phys. Rev. B* **69,** 035204 (2004).
54. Wang, F., Duckovic, G., Knoesel, E., Brus, L. E. & Heinz, T. F. Observation of rapid Auger recombination in optically excited semiconducting carbon nanotubes. *Phys. Rev. B* **70,** 241403 (R) (2004).
55. Schaller, R. D. & Klimov, V. I. High efficiency carrier multiplication in PbSe nanocrystals: implications for solar energy conversion. *Phys. Rev. Lett.* **92,** 186601 (2004).
56. Wang, S., Khafizov, M., Tu, X., Zheng, M. & Krauss, T. D. Multiple exciton generation in single-walled carbon nanotubes. *Nano Lett.* **10,** 2381–2386 (2010).
57. Konabe, S. & Okada, S. Multiple exciton generation by a single photon in single-walled carbon nanotubes. *Phys. Rev. Lett.* **108,** 227401 (2012).
58. Polli, D., Lüer, L. & Cerullo, G. High-time-resolution pump-probe system with broadband detection for the study of time-domain vibrational dynamics. *Rev. Sci. Intrum.* **78,** 103108 (2007).
59. Jensen, S. A. *et al.* Ultrafast photoconductivity of graphene nanoribbons and carbon nanotubes. *Nano Lett.* **13,** 5925–5930 (2013).
60. Lorenc, M. *et al.* Artifacts in femtosecond transient absorption spectroscopy. *Appl. Phys. B* **74,** 19–27 (2002).
61. Fantini, C. *et al.* Optical transition energies for carbon nanotubes from resonant raman spectroscopy: environment and temperature effects. *Phys. Rev. Lett.* **93,** 147406 (2004).
62. Abdelsayed, V. *et al.* Photothermal deoxygenation of graphite oxide with laser excitation in solution and graphene-aided increase in water temperature. *J. Phys. Chem. Lett.* **1,** 2804–2009 (2010).
63. Giannozzi, P. *et al.* QUANTUM ESPRESSO: a modular and open-source software project for quantum simulations of materials. *J. Phys. Condens. Matter* **21,** 395502 (2009).
64. Rozzi, C. A., Varsano, D., Marini, A., Gross, E. K. U. & Rubio, A. Exact Coulomb cutoff technique for supercell calculations. *Phys. Rev. B* **73,** 205119 (2006).
65. Marini, A., Hogan, C., Grüning, M. & Varsano, D. Yambo: an *ab initio* tool for excited state calculations. *Comput. Phys. Commun.* **180,** 1392–1403 (2009).
66. Villegas, C. E. P., Mendoca, P. B. & Rocha, A. R. Optical spectrum of bottom-up graphene nanoribbons: towards efficient atom-thick excitonic solar cells. *Sci. Rep* **4,** 6579 (2014).
67. Thijssen, J. M. *Computational Physics* (Cambridge University Press, 2007).
68. Hüser, M. H. & Berne, B. J. Circumventing the pathological behavior of path-integral Monte Carlo for systems with Coulomb potentials. *J. Chem Phys.* **107,** 571–575 (1997).
69. Colombier, L. *et al.* Detection of a biexciton in semiconducting carbon nanotubes using nonlinear optical spectroscopy. *Phys. Rev. Lett.* **109,** 197402 (2012).
70. Watanabe, K. & Asano, K. Biexcitons in semiconducting single-walled carbon nanotubes. *Phys. Rev. B* **83,** 115406 (2011).

Acknowledgements

This work was funded by MIUR PRIN Grant No. 20105ZZTSE, MIUR FIRB Grant No. RBFR12SWOJ, and MAE Grant No. US14GR12. It has been supported in part by the Austrian Science Fund FWF, under the SFB F49 NextLite, by NAWI Graz, the European Research Council grant on NANOGRAPH, DFG Priority Program SPP 1459 and European Union Projects MoQuaS. G.C., A.N., X.F. and K.M. acknowledge support by the EC under Graphene Flagship (contract no. CNECT-ICT-604391). Computing time was provided by the Center for Functional Nanomaterials at Brookhaven National Laboratory, supported by the US Department of Energy, Office of Basic Energy Sciences, under contract number DE-SC0012704.

Authors contributions

A.N., Y. H., X.F. and K.M. prepared the GNR sample. G.S., S.D.C., D.V., C.M. and G.C. performed the pump-probe experiments. G.S. and D.V. performed the data analysis and simulation with the coupled-rate equations model. D.P., U.H. and E.M. performed the simulations on the exciton and biexciton binding energies. G.S., D.P. and G.C. wrote the manuscript. All authors designed the research and discussed the results.

Additional information

Competing financial interests: The authors declare no competing financial interests.

17

Functionalization mediates heat transport in graphene nanoflakes

Haoxue Han[1,*], Yong Zhang[2,3,*], Nan Wang[3,*], Majid Kabiri Samani[3], Yuxiang Ni[4], Zainelabideen Y. Mijbil[5,6], Michael Edwards[3], Shiyun Xiong[7], Kimmo Sääskilahti[8], Murali Murugesan[3], Yifeng Fu[3,9], Lilei Ye[9], Hatef Sadeghi[5], Steven Bailey[5], Yuriy A. Kosevich[1,10], Colin J. Lambert[5], Johan Liu[2,3] & Sebastian Volz[1]

The high thermal conductivity of graphene and few-layer graphene undergoes severe degradations through contact with the substrate. Here we show experimentally that the thermal management of a micro heater is substantially improved by introducing alternative heat-escaping channels into a graphene-based film bonded to functionalized graphene oxide through amino-silane molecules. Using a resistance temperature probe for *in situ* monitoring we demonstrate that the hotspot temperature was lowered by $\sim$28 °C for a chip operating at 1,300 W cm^{-2}. Thermal resistance probed by pulsed photothermal reflectance measurements demonstrated an improved thermal coupling due to functionalization on the graphene-graphene oxide interface. Three functionalization molecules manifest distinct interfacial thermal transport behaviour, corroborating our atomistic calculations in unveiling the role of molecular chain length and functional groups. Molecular dynamics simulations reveal that the functionalization constrains the cross-plane phonon scattering, which in turn enhances in-plane heat conduction of the bonded graphene film by recovering the long flexural phonon lifetime.

[1] Laboratoire EM2C, CNRS, CentraleSupélec, Université Paris-Saclay, Grande Voie des Vignes, 92295 Châtenay-Malabry, France. [2] SMIT Center, School of Automation and Mechanical Engineering and Institute of NanomicroEnergy, Shanghai University, 20 Chengzhong Road, Shanghai 201800, China. [3] Electronics Materials and Systems Laboratory, Department of Microtechnology and Nanoscience, Chalmers University of Technology, Kemivägen 9, SE-412 96 Gothenburg, Sweden. [4] Department of Mechanical Engineering, University of Minnesota, 111 Church Street SE, Minneapolis, Minnesota 55455, USA. [5] Quantum Technology Center, Physics Department, Lancaster University, Lancaster LA1 4YB, UK. [6] Science Department, Veterinary Medicine College, Al-Qasim Green University, Babylon, Iraq. [7] Max Planck Institute for Polymer Research, Ackermannweg 10, D-55128 Mainz, Germany. [8] Department of Biomedical Engineering and Computational Science, Aalto University, FI-00076 Aalto, Finland. [9] SHT Smart High Tech AB, Ascherbergsgatan 46, SE-411 33 Gothenburg, Sweden. [10] Department of Polymers and Composite Materials, Semenov Institute of Chemical Physics, Russian Academy of Sciences, Kosygin Street 4, 119991 Moscow, Russia. * These authors contributed equally to this work. Correspondence and requests for materials should be addressed to C.J.L. (email: c.lambert@lancaster.ac.uk) or to J.L. (email: johan.liu@chalmers.se) or to S.V. (email: sebastian.volz@ecp.fr).

Anisotropic properties of two-dimensional (2D) layered materials make them promising in the application of next-generation electronic devices, among which graphene and few-layer graphene (FLG) have been most intensively studied for thermal management, due to their extraordinarily high in-plane thermal conductivity (κ)[1–5]. For instance, Yan *et al.*[6] reported that the maximum hotspot temperature can be lowered by $\sim$20 °C in transistors operating at $\sim$13 W mm^{-1} using FLG as a heat spreader for a gallium nitride (GaN) transistor. Gao *et al.*[7] reported that the maximum hotspot temperature decreased from 121 to 108 °C (ΔT = 13 °C) for a heat flux of 430 W cm^{-2} after the introduction of a single-layer graphene heat spreader. Moreover, the simulations of graphene heat spreaders were also reported for silicon-on-insulator integrated circuits[8] and three-dimensional (3D) integrated circuits[9]. The thermal conductivity of a graphene laminate film supported on substrate was also investigated and found to remain rather large[10]. However, in most practical applications, graphene/FLG will be supported by and integrated with insulators, both in electronic circuitry and heat-spreader applications[11]. Therefore, thermal energy flow will be limited both by the in-plane thermal conductivity (κ) of the supported graphene/FLG and by the thermal boundary resistance (R) at the graphene/FLG–substrate interface[12].

Owing to their exceedingly large surface-to-volume ratio, the properties of 2D layered materials are very sensitive to the interactions with external bodies. Indeed, when supported on an amorphous substrate, κ of suspended graphene decreased by almost one order of magnitude, from $\sim$4,000 (ref. 13) to $\sim$600 W m^{-1} K^{-1} (ref. 14). Such a striking discrepancy in κ significantly limits the thermal performance of graphene/FLG in real applications. It is reported that the different behaviours are due to the strong scattering of the important heat-carrying flexural acoustic (ZA) modes[15] to the substrate[16]. More specifically, it was identified that the phonon relaxation times of graphene ZA modes are suppressed when supported on a SiO_2 substrate. These studies have improved our fundamental understanding in the physics behind the problem, and it was suggested that making rational choice of the substrate material[14,17] and modulating its coupling to graphene[12] may be useful to improve κ of the supported graphene/FLG.

The thermal boundary resistance (R) of a graphene/FLG–substrate interface is another limiting factor to their thermal performance in devices. Covalent functionalization has been proved to efficiently promote heat transfer between interfaces by introducing additional thermal pathways through the functionalizing molecules[18–33]. For example, self-assembled monolayers (SAMs) were used to functionalize metallic surfaces to enhance heat transport across metal–water[20,27], metal–gas[28], metal–semiconductor[21] and metal–polymer[31] interfaces. Functionalization was used in graphene and carbon nanotube nanocomposites to mitigate the high thermal boundary resistance between the graphene/carbon nanotube fillers and the polymer matrices[18,25,26,30]. Functionalized molecules also assist to align and densely pack multilayer graphene sheets and reduce the interlayer thermal resistance of graphene[25]. Recently, it was shown that plasma-functionalized graphene raised the cross-plane thermal conductance between aluminium and its substrate by a factor of two[19]. Nevertheless, the functionalization-introduced point defects will further decrease κ of the supported graphene/FLG, as they introduce phonon-scattering centres[25,32,33]. To correct this drawback, a robust solution that maintains the high thermal conductivity of graphene/FLG when supported, while effectively reducing the interface thermal resistance is needed.

Here we demonstrate that thermal management of a micro heater is considerably improved via introducing alternative heat-spreading channels implemented with graphene-based film (GBF) bonded to functionalized graphene oxide (FGO) through amino-silane molecules. We probed interface thermal resistance by photothermal reflectance measurements to demonstrate an improved thermal coupling due to functionalization on the graphene–graphene oxide interface and the graphene oxide–silica interface. Molecular dynamics simulations and *ab initio* calculations reveal that the functionalization constrains the cross-plane scattering of low-frequency phonons, which in turn enhances in-plane heat conduction of the bonded graphene film by recovering the long flexural phonon lifetime. Our results provide evidence that a graphene film deposited on a FGO substrate provides a very attractive platform for thermal management applications.

Results

Graphene-based film and graphene oxide and device. A GBF bonded to the FGO substrate through silane molecules is shown in Fig. 1a,b. To synthesize GBF and FGO experimentally (Fig. 1c), we first prepared a graphene oxide (GO) dispersion (see Experimental method). The FGO was obtained by functionalizing GO with a silane-based chemistry suitable for reactive oxide-forming surfaces including the basal plane of GO and SiO_2. 3-Amino-propyltriethoxysilane (APTES) has three -Si–O- groups and one -NH_2 end, as shown in Fig. 1a. Owing to the simple chemistry and unique multifunctional nature of APTES, it can easily bind two different substrates. In our case, the –Si–O end of APTES binds to the GO substrate. -Si–OC_2H_5 groups of APTES hydrolyse in water and form crosslink bonds with each other. The crosslinked Si–O structure acts as a strong bonding layer between the substrate and GBFs. On the other hand, the -NH_2 end of APTES binds onto carboxyl groups on the functionalized graphene film. The FGO layer has a thickness of $\sim$5 nm. The graphene film was fabricated from chemically reduced GO and can recover relatively high in-plane thermal conductivity after thermal annealing[34]. The graphene film was then spin-coated[35] with the FGO and the resulting bundle was transferred to a thermal evaluation device[7], resulting in the formation of molecular bridges between the graphene surface and the device's SiO_2 substrate. The thermal evaluation device was integrated with micro Pt-based heating resistors as the hotspot and temperature sensors[7], as shown in Fig. 1d, acting as a simulation platform of an electronic component to demonstrate the heat-spreading capability of the supported graphene film.

***In situ* temperature measurement with resistance thermometry.** A direct current I was input into the circuit by applying an outer voltage V in Fig. 1e, and hence the generated power is calculated as $P = V \times I$. Since the lateral dimension of the hotspot (A = 400 × 400 μm^2) is much larger than its thickness (260 nm), most of the heat is dissipated through the lateral direction of the hotspot. Hence the heat flux is defined as $Q = P/A$, and the direction is parallel to the substrate. The calibration relationship between the resistance R (Ω) and the temperature T (°C) of the thermal evaluation chip is $R(T) = 0.21T + 112$. The temperature measurement uncertainty is $\epsilon = \pm 0.5$ °C. Figure 1e shows the temperature measured *in situ* at the hotspot and compares the thermal performance of the graphene film with and without the functionalization. With a constant heat flux of 1,300 ± 2.3 W cm^{-2} at the Pt chip, the hotspot temperature decreased from 135 ± 1.2 to 118 ± 1.1 °C (ΔT = 17 ± 2.3 °C) with a GBF deposited on non-functionalized GO compared with the case of a bare chip. Such a remarkable temperature decrease is far beyond the measurement uncertainty $\Delta T \gg \epsilon$ (see Supplementary Fig. 1 and Supplementary Note 1 for an uncertainty analysis).

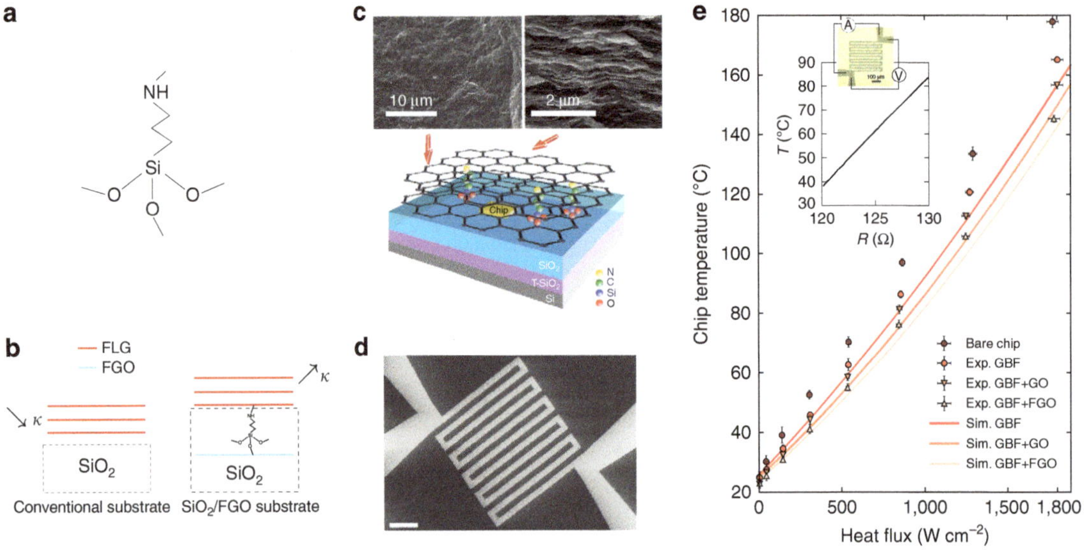

Figure 1 | Graphene-based film on FGO as heat spreader for hotspot. (**a**) Sketch of the chemical bonds of the silane molecule. (**b**) Schematic of a graphene film on different supports. Left: conventional silica substrate. Right: the proposed silica/FGO substrate. (**c**) Schematic of the measurement set-up. The chip is embedded in the SiO_2 substrate. T-SiO_2 stands for thermally grown SiO_2. Scanning electron microscopy (SEM) image of the in-plane and the cross-section of the GBF. (**d**) SEM image of the chip as the hotspot. Scale bar, 100 μm. (**e**) Measured (filled markers) and finite-element-simulated (lines) chip temperatures versus the in-plane heat fluxes dissipated in a bare hotspot (rectangles), a hotspot covered by a GBF (circles), a chip covered by a GBF with non-functionalized GO (up triangle) and a chip covered by a GBF with APTES FGO (down triangle). 'Exp.' and 'Sim.' stand for experiments and simulations, respectively. Inset: calibration relationship between the resistance R (Ω) and the temperature T (°C) of the thermal evaluation chip. Scale bar on the chip, 100 μm. The error bars correspond to the s.d.'s from the measurements on five samples.

Furthermore, with the same heat flux input, the hotspot temperature further decreased from 118 ± 1.1 to $107 \pm 0.8\,°C$ ($\Delta T = 11 \pm 1.9\,°C$) thanks to the presence of the APTES functionalization. The heat-spreading performance is thus enhanced by ~57% via the functionalization. We have implemented a finite-element model (Supplementary Figs 2–4, Supplementary Table 1 and Supplementary Note 2) of the heat-spreading device by taking the results of atomistic simulation as input parameters. As shown in Fig. 1e, the heat-spreading performance of the equivalent macroscopic finite-element model agrees reasonably well with the one measured by experiments.

Thermal resistance measurement with photothermal reflectance. To further confirm the enhanced heat spreading assisted by molecular functionalization, we measured the interface thermal resistance by using the pulsed photothermal reflectance (PPR) method[36,37]. To enhance heat absorption, a gold layer was evaporated on the surface of the GO and FGO layers after drop coating. The sample was first excited by a Nd:YAG laser pulse. This caused a fast rise in the surface temperature followed by a relaxation. The change of surface temperature was monitored by a probe laser, which reflects off from the samples' surface. Since the relaxation time is governed by the thermal properties of the underlying layers and interfacial thermal resistance between the layers, by obtaining the temperature profile one can extract the thermal properties of the layers and interface thermal resistance between the layers through a heat conduction model. Four sets of samples were fabricated, as shown in Fig. 2 and the thermal resistance R_1 between the Au-Cr film and the (functionalized) GO layer, and the resistance R_2 between (functionalized) GO layer and GBF or SiO_2 were measured. The experimental set-up and the procedure of thermal resistance extraction by fitting the photothermal response to the model (Supplementary Figs 5 and 6 and Supplementary Note 3). The normalized surface temperatures of the four sample sets are shown in Fig. 2 and the extracted thermal resistances are reported in Table 1. A fourfold reduction was achieved in the thermal resistance between GO and GBF from 3.8×10^{-8} to $0.9 \times 10^{-8}\,m^2\,K\,W^{-1}$. On the GO–$SiO_2$ interface, the functionalization remarkably reduced the thermal resistance by a factor of almost three, from 7.5×10^{-8} to $2.6 \times 10^{-8}\,m^2\,K\,W^{-1}$. We also observed a better thermal coupling on the metal–dielectric interface between Au-Cr and GO due to the surface chemistry.

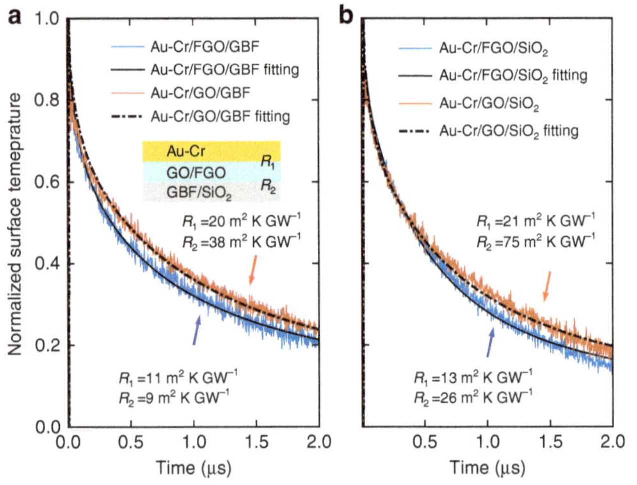

Figure 2 | Normalized surface temperature of PPR measurements of the thermal interface resistance. (**a**) Au-Cr/FGO/GBF and Au-Cr/GO/GBF samples and (**b**) Au-Cr/FGO/SiO_2 and Au-Cr/GO/SiO_2 samples. Inset: sample geometry for the PPR measurement. R_1 and R_2 refer to the thermal interface resistance between the Au-Cr film and the (APTES functionalized) GO layer, and that between (functionalized) GO layer and GBF or SiO_2. The thermal resistances are also reported in Table 1.

Table 1 | Au/(F-)GO and (F-)GO/X interface thermal resistances.

Interface	R_1 ($m^2 K W^{-1}$)	R_2 ($m^2 K W^{-1}$)
Au-Cr/GO/GBF	2.0E-8	3.8E-8
Au-Cr/FGO/GBF	1.1E-8	0.9E-8
Au-Cr/GO/SiO_2	2.1E-8	7.5E-8
Au-Cr/FGO/SiO_2	1.3E-8	2.6E-8

X = GBF or SiO_2. The fitting uncertainty is 20% (Supplementary Fig. 7).

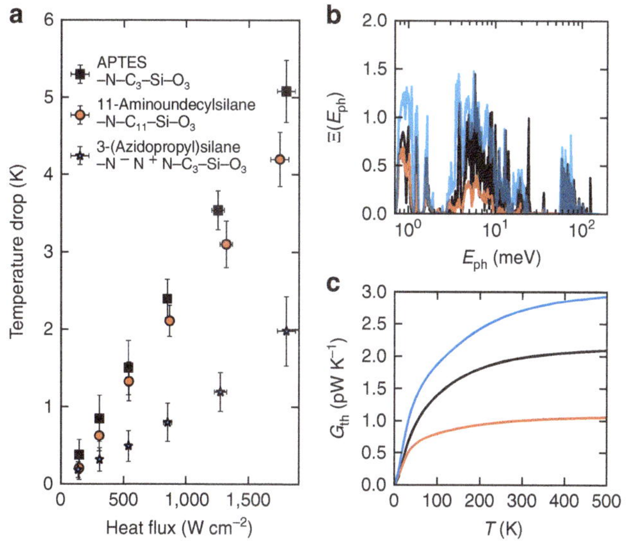

Figure 3 | Heat-spreading performance of GBF on FGO with different functional agents. The molecules are APTES, 11-Aminoundecyltriethoxysilane and 3-(Azidopropyl)triethoxysilane. (**a**) Measured temperature drop of heat spreaders with different functionalization molecules compared with that with non-functionalized GO. The error bars correspond to the s.d.'s of measurements on five samples for each molecule type. (**b**) Phonon transmission function $\Xi(E_{ph})$ for three different types of molecules used in the experiment. (**c**) Phonon thermal conductance through the molecules as a function of temperature. The colour code of the data curves in **b** and **c** is the same as in **a**.

Heat spreading through different functional molecules. To gain a deeper insight into the impact of molecular structure on the thermal transport along the molecules, we used APTES, 11-Aminoundecyltriethoxysilane and 3-(Azidopropyl)-triethoxysilane as different functional agent on the GO to evaluate their heat-spreading performance on the same thermal test platform. The same concentration of 0.1492 mol kg^{-1} was used for all three types of molecules. Figure 3 shows the temperature reduction of the hotspot covered by GBFs with FGO using three different molecules. The results show that the heat spreader of GO functionalized with APTES has the best cooling performance. To properly understand this difference, an exploration of the internal vibrational properties of the molecule is crucial[38]. We hence investigated how the differences in the phonon transmission impact the interfacial thermal transport. With this aim, we probed the phonon transmission $\Xi(E_{ph})$ by atomistic Green's function to characterize the local heat conduction with and without the presence of the molecule (Supplementary Note 4). $\Xi(E_{ph})$ enables a precise measurement of the atomic-scale molecule-graphene heat transport that the conventional models fail to provide. The phonon transmission functions through different molecules and the associated thermal

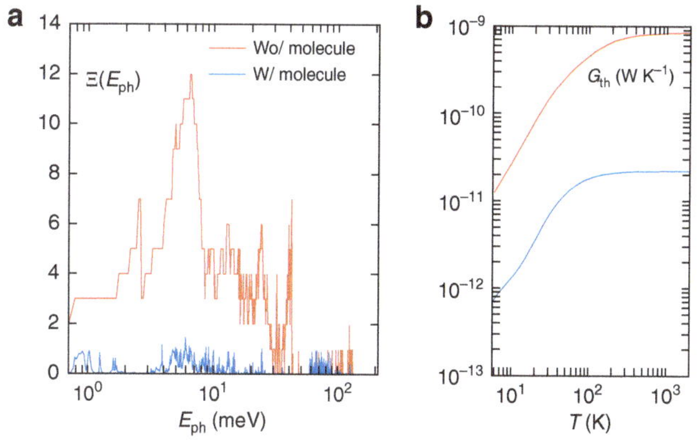

Figure 4 | Phonon transport through the APTES molecule. (**a**) Phonon transmission $\Xi(E_{ph})$ versus phonon energy $E_{ph} = \hbar\omega$ (red curve) between two adjacent graphene layers and (blue curve) through the APTES molecule bonding the two graphene layers. 'w/' and 'wo/' stand for with and without, respectively. (**b**) Thermal conductances G_{th} contributed by phonons versus temperature for the two cases.

conductances versus temperature are shown in Fig. 3b,c. The phonon transmission at low phonon energies across the 11-Aminoundecyltriethoxysilane molecule is comparable to that across the APTES molecule, whereas at high phonon energies (>4 meV), the phonon transmission is considerably suppressed. Such a distinct phonon transport behaviour is determined by the molecule chain length. By comparing the chemical structures, we can see that 11-Aminoundecyltriethoxysilane (–N–C_{11}–Si–O_3) has the same functional groups as APTES (–N–C_3–Si–O_3) but has a longer carbon backbone. Such a long chain length has a rather weak impact on the low-frequency phonons due to their very long wavelength but can strongly suppress the phonon transport at high frequencies. On the other hand, when comparing the phonon transmission through the junction of APTES and 3-(Azidopropyl)triethoxysilane (–N^-N^+N–C_3–Si–O_3), it is evident to identify a stronger transmission at all frequencies. By comparing the chemical structures in Fig. 3a, we can see that 3-(Azidopropyl)triethoxysilane has the same carbon backbone as APTES but has an azido group instead of an amino group. The azido group has strong interaction with the sp^2-bonded carbon lattice of graphene to form three-membered heterocycle structures. Given the similar backbone structure, the phonon eigenmodes in the molecule have not been significantly altered, but the transmission is enhanced due to the stronger thermal coupling to the graphene reservoirs. However, the introduction of nitrogen atoms into the sp^2 carbon structure can strongly scatter phonons by distorting the structure of the graphene sheet, which leads to a poor thermal performance of the heat spreader.

Phonon transport analysis in APTES. We now investigate the detailed vibrational and electronic transport properties of the APTES molecule since it presents the best performance in heat spreading. We probed the phonon transmission $\Xi(E_{ph})$ to characterize the local heat conduction with and without the presence of the APTES molecule. As shown in Fig. 4, the transmission function $\Xi(E_{ph})$ for the two adjacent graphene layers without any molecule displays a clear stepwise structure that provides the number of phonon channels. Low-energy phonons ($E_{ph} \leq 10$ meV) dominate heat conduction since the adjacent graphene flakes interact only through weak van der Waals (vdW) forces that inhibit the transmission of high-frequency phonons[39]. When the graphene layers are bridged by a amino-silane molecule, the high-frequency phonons act as the major contributors in the heat conductance G_{ph}, creating more phonon channels through the covalent bond vibrations. This is in line with the transmission calculation of Segal *et al.*[40] who observed a contribution to heat conduction by the higher-frequency phonons within the molecule coupled to the low-frequency phonons responsible for heat transport in the thermal reservoirs. The oscillations in the transmission spectrum may originate from phonon interferences within the alkane chain[41–43]. Fabry–Pérot-like interference effect occurs in the frequency region of $E_{ph} = 20–100$ meV, as was previously observed in an alkane SAM interface[41]. Such Fabry–Pérot-like interferences originate from the multiple reflected phonons interfering constructively within the alkane chain, as the local maxima in the transmission (Fig. 4) through the molecule can attain the same intensity of that through pristine graphene films at given frequencies. Although destructive quantum interference was believed unlikely to occur in a linear alkane chain[42], we observe strong destructive interference patterns in the high-frequency range ($\omega/2\pi = 40 - 60$ THz), which may correspond to two-path destructive phonon interferences[43,44]. We also investigated the interlayer electron transport in the graphene and the effect of the silane intercalation in such a hybrid nanostructure through a *ab initio* calculation combined with Green's function (Supplementary Fig. 8 and Supplementary Note 5). The main heat carrier in this system is phonon, as the thermal conductance due to the electron contributes to the total thermal conductance by ~4% at room temperature with functionalization and by ~2% without molecules.

Thermal conductance and conductivity calculations. To explore the effect of functional APTES molecules on the in-plane thermal conductance of the graphene film, we first perform molecular dynamics simulations to study a nanoscale molecular junction between two stacks of multilayer graphene nanoflakes. For a weak oxidation, the thermal resistance of the graphene film and its oxidized substrate on one hand, and on the other hand its thermal conductivity are very close to those with a non-oxidized graphene support. The dependence of thermal resistance and conductivity of graphene film to the oxidation rate of the GO substrate is reported in Supplementary Fig. 9 and Supplementary Note 6. We consider defect-free and isotopically pure graphene flakes of $10 \times 10\,nm^2$ thermalized at 300 K (Methods, Supplementary Fig. 10 and Supplementary Note 7). The FGO substrate has a 5-nm thickness as in the experiments. Conventional silica substrate results in a substantial decrease of the basal-plane thermal conductivity of graphene due to the non-conformality of the substrate–graphene interface[14]. To compare, the FGO substrate proposed herein minimize the perturbation of substrate on the morphology of graphene (Fig. 1c), thus maintaining its high thermal conductivity. Periodic boundary conditions were used in the in-plane directions so that the molecular dynamics system corresponds to two thin films connected through silane molecules with the number density ρ, defined as the molecule number per graphene unit area.

To illustrate the intriguing role of the functionalizing molecule, the in-plane thermal conductivity κ of the film and its interfacial thermal resistance R (Supplementary Fig. 10 and Supplementary Note 7) with the FGO substrate is plotted as a function of the graphene layer number l_G in the film in Fig. 5a,b, respectively. First, a supercell containing a single molecule is studied, which corresponds to $\rho = 0.081\,nm^{-2}$. For $l_G \geq 2$, the presence of the APTES molecule results in an unexpected increase both in the graphene film thermal conductivity κ and in R. An overall decaying trend of the in-plane thermal conductivity κ of the

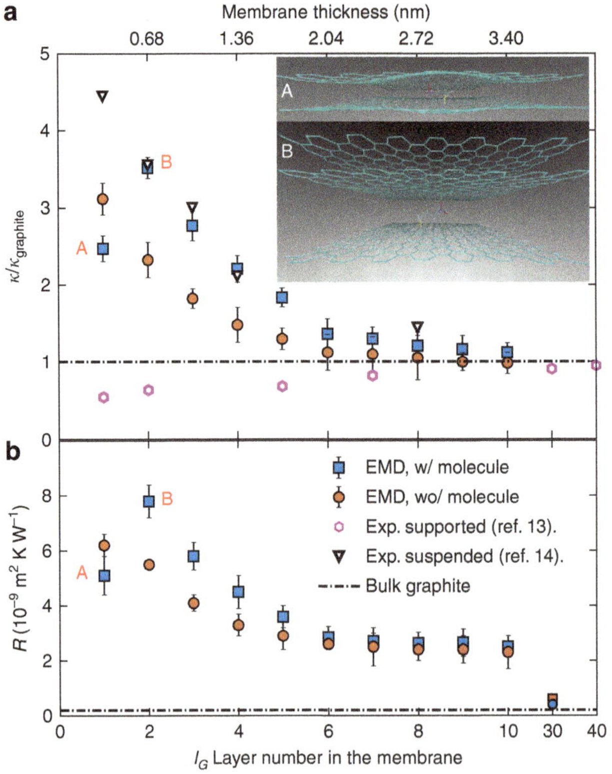

Figure 5 | Thermal resistance and in-plane thermal conductivity of the GBF versus the graphene layer number. (**a**) Molecular dynamics simulation results of in-plane thermal conductivity κ of the graphene film and (**b**) interfacial thermal resistance R between the FGO substrate and the graphene film versus the graphene layer number l_G in the film. The molecule density is $\rho = 0.081\,nm^{-2}$. Cases with (red circles) and without (blue squares) the APTES molecule are compared. 'w/' and 'wo/' stand for with and without, respectively. The values of thermal conductivity are normalized to that of the single-layer graphene. Inset A illustrates a saddle-like surface generated by the APTES molecule for $l_G = 1$. Inset B shows that the saddle-like curvature disappears for $l_G \geq 1$. The measurements of suspended and supported graphene are from refs 13,14, respectively. EMD stand for equilibrium molecular dynamics. 'Exp.' and 'Sim.' stand for experiments and simulations, respectively. The simulated (respectively measured) thermal conductivity is normalized with respect to that of the simulated (respectively measured) graphite to allow a reasonable comparison. The error bar corresponds to the s.d. of the ensemble average on 50 independent trajectories.

graphene film and its resistance R with the substrate versus the layer number l_G is observed until approaching the value of bulk graphite. This is due to the increased cross-plane coupling of the low-energy phonons. A similar decay was found both in experimental measurements of κ (ref. 13) of suspended graphene and in simulation-based estimation of R (ref. 45). Unlike silica-supported graphene[14], the FLG on FGO support recovers the high thermal conductivity and follows the same decaying trend versus l_G as the suspended graphene[13]. Therefore, the proposed FGO substrate better conserves the high thermal conductivity of graphene compared with the silica substrate. Interestingly, for $l_G = 1$, the presence of the molecule reduces κ, which goes against the case where no molecule interconnects the graphene film and the substrate. This breakdown of the thermal conductivity enhancement is due to a saddle-like surface generated around the molecule's chemical bonds of amino and silano groups connecting the graphene, with the bond centre as the saddle point, as shown in the inset A of Fig. 5a. The saddle-like surface strongly scatters all phonon modes, thus decreasing κ of the graphene film. Such a curved surface was found quite common in defective graphene[46,47], resulting from the Jahn-Teller effect to lower the energy by geometrical distortion[48]. The thermal conductance of a single silane molecule is determined to be 82 pW K^{-1} through the molecular footprint (Supplementary Fig. 11 and Supplementary Note 8). Our simulated result is comparable to recent measurements of the thermal conductance of alkane thiols SAM at a silica–gold interface[38].

It is clearly seen from Fig. 5a that the critical layer number for the in-plane thermal conductivity switch is $l_c = 2$. This critical number depends on the level of conformation distortion of the functionalized graphene sheet. Several factors can play a role (Supplementary Figs 12–14 and Supplementary Note 9): (i) molecule density: a diluted molecule distribution creates segregated ripples on the functionalized graphene sheet and even on the sheets further above thus interrupting the phonon mean free path. In this case more graphene sheets are required to recover the flat conformation, that is, $l_c > 2$. For a high molecule density, the ripples tend to merge with themselves, recovering the flat surface in an extended area. Such merging hence recovers the interrupted phonon MFP and alleviates the detrimental influence of the separate ripples on the basal-plane thermal conductivity of graphene. In this case the two layers of graphene with functionalization have higher basal-plane thermal conductivity compared with that without functionalization, that is, $l_c = 2$. (ii) Functional group of the molecule: when the functional agent forms a stronger bond with its substrate than that of the amino-silane molecule, more severe static graphene distortion occurs thus increasing the critical number l_c. (iii) Substrate of the graphene film: a highly mismatched substrate yields remarkable surface distortion of the graphene sheets in such a way that two layers of graphene are not sufficient to recover the high basal-plane thermal conductivity.

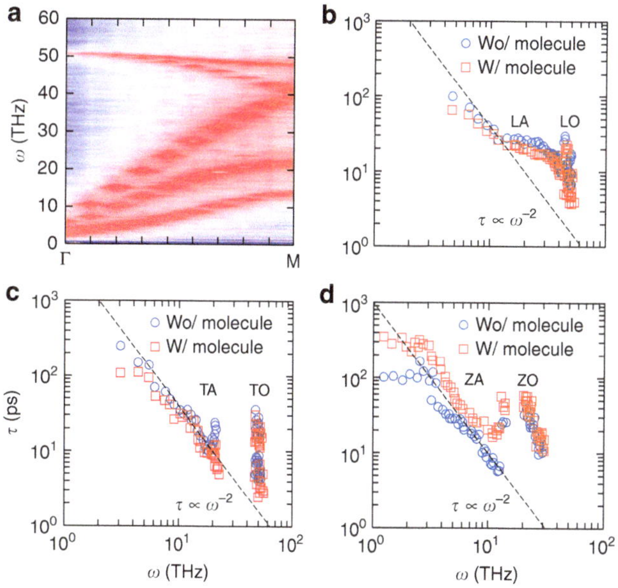

Figure 6 | Phonon dispersion and lifetimes of the graphene film. (**a**) Phonon dispersion of the graphene film bonded to the APTES-FGO substrate from molecular dynamics simulations, for $\rho = 0.081\,\mathrm{nm}^{-2}$ and $l_G = 2$. (**b–d**) Mode-wise phonon relaxation time for longitudinal modes including longitudinal acoustic (LA) and optical (LO) branches, for transverse modes including transverse acoustic (TA) and optical (TO) branches, and for flexural modes including flexural acoustic (ZA) and optical (ZO) branches, respectively. 'w/' and 'wo/' stand for with and without, respectively. For low-frequency phonons, τ follows the ω^{-2} scaling rule of the Umklapp processes at low frequency and high-temperature limits.

Functionalization effects on phonon lifetime of graphene. We investigate the microscopic origin of the thermal conductivity κ enhancement in the graphene film by probing the mode-wise phonon relaxation time (Supplementary Fig. 15 and Supplementary Note 10). The phonon relaxation time τ measures the temporal response of a perturbed phonon mode to relax back to equilibrium due to the net effect of different phonon-scattering mechanisms. τ can be defined as[49] $\partial n/\partial \tau = (n - n_0)/\tau$ where n and n_0 are the phonon occupation numbers out of and at thermal equilibrium. Under the single-mode-relaxation-time approximation, the thermal conductivity is given by $\kappa = \sum C_i v_i^2 \tau_i$ where C_i and v_i are the specific heat per volume unit and the group velocity of the i-th phonon mode. The phonon dispersion of the supported graphene film for $\rho = 0.081\,\mathrm{nm}^{-2}$ and $l_G = 2$, and the extracted relaxation time τ for all the phonon modes are shown in Fig. 6. By inserting the APTES molecule, the relaxation time of the acoustic flexural modes τ_{ZA} largely increase at low frequencies ($\omega < 10$ THz), whereas the longitudinal and transverse modes undergo a slight decrease in $\tau_{LA,LO,TA,TO}$. The notable increase in τ_{ZA} accounts for the enhancement in κ of the graphene film bonded to the substrate since the ZA modes contribute considerably to κ as much as 77% at 300 K (ref. 15). We attribute the increase in τ_{ZA} to the weakened coupling between the graphene film and the substrate, which is reflected as the increased thermal resistance R for $\rho = 0.081\,\mathrm{nm}^{-2}$ and $l_G \geq 2$, as is shown in Fig. 5b. An approximate expression from the perturbation theory for the relaxation time due to phonon leakage towards the contact interface yields $\tau^{-1} \propto g(\omega)K^2/\omega^2$, where $g(\omega)$ depends on the phonon density of states and K is the average vdW coupling constant between the graphene film and its substrate. A previous calculation[16] shows that K_{ZA} largely depends on the substrate morphology. The presence of the molecule strongly decreases the film–substrate coupling K_{ZA}, thus yielding a longer time τ_{ZA}. This demonstrates that the amino-silane molecules refrain the film–substrate phonon scattering, which in turn enhances in-plane heat conduction in the bonded graphene film. Note that ZA mode becomes 'massive' as shown in Fig. 6a, that is, the ZA branch does not reach all the way to zero frequency but shifts to higher frequencies. The reduced inter-plane scattering compensates the reduction in the phonon group velocity at Brillouin zone centre, so that the in-plane thermal conductivity increases after all (Supplementary Fig. 16 and Supplementary Note 11).

The cross-plane heat conduction between the top graphene films and the lower FGO layers is governed by the competing effects of (i) the intercalation of the molecules that tends to weaken the interlayer coupling of graphene therefore increasing the cross-plane thermal resistance and (ii) the additional heat channels introduced by the molecules covalently bonding the graphene sheets that tend to facilitate the cross-plane thermal coupling. We investigated the in-plane thermal conductivity κ of the graphene film and its thermal resistance R with the functionalized substrate versus the equivalent APTES molecule number density ρ in Fig. 7. At a low molecule density, the molecule population is too small in such a way that the additional thermal phonon transport through the molecules cannot compensate the weakened cross-plane thermal coupling through the interlayer vdW interactions due to the intercalation effect by the molecules (also see Supplementary Fig. 17 and Supplementary Note 12 for a detailed analysis on the κ–ρ relation at low molecular densities). Therefore, the net effect is an increase in the

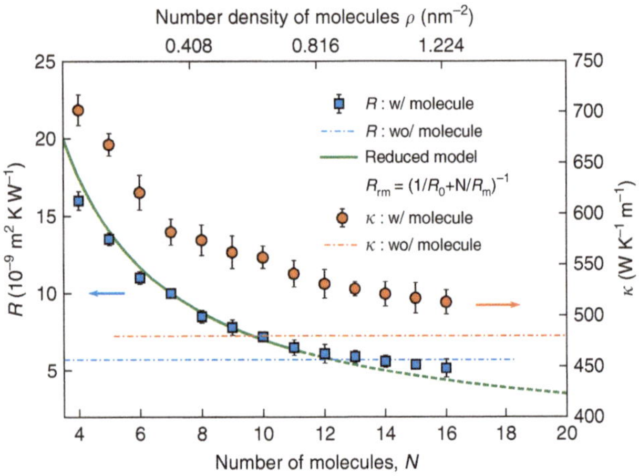

Figure 7 | Thermal resistance and in-plane thermal conductivity of the graphene film versus the molecule density. Thermal conductivity κ (red circles) of the graphene film and its thermal resistance R (blue squares) with the APTES functionalized substrate versus the equivalent molecule number density ρ. Green line: resistance R_{rm} predicted by a reduced model of parallel thermal resistors is to compare with the molecular dynamics (MD) results for small ρ. $R_{rm} = (1/R_0 + N/R_m)^{-1}$, where R_0 is the thermal resistance between two adjacent graphene layers and R_m is the additional resistance induced by a single molecule. The values of R_0 and R_m are determined in MD simulations: $R_0 = 0.203\,mm^2\,K\,W^{-1}$ and $R_m = 0.069\,mm^2\,K\,W^{-1}$. Red and blue dashed lines represent, respectively, κ and R without the interconnecting molecule at the film-substrate interface. 'w/' and 'wo/' stand for with and without, respectively. The error bar corresponds to the s.d. of the ensemble average on 50 independent trajectories.

cross-plane thermal resistance between the top graphene films and the lower GO layers compared to the non-functionalization case. In the limit of small number density, we consider the molecules as independent heat conductors connecting the film and the substrate. R agrees well with a reduced model of parallel thermal resistors. Thus, R decreases with the number of molecules. κ also decreases with the molecule number since the molecules increase the cross-plane phonon scattering in the graphene film. For larger ρ, the interactions among the molecules gain importance and R starts to deviate from the prediction of the parallel-thermal-resistor model, as shown in Fig. 7. The molecule population is large enough so that the additional thermal phonon transport through the molecules becomes very strong and compensates the weakened cross-plane thermal coupling through the interlayer interactions due to the intercalation effect by the molecules. Therefore, the net effect is that the cross-plane thermal resistance between the top graphene films and the lower FGO layers falls lower than that of the case without functionalization in the GO. For large ρ, the thermal resistance R is lower than that without molecules. However κ remains enhanced by a factor of $\sim$15% compared with its value for the non-functionalized graphene substrate. The molecule density ρ effectively tunes the thermal conductivity of the supported graphene.

Discussion

The correlation between the in- and cross-plane thermal transport of 2D materials has been recently studied from both theoretical[15,17] and experimental[13,14,16] points of view. Among them, Ghosh *et al.*[13] studied the dimensional crossover from a 2D graphene to a 3D graphite system. The in-plane thermal conductivity decreased with the increase of the layer number in the FLG. Wei *et al.*[17] used a 2D material model to show a negative correlation between the in-plane thermal conductivity and the cross-plane interaction. These studies have considered different mechanisms of the correlation, yet no general conclusion is applicable to any quasi-2D system. We show that the functionalization molecules mediate the cross-plane phonon scattering and in turn effectively control the in-plane thermal transport in the graphene-based heat spreader.

Our work tackles the key technological challenge of efficient thermal management in the industry of the next-generation integrated circuit. The excessive heat from the hotspots accelerates the failure rate and slows down the operating speed of microelectronic devices[50,51]. Extremely high power density beyond 1,000 W cm^{-2} can be observed in microprocessors[51]. Coolers are most commonly based on metallic materials due to their relatively high thermal conductivities, for example, 200–400 W m^{-1} K^{-1} for Al and Cu, and the low cost. However, for these technologically important metals, the thermal conductivity of a 100-nm-thick film constitutes only $\sim$20% of the thermal conductivity of the bulk[52,53]. On the other hand, GBF used in the present work can have an in-plane thermal conductivity as high as 1,600 W m^{-1} K^{-1} (ref. 54), which is more than two orders of magnitude higher than the ones of the metal thin films. Such a superior thermal conductivity significantly facilitates the heat conduction when GBFs are used as heat spreaders together with a better cross-plane thermal coupling with FGO. The APTES functionalization yields an additional temperature decrease ΔT of the hotspot of 11 °C compared with the case of GBF with non-functionalized GO ($\Delta T = 17$ °C), resulting in a large total temperature decrease of 28 °C when operating at 1,300 W cm^{-2}. In comparison, the highest ΔT achieved so far at a similar high heat flux of 1,250 W cm^{-2} is 19 °C by using a miniaturized thermoelectric cooler[51]. The contribution of functionalization to the overall ΔT is robust especially at higher heat flux beyond 1,000 W cm^{-2}, as shown in Fig. 1e. Therefore, we propose a significant package-level solution for the thermal management of hotspots in high-power electronics at the micro- and nanometre length scale.

Methods

Experimental method. (i) Sample synthesis: graphite (Sigma, 4 g); H_2SO_4 (92 ml, 98%); $NaNO_3$ (2 g); and $KMnO_4$ (12 g) were used to prepare GO dispersion by following Hummers' method[55]. The obtained GO dispersion was reduced by L-ascorbic acid, and polyvinyl alcohol was also added for better suspension. The GBF was prepared via vacuum filtration with polycarbonate filter paper with a pore size of 3 µm. The film thickness was controlled by the filtration volume and the graphene concentration in the suspension. After dissolving the filter paper in pure acetone, a freestanding GBF was obtained. The thickness of the GBF was measured as $\sim$20 µm. Raman spectroscopy data of the GBF before and after the spin-coating with FGO are shown in Supplementary Fig. 18. (ii) Functionalization: GO powder (20 mg) and dicyclohexylcarbodiimide (5 mg) were mixed with APTES (30 ml) by ultrasonication for 2 h to produce a homogeneous suspension. Then the suspension was heated up to 100 °C for 3 h, with continuous stirring to realize the functionalization. Fourier transform infrared spectroscopy data provide evidence for the functionalization (Supplementary Fig. 19). (iii) Transfer process: the GBF was first transferred onto a thermal release tape, and then spin-coated[35] with the FGO layer at 4,000 r.p.m. for 2 min onto the film. The thermal release tape was removed by heating the device.

Simulation set-up. Classical molecular dynamics simulations were performed using LAMMPS[56]. Adaptive intermolecular reactive empirical bond order potential[57] was used to simulate the graphene's C–C interactions. The intramolecular forces are taken into account through the ReaxFF potential[58], which uses distance-dependent bond-order functions to represent the contributions of chemical bonding to the potential energy. Periodic boundary conditions are applied in the in-plane directions and free boundary condition in the cross-plane direction of the graphene system. First, each supercell was relaxed at the simulation temperature to achieve zero in-plane stress. Then the systems were thermalized by using a Langevin heat bath. The system reached thermal equilibrium after 2 ns computed in the microcanonical ensemble. Temperature and heat flux were sampled in the microcanonical ensemble in the in the following 1 ns.

Autocorrelation functions for the resistance and the thermal conductivity were calculated over this latter duration.

Thermal conductance. Thermal conductance $\kappa = \kappa_e + \kappa_{ph}$, where κ_{ph} (κ_e) is the thermal conductance due to the phonons (electrons). From the phonon transmission $\Xi(E_{ph})$ the thermal conductance due to the phonon could be calculated as $\kappa_{ph}(T) = \int_0^\infty d\omega(-\frac{\partial f_{BE}}{\partial E})\hbar\omega\Xi(\omega)/2\pi$, where ω is the frequency, T refers to the mean temperature of the system, f_{BE} is the Bose–Einstein phonon statistics, and $\hbar$ represents the reduced Plancks constant. The thermal conductance due to the electrons[59] could be calculated from the electron transmission coefficient T_{el} as $\kappa_e(T) = \frac{L_0 L_2 - L_1^2}{hTL_0}$, where $L_n = \int_{-\infty}^{\infty} dE(E - E_F)^n \, T_{el}(E)(-\frac{\partial f_{FD}}{\partial E})$ in which f_{FD} is the Fermi–Dirac electron statistics.

References

1. Yan, Z., Nika, D. L. & Balandin, A. A. Thermal properties of graphene and few-layer graphene: applications in electronics. *IET Circ. Device. Syst.* **9,** 4–12 (2015).
2. Geim, A. K. & Novoselov, K. S. The rise of graphene. *Nat. Mater.* **6,** 183–191 (2007).
3. Balandin, A. A. *et al.* Superior thermal conductivity of single-layer graphene. *Nano Lett.* **8,** 902–907 (2008).
4. Balandin, A. A. Thermal properties of graphene and nanostructured carbon materials. *Nat. Mater.* **10,** 569 (2011).
5. Kong, Q. Q. *et al.* Hierarchical graphenecarbon fiber composite paper as a flexible lateral heat spreader. *Adv. Funct. Mater.* **24,** 4222–4228 (2014).
6. Yan, Z., Liu, G., Khan, J. M. & Balandin, A. A. Graphene quilts for thermal management of high-power GaN transistors. *Nat. Commun.* **3,** 827 (2012).
7. Gao, Z., Zhang, Y., Fu, Y., Yuen, M. M. & Liu, J. Thermal chemical vapor deposition grown graphene heat spreader for thermal management of hot spots. *Carbon* **61,** 342–348 (2013).
8. Subrina, S., Kotchetkov, D. & Balandin, A. A. Heat removal in silicon-on-insulator integrated circuits with graphene lateral heat spreaders. *IEEE Electron Device Lett.* **30,** 1281–1283 (2009).
9. Subrina, S. Modeling based design of graphene heat spreaders and interconnects in 3-D integrated circuits. *J. Nanoelectron. Optoelectron.* **5,** 281–286 (2010).
10. Malekpour, H. *et al.* Thermal conductivity of graphene laminate. *Nano Lett.* **14,** 5155 (2014).
11. Prasher, R. Graphene spreads the heat. *Science* **328,** 185–186 (2010).
12. Ong, Z. Y. & Pop, E. Effect of substrate modes on thermal transport in supported graphene. *Phys. Rev. B* **84,** 075471 (2011).
13. Ghosh, S. *et al.* Dimensional crossover of thermal transport in few-layer graphene. *Nat. Mater.* **9,** 555 (2010).
14. Sadeghi, M. M., Jo, I. & Shi, L. Phonon-interface scattering in multilayer graphene on an amorphous support. *Proc. Natl Acad. Sci. USA* **110,** 16321 (2013).
15. Lindsay, L., Broido, D. A. & Mingo, N. Flexual phonons and thermal transport in graphene. *Phys. Rev. B* **82,** 115427 (2010).
16. Seol, J. H. *et al.* Two-dimensional phonon transport in supported graphene. *Science* **328,** 213 (2010).
17. Wei, Z., Chen, Y. & Dames, C. Negative correlation between in-plane bonding strength and cross-plane thermal conductivity in a model layered material. *Appl. Phys. Lett.* **102,** 011901 (2013).
18. Luo, T. & Lloyd, J. R. Enhancement of thermal energy transport across graphene/graphite and polymer interfaces: a molecular dynamics study. *Adv. Mater.* **22,** 2495 (2012).
19. Hopkins, P. E. *et al.* Manipulating thermal conductance at metal-graphene contacts via chemical functionalization. *Nano Lett.* **12,** 590 (2012).
20. Ge, Z., Cahill, D. G. & Braun, P. V. Thermal conductance of hydrophilic and hydrophobic interfaces. *Phys. Rev. Lett.* **96,** 186101 (2006).
21. Wang, R. Y., Segalman, R. A. & Majumdar, A. Room temperature thermal conductance of alkanedithiol self-assembled monolayers. *Appl. Phys. Lett.* **89,** 173113 (2006).
22. Ramanathan, T. *et al.* Functionalized graphene sheets for polymer nanocomposites. *Nat. Nanotechnol.* **3,** 327–331 (2008).
23. Konatham, D. & Striolo, A. Thermal boundary resistance at the graphene-oil interface. *Appl. Phys. Lett.* **95,** 163105 (2009).
24. Collins, K. C., Chen, S. & Chen, G. Effects of surface chemistry on thermal conductance at aluminum-diamond interfaces. *Appl. Phys. Lett.* **97,** 083102 (2010).
25. Liang, Q., Yao, X., Wang, W., Liu, Y. & Wong, C. P. A three-dimensional vertically aligned functionalized multilayer graphene architecture: an approach for graphene-based thermal interfacial materials. *ACS Nano* **5,** 2392–2401 (2011).
26. Ni, Y. *et al.* Highly efficient thermal glue for carbon nanotubes based on azide polymers. *Appl. Phys. Lett.* **100,** 193118 (2012).
27. Harikrishna, H., Ducker, W. A. & Huxtable, S. T. The influence of interface bonding on thermal transport through solid-liquid interfaces. *Appl. Phys. Lett.* **102,** 251606 (2013).
28. Liang, Z., Evans, W., Desai, T. & Keblinski, P. Improvement of heat transfer efficiency at solid-gas interfaces by self-assembled monolayers. *Appl. Phys. Lett.* **102,** 061907 (2013).
29. OBrien, P. J. *et al.* Bonding-induced thermal conductance enhancement at inorganic heterointerfaces using nanomolecular monolayers. *Nat. Mater.* **12,** 118 (2013).
30. Taphouse, J. H., Smith, O. N. L., Marder, S. R. & Cola, B. A. A pyrenylpropyl phosphonic acid surface modifier for mitigating the thermal resistance of carbon nanotube contacts. *Adv. Funct. Mater.* **24,** 465 (2014).
31. Sun, F. *et al.* Molecular bridge enables anomalous enhancement in thermal transport across hard-soft material interfaces. *Adv. Mater.* **26,** 6093 (2014).
32. Chien, S. K. & Yang, Y. T. Influence of hydrogen functionalization on thermal conductivity of graphene: Nonequilibrium molecular dynamics simulations. *Appl. Phys. Lett.* **98,** 033107 (2011).
33. Kim, J. Y., Lee, J.-H. & Grossman, J. C. Thermal transport in functionalized graphene. *ACS Nano* **6,** 9050–9057 (2012).
34. Xin, G. *et al.* Large-area freestanding graphene paper for superior thermal management. *Adv. Mater.* **26,** 4521–4526 (2014).
35. Pasternack, R. M., Amy, S. R. & Chabal, Y. J. *Langmuir* **24,** 12963–12971 (2008).
36. Chen, G. & Hui, P. Pulsed photothermal modeling of composite samples based on transmission-line theory of heat conduction. *Thin Solid Films* **339,** 58–67 (1999).
37. Zhao, Y. *et al.* Pulsed photothermal reflectance measurement of the thermal conductivity of sputtered aluminum nitride thin films. *J. Appl. Phys.* **96,** 4563 (2004).
38. Meier, T. *et al.* Length-dependent thermal transport along molecular chains. *Phys. Rev. Lett.* **113,** 060801 (2014).
39. Yang, J. *et al.* Phonon transport through point contacts between graphitic nanomaterials. *Phys. Rev. Lett.* **112,** 205901 (2014).
40. Segal, D., Nitzan, A. & Hänggi, P. Thermal conductance through molecular wires. *J. Chem. Phys.* **119,** 6840 (2003).
41. Hu, L. *et al.* Phonon interference at self-assembled monolayer interfaces: Molecular dynamics simulations. *Phys. Rev. B* **81,** 235427 (2010).
42. Markussen, T. Phonon interference effects in molecular junctions. *J. Chem. Phys.* **139,** 244101 (2013).
43. Han, H. *et al.* Phonon interference and thermal conductance reduction in atomic-scale metamaterials. *Phys. Rev. B* **89,** 180301(R) (2014).
44. Han, H., Li, B., Volz, S. & Kosevich, Y. A. Ultracompact interference phonon nanocapacitor for storage and lasing of coherent terahertz lattice waves. *Phys. Rev. Lett.* **114,** 145501 (2015).
45. Ni, Y., Chalopin, Y. & Volz, S. Significant thickness dependence of the thermal resistance between few-layer graphenes. *Appl. Phys. Lett.* **103,** 061906 (2013).
46. Lusk, M. T. & Carr, L. D. Nanoengineering defect structures on graphene. *Phys. Rev. Lett.* **100,** 175503 (2008).
47. Jiang, J.-W., Wang, B.-S. & Wang, J.-S. First principle study of the thermal conductance in graphene nanoribbon with vacancy and substitutional silicon defects. *Appl. Phys. Lett.* **98,** 113114 (2011).
48. Hughbanks, T. & Hoffmann, R. Chains of trans-edge-sharing molybdenum octahedra: metal-metal bonding in extended systems. *J. Am. Chem. Soc* **105,** 3528 (1983).
49. Ziman, J. M. *Electrons and Phonons* (Oxford Univ. Press, 1960).
50. Trew, R. J., Green, D. S. & Shealy, J. B. AlGaN/GaN HFET reliability. *IEEE Microw. Mag.* **10,** 116127 (2009).
51. Yang, B., Wang, P. & Bar-Cohen, A. 'Mini-contact enhanced thermoelectric cooling of hot spots in high power devices'. *IEEE Trans. Compon. Packag. Technol.* **30,** 432–438 (2007).
52. Nath, P. & Chopra, K. L. Thermal conductivity of copper films. *Thin Solid Films* **20,** 5362 (1974).
53. Langer, G., Hartmann, J. & Reichling, P. Thermal conductivity of thin metal films measured by photothermal profile analysis. *Rev. Sci. Instrum* **68,** 15101513 (1997).
54. Kumar, P. *et al.* Large-area reduced graphene oxide thin film with excellent thermal conductivity and electromagnetic interference shielding effectiveness. *Carbon* **94,** 494–500 (2015).
55. Hummers, W. S. & Offeman, R. E. Preparation of graphitic oxide. *J. Am. Chem. Soc.* **80,** 1339–1339 (1958).
56. Plimpton, S. Fast parallel algorithms for short-range molecular dynamics. *J. Comput. Phys.* **117,** 1 (1995).
57. Stuart, S. J., Tutein, A. B. & Harrison, J. A. A reactive potential for hydrocarbons with intermolecular interactions. *J. Chem. Phys.* **112,** 6472 (2000).
58. Chenoweth, K., van Duin, A. C. & Goddard, W. A. ReaxFF reactive force field for molecular dynamics simulations of hydrocarbon oxidation. *J. Phys. Chem. A* **112,** 1040–1053 (2008).
59. Sadeghi, H., Sangtarash, S. & Lambert, C. Enhancing the thermoelectric figure of merit in engineered graphene nanoribbons. *Beilstein J. Nanotechnol.* **6,** 1176 (2015).

Acknowledgements

The work is sponsored by EU FP7 programme 'Nanotherm' under the Grant Agreement No 318117. This work is also supported by the National Chinese Science Foundation Projects (51272153 and 61574088) and the Science and Technology Commission of Shanghai Municipality programme (12JC1403900), as well as Shanghai Education Commission programme (Shanghai University Peak Discipline Construction Project), by the Swedish Foundation for Strategic Research (SSF) Frame Project Contract No: SE13-0061 and SSF Project Contract No: EM11-0002. This work is supported by EPSRC project EP/N017188/1 and by the EU ITN MOLESCO project number 606728. We acknowledge the financial support from The Production Area of Advance programme, Chalmers University of Technology, Sweden.

Author contributions

S.V. and J.L. proposed and designed the project; H.H., Y.Z. and N.W. contributed equally to this work; H.H. performed phonon and electron Green's function calculations and molecular dynamics simulations; Y.Z. characterized the samples with Raman spectroscopy, Fourier transform infrared spectroscopy and RTD measurement; N.W. and M.M. prepared GBF and functional materials; M.K.S. conducted the PPR measurement with assistance from Y.Z.; M.E. and H.H. performed FE simulations; Y.F. designed the experiments and analysed the experimental data; L.Y. supervised the experiments on functional materials; H.H., Y.N., K.S., S.Y., Y.A.K. and S.V. analysed and interpreted the phonon transport calculations; H.H. and Z.Y.M. conducted first-principle and Green's function calculations with assistance from S.X., H.S., S.B. and C.L.; H.H. integrated results among different research units and wrote the manuscript.

Additional information

Competing financial interests: The authors declare no competing financial interests.

18

Non-equilibrium induction of tin in germanium: towards direct bandgap $Ge_{1-x}Sn_x$ nanowires

Subhajit Biswas[1], Jessica Doherty[1], Dzianis Saladukha[2,3], Quentin Ramasse[4], Dipanwita Majumdar[5], Moneesh Upmanyu[6], Achintya Singha[5], Tomasz Ochalski[2,3], Michael A. Morris[7] & Justin D. Holmes[1,7]

The development of non-equilibrium group IV nanoscale alloys is critical to achieving new functionalities, such as the formation of a direct bandgap in a conventional indirect bandgap elemental semiconductor. Here, we describe the fabrication of uniform diameter, direct bandgap $Ge_{1-x}Sn_x$ alloy nanowires, with a Sn incorporation up to 9.2 at.%, far in excess of the equilibrium solubility of Sn in bulk Ge, through a conventional catalytic bottom-up growth paradigm using noble metal and metal alloy catalysts. Metal alloy catalysts permitted a greater inclusion of Sn in Ge nanowires compared with conventional Au catalysts, when used during vapour-liquid-solid growth. The addition of an annealing step close to the Ge-Sn eutectic temperature (230 °C) during cool-down, further facilitated the excessive dissolution of Sn in the nanowires. Sn was distributed throughout the Ge nanowire lattice with no metallic Sn segregation or precipitation at the surface or within the bulk of the nanowires. The non-equilibrium incorporation of Sn into the Ge nanowires can be understood in terms of a kinetic trapping model for impurity incorporation at the triple-phase boundary during growth.

[1] Materials Chemistry & Analysis Group, Department of Chemistry, Tyndall National Institute, University College Cork, Cork T12 YF78, Ireland. [2] Department of Photonics, Tyndall National Institute, University College Cork, Cork T12 R5CP, Ireland. [3] CAPPA, Cork Institute of Technology, Cork T12 T66T, Ireland. [4] SuperSTEM Laboratory, SciTech Daresbury Campus, Daresbury WA4 4AD, UK. [5] Department of Physics, Bose Institute, Kolkata 700009, India. [6] Group for Simulation and Theory of Atomic-Scale Material Phenomena (stAMP), Department of Mechanical and Industrial Engineering and Department of Bioengineering, Northeastern University, Boston, Massachusetts 02115, USA. [7] AMBER, CRANN, Trinity College Dublin, Dublin D02 R590, Ireland. Correspondence and requests for materials should be addressed to S.B. (email: s.biswas@ucc.ie) or to J.D.H. (email: j.holmes@ucc.ie).

Direct bandgap semiconductor materials are needed for new device architectures such as band-to-band tunnelling tunnel FETs[1], optical interconnects[2] and for the development of group IV photonics[3,4] because these technological modules are based on the direct transition of carriers between energy bands. However, a major problem arises with bulk Si and Ge in photonics, optoelectronics and tunnel FETs devices as they are indirect bandgap semiconductors, that is, the lowest-energy transition from the valence to the conduction band involves a change in crystal momentum[5]. Although highly doped, tensile strained Ge results in enhanced direct gap light emission, due to raising of the Fermi level, the doping levels and induced strain required are not practical for many post-CMOS devices. III–V compound semiconductors, for example InP, GaAs, InAs and so on, offer a solution for integrating direct bandgap materials as on-chip photonic and electronic components. However, the monolithic integration of direct bandgap group IV semiconductor materials is expected to lead to lower production costs and higher reliability than hybrid III–V-on-Si approaches[6]. Sn-based group IV alloys are predicted to be tunable direct gap semiconductor materials[7]. Apart from the direct transition of carriers, group IV alloy systems have also been predicted to exhibit high electron and hole mobilities, and low-carrier effective masses, making them ideal material platforms for co-integration of optoelectronic and high-speed electronic devices[7].

Considering group IV elements, on moving from Si to Ge to Sn, the conduction band at $k=0$ drops in energy until, in grey tin, the material acquires a direct (and vanishing) bandgap at $k=0$ (ref. 5). A direct band system from group IV elements is likely to require the presence of Sn[8,9]. Especially for Ge, the small energy separation of 140 meV between the indirect (L) and direct (Γ) conduction band valleys can be overcome by alloying with Sn. Theoretical modelling[10] as well as photoluminescence experimental studies[11] have found unstrained $Ge_{1-x}Sn_x$ to transition to a direct bandgap material at an alloy composition of no less than 6.5 at.% Sn, although some contradiction regarding this value exists[6,12,13], with the range of Sn incorporation to obtain a direct band transition predicted to be between 6.5 and 10 at.%. However, fundamental challenges (low solubility, metallic Sn segregation, lattice mismatch and so on) restrict the growth of Sn-based Si and Ge alloys with a high Sn content (>8 at.%) in any nanoform, for example, thin film, nanowire and so on[14,15].

In recent years, considerable effort has been used to grow $Ge_{1-x}Sn_x$ films on Si substrates, where the lattice mismatch with Si is fully relieved by periodic misfit dislocations at the interface with no dislocations into the films[16]. Recent advances in chemical vapour deposition (CVD) techniques have made it possible to grow binary $Ge_{1-x}Sn_x$ and ternary $Si_xGe_ySn_{1-x-y}$ group IV thin film semiconductor alloys using low-temperature epitaxy[17–19]. However, minimal effort has been applied to fabricate group IV direct bandgap materials in one-dimensional (1D) nanoform to keep on track with the miniaturization of Si-based nanoelectronics and to take advantage of their 1D geometry for new age field-effect transistor (FET) devices (finFET, gate-all-around FET and so on). Top-down processing to fabricate good quality (single crystalline, straight, uniform diameter nanowire with no Sn segregation) $Ge_{1-x}Sn_x$ nanowires is limited due to under-developed surface and etch chemistries, although encouraging results (with ~8 at.% Sn incorporation) were recently reported on the fabrication of suspended GeSn nanowires through competitive etching between $Ge_{1-x}Sn_x$ and Ge layers[20,21]. Using bottom-up growth paradigms, Ge and GeSn nanowires were synthesized by utilizing low-melting point Sn metal catalysts, but these techniques produced nanowires either with insufficient Sn incorporation[22] or low quality (bending and kinking) crystals with non-significant luminescence[23].

On the basis of thermodynamic limitations, a non-equilibrium growth scenario influenced by the kinetics of the system is required to incorporate a sufficient amount, far from equilibrium, of Sn into a 1D Ge lattice to achieve a direct bandgap transition. The triple-phase boundary at the catalyst-nanowire interface in a bottom-up growth process is known to be a feasible pathway for impurity incorporation in a 1D lattice and can act as a localized non-equilibrium centre for excessive impurity dissolution[24,25]. A kinetics-dependent framework was predicted to be responsible for the extraordinary incorporation of impurity adatoms from the catalyst tip.

Here, we report the application of a three-phase bottom-up growth protocol to fabricate highly crystalline, uniform diameter, direct bandgap $Ge_{1-x}Sn_x$ nanowires with considerable ($x>0.09$) Sn incorporation; around 10 times the equilibrium solubility. Third-party metal catalysts (Au or AuAg alloy) were used to guide the non-equilibrium incorporation of Sn adatoms into the precipitated Ge bi-layers, where the impurity Sn atoms become trapped with the deposition of successive layers, thus giving an extraordinary Sn content in the alloy nanowires.

Results

Growth of group IV alloy nanowires. Participation of Au and AuAg alloy seeds in the bottom-up growth of Ge nanowires has been well-documented by our group in previous reports[26–28]. Similarly, for the growth of $Ge_{1-x}Sn_x$ nanowires, we have used dodecanethiol-stabilized phase pure Au and $Au_{0.90}Ag_{0.10}$ alloy nanoparticles[29]. These small colloidal alloy nanoparticles were deposited onto silicon (001) substrates (with native oxide) and dried at 180 °C under vacuum, leading to the desorption of the surfactant molecules from the surface of the particles[30]. A liquid-injection CVD technique, using toluene as the solvent phase, was adopted for growing the $Ge_{1-x}Sn_x$ nanowires at 440 °C on the surface of Si(001) substrates. Diphenylgermane was used as the Ge source, whereas allyltributylstannane was used as the tin precursor. Similar decomposition kinetics and solubility of the tin and germanium precursors provoked the choice of diphenylgermane and allyltributylstannane, where the Sn precursor has a slightly higher boiling point (360 °C at atm. pressure) than the Ge precursor (325 °C at atm. pressure). The choice of Au and AuAg catalysts and the growth temperature was driven by the Au-Ge and Au-Ag-Ge phase diagrams[30], where a faster growth rate of Ge nanowires is expected using AuAg catalysts[27]. A faster Ge growth rate provides the opportunity to incorporate more Sn into the Ge lattice, with less chance of segregating on the surface or within the bulk[31]. At our growth temperature (440 °C) Au-Sn or Ag-Sn phase diagrams predict the formation of eutectic liquid alloys (Au-Sn-Ge or AuAg-Sn-Ge) with enormous Sn intakes in the catalyst, without any window for the precipitation of Sn layers[32].

Both Au and AuAg nanoparticles successfully catalysed the growth of Ge nanowires after a 2 h time period, as determined by scanning electron microscopy (see Fig. 1a,b). The absence (or very little amount) of particulate deposits, as a byproduct, on the nanowire surfaces and within the samples in general, verifies the controlled growth of the nanowires. The grown nanowires were straight without any observed kinks, bends or curling. A Ge and Sn precursor mixture containing 15 at.% Sn was used as the injection solution for the growth of the nanowires shown in Fig. 1a,b. The lengths of the nanowires grown from both Au and AuAg seeds were in the order of 1–3 μm, whereas their diameters were between 30–70 nm; with a mean diameter of 45.3 and 38.5 nm for Au and AuAg-seeded nanowires, respectively.

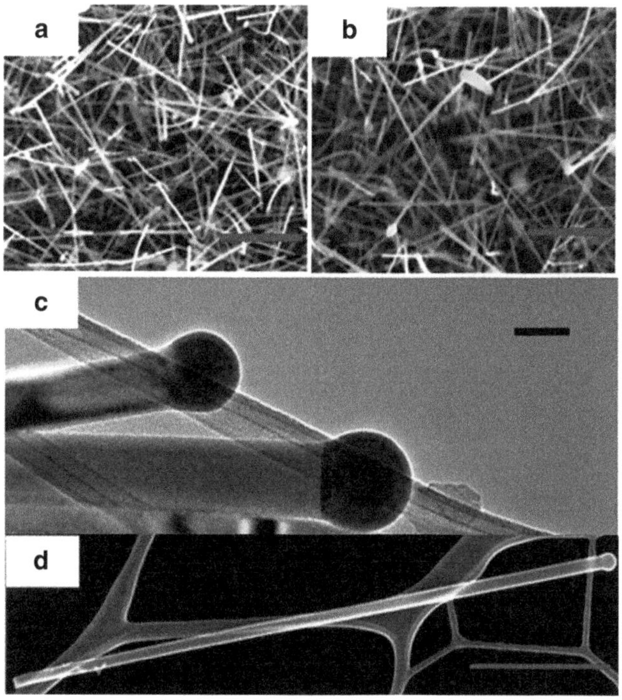

Figure 1 | Morphological examination of alloy nanowires by electron microscopy. SEM images of catalysed $Ge_{1-x}Sn_x$ nanowires grown using 15 at.% of Sn containing solution with: (**a**) Au (**b**) $Au_{0.90}Ag_{0.10}$ catalysts (scale bar, 1 μm). TEM image in **c** confirms the participation of VLS growth mechanism with dark-contrast spherical seed at the tip of the nanowire with AuAg catalysts from precursor solution containing 15 at.% Sn. Scale bar, 100 nm. HAADF STEM image in **d** confirms uniform nanowire diameter along the length with negligible tapering. Scale bar, 1 μm.

The bright-field transmission electron microscopy (TEM) image in Fig. 1c confirms the participation of catalytic vapour–liquid–solid (VLS) nanowire growth, as the dark-contrasted partially spherical metal seed can be seen at the tip of the nanowire in the image. A thin amorphous shell can also be observed on top of the nanoparticle seed, but the nanowire diameter is determined by the dimension of the metal seed at the tip. A flat interface was observed after growth between the nanoparticle seed and nanowire, without any lateral side facets at the tri-junction. The nanowires were fairly straight, with no or very little indication of tapering from the seed to the end of the nanowire as shown in the dark-field scanning TEM (STEM) image in Fig. 1d. Metal seeds at the nanowire tips were pinned at the interface, where the contact angle between the interface and the edge of the seed was larger than in the growth of Ge nanowires with Au or AuAg seeds[27,28], due to the relatively low surface energy of Sn-rich metal catalyst seeds. Participation of different growth regime with foreign Au and AuAg metal catalysts was confirmed by the fact that nanowires synthesized without any noble metal seeds, that is, self-seeded growth from Sn seeds, resulted in the formation of very short nanowires (200–300 nm in length), with prominent tapering from the seed-nanowire interface to the very end of the nanowire (Supplementary Fig. 1).

Elemental analysis of alloy nanowires through EDX. As the primary objective of this work was to fabricate direct bandgap $Ge_{1-x}Sn_x$ nanowires with a high Sn content, it is essential to explore the quantitative and qualitative incorporation of Sn in the nanowire body. Energy-dispersive x-ray (EDX) analysis and high-resolution electron energy loss spectroscopy (EELS) provide the means to investigate the chemical environment in the bulk of the alloy nanowire and also at the atomic scale. The composition of the nanowires and the distribution of elements within them were estimated through EDX point measurements and elemental mapping by STEM. For nanowires grown from pure Au catalysts, the amount of Sn in the injecting solution was varied from 10–20 at.%, resulting in a gradual increase in the actual Sn concentration in the nanowires. Very low amounts of Sn (mean concentration of 1.5 at.%) was determined in nanowires using an injecting solution of 10 at.% Sn. Increasing the Sn concentration in the injecting solution to 15 and 20 at.% resulted in the incorporation of Sn in the nanowires at levels of 6 and 9 at.%, respectively. These values of Sn concentrations are much higher (almost 6 and 9 times) than the extrapolated bulk equilibrium solid solubility of Sn in Ge[31]. Although a higher assimilation of Sn in the Ge nanowires was achieved with an injection solution of 20 at.% Sn, this high Sn concentration also resulted in homogeneous nucleation of metallic Sn as spherical clusters (Supplementary Fig. 2). Hence, under our reaction conditions, an initial Sn concentration of 15 at.% was determined to be ideal to obtain $Ge_{1-x}Sn_x$ nanowires with substantial Sn incorporation and with negligible secondary nucleation of unwanted spherical particulates in the sample. With the aim of including more Sn into the 1D Ge lattice, $Au_{0.90}Ag_{0.10}$ alloy nanoparticle catalysts were used as seeds rather than pure Au, as the alloy seeds have been previously shown to favour faster growth kinetics for phase pure Ge nanowire growth[27]. To avoid spherical metallic Sn clusters in the sample, injection solutions with 15 at.% of Sn were used. A slight increase in the Sn incorporation in the Ge nanowires from 6.0 (±0.5) to 6.6 (±0.6) at.% (error bars are defined in Method section) was observed when using AuAg alloy rather than pure Au seeds, respectively. For an accurate estimation of the amount of Sn included in each nanowire sample, EDX point measurements were performed on 50 different nanowires and mean values computed. An example of a point EDX measurement taken of a $Ge_{1-x}Sn_x$ nanowire grown from AuAg seed is shown in Supplementary Fig. 3. The distribution of Sn in the alloy nanowires grown with AuAg seeds was uniform along the length and width of the nanowires, without any segregation near the catalyst-nanowire interface or at the nanowire surfaces (Supplementary Figs 3 and 4). Elemental EDX mapping from a AuAg-seeded $Ge_{1-x}Sn_x$ nanowire also confirmed uniform Sn distribution in the entire nanowire volume (Supplementary Fig. 5). The uniform axial and radial distribution of Sn and the minimal tapering of our nanowires also rules out the diffusion of Sn through the nanowire sidewalls as a possible incorporation mechanism. The strong incorporation and uniform distribution of Sn atoms confirms the continuous dissolution of Sn atoms throughout the growth process at the seed-nanowire growth interface.

To achieve a higher concentration of Sn in the $Ge_{1-x}Sn_x$ nanowires, a step cooling method was utilized, where the initial injection of the solution at the growth temperature (440 °C) was followed by an annealing step for 2 h at 230 °C during the cool-down. Motivation to introduce a step cooling at 230 °C was driven by two reasons: a small window in the Sn-rich side of the Au-Sn phase diagram at 230 °C and the bulk Ge-Sn eutectic temperature at around 230 °C. This step cooling technique further forces a colossal amount of Sn (an example is shown in Fig. 2a), with an average concentration of 9.2 (±0.8) at.% (with a AuAg growth promoter and 15 at.% Sn injecting solution), into the nanowire while keeping the nanowire morphology intact (scanning electron microscopy image in Supplementary Fig. 6). The extraordinary amount of Sn incorporation as measured via EDX analysis was also supported through x-ray diffraction

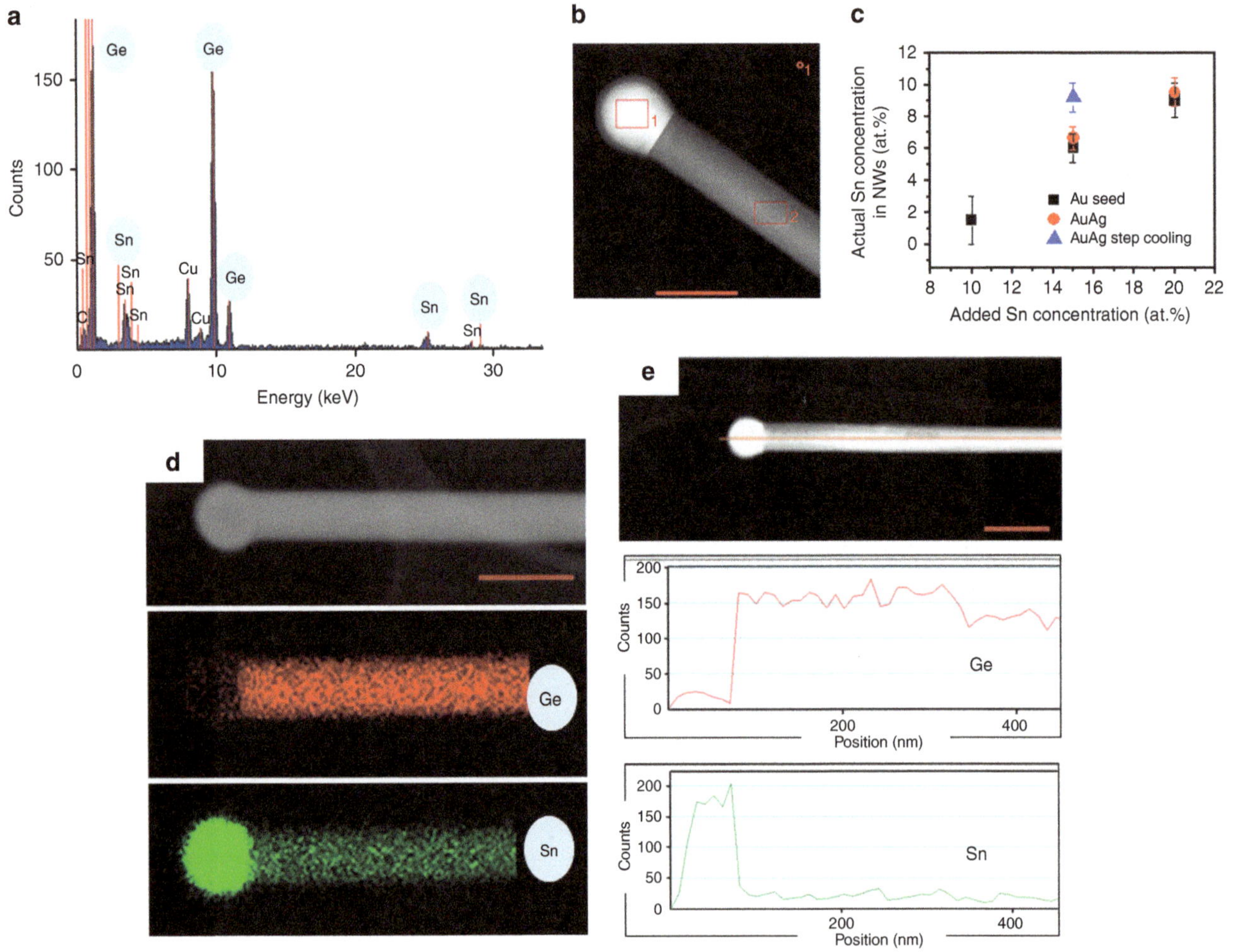

Figure 2 | EDX analysis of $Ge_{1-x}Sn_x$ nanowires. (**a**) EDX spectrum recorded from the body of an alloy nanowire (selected from the sample with 9.2 at.% average Sn incorporation) showing the presence of both Ge and Sn. This particular nanowire is shown in **b**. Scale bar, 50 nm. Variations in Sn concentration with different catalysts and growth conditions are demonstrated in **c**. Error bar indicates s.d. in EDX measurements measured over 50 nanowires. (**d**) Dark-field HAADF image and EDX mapping for Ge and Sn in a $Ge_{1-x}Sn_x$ nanowire with 9.4 at.% of Sn. HAADF image with the uniform distribution of Ge and Sn and Sn-rich catalyst is confirmed from EDX mapping and also from HAADF image and EDX linescan in **e**. Red curve denotes linescan for Ge, whereas green curve for Sn. Scale bar, 50 and 100 nm in **d** and **e**, respectively.

measurements (Supplementary Fig. 7). The amount of Sn in the nanowires was calculated as 9.8% from Vegard's law which is an empirical law that relates the substitution of a guest ion into the host lattice with the experimentally observed degree of lattice change. A plot showing the mean Sn concentration in $Ge_{1-x}Sn_x$ alloy nanowires as a function of different growth conditions can be seen in Fig. 2b. To confirm the homogeneity of Sn dissolution in the nanowire, that is, to rule out the formation of Sn precipitates or cluster formation in the core or on the surfaces of the nanowires after step annealing, EDX elemental mapping was performed on $Ge_{1-x}Sn_x$ nanowires with the highest Sn incorporation, that is, a mean concentration of 9.2 at.%. Elemental mapping of a particular nanowire with a Sn concentration of 9.4 at.% is shown in Fig. 2c. The elemental maps show a homogeneous distribution of Sn in the core of the nanowires without any surface segregation or precipitation near the seed-nanowire interface after step annealing at 230 °C. A high density of Sn was observed at the spherical tips of the nanowires, as confirmed from EDX mapping in Fig. 2c and the linescan in Fig. 2d, confirming the participation of a Sn-rich alloy seed (Sn alloyed with Au or AuAg) in VLS nanowire growth. An elemental EDX linescan of Ge and Sn along the nanowire axis clearly demonstrated the homogeneity of Sn incorporation along the nanowire length even after the step annealing process, thus confirming the continuous dissolution of Sn throughout the nanowire length (part of Fig. 2d). The uniformity of Sn dissolution in the alloy nanowires at the highest average Sn concentration was further confirmed through EDX point scans at different lengths along the nanowires (Supplementary Fig. 8). Even the radial Sn concentration detected by EDX demonstrates (Supplementary Fig. 9 shows the radial line profile for a nanowire selected from the sample with highest Sn content, that is, with step-down cooling) a flat profile (U-shape profile indicates surface segregation), thus again indicating uniform distribution of Sn without any clustering of Sn near nanowire side facets.

Atomic resolution Sn mapping through EELS. The uniform distribution of Sn atoms throughout the nanowires also suggests a single-atomic pathway for Sn impurity incorporation. The local distribution of Sn in the nanowires is a fine criterion to determine impurity incorporation and diffusion modes in the nanowires. The possible formation of Sn precipitates in the nanowire bulk or near the nanowire surface suggests multiple impurity incorporation pathways where the impurities are diffused to the preferred lattice sites such as crystal defects. Also, the formation of local metallic Sn segments and Sn–Sn dimers could quench efficient emission from these materials due to the creation of dark trapping sites for charge carriers. To confirm the sparse distribution of Sn in the Ge lattice of the nanowires, we have probed the spatial arrangement of Sn through high-resolution EELS, in a STEM. The spatial arrangement of dissolved Sn in a nanowire sample with the highest Sn incorporation (mean concentration of 9.2 ($\pm$0.8) at.%), as determined by EDX measurements were traced (error bars are defined in the Methods section). Two EELS maps were acquired at two vastly different heights along the wires, to make sure Sn was not only present at certain areas of each nanowire, such as close to the catalyst. Maps were de-noised by principal components analysis and the background was removed by fitting a power law over a region immediately in front of the core loss edges. The signal was then integrated over a 120 eV window above the onset of the Sn $M_{4,5}$ and Ge $L_{2,3}$ edges. HAADF images and the corresponding EELS chemical profile recorded from the rectangular box region are depicted in Fig. 3a,b. The HAADF survey image in Fig. 3a was acquired from the bulk of the nanowire to avoid strongly oxidized edges (and the overlap with the Sn edge) and the HAADF image shown in Fig. 3b was acquired closer to the edge of the wire. Both data sets were representative of the whole nanowire. Atomically resolved EELS spectral images highlight the incorporation of Sn in the core of the Ge nanowires. The sparse distribution of Sn in the Ge host lattice is clearly observed in the lattice-resolved EELS maps. No apparent sign of Sn precipitation was detected in the nanowires from EELS mapping, thus confirming the distribution of Sn atoms throughout the $Ge_{1-x}Sn_x$ 1D lattice. The Sn EELS M-edge is quite delocalized, thus making it very difficult to resolve Sn as part of the Ge-Sn dumb-bell due to inelastic scattering. A single Sn atom will look blurry and delocalized, especially when the maps are taken over relatively thick regions of the wires as the Sn atom may be buried deep inside the lattice and further scattering will give the impression of a poor image. So the proximity effect of Sn atoms in the lattice may represent as Sn clusters in high-resolution mapping (Fig. 3a,b) with a smaller field of view. Low-resolution EELS mapping from $Ge_{1-x}Sn_x$ nanowires also confirmed the sparse distribution of Sn without any formation of metallic Sn hotspots (Supplementary Fig. 10). Hence Sn was distributed uniformly throughout the lengths of the nanowires, but randomly at the atomic scale (as seen in Fig. 2 and Supplementary Fig. 10), without any phase separation. To assess precisely the catalyst-nanowire interface sharpness, EELS chemical maps (recorded from the rectangular red box denoted in the HAADF image attached to the map) and profiles were recorded by moving the electron probe serially across the interface along the line indicated by the black arrow and recording the Ge $L_{2,3}$ and Sn $M_{4,5}$ EELS edges (Fig. 3c). The red shaded area in the linescan (Fig. 3d) corresponds to the same spatial extent indicated on the HAADF image. EELS spectral images for Ge an Sn and chemical line-profiles confirm the very Sn-rich composition of the catalyst seed with sharp composition variation at the seed-nanowire interface. The oscillations of the integrated EELS intensities follow the oscillations of the simultaneously recorded HAADF signal, in both the nanowire

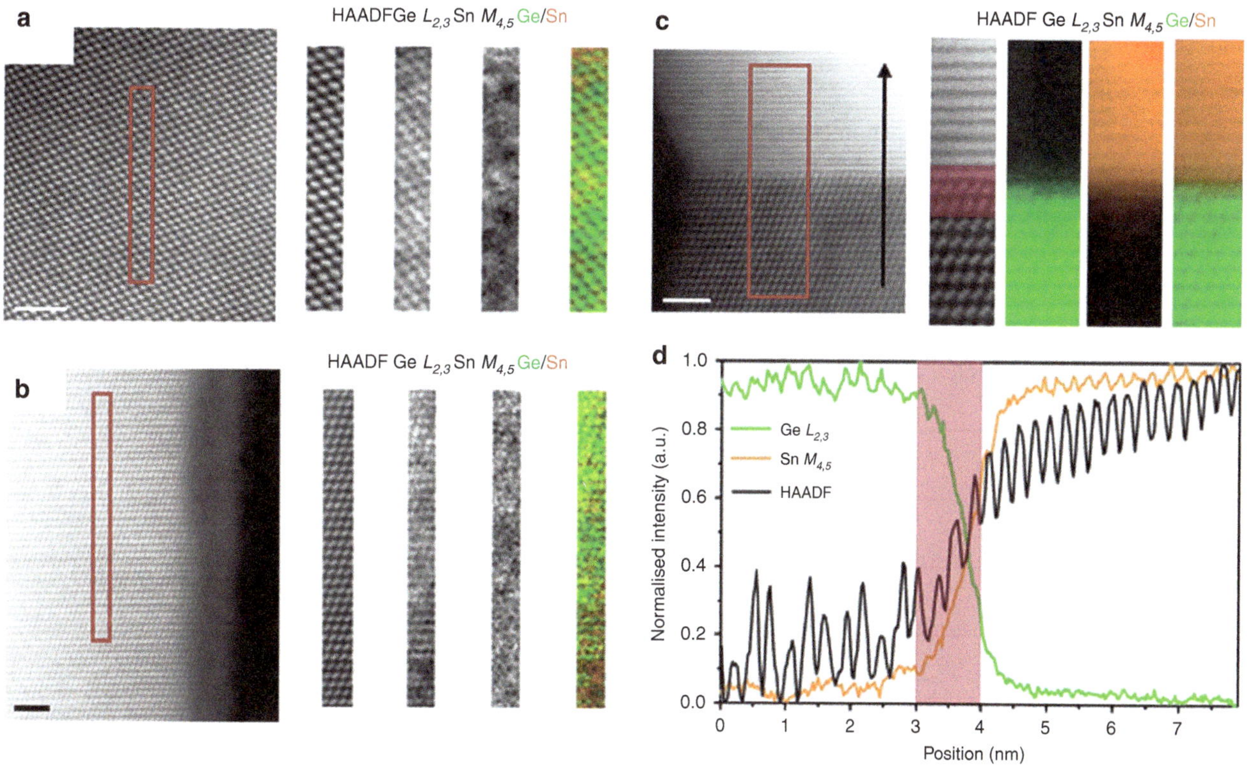

Figure 3 | High-resolution EELS mapping of $Ge_{1-x}Sn_x$ nanowires. Unprocessed HAADF survey image recorded from the centre (**a**) and near the edge (**b**) of a $Ge_{1-x}Sn_x$ nanowire with a Sn incorporation of 9 at.% (area of interest highlighted). Corresponding EELS map for Ge and Sn is also attached along with the simultaneously acquired HAADF image (green: Ge and orange: Sn). For the EELS map, after de-noising by principal components analysis, the background was removed by fitting a power law over a region immediately in front of the core loss edges. The signal was then integrated over a 120 eV window above the onsets of the Sn $M_{4,5}$ and Ge $L_{2,3}$ edges. (**c**) HAADF survey image of a seed-nanowire interface region with the Ge and Sn EELS map recorded from the highlighted region. Panel **d** shows a linescan acquired subsequently in the same region. The red shaded area in the linescan corresponds to the same spatial extent indicated on the HAADF image. Scale bar, 2 nm for all the HAADF images.

and in the seed regions. Random, non-uniform fluctuation of the Sn signal in the line-profiles of the nanowire region also suggests random $Ge_{1-x}Sn_x$ alloy formation with high Sn incorporation. The frequency of the HAADF oscillation signal increased in the catalyst thus confirming a much narrower interplanar spacing in the lattice of the catalyst than the nanowire. Abrupt composition fluctuations at the seed-nanowire interface confirmed a continuous trapping and dissolution process for Sn impurity incorporation rather than a Sn layer precipitation and diffusion process.

Structural characterization of nanowires via STEM and HRTEM. Impurity atoms (Sn in our case) in nanowires can induce structural defects, such as twins and stacking faults and these defects can act as preferential sites for subsequent impurity accumulation[33]. In other scenarios, pre-formed stacking faults in nanowires due to interface engineering, can also act as preferred sites for the segregation of foreign atoms from catalyst nanoparticles[34]. Hence, it is very important to probe the structural quality of the alloy nanowires to estimate the mode for impurity incorporation in Ge. Also nanowires with defects are not suitable for nanoelectronic devices as stacking faults and twin boundaries can encourage electron scattering[35]. A bright-field high-resolution TEM (HRTEM) image (Fig. 4a) confirms the high crystallinity of a single $Ge_{1-x}Sn_x$ nanowire with a 9.2 at.% Sn incorporation and with a 2–3 nm amorphous oxide coating. Fast Fourier transform (FFT) analysis showed a pseudo hexagonal symmetry and the reflections can be assigned to the high-order Laue zone diffraction of {111} and {002} planes in group IV crystals[36]. FFT and HRTEM images depict an interplanar spacing (*d*) between {111} planes in the nanowire to be 0.323 nm, which is

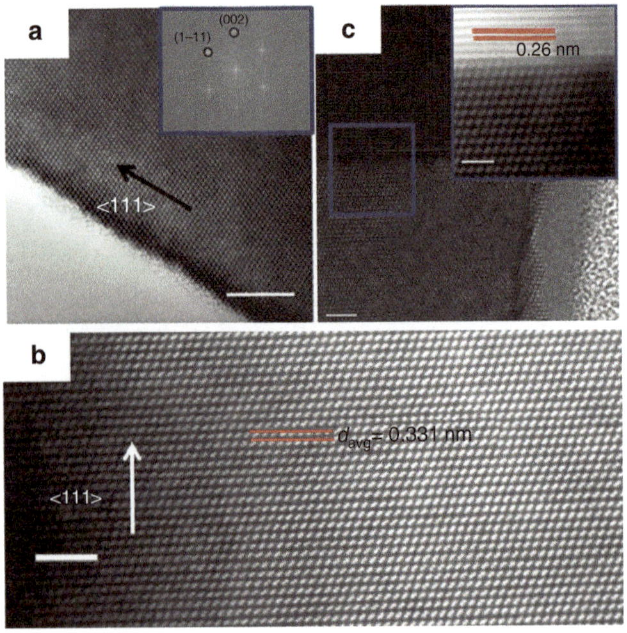

Figure 4 | Structural study of alloy nanowires with HRTEM and STEM. (**a**) HRTEM image of highly crystalline $Ge_{1-x}Sn_x$ nanowire (9.2 at.% Sn incorporation). Scale bar, 5 nm. FFT pattern in the inset confirms the crystallinity and growth orientation of the alloy nanowire. (**b**) Lattice-resolved STEM HAADF image recorded from the core of the alloy nanowire showing the single crystalline nature with an inter-planer spacing of 0.33 nm. Scale bar, 2 nm. (**c**) High-resolution HAADF image of a seed-nanowire interface (magnified image in the inset) shows abrupt catalyst-nanowire interface with no precipitation of metallic Sn. Scale bar, 2 and 1 nm for the inset.

very close to the *d* value for bulk diamond Ge crystal (JCPDS 04-0545). An increase in the *d* value from bulk Ge is expected with the incorporation of large amounts of Sn in Ge lattice. However this discrepancy may arise from the fact that this particular nanowire could only be aligned to a relatively minor zone axis. Fig. 4b shows a high-resolution STEM image of another nanowire from the sample with the highest Sn incorporation using the HAADF mode. The image was recorded with <110> zone axis alignment. Generally, the crystal structure of the $Ge_{1-x}Sn_x$ alloy nanowires, with various Sn incorporations, exhibited a bulk diamond cubic crystal structure with a 3C lattice arrangement without any stacking faults and twin boundaries, with <111> being the dominant growth direction. Although there is a large lattice mismatch between the components (Ge and Sn) of the alloy[31], the epitaxial mismatch in the nanowires is compensated by elastic deformation near the hetero-interface and relieved at the nanowire surfaces[37], thus maintaining highly crystalline nanowires. The liquid eutectic catalyst at the tip of the nanowires can also naturally accommodate elastic strain. Atomic-scale randomness in Sn incorporation in the Ge lattice, as observed through EELS mapping, can generate varied local lattice distortion and spacing at an Ångström scale. To compensate the effect of random alloying on the *d* value, we have calculated the inter planner spacing of 50 successive lattice planes (over >15 nm length) and determined the average *d* value to be 0.331 nm, which is slightly above the bulk 3C-Ge value (Supplementary Fig. 11). Bright-field STEM imaging (Fig. 4c) of the interface between the $Ge_{1-x}Sn_x$ nanowires and the metallic tips confirmed the sharp nature of the interface, with no tailing effect or segregation of metal at the interface. For STEM imaging, stacks of images were acquired sequentially at high scanning speed to minimize drift and instabilities and were aligned and summed for a high signal-to-noise ratio. An atomic resolution view of a catalyst-nanowire interface area, indicated by the blue box in Fig. 4c, is depicted in the HAADF image in the inset of Fig. 4c. The sharp contrast in the HAADF intensity at the interface clearly suggests the abrupt nature of the interface. A lattice spacing of 0.26 nm was measured at the metallic tip which is relatively close to metallic Sn (JCPDS cards #04-0673), thus further confirming the formation of a Sn-rich alloy at the tip.

Raman spectroscopy on alloy nanowires. Raman scattering is an effective tool to estimate the structural and chemical environment in the core of a nanowire. Raman spectroscopy was used to accurately probe the local chemical bonding environment and also to estimate the amount of Sn in the alloy nanowire samples. Figure 5a shows the Raman spectra of alloy nanowires with different Sn concentrations and for reference a spectrum from bulk Ge. The strong peak around 302 cm^{-1} in bulk Ge is attributed to the Ge-Ge LO mode. The Ge-Ge Raman peak progressively shifts to a lower energy with increasing Sn concentration (as determined by EDX measurements). A red shift of 1.2 to 5.9 cm^{-1} of the Ge-Ge LO mode was observed for a variation in the Sn concentration from 1.5 to 9.2 at.%, compared with bulk Ge. We could not compare the Raman shift with a sample of pure Ge nanowires, as the similar growth conditions without any Sn yielded Ge nanowires of entirely different dimensions and morphology[27]. In the alloy nanowires, apart from the Ge-Ge LO peak, additional modes due to Ge-Sn bonds appeared at around 260 cm^{-1}, as shown in Fig. 5b. The presence of a Ge-Sn vibrational mode indicates the formation of $Ge_{1-x}Sn_x$ alloys, where an increase in the intensity ratio between Ge-Sn and Ge-Ge LO modes with increasing Sn content, implies a larger substitution of Sn in the Ge lattice for $Ge_{1-x}Sn_x$ nanowires. The origin of the Ge-Ge frequency shift in the Raman spectra of

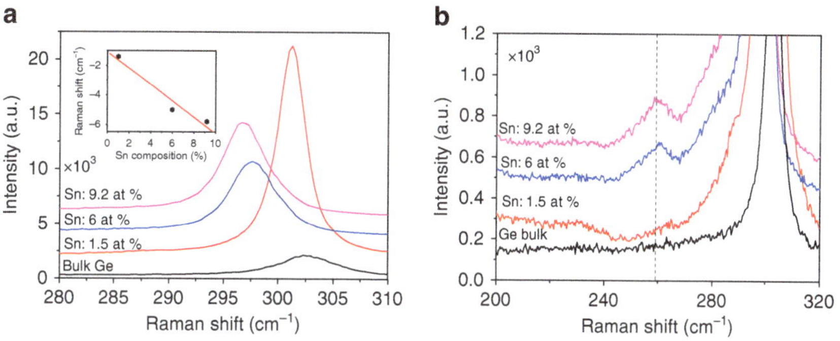

Figure 5 | Confirmation of $Ge_{1-x}Sn_x$ alloy formation with Raman analysis. Raman spectra of bulk Ge and $Ge_{1-x}Sn_x$ nanowires (where $x = 0.06$ and 0.092) within the range (**a**) 280-310 cm^{-1} and (**b**) 200-320 cm^{-1}. Vertical line in **b** represents position of Ge-Sn vibration. The inset of **a** shows the downshift of Ge-Ge LO mode as a function of Sn percentage. Experimental data are represented with dots which fits (straight line) well with the linear expression, $\omega(x) = \omega_0 + \Delta\omega x$.

the $Ge_{1-x}Sn_x$ alloys is due to compositional variations and strain effects. Participation of compressive and tensile strain towards the Raman shift is not justified for nanowire samples, as due to the large surface area, strain can be effectively released for these nanostructures. Compositional variations can originate from two factors: (i) mass disorder and (ii) bond distortion. The Ge-Ge LO mode progressively shifts towards a lower frequency with an increasing Sn concentration as displayed in the inset of Fig. 5a. We have fitted the Raman peak shift ($\Delta\omega$) against Sn composition (x), as determined through EDX analysis, with a linear expression, $\omega(x) = \omega_0 + \Delta\omega x$, and the obtained value of $\Delta\omega$ was found to be $-(64.3 \pm 0.1)$ cm^{-1}. This value is consistent with the value of—(68 ± 5) cm^{-1} reported by Li *et al.*, who assumed that their alloy films were completely strain free[38]. The linear correlation between the Raman peak shift and the Sn concentration (determined by EDX) in the alloy nanowires further validates the high Sn content in $Ge_{1-x}Sn_x$ nanowires. Compared with a few other recent reports, a discrepancy in the $\Delta\omega$ value and Raman shift is observed for the 9.2 at.% Sn containing nanowire sample with a relatively smaller shift in the Ge-Ge LO peak[21,39]. This downshift in Raman frequency may arise from the random alloying affect[38], instead of a spontaneous ordering, as observed in our nanowire sample through high-resolution EELS and TEM measurement.

Photoluminescence study of $Ge_{1-x}Sn_x$ nanowires. As there are contradictions regarding the amount of Sn needed in $Ge_{1-x}Sn_x$ thin films and bulk alloy to obtain a direct bandgap, it is essential to investigate the emission characteristics of strain free bottom-up grown $Ge_{1-x}Sn_x$ alloy nanowires. Bandgap information on $Ge_{1-x}Sn_x$ alloys was extracted through photoluminescence measurements at low temperature. A photoluminescence study to probe the bandgap characteristics of $Ge_{1-x}Sn_x$ nanowire samples was conducted on samples cooled to 7 K, using a He cryostat. Photoluminescence spectra of two nanowire samples (with an average Sn concentration of 6 and 9.2 at.%) recorded at 7 K is shown in Fig. 6a. The photoluminescence spectrum for the nanowire sample with the relatively low Sn content (6 at.%) exhibited a main peak which corresponds to the direct energy gap emission, at a wavelength of around 2,200 nm, with a broad line-width (232 nm) of the emission spectrum. The direct peak is due to the strong radiative recombination of the direct bandgap transition. At this Sn content, separate peaks due to direct and indirect transitions cannot be clearly identified due to the reduced energy difference between the direct and indirect bandgap, resulting in only a single peak with broad line-width (232 nm) and tailing. A large amount of Sn in the $Ge_{1-x}Sn_x$ alloy nanowires resulted in the reduction of the bandgap energy difference between the direct and indirect transition, which was approximately 0.14 eV in bulk Ge. Nanowire samples containing a high Sn content (9.2 at.%) exhibited a single photoluminescence emission peak centred at 2,233 nm (band gap (E_g) around 0.55 eV), with a relatively narrow line-width (202 nm) compared with the photoluminescence plot from the low Sn content $Ge_{1-x}Sn_x$ nanowire sample. The relatively narrower line-width of the photoluminescence emission confirms the single energy emission at the Γ point[40]. Typically, a photoluminescence emission with a broad line-width is observed in indirect bandgap alloys with a high Sn content due to the amalgamation of the indirect valley and the direct peak into a single broad emission. However, for our $Ge_{1-x}Sn_x$ nanowires, with 9.2 at.% Sn incorporation, the direct band-to-band transition resulted in relatively narrow photoluminescence emission, compared with the broad emission from $Ge_{1-x}Sn_x$ alloy nanowires incorporating 6 at.% Sn. A single peak with a relatively narrow line-width could signify emission from only the direct bandgap transition rather than unification of both direct and indirect transitions. The low photoluminescence emission intensity observed from the nanowire samples could be due to the luminescence quenching from metallic Sn impurities, which are present in the catalysts at the tip of the nanowires and also in negligible amounts as spherical particles in samples. Also the high surface-to-volume ratio of nanowires compared with thin films can account for the lower luminescence intensity. Relatively broad photoluminescence spectra for both nanowire samples, compared with previous reports, could result from the random distribution of Sn in the alloys. Sn incorporation in the nanowire samples also showed a standard deviation (around 1%) which may also account for the broadness of the photoluminescence spectra obtained.

The position of the maximum in the direct energy emission from the $Ge_{1-x}Sn_x$ nanowires matched well with the reported emission from alloy thin films and disks with similar Sn incorporation (8–10 at.%)[11,21,40,41]. Specifically, the emission energy matched very well with reported data for unstrained GeSn disks[21]. Emission from the $Ge_{1-x}Sn_x$ nanowires at low temperature also complements the low temperature photoluminescence observation from GeSn thin films[40]. Thin film samples with 8 and 9 at.% Sn showed similar broad peaks at 10 K, which transformed to a single emission narrow peak with 10 at.% Sn in the film samples[40]. A shift (around 0.01 eV) in the photoluminescence maximum to a lower energy was also observed in the photoluminescence plots of our nanowire samples with an

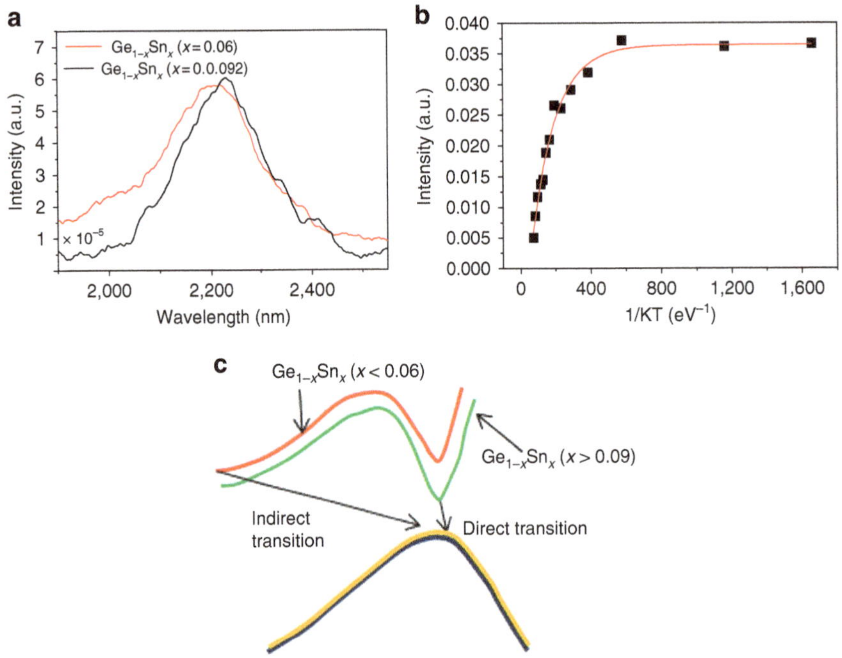

Figure 6 | Optical emission characteristics of alloy nanowires. (**a**) Photoluminescence spectra of $Ge_{1-x}Sn_x$ nanowires ($x = 0.06$ and 0.092) recorded at 7 K. Broadened photoluminescence peak is observed for alloy nanowires with $x = 0.06$. Single emission with narrow line-width is observed for $Ge_{1-x}Sn_x$ nanowires with $x = 0.092$. (**b**) The Arrhenius plot from 9.2% $Ge_{1-x}Sn_x$ nanowires in agreement with single exponential decay of photoluminescence intensity with temperature with the coefficient of determination close to unity ($R^2 = 0.986$). (**c**) A projection (lines are only guide to eyes) of possible direct and indirect transition pathway for different Sn compositions.

increase in the average Sn concentration from 6 to 9.2 at.%. The shift in the photoluminescence peak energy we observed for nanowire samples incorporating 6 and 9.2 at.% Sn was lower than has been reported for thin film samples[11,41]. However, the change in the bandgap of the nanowire alloys with different compositions and morphologies will depend on hole splitting, changes in effective mass, alloy broadening, band-tail states, carrier lifetime and steady state carrier occupation. Also, the different degrees of randomness in the alloys, which may be present between the 9.2 and 6 at.% samples (9.2 at.% samples had an additional step cooling, leading to high degree of randomness as observed in the EELS maps) can affect the band-structure and bandgap tuning. A comparison of photoluminescence spectra at 77 K between pure Ge and $Ge_{1-x}Sn_x$ nanowires (9.2 at.% Sn) show a massive red shift in the emission wavelength with the inclusion of Sn (Supplementary Fig. 12). In order to achieve similar emission intensity the excitation power in the case of Ge nanowires was 700 mW in comparison with 30 mW for GeSn nanowire samples. The peak position for the Ge nanowire samples was around 1,750 nm, which matches well with the indirect bandgap of Ge, while for GeSn samples emission was at 2,150 nm.

To predict the direct band transition, a temperature-dependent photoluminescence study between 7 and 160 K was performed. An Arrhenius plot, depicting integrated photoluminescence intensity as a function of inverted temperature is shown in Fig. 6b (for 9.2 at.%) and in Supplementary Fig. 13 (for 6 at.%). The photoluminescence of direct bandgap semiconductors generally decreases in intensity with increasing temperature, which can be attributed to a reduced transfer of electrons from the Γ to L valleys by thermal activation[4,42]. Thus the increase in the intensity of the photoluminescence peak with decreasing temperature for GeSn nanowires is attributed to the higher population of the Γ valley. With increasing temperature the fast diffusion of photocarriers toward surfaces and interfaces leads to non-radiative surface and interface recombination, respectively, reducing the radiative transition rate[42]. Furthermore, we also observed broadening of the photoluminescence peak (Supplementary Fig. 12b) with increasing temperature for $Ge_{1-x}Sn_x$ nanowires, which can be ascribed to the temperature-dependent broadening of the Fermi distribution of carriers within electron bands[42]. A methodology to discriminate a direct from an indirect fundamental bandgap using temperature-dependent photoluminescence measurements has been presented recently[4]. On the basis of the same arguments, nanowire samples with 6 and 9.2 at.% Sn, manifested by monotonically increasing photoluminescence intensity with decreasing temperature, is similar to photoluminescence observed from direct bandgap III – V alloys or dichalcogenides[43,44]. A power-dependent photoluminescence measurement of $Ge_{1-x}Sn_x$ nanowires (9.2 at.% Sn) (Supplementary Fig. 12c) depicts the evolution of photoluminescence spectra from the GeSn nanowires under different excitation power densities at 77 K. For the lowest power ($P_0 = 30$ mW, enhanced in figure by a factor of 300 for clarity) the photoluminescence peak position was around 2,150 nm. With increasing excitation (up to 16 P_0) spectral broadening was observed, coupled with a blue-shift of the photoluminescence peak position. We attribute this change to carrier filling of closely-spaced Γ and L energy bands. This shift was not observed in the case of Ge nanowires within the scope of available excitation powers between 350 and 700 mW. In addition, in order to achieve similar photoluminescence intensity from both Ge and $Ge_{1-x}Sn_x$ nanostructures (Supplementary Fig. 12a), Ge nanowires had to be excited with a laser power of 700 mW, while for $Ge_{1-x}Sn_x$ nanowires 30 mW excitation power was sufficient. This enhancement of a factor greater than 20 yields provides further evidence to support the transition from indirect to direct bandgap nanowires with increasing Sn incorporation.

We have calculated the activation energies for non-radiative processes from Arrhenius plots. The activation energy was found to increase with increasing directness of the bandgap[42]. The decrease in the photoluminescence intensity at high temperatures is due to an increase of the non-radiative recombination affects, that is, the activation (deactivation) energy values[42]. Experimental activation energy for 9.2 at.% nanowire samples, calculated from Arrhenius plots, was 7 meV and for 6 at.% nanowire samples was 3 meV. The activation energy value for the 9.2 at.% nanowire samples matches well with the reported value for 12 at.% Sn incorporated thin film samples[42], which have been designated in previous reports as direct transitions. Furthermore, Arrhenius plots have been fitted with a single exponential function. A coefficient of determination (R^2) close to unity, that is, for a good fit, indicates a single channel of recombination, while poor fit indicates competitive transition channels. For 6 at.% Sn containing nanowire samples the value of R^2 was calculated to be 0.935 (Supplementary Fig. 13), while for the 9.2 at.% Sn incorporated nanowire samples R^2 values of 0.986 were obtained (Fig. 6b). This result indicates that a single charge carrier transition mechanism dominates only for samples with a high Sn content (Fig. 6c). Although the current photoluminescence measurements indicate a direct bandgap, further confirmation regarding the nature of the emission is required, for example, including radiative rates and quantum efficiencies. Spatial orientation, quantum confinement effects and the internal strain in nanowires can strongly impact the electronic band-structure and bandgap bowing[45] of alloys at the nanoscale.

Discussion

$Ge_{1-x}Sn_x$ nanowires fabricated with Au and AuAg catalysts at 440 °C showed considerable incorporation of Sn in the range around 6–9 at.%, much beyond the bulk equilibrium solubility (around 1%)[45]. The incorporation of Sn in the Ge nanowires through nanowire sidewalls due to homoepitaxy and vapour–solid growth is negligible, as the nanowires were not tapered and demonstrated a uniform radial Sn distribution, as determined by EDX line-profiles (Supplementary Fig. 4). A U-shape line profile with a larger Sn concentration at the edge of the nanowires would have been observed for sidewall Sn incorporation. Size-dependent corrections to the bulk phase diagram due to the influence of capillary forces and stress at the nanoscale[28,46,47] results in significant undercooling of the liquid droplet and can in principle alter the equilibrium content of Sn in Ge. Calculations of the nanoscale equilibrium content of a solid impurity in a 1D lattice, taking account of surface anisotropy and elastic stress, do not support a large dissolution of impurity atoms much beyond equilibrium solubility[25]. Careful analysis of the Sn-rich portion of the ternary Au-Ge-Sn phase diagram[48] shows that for our growth conditions, at equilibrium the Sn-rich (more than 90%) droplet has a Ge:Au ratio of close to unity and the growth should occur via the invariant reaction U4: $L + AuSn_2 \leftrightarrow$ diamond $A4 + AuSn_4$ mediated by the formation of AuSn intermetallic phases. We did not observe the presence of these intermetallic phases within the nanowire (EDX analysis in Supplementary Fig. 14 shows no traceable amount of Au or Ag in nanowires) or at the hetero-interface through EDX and EELS observations, effectively ruling out equilibrium growth. The droplet morphology/volume also did not change significantly on varying the amount of Sn in the injecting solutions, as would have been expected for near-equilibrium growth[49]. Also, the amount of Sn incorporated in the nanowires increased on using AuAg catalysts which promotes faster Ge nanowire growth kinetics than Au seeds[27], thus underscoring the role of kinetic factors in the non-equilibrium incorporation of Sn in Ge.

Non-equilibrium induction of Sn impurity in the Ge host is justified through diffusionless solute trapping at a finite growth velocity of the crystals. Solute trapping is a process of solute redistribution at the interface resulting in an increase of chemical potential and deviation of partition coefficient[50]. Local chemical equilibrium at the alloy solidification front at the liquid (seed)–solid (nanowire) interface is relaxed due to a large interface velocity resulting in kinetic interface undercooling. At a high solidification rate at the catalyst-nanowire interface impurity adatoms can be trapped on the high energy sites of the crystal lattice, leading to the formation of metastable solids; for example $Ge_{1-x}Sn_x$ with non-equilibrium Sn content, at the nanowire growth front. The kinetic incorporation of Sn is aided by the following factors: (i) Sn diffusion in Ge at the growth conditions is negligible, (ii) the epitaxial mismatch between Sn and Ge results in elastic strains at and near the catalyst-nanowire interface and (iii) the lack of truncating side facets at the catalyst-particle interface. Assuming that growth of the nanowire is layer by layer, the step flow kinetics can result in solute trapping of Sn from the Sn-rich droplet. The deviation of chemical equilibrium at the interface is influenced by the kinetic parameter, that is, interfacial diffusion speed in this case. For bulk metal-semiconductor systems impurity trapping at the liquid–solid interface is highly probable at a very high interface velocity in the order of m sec^{-1} (ref. 51). However, in the 1D $Ge_{1-x}Sn_x$ nanoscale systems the growth rate at the liquid–solid interface is only of the order of nm sec^{-1}. The growth velocity of $Ge_{1-x}Sn_x$ nanowires is much lower (~0.5–1 nm sec^{-1}) than the growth velocity required for kinetic driven solute trapping. However, for our particular system a much higher Sn concentration in the catalyst seed (>90 at.%) than the impurity concentration in a typical bulk solidification process and a continuous Sn flux throughout nanowire growth could account for the high Sn incorporation[34]. In the nanoscale system, where the crystal growth proceeds with the formation of steps at the interface, impurity atoms remains frozen at the step edges on the formation of new row of atoms[25]. Hence, impurity incorporation during nanowire growth depends on the step velocity rather than on the interface velocity. With a high step velocity, the time required for local impurity exchange at the catalyst-nanowire interface decreases thus the rate of solute trapping increases in the nanowire. We delegate a detailed model to a later study, but it is important to note that solute trapping has been implicated in the catalyst incorporation of Al-catalysed growth of Si nanowires[25]. A key difference, although, is that the equilibrium Al-solubility in the Al-Si droplet is much smaller (<5%), unlike the Sn-rich droplet that catalyses the growth of $Ge_{1-x}Sn_x$ nanowires. The high Sn concentration around the growing steps is further aided by low Sn diffusivity within the nanowire and along its sidewalls, and the kinetic pathway favors Sn incorporation into the newly forming layer at the expense of elastic strains relative to the pristine Ge crystal. The elastic strains can be effectively accommodated by the Sn-rich droplet. Finally, at equilibrium the catalyst-nanowire interface also involves truncating side facets[52]. These truncating facets are absent in $Ge_{1-x}Sn_x$ nanowires and the interface is fully faceted which is indirect confirmation of the elastic strains due to non-equilibrium Sn incorporation. As a result, the Sn incorporation is uniform through the nanowire, as opposed to being localized at the core or within a surface shell.

Induction of Sn in the alloy nanowires is further encouraged with a 2 h annealing at 230 °C, during the cooling down of nanowires (Supplementary Fig. 6). The choice of the step cool-down process and temperature was driven by the existence of a small Sn precipitation window at the lowest eutectic, near the

Sn-rich side of the binary bulk Au-Sn phase diagram[32], at around 215–230 °C. The position and width of the Sn precipitating window in the AuAg-Sn pseudo binary phase diagram was assumed from the Au-Sn and Ag-Sn phase diagrams. This small window encourages further precipitation of Sn from non-equilibrium Sn-rich eutectic Au-Sn and AuAg-Sn catalysts during the annealing process. A deposition and dissolution based process could be responsible for the increase in the Sn amount in the alloy nanowire, where a Sn precipitation from the super-saturated catalyst drop is encouraged at 230 °C. Precipitated tin from the supersaturated catalyst gets further dissolved into the Ge nanowire host lattice at 230 °C due to the eutectic solubility. Metastability and continuous dissolution of Sn in the Ge host is expected at the eutectic temperature. A very large amount of Sn could be dissolved in the Ge lattice in the metastable state as projected in the Ge-Sn phase diagram. So the coincidence of the Sn precipitation from the seed droplet at 230 °C and dissolution of this Sn in the Ge nanowires at the eutectic temperature (at 230 °C) encourages large homogeneous Sn influx into the nanowire. Sn diffusion in Ge at 230 °C is negligible. So a diffusion mediated incorporation process would have a large concentration of Sn near the seed-nanowire interface, with a continuous drop in Sn concentration along the nanowire length. Compositional analysis of 9.2 at.% sample (grown following the step cool-down process) does not demonstrate this trend but shows similar distribution of Sn along the length of nanowires (Supplementary Fig. 8). To confirm the participation of the particular step-down temperature of 230 °C in large Sn incorporation, nanowires were annealed during cool-down at four different temperatures of 210, 220, 230 and 250 °C. However, only those subsequently annealed at 230 °C displayed an increased Sn incorporation, whereas no or negligible increase was observed for other temperatures (Supplementary Fig. 15). So in the step cool-down process, the coincidence of the lowest eutectic in AuAg-Sn system and Ge-Sn system assists to increase the amount of Sn (~2.5 at.%) further beyond the capability of kinetic trapping. Undercooling and the shift in the liquidus from their bulk counterpart for nanoscale Ge-Sn systems may also be expected for Au-Sn and Ge-Sn binary phase diagram.

In summary, we have successfully fabricated a stable direct bandgap 1D $Ge_{1-x}Sn_x$ nanosystem, usually metastable under thermodynamic equilibrium, compatible with existing Si electronics platforms. Kinetic driven solute trapping model in a catalytic bottom-up growth with noble metal catalysts facilitates the dissolution of high amounts of Sn into the alloy nanowires. The sparse spatial distribution of Sn in the nanowires and the formation of GeSn alloys in the nanowire cores were confirmed through atomic resolution EELS mapping and Raman spectroscopy. The use of innovative AuAg alloy catalysts, which enhances nanowire growth kinetics, facilitated larger incorporation of Sn into the $Ge_{1-x}Sn_x$ lattice than pure Au catalyst. Calcination of the nanowires near Ge-Sn eutectic temperature during the cool-down further aided Sn precipitation and dissolution to achieve direct bandgap $Ge_{1-x}Sn_x$ alloy ($x = 0.092$) nanowires. Further manipulation of growth temperature, choice of precursors and catalysts could lead towards $Ge_{1-x}Sn_x$ nanowires with even a higher Sn content. Three-phase bottom-up growth is a feasible way to incorporate large amount of impurities and dopants in semiconductor nanowires. Demonstration of colossal incorporation of foreign atoms in the host semiconductor lattice allows new or added functionalities (strain engineering, controlled defect formation, band-structure modulation and so on) in the existing semiconductor architecture. The demonstrated VLS process with its innovative catalyst and precursors contributes towards the ongoing research on dopant/impurity incorporation in nanowires by directly demonstrating kinetics-dependent tin incorporation in Ge nanowire. The protocol demonstrated here is general and could be applied to other doped or alloy nanowire systems (for example III–V nanowires, ternary group IV alloy nanowires and so on) to create new and innovative nanomaterials for novel physics and devices. This paper also addresses towards the ongoing demand for a nanoscale material for group IV photonics. The demonstration of direct bandgap silicon compatible nanowires will instigate a lot of initiative on fundamental research on band-structure engineering of binary and ternary group IV alloy nanowires and implementation of these nanoscale materials in photonic and electronic devices. The great success of III–V quaternary semiconductors in decoupling strain and band-structure effects suggests that ternary compounds should have a similar impact in the group IV arena. Direct bandgap $Ge_{1-x}Sn_x$ nanowires fabricated through a cheap and feasible bottom-up technique opens up unlimited possibilities in group IV photonics, nanoelectronics and optoelectronics.

Methods

Nanowire growth. Continuous-flow reactions for nanowire growth were carried out in a toluene medium using a liquid-injection CVD technique. Metal nanoparticles were spin-coated onto a Si (001) substrate and loaded into a stainless steel micro reactor cell, connected to metal tubing. The catalyst nanoparticle concentration in each case was fixed at 40 μM cm^{-3}. Solutions of diphenylgermane and allyltributylstannane in anhydrous toluene were prepared in an N_2 glove box with a typical Ge precursor concentration of 10 μM ml^{-1} and varying Sn concentrations. The concentration of diphenylgermane in toluene was fixed at 10 μM ml^{-1}, whereas tin precursor concentrations were varied from 1–2 μM ml^{-1} for the incorporation of different amounts of Sn in the $Ge_{1-x}Sn_x$ nanowires. A precursor solution was loaded into a Hamilton sample-lock syringe inside a nitrogen-filled glovebox. Before injection, the coated Si substrate was annealed for 15 min at 440 °C under a flowing H_2/Ar atmosphere inside a tube furnace. The precursor solution was then injected into the metal reaction cell using a high-pressure syringe pump at a rate of 0.025 ml min^{-1}. A H_2/Ar flow rate of 0.5 ml min^{-1} was maintained during the entire growth period. A typical nanowire growth time was 2 h. An additional annealing step was also introduced during the cool-down process where the substrate was kept at 230 °C for 2 h under a H_2/Ar flowing atmosphere. The reaction cell was allowed to cool to room temperature and disassembled to access the growth substrate. Nanowires were washed with dry toluene and dried under N_2 flow for further characterization.

Characterization. Bottom-up grown $Ge_{1-x}Sn_x$ nanowires were imaged on an FEI Helios NanoLab 600i scanning electron microscope. All EDX measurements were recorded in high-angle annular dark-field mode in the FEI Helios NanoLab 600i operating at 20 kV and 1.4 nA with an attached Oxford X-Max 80 detector. Error in the EDX measurements indicates standard deviation in EDX measurements measured over 50 nanowires. TEM analysis was done in a JEOL JEM-2100 operating at 200 kV in bright-field condition for imaging. High-resolution STEM imaging and EELS mapping was done using Nion UltraSTEM100 microscope, operated at 100 kV. Probe-forming optics were adjusted to deliver a 0.9 Å probe, with 120 pA beam current and 31 mrad convergence semi-angle. EELS data was acquired on a Gatan Enfina spectrometer, at 1 eV per channel to capture both the Sn and Ge edges simultaneously. As a result, the effective energy resolution was limited to 2.5 eV by the detector point spread function (~3 pixels), even though the cold field emission gun of the instrument had a native energy width of 0.35 eV in the operating conditions. Raman scattering measurements were performed in a backscattering geometry using a micro-Raman setup consisting of a spectrometer (model LabRAM HR, Jobin Yvon) and a Peltier-cooled charge-coupled device detector. An air cooled He-Ne laser of wavelength 633 nm was used as an excitation source. The photoluminescence measurements were performed using a confocal configuration. Samples were cooled to 7 K using a Helium cryostat. A pulsed titanium-sapphire 800 nm laser was used as an excitation source. Laser frequency was 76 MHz and pulse width was 300 fs. The laser beam was focused down to a 50 μm spot and the power was measured to be 500 mW. For temperature- and power-dependent experiments, the structures were encased in a liquid He and liquid nitrogen cryostat, respectively equipped with KBr window and cooled to desired temperature. The photoluminescence emission was collected by a monochromator and then sent to a thermoelectrically cooled, photoconductive, extended-range InGaAs detector, sensitive in the mid-IR spectral range from 1.2 to 2.6 μm and facilitated by CaF2 optical components. Lock-in and chopper was used as a standard noise-cancellation tool.

References

1. Hu, C. Green transistor as a solution to the IC power crisis. *Proc. 9th Int. Conf. Solid-State Integrated-Circuit Technol* 16–20 (IEEE, 2008).
2. Brubaker, M. D. *et al.* On-chip optical interconnects made with gallium nitride nanowires. *Nano Lett.* **13,** 374–377 (2013).
3. Soref, R. Mid-infrared photonics in silicon and germanium. *Nat. Photon.* **4,** 495–497 (2010).
4. Wirths, S. *et al.* Lasing in direct-bandgap GeSn alloy grown on Si. *Nat. Photon.* **9,** 88–92 (2015).
5. Zhang, P. H., Crespi, V. H., Chang, E., Louie, S. G. & Cohen, M. L. Computational design of direct-bandgap semiconductors that lattice-match silicon. *Nature* **409,** 69–71 (2001).
6. Moontragoon, P., Soref, R. A. & Ikonic, Z. The direct and indirect bandgaps of unstrained $Si_xGe_{1-x-y}Sn_y$ and their photonic device applications. *J. Appl. Phys.* **112,** 073106 (2012).
7. Sau, J. D. & Cohen, M. L. Possibility of increased mobility in Ge-Sn alloy system. *Phys. Rev. B* **75,** 045208 (2007).
8. Jenkins, D. W. & Dow, J. D. Electronic-properties of metastable Ge_xSn_{1-x} alloys. *Phys. Rev. B* **36,** 7994–8000 (1987).
9. He, G. & Atwater, H. A. Interband transitions in Sn_xGe_{1-x} alloys. *Phys. Rev. Lett.* **79,** 1937–1940 (1997).
10. Yin, W.-J., Gong, X.-G. & Wei, S.-H. Origin of the unusually large band-gap bowing and the breakdown of the band-edge distribution rule in the Sn_xGe_{1-x} alloys. *Phys. Rev. B* **78,** 161203 (2008).
11. Chen, R. *et al.* Increased photoluminescence of strain-reduced, high-Sn composition $Ge_{1-x}Sn_x$ alloys grown by molecular beam epitaxy. *Appl. Phys. Lett.* **99,** 181125 (2011).
12. Moontragoon, P., Ikonic, Z. & Harrison, P. Band structure calculations of Si-Ge-Sn alloys: achieving direct band gap materials. *Semicond. Sci. Technol.* **22,** 742–748 (2007).
13. Gupta, S., Magyari-Koepe, B., Nishi, Y. & Saraswat, K. C. Achieving direct band gap in germanium through integration of Sn alloying and external strain. *J. Appl. Phys.* **113,** 073707 (2013).
14. Kouvetakis, J., Menendez, J. & Chizmeshya, A. V. G. Tin-based group IV semiconductors: New platforms for opto- and microelectronics on silicon. *Annu. Rev. Mater. Res.* **36,** 497–554 (2006).
15. Johnson, K. A. & Ashcroft, N. W. Electronic structure of ordered silicon alloys: direct-gap systems. *Phys. Rev. B* **54,** 14480–14486 (1996).
16. Bauer, M. R. *et al.* Tunable band structure in diamond-cubic tin-germanium alloys grown on silicon substrates. *Solid State Commun.* **127,** 355–359 (2003).
17. Bauer, M. *et al.* Ge-Sn semiconductors for band-gap and lattice engineering. *Appl. Phys. Lett.* **81,** 2992–2994 (2002).
18. Bauer, M. R. *et al.* SnGe superstructure materials for Si-based infrared optoelectronics. *Appl. Phys. Lett.* **83,** 3489–3491 (2003).
19. Bauer, M. *et al.* Synthesis of ternary SiGeSn semiconductors on Si(100) via Sn_xGe_{1-x} buffer layers. *Appl. Phys. Lett.* **83,** 2163–2165 (2003).
20. Ragan, R., Ahn, C. C. & Atwater, H. A. Nonlithographic epitaxial Sn_xGe_{1-x} dense nanowire arrays grown on Ge(001). *Appl. Phys. Lett.* **82,** 3439–3441 (2003).
21. Gupta, S. *et al.* Highly selective dry etching of germanium over germanium-tin ($Ge_{1-x}Sn_x$): a novel route for $Ge_{1-x}Sn_x$ nanostructure fabrication. *Nano Lett.* **13,** 3783–3790 (2013).
22. Mullane, E., Kennedy, T., Geaney, H., Dickinson, C. & Ryan, K. M. Synthesis of tin catalyzed silicon and germanium nanowires in a solvent-vapor system and optimization of the seed/nanowire interface for dual lithium cycling. *Chem. Mater.* **25,** 1816–1822 (2013).
23. Barth, S., Seifner, M. S. & Bernardi, J. Microwave-assisted solution-liquid-solid growth of $Ge_{1-x}Sn_x$ nanowires with high tin content. *Chem. Commun.* **51,** 12282–12285 (2015).
24. Chen, W. *et al.* Incorporation and redistribution of impurities into silicon nanowires during metal-particle-assisted growth. *Nat. Commun.* **5,** 4134 (2014).
25. Moutanabbir, O. *et al.* Colossal injection of catalyst atoms into silicon nanowires. *Nature* **496,** 78–82 (2013).
26. Biswas, S., Singha, A., Morris, M. A. & Holmes, J. D. Inherent control of growth, morphology, and defect formation in germanium nanowires. *Nano Lett.* **12,** 5654–5663 (2012).
27. Biswas, S., O'Regan, C., Petkov, N., Morris, M. A. & Holmes, J. D. Manipulating the growth kinetics of vapor-liquid-solid propagated ge nanowires. *Nano Lett.* **13,** 4044–4052 (2013).
28. Biswas, S., O'Regan, C., Morris, M. A. & Holmes, J. D. *In-situ* observations of nanoscale effects in germanium nanowire growth with ternary eutectic alloys. *Small.* **11,** 103–111 (2015).
29. He, S. T., Xie, S. S., Yao, J. N., Gao, H. J. & Pang, S. J. Self-assembled two-dimensional superlattice of Au-Ag alloy nanocrystals. *Appl. Phys. Lett.* **81,** 150–152 (2002).
30. Prince, A., Liang, P. & Fabrichnaya, O. *Springer Materials The Landolt-Bornstein New Series.* IV/IIB (Springer, 2012).
31. Kasper, E., Kittler, M., Oehme, M. & Arguirov, T. Germanium tin: silicon photonics toward the mid-infrared. *Photon. Res.* **1,** 69 (2013).
32. Ciulik, J. & Notis, M. R. The Au-Sn phase-diagram. *J. Alloy. Comp.* **191,** 71–78 (1993).
33. Hemesath, E. R. *et al.* Catalyst incorporation at defects during nanowire growth. *Nano Lett.* **12,** 167–171 (2012).
34. Allen, J. E. *et al.* High-resolution detection of Au catalyst atoms in Si nanowires. *Nat. Nanotechnol.* **3,** 168–173 (2008).
35. Caroff, P. *et al.* Controlled polytypic and twin-plane superlattices in III-V nanowires. *Nat. Nanotechnol.* **4,** 50–55 (2009).
36. Bell, D. C. *et al.* Imaging and analysis of nanowires. *Microsc. Res. Tech.* **64,** 373–389 (2004).
37. Jiang, X. *et al.* Rational growth of branched nanowire heterostructures with synthetically encoded properties and function. *Proc. Natl Acad. Sci. USA* **108,** 12212–12216 (2011).
38. Li, S. F., Bauer, M. R., Menendez, J. & Kouvetakis, J. Scaling law for the compositional dependence of Raman frequencies in SnGe and GeSi alloys. *Appl. Phys. Lett.* **84,** 867–869 (2004).
39. Lin, H., Chen, R., Huo, Y., Kamins, T. I. & Harris, J. S. Raman study of strained $Ge_{1-x}Sn_x$ alloys. *Appl. Phys. Lett.* **98,** 261917 (2011).
40. Ghetmiri, S. A. *et al.* Direct-bandgap GeSn grown on silicon with 2230 nm photoluminescence. *Appl. Phys. Lett.* **105,** 151109 (2014).
41. Senaratne, C. L., Gallagher, J. D., Aoki, T., Kouvetakis, J. & Menendez, J. Advances in light emission from group-IV alloys via lattice engineering and n-type doping based on custom-designed chemistries. *Chem. Mater.* **26,** 6033–6041 (2014).
42. Stange, D. *et al.* Optical transitions in direct-bandgap $Ge_{1-x}Sn_x$ alloys. *ACS Photon.* **2,** 1539–1545 (2015).
43. Lezama, I. G. *et al.* Indirect-to-direct band gap crossover in few-layer $MoTe_2$. *Nano. Lett.* **15,** 2336–2342 (2015).
44. Peng, X., Wei, Q. & Copple, A. Strain-engineered direct-indirect band gap transition and its mechanism in two-dimensional phosphorene. *Phys. Rev. B* **90,** 085402 (2015).
45. Harris, C. & O'Reilly, E. P. Nature of the band gap of silicon and germanium nanowires. *Physica E Low Dimens. Syst. Nanostruct.* **32,** 341–345 (2006).
46. Sutter, E. A. & Sutter, P. W. Size-dependent phase diagram of nanoscale alloy drops used in vapor-liquid-solid growth of semiconductor nanowires. *ACS Nano* **4,** 4943–4947 (2010).
47. Schwalbach, E. J. & Voorhees, P. W. Phase equilibrium and nucleation in VLS-grown nanowires. *Nano Lett.* **8,** 3739–3745 (2008).
48. Jin, S., Valenza, F., Novakovic, R. & Leinenbach, C. Wetting behavior of ternary Au-Ge-X (X = Sb, Sn) alloys on Cu and Ni. *J. Electron. Mater.* **42,** 1024–1032 (2013).
49. Lim, S. K., Crawford, S., Haberfehlner, G. & Gradečak, S. Controlled modulation of diameter and composition along individual III-V nitride nanowires. *Nano. Lett.* **13,** 331–336 (2012).
50. Galenko, P. Solute trapping and diffusionless solidification in a binary system. *Phys. Rev. E* **76,** 031606 (2007).
51. Baker, J. & Gahn, J. Solute trapping by rapid solidification. *Acta Metall.* **17,** 575–578 (1969).
52. Wang, H., Zepeda-Ruiz, L. A., Gilmer, G. H. & Upmanyu, M. Atomistics of vapour-liquid-solid nanowire growth. *Nat. Commun.* **4,** 1956–1956 (2013).

Acknowledgements

We acknowledge financial support from Science Foundation Ireland (SFI Grants: 14/IA/2513 and 12/RC/2278) and the SFI International Strategic Co-operation Award (ISCA) India-Ireland programme. D.S. and T.O. acknowledge SFI US-Ireland R&D Partnership Program Grant No. SFI/14/US/I3057.

Author contributions

S.B. and J.D.H. designed the experiments. S.B. and J.D. conducted the experiments. S.B., M.U., M.A.M. and J.D.H. contributed to manuscript writing. Q.R. performed the STEM and EELS experiments. A.S. and D.M. participated in Raman measurements and analysis, while T.O. and D.S. measured photoluminescence spectroscopy and analysed the PL data. All authors participated in discussion to review the manuscript.

Additional information

Competing financial interests: The authors declare no competing financial interests.

19

Thermoelectric effect and its dependence on molecular length and sequence in single DNA molecules

Yueqi Li[1,2], Limin Xiang[1,2], Julio L. Palma[1,2], Yoshihiro Asai[3] & Nongjian Tao[1,4]

Studying the thermoelectric effect in DNA is important for unravelling charge transport mechanisms and for developing relevant applications of DNA molecules. Here we report a study of the thermoelectric effect in single DNA molecules. By varying the molecular length and sequence, we tune the charge transport in DNA to either a hopping- or tunnelling-dominated regimes. The thermoelectric effect is small and insensitive to the molecular length in the hopping regime. In contrast, the thermoelectric effect is large and sensitive to the length in the tunnelling regime. These findings indicate that one may control the thermoelectric effect in DNA by varying its sequence and length. We describe the experimental results in terms of hopping and tunnelling charge transport models.

[1]Center for Bioelectronics and Biosensors, Biodesign Institute, Arizona State University, Tempe, Arizona 85287-5801, USA. [2]School of Molecular Sciences, Arizona State University, Tempe, Arizona 85287-5801, USA. [3]Research Center for Computational Design of Advanced Functional Materials (CD-FMat), National Institute of Advanced Industrial Science and Technology (AIST), Central 2, Umezono 1-1-1, Tsukuba, Ibaraki 305-8568, Japan. [4]School of Chemistry and Chemical Engineering, Nanjing University, Nanjing, Jiangsu 210093 China. Correspondence and requests for materials should be addressed to N.T. (email: Nongjian.Tao@asu.edu).

The thermoelectric effect is a basic property of materials and has found applications in energy conversion, temperature sensing and regulation. This important effect in single molecules is expected to be distinctly different from that in bulk materials[1]. In addition to potential applications, studying the thermoelectric effect in single molecules helps assess the molecular orbital level alignment, understand the energy conversion mechanism associated with charge transport and determine whether electrons or holes are responsible for the charge transport. The work by Reddy *et al.*[2] has stimulated many efforts in investigating the thermoelectric effect in molecules[3–6]. However, to date, research has been limited to the regime where the Fermi levels of the electrodes are in the highest occupied molecular orbital (HOMO)–lowest unoccupied molecular orbital (LUMO) gaps of the molecules, and electrons or holes transport through molecules via coherent tunnelling, a quantum mechanical process with a characteristic exponential dependence of the molecular resistance with length.

Here we report a study of the thermoelectric effect in double-stranded DNA (dsDNA) molecules. By selecting different DNA sequences, we study the thermoelectric effect not only in the tunnelling regime but also in the hopping regime, in which electrons or holes hop along a molecule from one end to another sequentially via multiple steps, leading to a linear dependence of the molecular resistance with length. The work demonstrates that DNA thermoelectricity can be tuned by its sequences and length, which provides new insights into the thermoelectric effect in single molecules, and basic knowledge for potential applications with programmable DNA nanostructures, a field that has been advancing rapidly in recent years[7].

Results

dsDNA sequences. To study the thermoelectric effect in the hopping regime, we selected DNA molecules with sequences of 5′-A(CG)$_n$T-3′ ($n = 3$, 4, 5, 6 and 7; Fig. 1b), where A, C, G and T denote the four DNA bases, adenine, cytosine, guanine and thymine, respectively. Note that this sequence notation shows only one strand of the dsDNA, and the second strand has the complementary sequence. The complete sequences of all the dsDNA studied in this work are listed in Table 1. Previous studies have shown that charge transport in GC sequences is dominated by hopping of holes along the molecules with G as hopping sites[8–13]. To investigate the thermoelectric effect in the tunnelling regime, we studied DNA sequences, 5′-ACGC(**AT**)$_m$ GCGT-3′, where $m = 1$, 2, 3 and 4, and 5′-ACGC(**AT**)$_{m-1}$ AGCGT-3′, where $m = 1$, 2 and 3 (Fig. 1b; Table 1). The insertion of a block of AT bases in the middle of the GC sequences introduces a tunnelling barrier into the DNA molecules, as shown in literature[14–19], allowing us to investigate the thermoelectric effect in the tunnelling regime. To measure the DNA conductance and thermal electric effect, we modified T base at the 3′ end with an amino group, which binds to gold electrodes to establish electrical contact between the electrodes and the molecule[20].

Conductance. We measured the DNA conductance and thermoelectric effect with a temperature controlled scanning tunnelling microscopy (STM) set-up shown in Fig. 1a, following a procedure described in detail elsewhere[21] (see also the Methods section). The experiment generated conductance versus STM tip–substrate distance traces, showing plateaus associated with the formation and breakdown of single-molecule junctions (marked with arrows in the red traces in Fig. 1c). As a control experiment, we performed the experiment without DNA and observed no characteristic plateaus in the conductance traces (black traces). By collecting thousands of conductance traces, we constructed a conductance histogram (Fig. 1d), which shows a broad peak, where the peak position measures the average conductance of a single dsDNA molecule. The width of the histogram peak is not due to experimental error, but rather it reflects the intrinsic variability and distribution in the molecule–electrode binding geometry[22]. To determine the experimental error, we repeated the above experiment more than three times, each using a new sample to generate thousands of conductance traces, for every DNA sequence, and used the s.d. in the conductance values as the error bar.

We characterized the DNA samples by performing non-denaturing polyacrylamide gel electrophoresis, circular dichroism spectroscopy and melting temperature experiments in phosphate buffer (Supplementary Figs 1, 2 and 3; Supplementary Methods 1, 2 and 3). These experiments confirmed that each DNA sequence studied here was in B-form double-helical structure in phosphate buffer. To ensure the DNA molecule retained its double-helical structure in the experiments (~20% relative humidity), we compared the conductance of DNA measured in humidified air with that measured in phosphate buffer and found similar values within the experimental uncertainty (Supplementary Figs 4 and 5). Another parameter that characterizes molecular junction stability is the stretching length, which is the average distance that one can stretch the junction before it breaks. We also found similar stretching lengths measured in phosphate buffer and in humidified air (Supplementary Fig. 6; Supplementary Discussion 1). These facts led us to conclude that the DNA molecules in humidified air retained its double-helical structure, possibly due to a thin absorbed water layer on the substrate surface. This conclusion is further supported by atomic force microscopy[23], STM[24], mechanical break junction[25] and polarization modulation infrared[26] measurements of dsDNA on a surface in the air.

The conductance (or resistance) of DNA depends on both the molecular length and sequence. For A(CG)$_n$T ($n = 3$–7), the resistance increases linearly with the molecular length (Fig. 2a), which is expected for hopping transport[20,27,28]. According to the hopping model[29,30], holes hop along a DNA molecule from one end to another, where the individual G bases act as hopping sites such as stepping stones. Consequently, the total resistance is proportional to the number of hopping sites (G bases), and thus the length of the molecule (see more discussions in Supplementary Fig. 7 and Supplementary Discussion 2). When DNA is connected to two electrodes (tip and substrate), an additional contribution to the resistance arises from the two contacts. The total resistance of the DNA is given by[29,30]

$$R = R_c + R_h = R_c + (N-1)R_{G-G}, \qquad (1)$$

where R_c is contact resistance, R_h is intrinsic DNA resistance, N is number of base pairs and R_{G-G} is resistance of a hopping step (one GC base pair). Table 2 summarizes R_c and R_h of all the A(CG)$_n$T sequences. By fitting the experimental data with equation (1), we found that the contact resistance, $R_c = 0.48 \pm 0.16\,\mathrm{M\Omega}$, and the resistance of one step hopping, $R_{G-G} = 0.28 \pm 0.02\,\mathrm{M\Omega}$ per GC base pair. Note that the contact resistance is small compared with the total resistance for A(CG)$_n$T (see Table 2), indicating efficient electronic coupling between the electrodes and the molecule, and the DNA resistance is dominated by the molecule itself, rather by the contact.

In contrast to A(CG)$_n$T described above, the resistance length dependence of ACGC(**AT**)$_m$GCGT/ACGC(**AT**)$_{m-1}$**A**GCGT is significantly different (Fig. 2b). For $m = 0$, 1 and 2, the resistance increases more rapidly with length (L) and can be best fitted with an exponential function, rather than with a shallow linear dependence found for A(CG)$_n$T. For $m > 2$, however, the resistance becomes weakly length dependent, indicating a

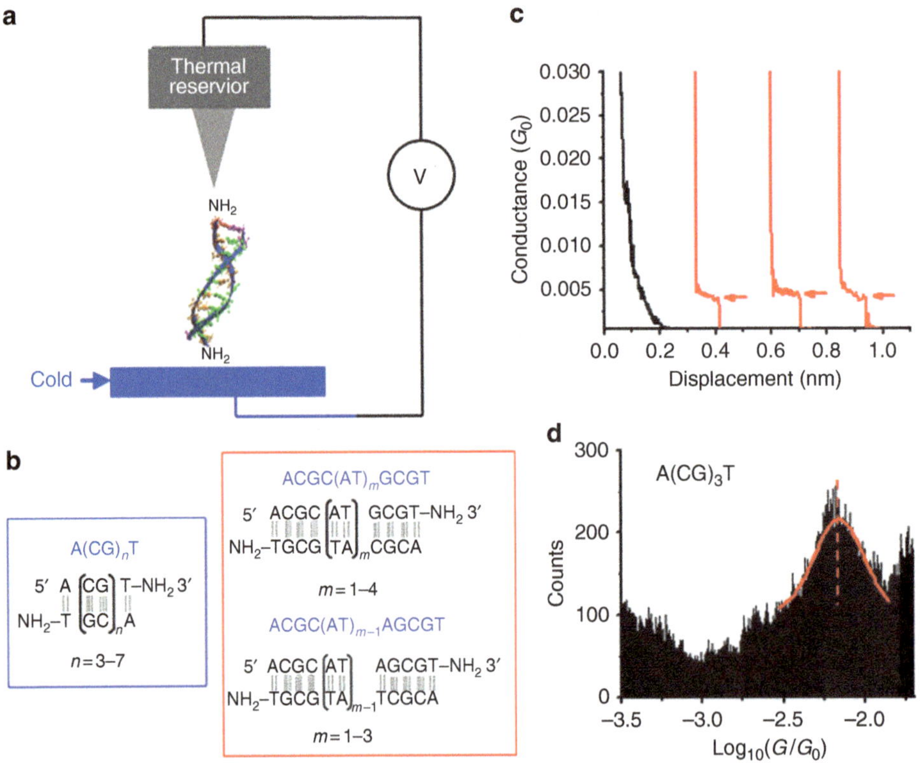

Figure 1 | Thermoelectric and conductance measurements of DNA. (**a**) Conductance and thermoelectric effect of a DNA molecule bridged between STM tip (kept at 295 K) and substrate (cold). (**b**) Two families of DNA sequences are studied in this work, denoted as (1) A(CG)$_n$T, (2) ACGC(**AT**)$_m$GCGT and ACGC(**AT**)$_{m-1}$**A**GCGT (see Table 1 for a full list). (**c**) Conductance traces without (black) and with (red) dsDNA A(CG)$_3$T, where plateaus are marked with red arrows. (**d**) Conductance histogram of A(CG)$_3$T, where the solid red curve is Gaussian fit to the conductance peak. Note: $G_0 = 2e^2/h = 77.48\,\mu$S.

Table 1 | Sequences of dsDNA studied in this work.

Name	Duplex	Abbreviation
A(CG)$_3$T	5′-ACGCGCGT-3′; 3′-TGCGCGCA-5′	A(CG)$_n$T
A(CG)$_4$T	5′-ACGCGCGCGT-3′; 3′-TGCGCGCGCA-5′	A(CG)$_n$T
A(CG)$_5$T	5′-ACGCGCGCGCGT-3′; 3′-TGCGCGCGCGCA-5′	A(CG)$_n$T
A(CG)$_6$T	5′-ACGCGCGCGCGCGT-3′; 3′-TGCGCGCGCGCGCA-5′	A(CG)$_n$T
A(CG)$_7$T	5′-ACGCGCGCGCGCGCGT-3′; 3′-TCGGCGCGCGCGCGCA-5′	A(CG)$_n$T
ACGC**A**GCGT	5′-ACGCAGCGT-3′; 3′-TGCGTCGCA-5′	ACGC(AT)$_{m-1}$AGCGT
ACGC**AT**GCGT	5′-ACGCATGCGT-3′; 3′-TGCGTACGCA-5′	ACGC(AT)$_m$GCGT
ACGC**ATA**GCGT	5′-ACGCATAGCGT-3′; 3′-TGCGTATCGCA-5′	ACGC(AT)$_{m-1}$AGCGT
ACGC(**AT**)$_2$GCGT	5′-ACGCATATGCGT-3′; 3′-TGCGTATACGCA-5′	ACGC(AT)$_m$GCGT
ACGC(**AT**)$_2$**A**GCGT	5′-ACGCATATAGCGT-3′; 3′-TGCGTATATCGCA-5′	ACGC(AT)$_{m-1}$AGCGT
ACGC(**AT**)$_3$GCGT	5′-ACGCATATATGCGT-3′; 3′-TGCGTATATACGCA-5′	ACGC(AT)$_m$GCGT
ACGC(**AT**)$_4$GCGT	5′-ACGCATATATATGCGT-3′; 3′-TGCGTATATATACGCA-5′	ACGC(AT)$_m$GCGT

dsDNA, double-stranded DNA.

transition in the charge transport mechanism at $m = 2$ (or 4 A–T base pairs). This observation agrees with previous photochemical and direct conductance measurements of DNA, which have shown that inserting a short AT block in the middle of a GC sequence introduces a tunnelling barrier, leading to an exponential increase in the DNA resistance with the AT block length[14–19], and inserting a longer AT block results in the transition in the charge transport mechanism from tunnelling to hopping[14,16]. The tunnelling–hopping transition has also been predicted by theoretical calculations[14–19]. See Supplementary Discussion 2 for further discussion.

On the basis of the above results, we modelled the ACGC(**AT**)$_m$GCGT and ACGC(**AT**)$_{m-1}$**A**GCGT resistance as a sum of three contributions, electrode–DNA contact, hopping via GC sequences and tunnelling via the AT block, given by $R = R_c + R_h + R_t$, where R_c is the contact resistance, R_h is the GC hopping resistance and R_t is the AT tunnelling resistance. Since both R_c and R_h are known from the A(CG)$_n$T measurement, the above relation allowed us to determine the resistance of the inserted AT block, which is listed in Table 2. Semi-logarithmic plot of the AT block resistance (R_t) versus the AT block length shows an exponential dependence, following $R_t = R_0\exp(\beta L)$, where R_0 is the contact resistance between the GC and AT block, and β is the decay constant and L is the AT block length (Fig. 2c). Fitting the experimental data with the above exponential relation leads to a β of $2.03 \pm 0.12\,\text{nm}^{-1}$ (Fig. 2c). This decay constant is smaller than $\sim 4\,\text{nm}^{-1}$ reported by Xu *et al.*[15] for thiolated DNA molecules, but is within the

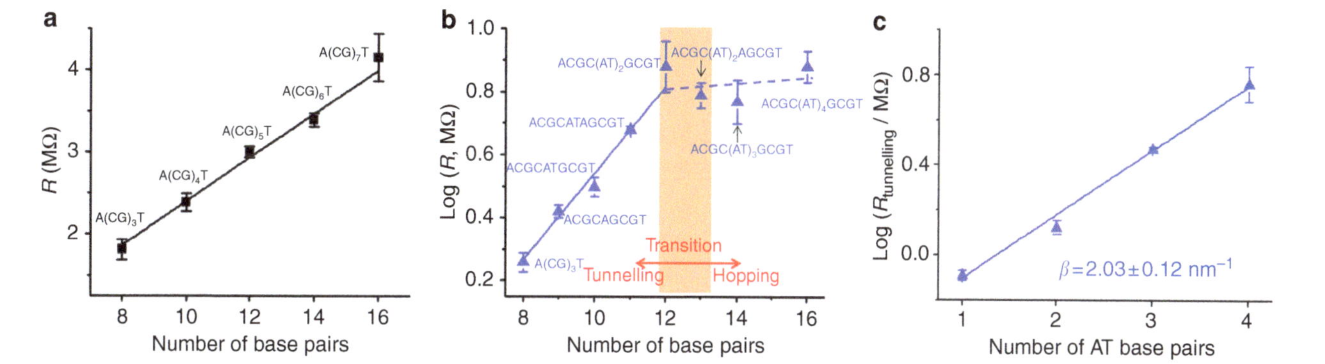

Figure 2 | DNA resistance versus molecular length. (**a**) A(CG)$_n$T ($n=3$-7), (**b**) ACGC(**AT**)$_m$GCGT ($m=0$-4) and ACGC(**AT**)$_{m-1}$**A**GCGT ($m=1$-3) sequences, and (**c**) the AT blocks ((AT)$_m$ and (AT)$_{m-1}$A) in ACGC(AT)$_m$GCGT and ACGC(AT)$_{m-1}$AGCGT sequences. The solid lines in **a**,**b** and **c** are linear fits to the data, and the dashed line in **b** is a guide to eye. The transition from tunnelling to hopping occurs the AT block is longer than 4 base pairs, which is markers with a shaded orange region.

Table 2 | Conductance and Seebeck coefficients of DNA with different sequences.

Sequence	Conductance peak position (log(G/G$_0$))*	Conductance peak width	Resistance (MΩ)	R_c (MΩ)†	R_h (MΩ)†	R_t (MΩ)†	Seebeck coefficient (μV K^{-1})
A(CG)$_3$T	−2.17 ± 0.03	0.20	1.9 ± 0.1	0.48	1.40	NA	1.4 ± 0.2
A(CG)$_4$T	−2.27 ± 0.02	0.18	2.4 ± 0.1	0.48	1.96	NA	0.8 ± 0.2
A(CG)$_5$T	−2.37 ± 0.01	0.17	3.0 ± 0.1	0.48	2.52	NA	1.0 ± 0.2
A(CG)$_6$T	−2.42 ± 0.01	0.18	3.4 ± 0.1	0.48	3.08	NA	0.6 ± 0.2
A(CG)$_7$T	−2.51 ± 0.03	0.20	4.2 ± 0.3	0.48	3.64	NA	0.8 ± 0.2
ACGC**A**GCGT	−2.31 ± 0.02	0.25	2.6 ± 0.1	0.48	1.12	1.0	5.0 ± 0.3
ACGC**AT**GCGT	−2.36 ± 0.03	0.35	3.0 ± 0.2	0.48	1.12	1.4	5.9 ± 0.4
ACGC**ATA**GCGT	−2.57 ± 0.01	0.30	4.8 ± 0.1	0.48	1.12	3.2	6.5 ± 0.8
ACGC(**AT**)$_2$GCGT	−2.76 ± 0.08	0.30	7.4 ± 1.4	0.48	1.12	5.8	7.9 ± 0.5
ACGC(**AT**)$_2$**A**GCGT	−2.68 ± 0.04	0.50	6.1 ± 0.6	0.48	1.12	4.5	4.9 ± 0.2
ACGC(**AT**)$_3$GCGT	−2.65 ± 0.07	0.52	5.8 ± 0.9	0.48	1.12	4.2	2.0 ± 0.2
ACGC(**AT**)$_4$GCGT	−2.77 ± 0.05	0.40	7.6 ± 0.9	0.48	1.12	6.0	2.0 ± 0.3

NA, not applicable.
*G_0 ($=2e^2/h=77.48$ μS) is the conductance quantum.
†R_c, R_h and R_t are the contact resistance, hopping part resistance (CG sequences) and tunnelling part (AT sequences) resistance, respectively.

range reported by various other works[8–13] and consistent with the calculations by Torisi *et al.*[19] and Voityuk *et al.*[31] for dsDNA with alternating AT sequences.

Thermoelectric effect. The thermoelectric effect of a single DNA-electrode junction can be expressed by

$$S_{\text{junction}} = S_{\text{Au}} - \frac{\Delta U_{\text{TE,m}}}{\Delta T} = S_{\text{Au}} - \frac{-\Delta\left(I_{\text{TE,m}}/G_{\text{m}}\right)}{\Delta T}, \qquad (2)$$

where S_{Au} is the Seebeck coefficient of gold ($\sim 2\,\mu\text{V K}^{-1}$ (ref. 32)), $U_{\text{TE,m}}$ is the open-circuit voltage, $I_{\text{TE,m}}$ is the short-circuit current, G_{m} is the conductance of the molecular junction and ΔT is the temperature difference between the STM tip and substrate electrodes. We measured the thermoelectric effect by holding the tip at 295 K while cooling the substrate from 295 to 275 K (ref. 33). The selection of this temperature range ensures that all the dsDNA sequences are structurally stable, and their resistance does not change significantly with temperature[34].

The thermoelectric effect measurement started by detecting plateaus in the conductance traces during the retraction of the STM tip from the substrate (Fig. 3a). Once a plateau was detected, the tip retraction was halted and the bias voltage between the tip and substrate electrodes was swept over ±10 mV to obtain a current (I)–voltage (V) characteristic curve. After recording an I–V curve, the tip was further retracted to a new position, and the I–V measurement was repeated until the molecular junction broke down. Figure 3b shows several I–V curves for ACGC(AT)$_2$GCGT measured at different locations of a conductance plateau, where the colours of curves match the colours of the dots marked on the conductance plateau shown in Fig. 3a. These I–V curves are linear within the bias range.

From the I–V characteristics, we obtained the average Seebeck coefficient of a DNA molecule with the following procedure. First, each I–V curve was fitted with a linear function to obtain G, and then a corresponding I/G–V curve was obtained by dividing I by G. From thousands of I/G–V curves, a two-dimensional I/G–V histogram was constructed for each temperature difference (Fig. 3c,d). When the substrate is held at the same temperature as the tip (room temperature), the centre of the histogram passes through the origin of I/G–V plot (Fig. 3c). However, when the substrate is cooled below the tip temperature, there is an offset in the histogram in the I/G–V plot, due to the thermoelectric effect (Fig. 3d). By extracting the voltage offsets (at zero current) from the I/G–V histograms, we constructed the thermoelectric voltage histograms at different tip–substrate temperature differences (Fig. 3e; Supplementary Figs 8 and 9; Supplementary Table 1; Supplementary Discussion 3). The peak in each thermoelectric voltage histogram was fitted with a Gaussian distribution, from which $U_{\text{TE,m}}$ was determined. Figure 3f shows that $U_{\text{TE,m}}$ depends linearly on ΔT, and the slope was used to determine S_{junction}, the Seebeck coefficient of the dsDNA molecule according to equation (2).

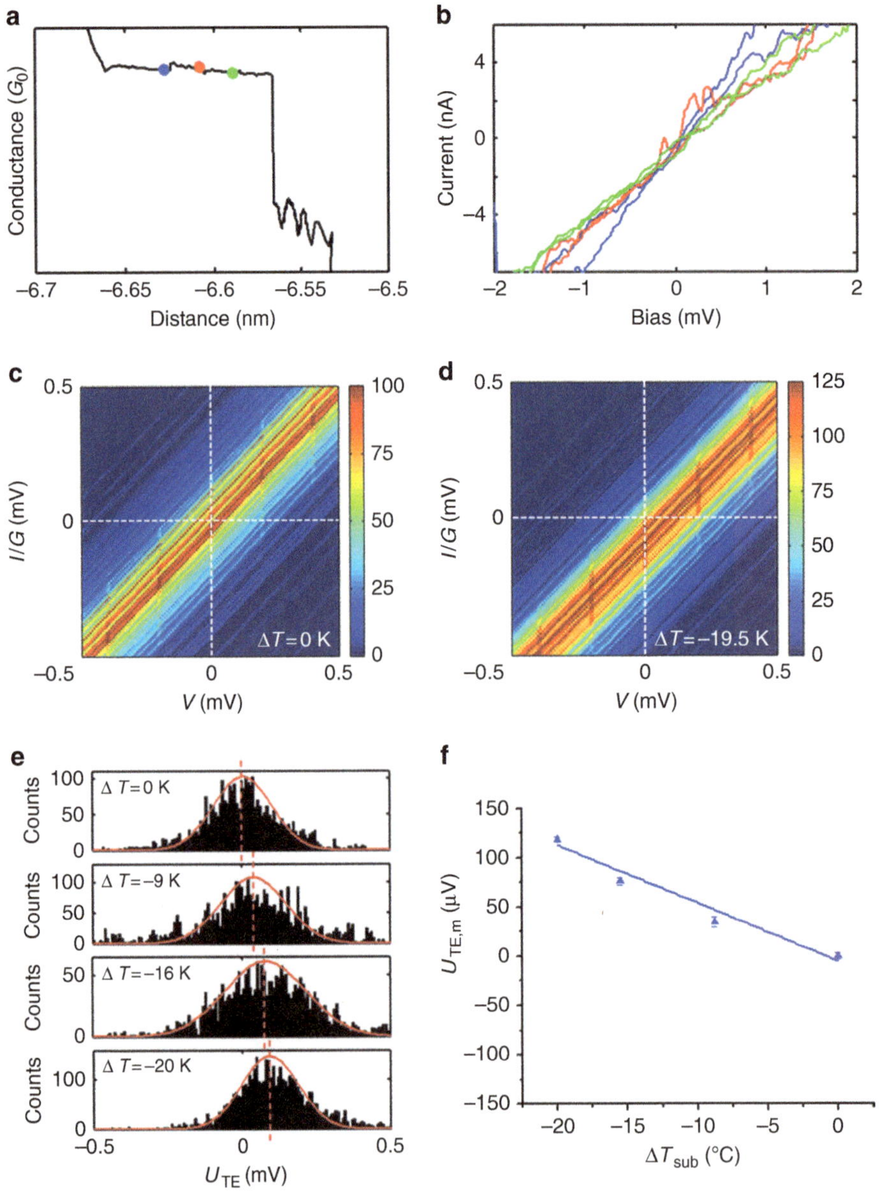

Figure 3 | Thermoelectric measurement of DNA. (**a**) Semi-logarithmic plot of conductance-distance trace (the coloured dots mark, where *I-V* curves are measured). (**b**) Individual *I-V* curves measured at the locations marked with corresponding colour in **a**. (**c**) *I/G-V* histogram at $\Delta T = 0$ K. (**d**) *I/G-V* histogram at $\Delta T = 19.5$ K, showing an offset due to the thermoelectric effect. (**e**) Thermal voltage histogram, where the red curves show Gaussian fits. (**f**) Thermoelectric voltage versus ΔT, where the straight line is linear fit to the data, and error bars are fitting errors. DNA sequence: ACGC(**AT**)$_2$GCGT.

Figure 4 summarizes the Seebeck coefficients of DNA with different lengths and sequences, from which we can draw several important conclusions. First, the Seebeck coefficients are positive for all the DNA sequences in both the tunnelling and hopping regimes studied here. It has been predicted that a positive Seebeck coefficient corresponds to hole-dominated charge transport and a negative Seebeck coefficient signals electron-dominated charge transport[2]. The observation of positive Seebeck coefficients here indicates that holes dominate charge transport in DNA. This is expected because the DNA HOMO level is close to the electrode Fermi level compared with its LUMO level, and is also consistent with the previous experiments[14,16,18,19,35]. However, the prediction of the Seebeck coefficient sign is based on a coherent tunnelling model, which is not necessarily applicable to hopping transport. The data shown in this indicate that this prediction appears to be valid also in the hopping regime. Second, the Seebeck coefficients of DNA in the hopping regime (in A(CG)$_n$T) are small, and weakly depend on the molecular length compared with other organic molecules[36,37]. Last, inserting a short AT block (shorter than 5 AT base pairs) into the middle of A(CG)$_n$T leads to a much greater Seebeck coefficient, and it increases with the AT block length. However, when the AT block is longer than 5 AT base pairs, the Seebeck coefficient drops to the level of A(CG)$_n$T and become insensitive the AT length. This transition coincides with the observed tunnelling–hopping transition near 4–5 AT base pairs, which strongly suggests that the thermoelectric effect is large in the tunnelling regime and small in the hopping regime. We discuss these observations below.

For bulk materials, the Seebeck coefficient can be expressed by the Mott formula[38],

$$S = \frac{\pi^2 k_B^2 T}{3e} \cdot \left(\frac{1}{R} \cdot \frac{\partial R}{\partial E}\right)_{E=E_F}, \tag{3}$$

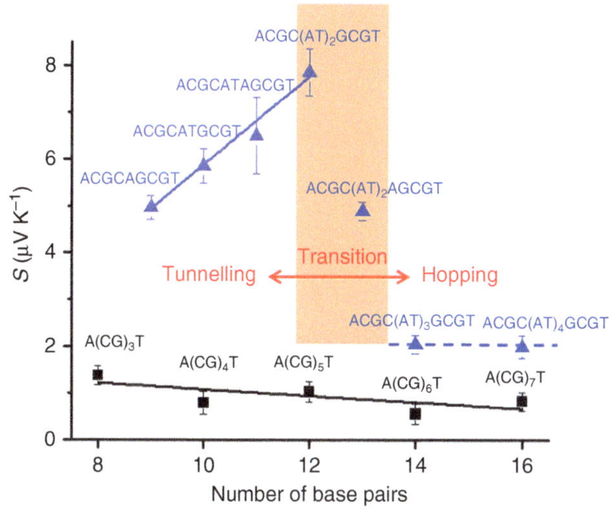

Figure 4 | Seebeck coefficients of DNA with different molecular lengths and sequences. The black squares are the Seebeck coefficients of A(CG)$_n$T (n=1, 2, 3). The blue triangles are the Seebeck coefficients of ACGC(**AT**)$_m$GCGT (m=1-4) and ACGC(**AT**)$_{m-1}$**A**GCGT (m=1−3) sequences, showing a transition when the AT block length is longer than 4 base pairs. The solid and dashed lines are guides to eye.

where R is the energy-dependent resistance function and E is the energy of the carriers. k_B is the Boltzmann constant, T is temperature and e is elementary charge. The Mott formula assumes incoherent transport, which is applicable to hopping transport in DNA. For coherent tunnelling transport, the Seebeck coefficient can be obtained with Landauer's formula, which leads to a similar expression as equation (3) except that R is replaced by the energy-dependent transmission probability function[5,39].

We analyse the experimental thermoelectric effect in the hopping regime with equation (3) by assuming that the resistance function, $R(E)$, can be expressed as a sum of the contact and hopping resistance functions (equation (1)), which is modelled as Fig. 5a. Consequently, the overall Seebeck coefficient (S) consists of contributions from the molecule–electrode contact (S_c) and hopping along the dsDNA sequences (S_h), given by

$$S[\mathrm{A(CG)}_n\mathrm{T}] = \frac{R_c}{R}S_c + \frac{R_h}{R}S_h, \quad (4)$$

where R, R_c and R_h in equation (4) are the total resistance, contact resistance and DNA resistance, respectively. We discuss the validity of the assumption and derivation of equation (4) in Supplementary Discussion 4. Both R_c and R_h are known from the linear fitting to the experimental data shown in Fig. 2a (Table 2).

In the case of ACGC(**AT**)$_m$GCGT and ACGC(**AT**)$_{m-1}$ **A**GCGT sequences, the total resistance R[AT] must include a contribution, R_t, arising from the inserted AT block in the middle of the GC sequence, as shown in Fig. 5b. The overall Seebeck coefficient of the DNA junction must also include the contribution from the inserted AT block, expressed by,

$$S[\mathrm{AT}] = \frac{R_c}{R}S_c + \frac{R_h}{R}S_h + \frac{R_i}{R}S_i, \quad (5)$$

where S_t is the Seebeck coefficient of the inserted AT part.

In equation (4), although R_h is proportional to the molecular length, S_h is independent of the length because it is proportional to the relative change in R_h (a normalized quantity). S_c is also independent of the molecular length. So, we conclude that the molecular length dependence of the overall Seebeck coefficient, S, arises from R in equation (4), which as shown in Fig. 2a is weakly dependent on molecular length. By fitting the measured Seebeck coefficients times R with R_h at different lengths (Fig. 5c), we obtained the slope, $S_h = 0.1 \pm 0.3\,\mu\mathrm{V\,K}^{-1}$, and from the intercept we found that $S_c = 4.8 \pm 1.7\,\mu\mathrm{V\,K}^{-1}$, showing that the contact plays a dominant role in the thermoelectric effect of A(CG)$_n$T. The above model explains the weak length dependence of the observed thermoelectric effect.

For equation (5), we plotted $S \times R$ versus R_t (Table 2) in Fig. 5d for short AT blocks (m=1–2), which follows a linear function. According to equation (5), the slope of the linear relation is S_t, from which we found that $S_t = 8.9 \pm 1.4\,\mu\mathrm{V\,K}^{-1}$. For longer AT blocks ($m>2$), we determined S_t from equation (5) (Table 3). Compared with the short AT blocks, these sequences have smaller Seebeck coefficients ($\sim 2\,\mu\mathrm{V\,K}^{-1}$).

To further understand the tunnelling Seebeck coefficient, S_t, from the AT block, we express the transmission function of the tunnelling barrier with a Lorentzian distribution by[33,36,37,40]

$$\tau(E) = \frac{4\Gamma^2}{4\Gamma^2 + \Delta E_H^2}, \quad (6)$$

where Γ is the width of the HOMO level of the AT block and ΔE_H is the energy difference between the AT block and GC HOMO levels. Using the above transmission function, the contributions of the AT block to the resistance and Seebeck coefficient are expressed as

$$\frac{1}{R_t} = G_0\frac{4\Gamma^2}{4\Gamma^2 + \Delta E_H^2}, \quad (7)$$

$$S_t = \frac{\pi^2 k_B^2 T}{3e}\cdot\frac{2\Delta E_H}{4\Gamma^2 + \Delta E_H^2}. \quad (8)$$

By fitting the experimental data with equations (7) and (8), we obtained Γ and ΔE_H (Table 3), which shows that $\Gamma \ll \Delta E_H$, indicating a weak coupling between the AT block and the GC sequences and electrodes. In this weak-coupling limit, S_t is proportional to $1/\Delta E_H$ (refs 33,41). Because the electronic coupling between alternating AT base pairs in the AT block is weak[31], ΔE_H is expected to be weakly dependent on the AT block length, which explains the constant slope (S_t) in Fig. 5d.

The above hopping and tunnelling models help understand the length dependence of the Seebeck coefficients in DNA. However, to explain that the insertion of a short AT block increases the Seebeck coefficient, a model including sufficient chemical and structural details of dsDNA is needed. We used an extended Su–Schrieffer–Heeger model with the Lennard–Jones potential to describe the dsDNA (see Supplementary Fig. 10 and Supplementary Discussion 5 for details). The theory reproduces sequence dependence, and also the observed large increase in the Seebeck coefficient associated with the insertion of an AT block in the middle of CG sequences (Supplementary Fig. 11). The increase can be attributed to the difference in the site energies between AT and GC base pairs. However, we emphasized that the calculation was performed at 8 K. As such, the theory cannot provide quantitative agreement with the experimental Seebeck coefficient nor it predicts the transition from tunnelling to hopping. Further investigation is thus still needed to quantitatively describe the measured Seebeck coefficients and the tunnelling–hopping transition.

In this paper we, have studied the charge transport and thermoelectric effect in single DNA molecules with different sequences and lengths. The transport mechanism in CG sequences (5′-A(CG)$_n$T-3′, with n=3–7) is hole hopping, as reflected by linear dependence of the resistance on the molecular length. The Seebeck coefficient that describes the thermoelectric effect is small and weakly dependent on the molecular length. Inserting a block of AT base pairs in the middle of a CG sequence

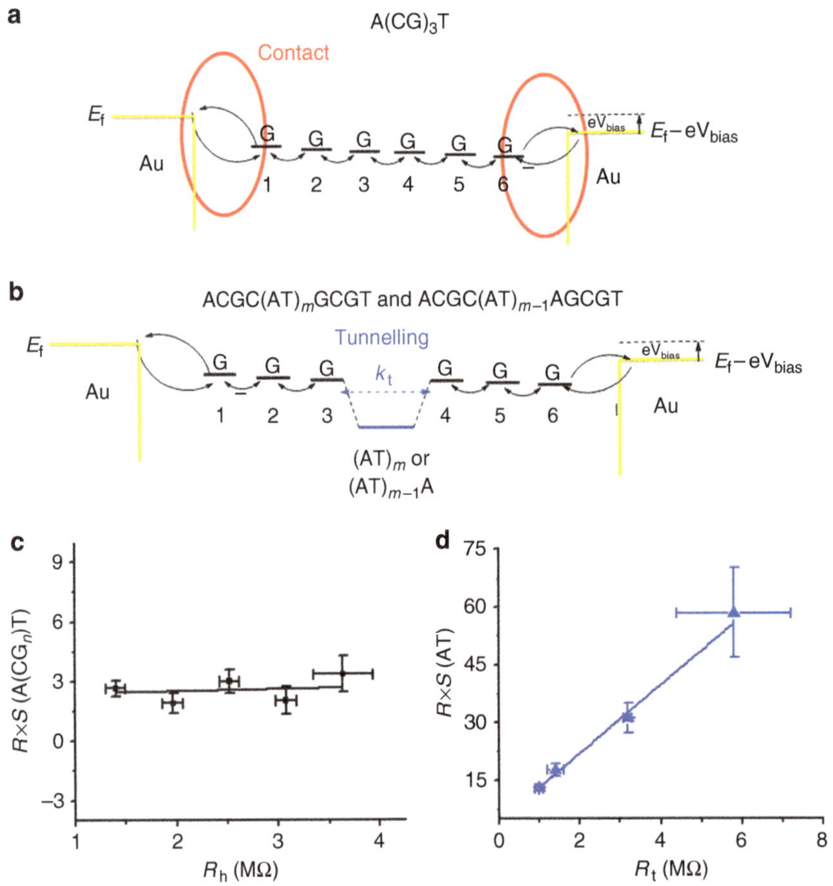

Figure 5 | Thermoelectric effect in DNA in terms of tunnelling and hopping models. (**a**) Charge transport in $A(CG)_nT$, where a hole is injected from left electrode into the first G, then hops along the molecule with each G as a hopping site, and eventually reaches the right electrode. The charge transfer rates at the contacts (red part) are energy dependent, which is a dominant contribution to the Seebeck coefficient. (**b**) Charge transport in ACGC(**AT**)$_m$GCGT and ACGC(**AT**)$_{m-1}$**A**GCGT, where a tunnelling barrier (marked blue) arises from the AT block. (**c**) $R \times S$ versus R_h for $A(CG)_nT$ sequences. (**d**) $R \times S$ versus R_t for ACGC(**AT**)$_m$GCGT and ACGC(**AT**)$_{m-1}$**A**GCGT ($m = 1$ and 2). Solid lines in **c** and **d** are linear fits to the data.

Table 3 | The resistance (R_t) and the Seebeck coefficients (S_t) of the AT blocks, from which the broadening of the HOMO levels (Γ), and the energy difference between the HOMO levels of AT and GC blocks were determined.

Sequences	R_t (MΩ)	S_t (μV K^{-1})	Γ (eV)	ΔE_H (eV)
ACGC**A**GCGT	1.0 ± 0.3	8.9 ± 1.4	0.07 ± 0.02	1.3 ± 0.4
ACGC**AT**GCGT	1.4 ± 0.4	8.9 ± 1.4	0.06 ± 0.02	1.3 ± 0.4
ACGC**ATA**GCGT	3.2 ± 0.3	8.9 ± 1.4	0.05 ± 0.01	1.6 ± 0.3
ACGC(**AT**)$_2$GCGT	5.8 ± 1.6	8.9 ± 1.4	0.03 ± 0.01	1.4 ± 0.5
ACGC(**AT**)$_2$**A**GCGT	4.5 ± 0.8	6.1 ± 1.3	NA	NA
ACGC(**AT**)$_3$GCGT	4.2 ± 1.1	2.2 ± 0.8	NA	NA
ACGC(**AT**)$_4$GCGT	6.0 ± 1.1	2.1 ± 0.7	NA	NA

HOMO, highest occupied molecular orbital; NA, not applicable.

(ACGC(**AT**)$_m$GCGT-3′, $m = 1$–4; and 5′-ACGC(**AT**)$_{m-1}$ **A**GCGT-3′, $m = 1$–3) changes both the charge transport and thermoelectric effect substantially. When the inserted AT block is shorter than 4 base pairs, it acts as a tunnelling barrier with resistance increases exponentially with the block length, and its Seebeck coefficient is large compared with the CG sequences. However, when the inserted AT block is longer than 4 base pairs, the charge transport mechanism changes from tunnelling to hopping, the resistance becomes weakly length dependent and the Seebeck coefficient drops to smaller values. The experimental results have been analysed in terms on tunnelling and hopping models, and compared with an extended Su–Schrieffer–Heeger model with the Lennard–Jones potential. The work demonstrates that the DNA thermoelectricity may be tuned by its length and sequence, and studying the thermoelectric effect provides new insights into charge transport in DNA.

Methods

Preparation of dsDNA SAM. DNA molecules with different sequences and lengths were purchased from Bio-Synthesis Inc. The T terminal of each DNA molecule was modified with an amine group[20], allowing binding to gold electrodes for electrical measurement. The name and structure of all the sequences are listed in Table 1. The DNA molecules were annealed using a thermal cycler (TC-050-18, Labnet Inc) to form the B-form double-helical structure in phosphate buffer, which was verified by gel electrophoresis, circular dichroism and melting curve (Supplementary Figs 1, 2 and 3; Supplementary Methods 1, 2 and 3). The phosphate buffer was prepared with $Na_2HPO_4.2H_2O$ (for high-performance liquid chromatography (HPLC) ≥98.5%) and NaH_2PO_4 (for HPLC ≥99%) purchased from Fluka. The DNA sample was immobilized on a gold substrate (gold with thickness of ~160 nm on mica) prepared in a vapour deposition system under ultrahigh vacuum. Before each experiment, the gold substrate was annealed with hydrogen flame briefly, and immediately immersed in phosphate buffer (5 mM, pH = 7) containing 5 μM double-helical DNA, incubated overnight. The substrate electrode was then rinsed with phosphate buffer and dried with nitrogen gas. The DNA-coated gold substrate was attached to the sample holder of the STM, which was placed in a glove box under nitrogen atmosphere (relative humidity ~20%).

STM set-up. The charge transport and thermoelectric measurements were carried out with the STM consisting of a controller (Nanoscope IIIA, Digital Instruments Inc.) and a STM scanner (Molecular Imaging). The STM tip was freshly cut from a gold wire (0.25 mm diameter, 99.95%, Alfa Aesar). A piece of electrically shielded copper strip (0.7 × 0.4 × 2 cm) was thermally connected to the STM tip holder to serve as a thermal reservoir. The STM sample holder was mounted on a semiconductor cooler with a thermistor to measure the temperature (controlled by Lake Shore 331 temperature controller).

Measurement of thermoelectricity. The STM break junction technique was used to measure the conductance of DNA with a procedure described in detail elsewhere[21] (Fig. 1a). Briefly, the STM tip was first brought into contact with the DNA molecules on the gold substrate, allowing the amine-terminated DNA to bind with the tip and substrate electrodes. The tip was then retracted from the substrate, during which a conductance–distance trace was recorded. The last plateau in the conductance trace corresponded to a single DNA molecule attached to the tip and substrate electrodes to form a molecular junction. The process was repeated thousands of times. A conductance histogram was constructed from the individual conductance traces, and the peak in the conductance histogram gave the average conductance of the DNA molecule.

To measure the thermoelectric effect of DNA molecules, a temperature gradient was created across the molecule by cooling the sample holder while maintaining the STM tip at the room temperature[33]. The temperature of the sample was monitored with thermistors during the experiment. When a molecular junction was formed in the STM break junction measurement as indicated by the formation of a plateau feature in the conductance–distance trace, the tip was held in position and an I–V curve was recorded. The tip was then retracted further away from the substrate by an additional distance. If the current did not drop abruptly, signalling that the molecular junction was still intact, another I–V curve was recorded. Otherwise, the measurement was started over again.

Thousands of I–V curves were collected for each temperature gradient. These I–V curves were normalized by conductance (G) and presented as a two-dimensional I/G versus V histogram. The voltage offsets in the I/G versus V histogram were used to construct a histogram of the thermoelectric effect, and Gaussian fitting of the thermoelectric effect histogram provided the Seebeck coefficient of DNA. Controlling the STM tip movement, recording and processing the experimental data described above were achieved with Labview 8.5 and data acquisition card.

References

1. Mahan, G. D. & Sofo, J. O. The best thermoelectric. *Proc. Natl Acad. Sci. USA* **93,** 7436–7439 (1996).
2. Reddy, P., Jang, S.-Y., Segalman, R. A. & Majumdar, A. Thermoelectricity in molecular junctions. *Science* **315,** 1568–1571 (2007).
3. Dubi, Y. & Di Ventra, M. Thermoelectric effects in nanoscale junctions. *Nano Lett.* **9,** 97–101 (2009).
4. Bergfield, J. P. & Stafford, C. A. Thermoelectric signatures of coherent transport in single-molecule heterojunctions. *Nano Lett.* **9,** 3072–3076 (2009).
5. Ke, S. H., Yang, M., Curtarolo, S. & Baranger, H. U. Thermopower of molecular junctions: an ab initio study. *Nano Lett.* **9,** 1011–1014 (2009).
6. Liu, Y., Chen, Y. & Chen, Y. Thermoelectric efficiency in nanojunctions: a comparison between atomic junctions and molecular junctions. *ACS Nano* **3,** 3497–3504 (2009).
7. Seeman, N. C. Nanomaterials Based on DNA. *Annu. Rev. Biochem.* **79,** 65–87 (2010).
8. Fukui, K. & Tanaka, K. Distance dependence of photoinduced electron transfer in DNA. *Angew. Chem. Int. Ed.* **37,** 158–161 (1998).
9. Kelley, S. O. & Barton, J. K. Electron transfer between bases in double helical DNA. *Science* **283,** 375–381 (1999).
10. Lewis, F. D. *et al.* Distance-dependent electron transfer in DNA hairpins. *Science* **277,** 673–676 (1997).
11. Meade, T. J. & Katten, J. F. Electron-transfer through DNA-site-specific modification of duplex DNA with Ruthenium donors and accepters. *Angew. Chem. Int. Ed.* **34,** 352–354 (1995).
12. Meggers, E., Kusch, D., Spichty, M. & Wille, U. Electron transfer through DNA in the course of radical-induced strand cleavage. *Angew. Chem. Int. Ed.* **37,** 460–462 (1998).
13. Brun, A. M. & Harriman, A. Dynamics of electron transfer between intercalated polycyclic molecules: effect of interspersed bases. *J. Am. Chem. Soc.* **114,** 3656–3660 (1992).
14. Giese, B., Amaudrut, J., Koehler, A.-K., Spormann, M. & Wessely, S. Direct observation of hole transfer through DNA by hopping between adenine bases and by tunnelling. *Nature* **412,** 318–320 (2001).
15. Zhang, X. u. & Tao, L. i. Direct conductance measurement of sngle DNA molecules in aqueous solution. *Nano Lett.* **4,** 1105–1108 (2004).
16. Lewis, F. D., Zhu, H., Daublain, P., Cohen, B. & Wasielewski, M. R. Hole mobility in DNA a tracts. *Angew. Chem. Int. Ed.* **45,** 7982–7985 (2006).
17. Jortner, J., Bixon, M., Voityuk, A. A. & Rösch, N. Superexchange mediated charge hopping in DNA. *J. Phys. Chem. A* **106,** 7599–7606 (2002).
18. Renaud, N., Berlin, Y. A., Lewis, F. D. & Ratner, M. A. Between superexchange and hopping: an intermediate charge-transfer mechanism in Poly(A)-Poly(T) DNA hairpins. *J. Am. Chem. Soc.* **135,** 3953–3963 (2013).
19. Troisi, A. & Orlandi, G. The hole transfer in DNA: calculation of electron coupling between close bases. *Chem. Phys. Lett.* **344,** 509–518 (2001).
20. Xiang, L. *et al.* Intermediate tunnelling–hopping regime in DNA charge transport. *Nat. Chem.* **7,** 221–226 (2015).
21. Xu, B. & Tao, N. J. Measurement of single-molecule resistance by repeated formation of molecular junctions. *Science* **301,** 1221–1223 (2003).
22. Williams, P. D. & Reuter, M. G. Level alignments and coupling strengths in conductance histograms: the information content of a single channel peak. *J. Phys. Chem. C* **117,** 5937–5942 (2013).
23. Uchihashi, T. *et al.* Identification of B-form dna in an ultrahigh vacuum by noncontact-mode atomic force microscopy. *Langmuir* **16,** 1349–1353 (1999).
24. Tanaka, H., Tanaka, H. & Kawai, T. Partial sequencing of a single DNA molecule with a scanning tunnelling microscope. *Nat. Nanotechnol.* **4,** 518–522 (2009).
25. Kang, N. Electrical characterization of DNA in mechanically controlled break-junctions. *New J. Phys.* **10,** 023030 (2008).
26. Brewer, S. H., Anthireya, S. J., Lappi, S. E., Drapcho, D. L. & Franzen, S. Detection of DNA Hybridization on Gold Surfaces by Polarization Modulation Infrared Reflection Absorption Spectroscopy. *Langmuir* **18,** 4460–4464 (2002).
27. McCreery, R. L. Molecular electronic junctions. *Chem. Mater.* **16,** 4477–4496 (2004).
28. Luo, L., Choi, S. H. & Frisbie, C. D. Probing hopping conduction in conjugated molecular wires connected to metal electrodes. *Chem. Mater.* **23,** 631–645 (2010).
29. Segal, D., Nitzan, A., Ratner, M. & Davis, W. B. Activated conduction in microscopic molecular junctions. *J. Phys. Chem. B* **104,** 2790–2793 (2000).
30. Nitzan, A. The relationship between electron transfer rate and molecular conduction 2. The sequential hopping case. *Isr. J. Chem.* **42,** 163–166 (2002).
31. Voityuk, A. A., Jortner, J., Bixon, M. & Rösch, N. Electronic coupling between Watson–Crick pairs for hole transfer and transport in desoxyribonucleic acid. *J. Chem. Phys.* **114,** 5614–5620 (2001).
32. Blatt, F. J., Schroeder, P. A., Foiles, C. L. & Greig, D. *Thermoelectric Power of Metals* (Plenum Press, 1976).
33. Guo, S., Zhou, G. & Tao, N. Single molecule conductance, Thermopower, and Transition Voltage. *Nano Lett.* **13,** 4326–4332 (2013).
34. Joshua, H., Fang, C., Peiming, Z. & Nongjian, T. Thermal and electrochemical gate effects on DNA conductance. *J. Phys. Condens. Matter* **19,** 215202 (2007).
35. Di Felice, R., Calzolari, A., Molinari, E. & Garbesi, A. *Ab initio* study of model guanine assemblies: The role of pi-pi coupling and band transport. *Phys. Rev. B* **65,** 045104 (2001).
36. Malen, J. A. *et al.* Identifying the length dependence of orbital alignment and contact coupling in molecular heterojunctions. *Nano Lett.* **9,** 1164–1169 (2009).
37. Widawsky, J. R. *et al.* Length-dependent thermopower of highly conducting Au–C bonded single molecule junctions. *Nano Lett.* **13,** 2889–2894 (2013).
38. Mott, N. F. & Davis, E. A. *Electronic Processes in Non-Crystalline Materials* (Clarendon Press, 1979).
39. Paulsson, M. & Datta, S. Thermoelectric effect in molecular electronics. *Phys. Rev. B* **67,** 241403 (2003).
40. Kim, Y., Jeong, W., Kim, K., Lee, W. & Reddy, P. Electrostatic control of thermoelectricity in molecular junctions. *Nat. Nanotechnol.* **9,** 881–885 (2014).
41. Bâldea, I. Interpretation of stochastic events in single-molecule measurements of conductance and transition voltage spectroscopy. *J. Am. Chem. Soc.* **134,** 7958–7962 (2012).

Acknowledgements

We thank Hao Yan for the help on the gel electrophoresis experiments, Mark A. Ratner, Yuri A. Berlin and Vladimiro Mujica for the support and discussions. Financial support from Office of Naval Research (ONR N00014-11-1-0729) is gratefully acknowledged.

Author contributions

N.T. conceived the project. Y.L. designed, conducted and analysed the experiments. N.T., L.X., J.L.P. and Y.L. made the theoretical explanation with hopping model. Y.A. completed the SSH calculation. N.T., L.X. and Y.L. wrote the paper.

Additional information

Competing financial interests: The authors declare no competing financial interests.

Transformation of metal-organic frameworks for molecular sieving membranes

Wanbin Li[1], Yufan Zhang[2], Congyang Zhang[1], Qin Meng[3], Zehai Xu[1], Pengcheng Su[1], Qingbiao Li[4], Chong Shen[3], Zheng Fan[1], Lei Qin[1] & Guoliang Zhang[1]

The development of simple, versatile strategies for the synthesis of metal-organic framework (MOF)-derived membranes are of increasing scientific interest, but challenges exist in understanding suitable fabrication mechanisms. Here we report a route for the complete transformation of a series of MOF membranes and particles, based on multivalent cation substitution. Through our approach, the effective pore size can be reduced through the immobilization of metal salt residues in the cavities, and appropriate MOF crystal facets can be exposed, to achieve competitive molecular sieving capabilities. The method can also be used more generally for the synthesis of a variety of MOF membranes and particles. Importantly, we design and synthesize promising MOF membranes candidates that are hard to achieve through conventional methods. For example, our CuBTC/MIL-100 membrane exhibits 89, 171, 241 and 336 times higher H_2 permeance than that of CO_2, O_2, N_2 and CH_4, respectively.

[1] Institute of Oceanic and Environmental Chemical Engineering, State Key Lab Breeding Base of Green Chemical Synthesis Technology and Collaborative Innovation Center of Membrane Separation and Water Treatment of Zhejiang Province, Zhejiang University of Technology, Chaowang Road 18#, Hangzhou 310014, China. [2] Department of Materials Science and Engineering, College of Engineering, University of California, Berkeley, California 94720, USA. [3] Department of Chemical and Biological Engineering, College of Chemical and Biological Engineering, State Key Laboratory of Chemical Engineering, Zhejiang University, Hangzhou 310027, China. [4] Department of Chemical and Biochemical Engineering, College of Chemistry and Chemical Engineering, National Laboratory for Green Chemical Productions of Alcohols, Ethers and Esters, Key Lab for Chemical Biology of Fujian Province, Xiamen University, Xiamen 361005, China. Correspondence and requests for materials should be addressed to G.Z. (email: guoliangz@zjut.edu.cn).

Molecular sieving membranes composed of silica[1], zeolites[2,3], graphene[4,5] or metal-organic frameworks (MOFs)[6,7] have intrinsic advantages, including both larger permeability and better selectivity, compared with those of conventional polymeric membranes[8]. MOFs consisting of metal ions or metal clusters coordinated with organic linkers have great potential as molecular sieves because of their large surface area, diverse structures and tailorable pore sizes. Until now, tens of thousands of MOFs have been reported and investigated[9]. However, compared with the total number of MOFs reports, those that can be assembled into continuous membranes are limited because of the necessary complex fabrication and activation procedures[10]. Substitution reaction chemistry has been extensively employed in various MOF modification procedures, notably to expand pore aperture or enhance adsorption of MOFs, by linker exchange or transmetalation[11–17]. However, MOFs synthesized through substitution protocols usually possess the same topological structure as their precursor, and no such investigation has focused on MOF membranes. Therefore, if we can establish a synthetic connection between different series of MOFs, we may greatly increase their potential in filtration applications.

Here we report a methodology for realizing the connection and complete transformation of different series of MOF membranes and particles based on multivalent cation substitution. This strategy combines three key concepts: (i) facile transformation of unstable and easily fabricated MOF particles to obtain the stable MOFs with completely different topology structure, which are usually fabricated in a relatively harsh synthetic conditions; (ii) *in situ* transformation of one common MOF membrane to another MOF membrane, which is hard to be synthesized in conventional methods at present; (iii) reducing the pore size through immobilizing the metal salt residue in cavities and exposing the appropriate crystal facets of the MOFs to achieve competitive molecular sieving ability by the transformation of MOF membranes. Our strategy can be used more generally to various MOF membranes and particles, but we exhibit our key findings here with two examples, one is the transformation of CuBTC to MIL-100, which takes the advantages of easy preparation and material stability[18–20]. Another is the transformation of CuBTC membrane to CuBTC/MIL-100 membrane, which provides a facile route to design promising candidates of MOF membrane for molecular sieving.

Results

Transformation of CuBTC to MIL-100. We first conducted the transformation of CuBTC to MIL-100. CuBTC (also known as HKUST-1), a MOF with a cubic Pt_3O_4-type network, is composed of paddle wheel dimeric copper carboxylate units bridged by three-connecting 1,3,5-benzenetricarboxylate (BTC) linkers[18]. As one of the earliest reported and most studied MOF materials, it has been produced in large scale and used to fabricate continuous membranes[10,19] despite its relatively poor chemical stability[20]. The MIL series of MOFs possess the excellent chemical stability, yet their synthetic conditions are much more rigorous compared with those of CuBTC[21]. For example, MIL-100(Fe) (Supplementary Fig. 1), composed of BTC and Fe-based centre with zeotype architecture[22], is typically synthesized at 150 °C for 6 days and the precursor solution contains hazardous hydrofluoric acid, which is harmful to the environment. Although the preparation process of MIL-100(Fe) has been simplified considerably by reflux and microwave-assisted method at about 100 °C recently[23,24] compared with CuBTC, which can be straightforwardly fabricated at low temperature[19,25–27], the synthetic conditions and experimental equipment are still more rigorous. Moreover, new synthetic methods are desired highly to formation of MOF at room temperature[27]. Thus, we selected CuBTC and MIL-100 as main precursor and product to demonstrate the transformation between the MOFs. Scanning electron microscopy (SEM) image of CuBTC particle and transformation procedure in metal centre and second building unit are shown in Fig. 1a,b. The transformation was implemented by immersing CuBTC into $FeCl_3 \bullet 6H_2O$ solution at room temperature. All water, methanol, ethanol and *N*,*N*-dimethylformamide (DMF) were employed as the solvents. From the comparison between the simulated X-ray diffraction (XRD) of MIL-100 and XRD patterns of the prepared materials, we found that only the pattern of MIL-100 transformed in methanol showed all characteristic peaks similar to simulated XRD patterns, which demonstrated that the methanol was a relatively better solvent for achieving pure MIL-100 crystals with excellent crystallinity, but the three other solvents were not the same case (Supplementary Fig. 2). This phenomenon should be attributed to the balance among the diffusion rate of the solvent and Fe^{3+} ions in the pores of crystals, hydrolysis rate of CuBTC and substitution rate. CuBTC is unstable in water, and the Lewis acidic species of Fe^{3+} further accelerate the hydrolysis. Therefore, CuBTC was hydrolysed before the reaction was completed in aqueous solution. Compared with methanol, ethanol and DMF possess larger diameter and smaller polarity, which may conduce to much lower diffusion rate of reagents in CuBTC and substitution rate, and led to little cation substitution. Some monovalent and divalent cations were also employed to transform CuBTC, the result revealed that the MOF crystal structure had not been changed (Supplementary Fig. 3). Moreover, similar molar quantity of Cu^{2+} and BTC as CuBTC cannot be used to synthesize MIL-100 (Supplementary Fig. 4), demonstrating that the importance of the CuBTC framework and the transformation process was based on multivalent cation substitution.

Transformation mechanism. We characterized the CuBTC/MIL-100 particles with different substitution times to investigate the reaction processes of transformation. After substitution for 1 h, the number of peaks in XRD pattern showed no big change (Fig. 1c). When the duration time was extended to 21 h, peaks for MIL-100 emerged. With the passage of time, the peak intensities of {200} and {220} degraded rapidly, and the peak intensity ratios of {200}/{222} and {220}/{222} also decreased dramatically (Fig. 1e). Moreover, it should be noted that the peak intensity of {333} also displayed a much lower degradation rate compared with other peaks. CuBTC exhibits three window structures, largest pores with diameter of 9 Å, moderate side pores with the diameter of 5 Å and smallest triangular shaped windows with the diameter of 3.5 Å (ref. 28). Correspondingly, the window size of {111} facets is 3.5 Å, which is smaller than the pore size of 5 and 9 Å in {100} facets (Fig. 1d). The octahedron crystal morphology and the much higher peak intensity of {222} compared with {200} and {220} reveal that the CuBTC has mainly exposed {111} facet[29]. Thus, by combining above experimental results and CuBTC structures, we can deduce the following reaction processes: when CuBTC crystals were immersed into the substitution solutions, the exposed {111} facets would greatly hinder the diffusion of the solvent and Fe^{3+} ions, this may be the reason for the little cation substitution of CuBTC in above-mentioned three solvents. After the copper ions were replaced by the Fe^{3+} ions, some MIL-100 nanocrystals formed on the surface of the CuBTC crystals, and subsequently the channel for diffusion of the solvent and Fe^{3+} ion was produced due to the larger window of MIL-100 and the gap between the two kinds of MOFs.

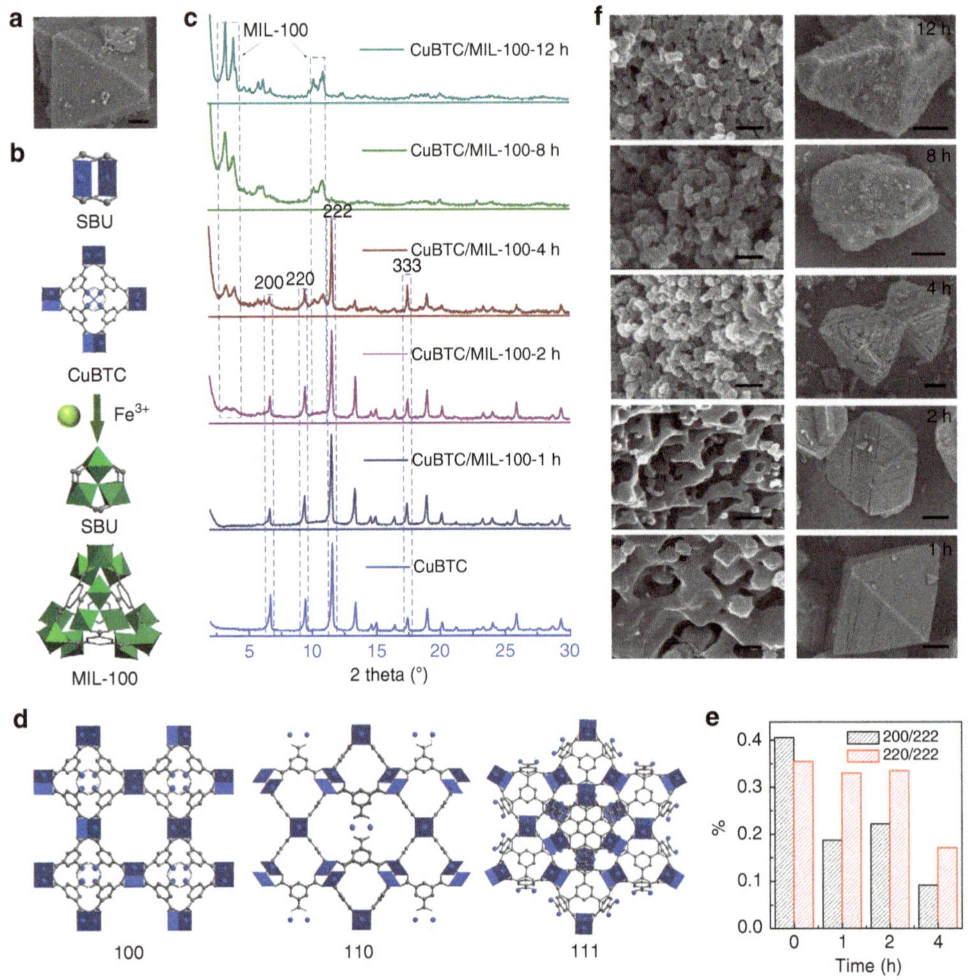

Figure 1 | Transformation of CuBTC to MIL-100. (**a**) SEM image of as-synthesized octahedron CuBTC crystal. Scale bar, 10 μm. (**b**) Transformation procedure in metal centre and second building unit of CuBTC and MIL-100. The Cu and Fe coordination polyhedra are represented in blue and green, respectively, whereas the BTC links are depicted by sticks. (**c**) XRD patterns of CuBTC/MIL-100 with different transformation time. (**d**) CuBTC structure viewed at different facet. (**e**) Peak intensity ratios of {200}/{222} and {220}/{222} with the transformation time increasing. (**f**) SEM images of CuBTC/MIL-100 with different transformation time. The octahedron CuBTC crystal (bottom, 1 h) transformed into nanoparticle MIL-100 crystal (top,12 h). Scale bar, 200 nm (left); 20 μm (right).

As a result, the solvent and the Fe^{3+} ion would pass through the {100} facets quickly and replace the adjacent Cu^{2+} ions in CuBTC crystals to form MIL-100. Although the window size of {110} facets was the same as that of {111} facets, the {111} facets possessed the large atomic density and accordion-like channel, which greatly decreased the substitution rate and increased the diffusion resistance. This also led to the decrease of the value of {220}/{222}. The morphologies of these CuBTC/MIL-100 crystals are consistent with the XRD results and the deduction (Fig. 1f and Supplementary Figs 5 and 6). After substitution for 1 h, the crystals displayed that CuBTC/MIL-100 had superstructural composite surfaces and three-dimensional CuBTC shelf with interior MIL-100 nanocrystals. When the substitution time increased, the superstructure extended to the middle of the CuBTC crystals. After reaction for 81 h, all CuBTC were transformed into MIL-100. For identification of the inside structure of CuBTC/MIL-100 in transformation, the optical microscopy and energy dispersion spectroscopy (EDS) mapping images of CuBTC/MIL-100 with different transformation time were collected and presented in Fig. 2a–c. We can find that, with the extension of transformation times, the blue part of CuBTC became small but the shape was still octahedron. The optical microscopy images showed the lamellar structure, which was parallel to the boundary of CuBTC crystal, this may be attributed to the untransformed CuBTC with preferred {111} crystal facet, which was consistent with the results of XRD and showed the important role of {111} facet in transformation. After transformation for 12 h, the whole blue CuBTC was changed to yellow MIL-100. EDS mapping showed the similar phenomenon. When transformation time was extended to 12 h, the Cu/Fe ratio was reduced to 0.4%. These results also revealed that the transformation was conducted from outside to inside. To further demonstrate the importance of the exposed {111} facet for transformation, we synthesized the cubic CuBTC with mainly exposed {100} facet as previous study (Fig. 2d)[30], and used it to obtain the MIL-100. The XRD pattern revealed that this transformation failed to obtain the MIL-100 (Fig. 2f). Compared with the above successful transformation, two differences may lead to the unsuccessful transformation, the difference in exposed facet and the difference in CuBTC particle size. We further prepared the small octahedron CuBTC by simply stirring the precursor at room temperature (Fig. 2e)[26], and found that this material could be transformed to be MIL-100 successfully. This results verified that the difference in exposed

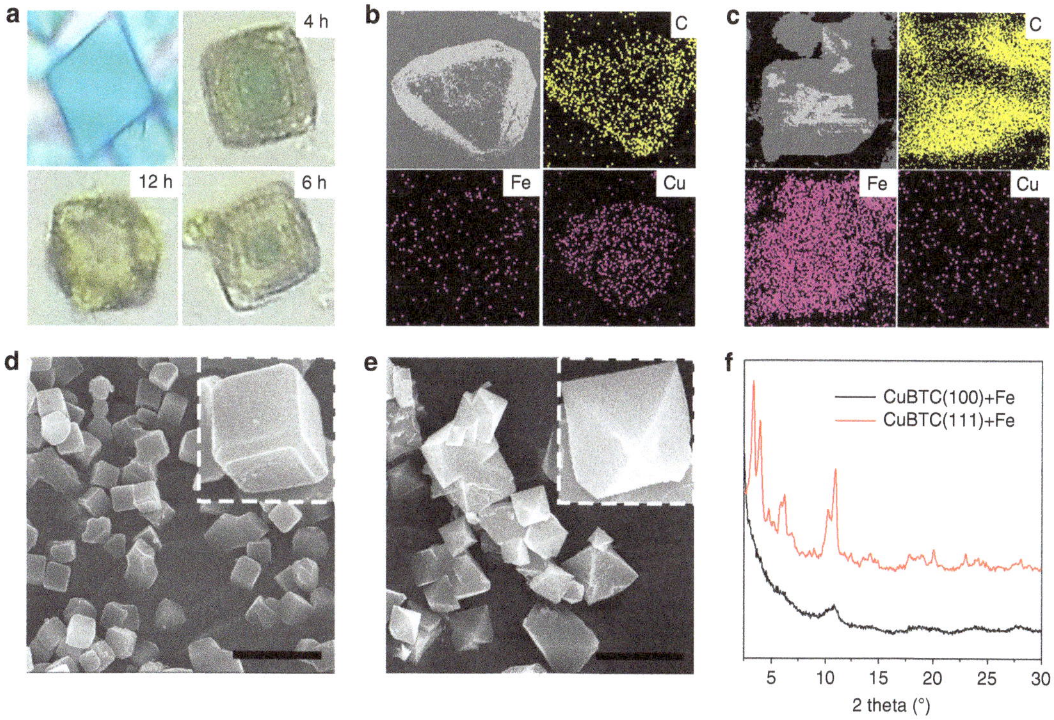

Figure 2 | Transformation process from CuBTC to MIL-100. (**a**) Optical microscopy images of CuBTC with different transformation time, the transformation was carried out under standing. (**b**) EDS mapping of the CuBTC/MIL-100 after transformation for 4 h and purification in methanol, Fe/Cu ratio is 23.5%. (**c**) EDS mapping of the CuBTC/MIL-100 after transformation for 12 h and purification in methanol and hot water, Cu/Fe ratio is 0.4%. (**d**) SEM images of the cubic CuBTC with mainly exposed {100} facet. Scale bar, 5 μm. (**e**) SEM images of the small octahedron CuBTC with mainly exposed {111} facet. (**f**) XRD patterns of cubic CuBTC and small octahedron CuBTC after transformation. Scale bar, 5 μm.

facet was the main reason of the failing transformation of the cubic CuBTC with mainly exposed {100} facet, rather than the difference in CuBTC particle size. The experiment also showed that the MIL-100 can be prepared by transformation of the small octahedron CuBTC in mild route at room temperature without using any special equipment (Supplementary Table 1).

Universality of transformation. To present the versatility of this strategy, we transformed the MOF-5 to CuBDC by Cu^{2+} substitution and transformed the CuBDC to MIL-53 by Fe^{3+} substitution. In this process, MOF-5 was a face-centred cubic material, formed by Zn-based metal cluster and 1,4-ben-zenedi-carboxylate (BDC) linker[31], and can be synthesized at room temperature by adding triethylamine[32]; CuBDC consisted of paddle wheel dimeric copper carboxylate units interconnected by BDC[33], and was usually fabricated in DMF at 110 °C or in acetonitrile system[33–36]; MIL-53 was formed by $FeO_4(OH)_2$ clusters with BDC[37], and prepared typically in DMF or in water at high temparature[38–40]. The crystalline structures of these MOFs are presented in Supplementary Fig. 7. The XRD patterns and typical SEM images reveal the successful transformation (Supplementary Figs 7, 8 and 9). Because of the mild and environment-friendly process, the complex synthesis condition of some MOFs can be simplified greatly by transformation, which is of benefit to the scalable production of these MOFs.

Property of transformed MIL-100. We further studied the features of the transformed MIL-100. Nitrogen adsorption experiments reveal that the CuBTC sample has Brunauer–Emmett–Teller (BET)-specific surface area of $960\ m^2\ g^{-1}$ and pore volume of $0.48\ cm^3\ g^{-1}$ (Fig. 3a). After transformation, BET-specific surface area and pore volume increase obviously to $1880\ m^2\ g^{-1}$ and $1.07\ cm^3\ g^{-1}$, respectively, and the hysteresis loop reveals the existence of mesopores in MIL-100 (ref. 41). These values are consistent with those reported in previous studies (Supplementary Table 1), and demonstrate the good porosity of the prepared MIL-100. The DFT pore size distribution curve displays that the prepared MIL-100 possesses the multi-scale pore structure (Fig. 3b), as similar results have been observed in many previous studies[24,42–44]. The small skewing in O 1s X-ray photoelectron spectroscopy demonstrates the successful transformation of the C–O–Cu bond in CuBTC to C–O–Fe bond in MIL-100 (Supplementary Fig. 10). Thermal analysis shows the same good thermal stability of the transformed MIL-100 and normal thermal degradation curve as those in previous reports[22] (Supplementary Fig. 11). From the SEM images (Fig. 1f), we can find that MIL-100 nanocrystals have a relatively small size in the range of 30–60 nm. The integrated MIL-100 particle after treated by ultrasonic treatment demonstrates that MIL-100 nanoparticles grew together tightly, rather than simple accumulation (Supplementary Fig. 12). These features exhibit that MIL-100 was hierarchically porous material. Moreover, we found that some Fe-based materials can be kept in cavities by insufficient purification to reduce the pore size of the obtained MOFs. With the decrease of purification cycle, the adsorption–desorption isotherms change from type-I/IV to type-II. The specific surface area and pore volume reduce to $122\ m^2\ g^{-1}$ and $0.18\ cm^3\ g^{-1}$ (Fig. 3a), respectively, and the pore size also decreases obviously (Fig. 3b). The relative intensities of XRD patterns of the as-synthesized MIL-100 also show a strong reduction compared with the MIL-100 after purification, especially at low angles (Supplementary Fig. 13), which may be attributed to the impregnation of amorphous $FeCl_3$ (refs 21,42).

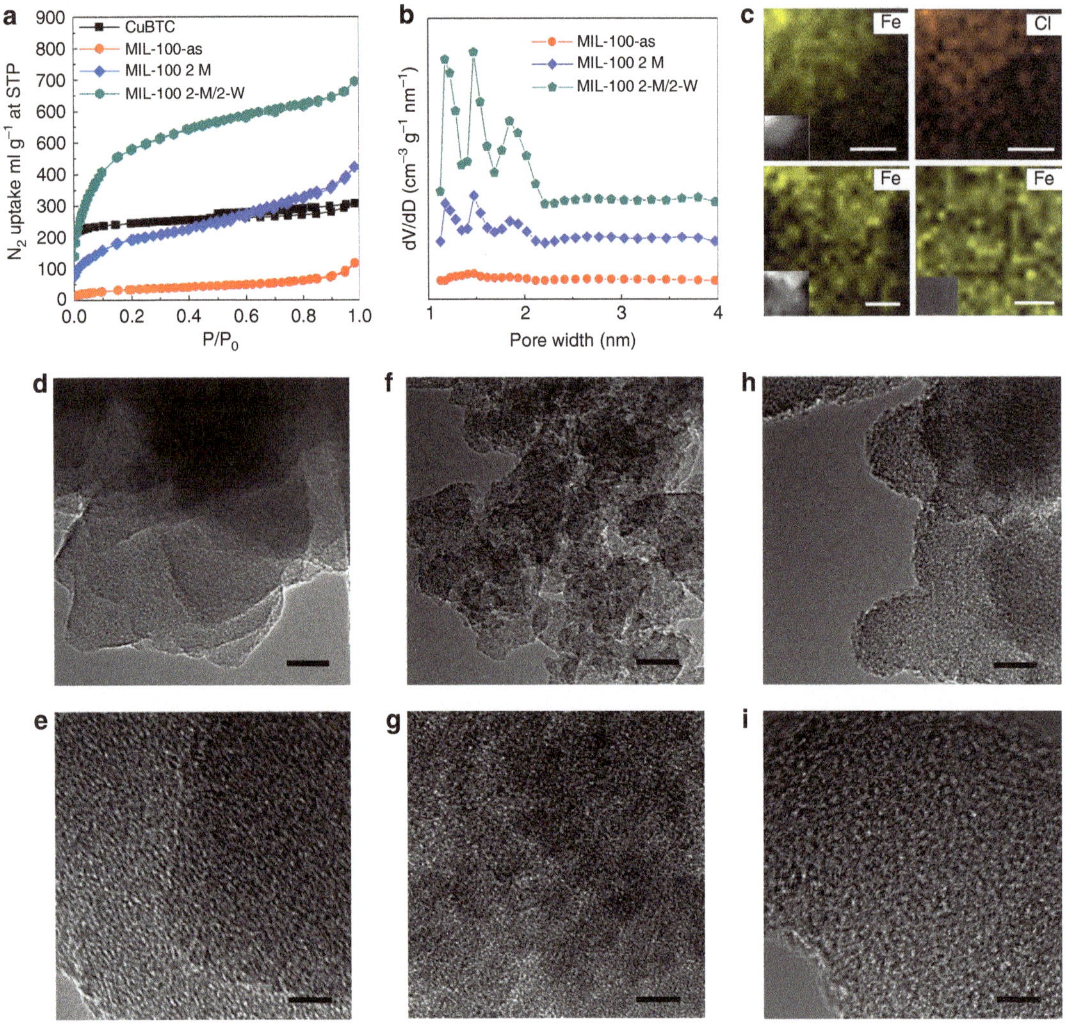

Figure 3 | Characterization of MIL-100. (**a**) N_2 adsorption-desorption isotherms on MIL-100 with various purification cycles, showing the specific surface area reduced with the purification cycle, two cycles in methanol (2-M), two cycles in methanol and then in hot water at 70 °C (2-M/2-W). (**b**) Corresponding pore size distributions calculated by DFT method. (**c**) EDS mapping of the MIL-100 with different post-processing. Top two images: as-synthesized MIL-100 and demonstrating the well-dispersed chlorine, scale bar, 20 nm; bottom left image: as-synthesized MIL-100 after calcination at 200 °C for 4 h, scale bar, 10 nm; bottom right image: MIL-100 after purification; scale bar, 5 nm. (**d,e**) TEM images of as-synthesized MIL-100, showing the amorphous $FeCl_3$ in the cavities. (**f,g**) TEM images of as-synthesized MIL-100 after calcination at 200 °C for 4 h, indicating the presence of nanoclusters in the cavities of the MIL-100. (**h,i**) TEM images of MIL-100 after purification, indicating clear porous structure and pure framework. Scale bar, 20 nm (in **d,f,h**); 5 nm (in **e,g,i**).

Combining with the vast existence of well-dispersed chlorine (10.2%) in the as-synthesized MIL-100 (Supplementary Table 2 and Fig. 3c), and the small amount of chlorine (0.2%) in MIL-100 after purification in methanol and hot water, we come to a conclusion that the cavities of as-synthesized MIL-100 have been occupied by highly dispersed amorphous $FeCl_3$, which can be utilized to artificially control the pores of the transformed MOFs. To further prove the existence of $FeCl_3$ in the cavities of as-synthesized MIL-100, it was calcined at 200 °C for 41 h. The chlorine displays a great reduction (74.9%) and the XRD patterns show some new peaks of Fe_2O_3 for as-synthesized MIL-100 (Supplementary Fig. 14). All these reveal that the amorphous $FeCl_3$ in the cavities has been converted to Fe_2O_3 nanoclusters. To further identify the formation of $FeCl_3$ and the structure of the MIL-100, we collected the transmission electron microscopy (TEM) images of as-synthesized MIL-100, as-synthesized MIL-100 after calcination at 200 °C for 4 h and MIL-100 after purification in methanol and hot water (Fig. 3d–i). TEM images clearly indicate that abundant of amorphous materials are left in the cavities of as-synthesized MIL-100 (Fig. 3d,e), however, the purified MIL-100 has clear porous structure and pure framework (Fig. 3h,i), there are almost no residues left in the cavities of purified MIL-100. For as-synthesized MIL-100 after calcination, the uniformly dispersed dark spots in TEM images demonstrates that the Fe_2O_3 nanoclusters are immobilized inside the cavities of insufficiently purified MIL-100 (Fig. 3f,g). The nanoclusters in cavities may be beneficial to improve the activity of MOFs. In addition, the reduced pore size will be very important for the performance of molecular sieve.

Transformation of MOF membrane. After demonstrating the transformation of MOFs, our strategy was further employed to transform MOF membranes. CuBTC membrane was synthesized on polymeric hollow fibre by solvothermal method and used as precursor[45] (Supplementary Fig. 15). Figure 4a–c presents the SEM images of CuBTC membrane, as-synthesized CuBTC/MIL-100 membrane and CuBTC/MIL-100 membrane after purification. The CuBTC layer with thickness of ~20 μm is

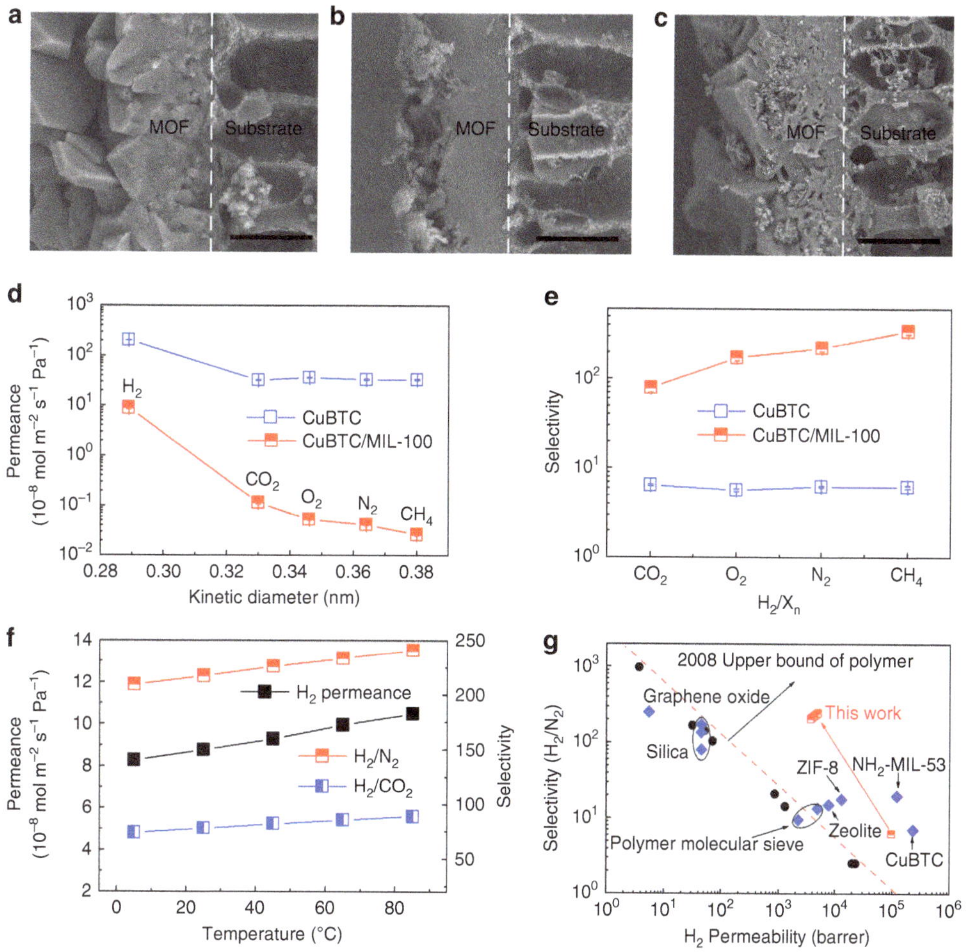

Figure 4 | Transformation of CuBTC membrane and their performance. (**a**-**c**) SEM images of original CuBTC membrane, transformed CuBTC/MIL-100 membrane and transformed CuBTC/MIL-100 membrane after purification, respectively. Scale bar, 20 μm. (**d**,**e**) Gas permeance and selectivities of the CuBTC and CuBTC/MIL-100 membranes. All the average permeation results with standard deviation were calculated from three measurement data. (**f**) Effect of temperature on H_2 permeance and H_2/CO_2 and H_2/N_2 selectivities for CuBTC/MIL-100 membrane. (**g**) Comparison of CuBTC/MIL-100 membrane with polymeric[8,47], silica[1], zeolite[3], other MOF[46,48,49] and graphene oxide membranes[5] for H_2/N_2 system. 1 barrer = 3.348×10^{-16} mol m^{-2} s^{-1} Pa^{-1}, the red dotted line is the Robeson's upper-bound reported in 2008 (ref. 8).

anchored to the substrate continuously. The octahedron morphology of crystal demonstrates that the CuBTC membrane also has mainly exposed {111} facet. After transformation, the as-synthesized CuBTC/MIL-100 membrane seems to become denser due to the $FeCl_3$ residue (Supplementary Figs 16 and 17). However, after the $FeCl_3$ was removed, the massive MIL-100 nanoparticles accompanied by CuBTC flake occurred on the cross-section of purified CuBTC/MIL-100 membrane (Fig. 4c and Supplementary Fig. 16). EDS mapping demonstrates the incomplete transformation (Supplementary Fig. 17). The existence of the peak with 2 theta at 11.6° in XRD pattern also shows the incomplete transformation and the residual CuBTC with exposed {111} facet (Supplementary Fig. 18). This is consistent with the phenomenon observed in incompletely transformed MOF particles, and can be explained by the smaller mass transfer-specific surface area exposed to the solution of membrane. When the duration time was extend to 48 h, all the CuBTC were transformed to MIL-100 (Supplementary Fig. 16).

Performance of transformed membrane. To investigate the performance of the prepared MOF membranes by transformation, we used the constant-pressure method to measure H_2 (kinetic diameter: 0.289 nm), CO_2 (0.33 nm), O_2 (0.346 nm), N_2 (0.364 nm) and CH_4 (0.38 nm) permeances through the two dense membranes[4]. The CuBTC membrane exhibited H_2 permeance and selectivities of H_2/X_n (X_n: other gases) in the range of 5.5–6.5, similar to those previously reported[46] (Fig. 4d,e). For the transformed CuBTC/MIL-100 membrane, the permeances of all gases were smaller than those of CuBTC membrane, and the largest H_2 permeance was 8.8×10^{-8} mol m^{-2} s^{-1} Pa^{-1}. However, the selectivities of H_2/CO_2, H_2/O_2, H_2/N_2 and H_2/CH_4 displayed great improvement and reached about 77.6, 170.6, 217.0 and 335.7, respectively. Moreover, the selectivities increased with temperature and reached 89.0 and 240.5 for H_2/CO_2 and H_2/N_2 at 85 °C, respectively, as the H_2 permeances grew faster than other gases. Meanwhile, the H_2 permeance also increased to 10.5×10^{-8} mol m^{-2} s^{-1} Pa^{-1} (Fig. 4f). The transformed membrane also presented excellent durability, which maintained good performance with only small fluctuation over a 192-h period (Supplementary Fig. 19). Our transformed molecular sieving membrane had much better performance than polymeric membranes and can easily exceed the Robeson's upper-bound reported in 2008 for all H_2/CO_2, H_2/N_2 and H_2/CH_4 systems[8,47]. With regard to porous zeolites[3] and conventional MOF membranes[46,48,49], the transformed CuBTC/MIL-100 membrane showed superior performance in selectivities (Fig. 4g and Supplementary Fig. 20). Even compared with the silica, graphene oxide and $Zn_2(bim)_4$ nanosheet membranes reported, which exhibited great selectivities for H_2/CO_2 or/and H_2/N_2 systems[1,4,5,7], our membrane also exhibited competitive performance. The transformed CuBTC/MIL-100 membrane possessed excellent H_2 permeability as high as ~4,000 barrer as well as good stability under high trans-membrane pressure of 0.25 MPa (Supplementary Fig. 21), while the reported membranes were usually operated under very low trans-membrane pressure of 0.1 or even 0 MPa to prevent possible cracks. In binary mixture separation, because of the competitive adsorption, the membrane showed a little smaller H_2 permeance and selectivities compared with the values measured by constant-pressure method, but the membrane still showed the competitive selectivities (Supplementary Table 3). Moreover, the synthetic procedure has good reproducibility, which has been demonstrated by the similar separation performance of two further additional membranes. We speculate that the good molecular sieving performance of transformed CuBTC/MIL-100 membrane is caused by the following three factors (Supplementary Fig. 22): First, the transformed membrane is continuous, which is the basis for good performance. Second, as mentioned above, the reagents first enter into of the {100} facets of CuBTC and then the transformation occurs, the peak at 11.6° in XRD pattern displays the exposed {111} facets of the residual CuBTC, so the triangular shaped window with a diameter of 3.5 Å in the residual and the exposed {111} facets is a main channel for gas to pass through. This window size is similar to the kinetic diameter of the CO_2, O_2 and N_2, and is much larger and smaller than that of H_2 and CH_4, respectively. Thus, H_2 molecules can penetrate the transformed membrane quickly. Third, the massive amorphous $FeCl_3$ can fill the gaps between MIL-100 and CuBTC and also occupy the cavities and pores of MOFs. As a result, the void interfaces can be eliminated, the gas channel can be reduced and the gas selectivities of the transformed membrane are increased. Because of the tailorable pore sizes of MOF materials and the diversity in membrane synthesis, we envisage there would be some optimal MOFs and proper preparation conditions, where the membranes with smaller thickness and better separation performance can be achieved by transformation in the future.

Discussion

We have developed a facile and general methodology for realizing the connection and complete transformation of different series of MOF membranes and particles based on multivalent cation substitution. Through this strategy, the unstable and easily fabricated MOF particles can be transformed to obtain the stable MOFs with completely different topology structure, which are usually fabricated in relatively harsh synthetic conditions. The common MOF membranes can also be *in situ* transformed to another MOF membrane, which is hard to be synthesized in conventional methods so far. The pore size can be controlled through immobilizing the metal salt residue in cavities and the appropriate crystal facets of the MOFs can be exposed to achieve competitive molecular sieving ability by the transformation. The typically transformed CuBTC/MIL-100 membrane with good stability exhibits 89, 171, 241 and 336 times higher H_2 permeance than that of CO_2, O_2, N_2 and CH_4, respectively. The method can be used more generally to various MOF membranes and particles with great potential in wide applications.

Methods

Synthesis of CuBTC. $Cu(NO_3)_2{\bullet}3H_2O$ and BTC were dissolved in water and ethanol, respectively. For crystallization, the two solutions were mixed and transferred into a Teflon-lined autoclave. Then, the reaction mixture was heated. After cooling to room temperature, the blue particles were isolated by centrifugation. Eventually, CuBTC particles were washed by ethanol and methanol and dried.

Synthesis of MOF-5. $Zn(NO_3)_2{\bullet}6H_2O$ and BDC were dissolved into DMF, respectively. For crystallization, the two solutions were mixed and heated in an autoclave[50]. After cooling to room temperature, the white particles were separated by centrifugation. Eventually, the prepared particles were washed by DMF and methanol and dried.

Synthesis of CuBDC. $Cu(NO_3)_2{\bullet}3H_2O$ and BDC were dissolved into the DMF, respectively. For crystallization, the two solutions were mixed and heated[34]. After cooling, the blue particles were isolated by centrifugation. Eventually, CuBTC particles were washed by DMF and methanol and dried.

Synthesis of CuBTC with mainly exposed {100} facet. $Cu(NO_3)_2{\bullet}3H_2O$ and lauric acid were dissolved into ethanol (10 ml), BTC was also dissolved in ethanol (10 ml). For crystallization, the two solutions were mixed and heated at 150 °C for 24 h (ref. 30). After natural cooling, the resulting blue powders were isolated by centrifugation and washed with ethanol.

Synthesis of small CuBTC with mainly exposed {111} facet. $Cu(NO_3)_2{\bullet}3H_2O$ and BTC were dissolved in water and ethanol, respectively, to obtain the solutions with concentration of 80 mmol l^{-1}. For crystallization, the BTC solution was added into the metal salt solution and stirred at room temperature for 18 h. After reaction, the blue CuBTC crystals were separated by centrifugation and washed by ethanol.

Transformation of CuBTC to MIL-100. $FeCl_3{\bullet}6H_2O$ (2.00 g) was dissolved in different solvents. CuBTC (0.1 g) was dispersed into the above solution. For transformation, the obtained suspension was fixed into a shaker. Reaction was carried out at room temperature. The reaction time was 1, 2, 4, 8 or 12 h. After reaction, the prepared particles were separated by centrifugation (5,000 r.p.m., 5 min) and purified with methanol. The obtained MIL-100 was purified with various purification cycles, two cycles in methanol (2-M), two cycles in methanol and then in hot water at 70 °C (2-M/2-W). To further identify the nanoclusters in MIL-100, the obtained particles were dried and calcined at 200 °C for 4 h. As comparison, water, ethanol and DMF were all employed as the solvent for transformation. To demonstrate the importance of the CuBTC and that the transformation was not a hydrolysis-recrystallization process, the $Cu(NO_3)_2{\bullet}3H_2O$ (0.124 g) and BTC (0.066 g) were applied to displace the CuBTC, which possesed the same molar quantity as the CuBTC. The result revealed that the CuBTC was vital for transformation.

Transformation of MOF-5 to CuBDC. MOF-5 (0.1 g) was dispersed into $Cu(NO_3)_2{\bullet}3H_2O$ methanol solution (0.05 g ml^{-1}). The above suspension was fixed into a shaker. Transformation was carried out at room temperature. The prepared blue particles were collected by centrifugation and purified by methanol and dried at 80 °C.

Transformation of CuBDC to MIL-53. CuBDC (0.1 g) was dispersed into $FeCl_3 \cdot 6H_2O$ methanol solution (0.05 g ml^{-1}). The above suspension was fixed into a shaker. Transformation was carried out at room temperature for 12 h. The prepared particles were collected by centrifugation and purified by methanol and dried at 80 °C.

Synthesis of CuBTC membrane. For dopamine modification[48], the polyvinylidene difluoride (PVDF) was washed using water and dried at room temperature to remove impurities on membrane surface, and levodopa was added in 10 mM Tris-HCl to obtained transparent solution. Then, the PVDF hollow fibre was immersed in the prepared solution for dopamine deposition. After deposition, the membrane was washed using water and dried at atmosphere. For non-activation ZnO array growth, 0.592 g of zinc nitrate hexahydrate, 0.135 g of sodium formate and 0.245 g of 2-methyl-imidazole were dissolved in methanol and transferred into an autoclave. The modified PVDF hollow fibre was immersed vertically into the mixture solution by using a self-made Teflon holder. Then, the autoclave was sealed and heated at 80 °C. After reaction for 12 h, PVDF hollow fibre with non-activation ZnO array was taken out and washed by ultrasound for 60 s to remove the loose powder. Ultimately, the hollow fibre was washed by methanol and dried at atmosphere. For synthesis, $Cu(NO_3)_2{\bullet}3H_2O$ (0.5 g) and BTC (0.25 g) were dissolved in water and ethanol, respectively. These two solutions were mixed to obtain a clear precursor solution. The mixed solution was poured into an autoclave, and the PVDF hollow fibre with non-activation ZnO array was also placed vertically in the autoclave by using a self-made Teflon holder. The crystallization was executed at 85 °C for 48 h. After crystallization, the hollow fibre membrane was taken out, washed with ethanol and dried at room temperature.

Transformation of CuBTC membrane. $FeCl_3 \bullet 6H_2O$ (1.00 g) was dissolved into the methanol. CuBTC hollow fibre membrane was soaked into the prepared solution at room temperature for 12 or 48 h. After reaction, the membrane was taken out and dried directly for obtaining the dense CuBTC/MIL-100 membrane. As comparison, the membrane after transformation was immersed in the pure methanol to remove the excess $FeCl_3$ component.

Gas permeation measurement. The gas permeation properties of the membrane were studied by constant-pressure, variable-volume method[4]. The dense membranes were put in a permeation module and sealed by epoxy glue. The effective area was calculated by the outer surface. The measured gas was used to rinse the permeation module. The feed gas and permeate gas were fed and collected at the shell side and tube side of the membrane, respectively. Upstream pressure and downstream pressure were 2 and 1 bar (atmosphere conditions), respectively, and transmembrane pressure was 1 bar. The experiment was carried out with different kinetic diameters in the following order: H_2 (0.289 nm), CO_2 (0.33 nm), O_2 (0.346 nm), N_2 (0.364 nm), CH_4 (0.38 nm), CH_4, N_2, O_2, CO_2 and H_2. Gas flow rates were measured by a bubble flow-metre. The gas permeation data were calculated by averaging the measured values of two cycles. The data were read and recorded until the system running stably. Gas permeance (P, mol m^{-2} s^{-1} Pa^{-1}) and gas permeability (P_G, *Barrer* = 3.348 × 10^{-16} mol m^{-2} s^{-1} Pa^{-1}) were calculated by using the following equations:

$$P = \frac{1}{p_u - p_d} \times \frac{1}{A} \times \frac{p_d \times \Delta V}{R \times (273.15 + T) \times \Delta t} \quad (1)$$

$$P_G = \frac{1}{p_u - p_d} \times \frac{l}{A} \times \frac{p_d \times \Delta V}{R \times (273.15 + T) \times \Delta t} \quad (2)$$

where p_u and p_d are the upstream pressure and downstream pressure, respectively, A and l are the membrane effective area and membrane thickness (thickness of the MOF layer), R and T are the gas constant value and temperature (Celsius), ΔV and Δt are the volume of the gas through the membrane and the corresponding time. The permselectivity (α) is defined as the ratio of two kinds of gas permeances.

$$\alpha = \frac{P_1}{P_2} \quad (3)$$

Characterization. XRD patterns were recorded by a PANalytical X' Pert PRO X-ray diffractometer with Cu Kα radiation ($\lambda = 0.154056$ nm) at 40 kV and 40 mA. A field-emission scanning electron microscope (S-4700, Hitachi) was used to observe the morphologies of the membranes. Accelerating voltage was 15 kV. The attached X-ray EDS (GENESIS4000, EDAX) was applied to analyse the element content of prepared MIL-100 particle after tabletting. To keep the cross-section morphology the MOF membranes, it was freeze-fractured in liquid nitrogen. All the prepared samples were coated with an ultrathin layer of platinum using an ion sputter coater to minimize charging effects. X-ray photoelectron spectroscopy experiments were performed on a RBD upgraded PHI-5000C ESCA system (Perkin Elmer) with an incident radiation of monochromatic Mg Kα X-rays ($h\nu = 1253.6$ eV) at 250 W. The spectra of all the elements were collected by using RBD 147 interface (RBD Enterprises). A JEM-2100 (JEOL Co.) operated with accelerating voltage of 200 kV was used for obtaining the TEM images. TG measurements were executed on a thermal gravimetric analyser (PERKIN ELMER, Model TGA 7). The samples were heated from 25 or 40 to 700 °C with a heating rate of 20 °C min^{-1} under a flow of synthetic air with a flow rate of 20 ml min^{-1}. N_2 adsorption–desorption isotherms were measured on a Micromeritics-Accelerated Surface Area and Porosimetry system (ASAP 2020M + C, Micromeritics Instrument Co.). Measurements were carried out at 77 K held using a liquid nitrogen bath. The samples were degassed in vacuum at 150 °C for 12 h before the analysis. BET method was used to calculate the specific surface areas in the P/P_0 range of 0.05–0.1. DFT was used to obtaining the pore-size distributions.

References

1. de Vos, R. M. & Verweij, H. High-selectivity, high-flux silica membranes for gas separation. *Science* **279,** 1710–1711 (1998).
2. Lai, Z. P. *et al.* Microstructural optimization of a zeolite membrane for organic vapor separation. *Science* **300,** 456–460 (2003).
3. Guan, G., Tanaka, T., Kusakabe, K., Sotowa, K. & Morooka, S. Characterization of AIPO(4)-type molecular sieving membranes formed on a porous alpha-alumina tube. *J. Membr. Sci.* **214,** 191–198 (2003).
4. Kim, H. W. *et al.* Selective gas transport through few-layered graphene and graphene oxide membranes. *Science* **342,** 91–95 (2013).
5. Li, H. *et al.* Ultrathin, molecular-sieving graphene oxide membranes for selective hydrogen separation. *Science* **342,** 95–98 (2013).
6. Brown, A. J. *et al.* Interfacial microfluidic processing of metal-organic framework hollow fiber membranes. *Science* **345,** 72–75 (2014).
7. Peng, Y. *et al.* Metal-organic framework nanosheets as building blocks for molecular sieving membranes. *Science* **346,** 1356–1359 (2014).
8. Robeson, L. M. The upper bound revisited. *J. Membr. Sci* **320,** 390–400 (2008).
9. Furukawa, H., Cordova, K. E., O'Keeffe, M. & Yaghi, O. M. The chemistry and applications of metal-organic frameworks. *Science* **341,** 1230444 (2013).
10. Qiu, S., Xue, M. & Zhu, G. Metal-organic framework membranes: from synthesis to separation application. *Chem. Soc. Rev.* **43,** 6116–6140 (2014).
11. Han, Y., Li, J. R., Xie, Y. & Guo, G. Substitution reactions in metal-organic frameworks and metal-organic polyhedral. *Chem. Soc. Rev.* **43,** 5952–5981 (2014).
12. Brozek, C. K. & Dincă, M. Cation exchange at the secondary building units of metal-organic frameworks. *Chem. Soc. Rev.* **43,** 5456–5467 (2014).
13. Dincă, M. & Long, J. R. High-enthalpy hydrogen adsorption in cation-exchanged variants of the microporous metal-organic framework $Mn_3[(Mn_4Cl)_3(BTT)_8(CH_3OH)_{10}]_2$. *J. Am. Chem. Soc.* **129,** 11172–11176 (2007).
14. Nouar, F., Eckert, J., Eubank, J. F., Forster, P. & Eddaoudi, M. Zeolite-like metal-organic frameworks (ZMOFs) as hydrogen storage platform: lithium and magnesium ion-exchange and H_2-(rho-ZMOF) interaction studies. *J. Am. Chem. Soc.* **131,** 2864–2870 (2009).
15. Kim, M., Cahill, J. F., Fei, H., Prather, K. A. & Cohen, S. M. Postsynthetic ligand and cation exchange in robust metal-organic frameworks. *J. Am. Chem. Soc.* **134,** 18082–18088 (2012).
16. Li, T., Kozlowski, M. T., Doud, E. A., Blakely, M. N. & Rosi, N. L. Stepwise ligand exchange for the preparation of a family of mesoporous MOFs. *J. Am. Chem. Soc.* **135,** 11688–11691 (2013).
17. Brozek, C. K. & Dincă, M. Ti^{3+}-, $V^{2+/3+}$-, $Cr^{2+/3+}$-, Mn^{2+}-, and Fe^{2+}-Substituted MOF-5 and redox reactivity in Cr- and Fe-MOF-5. *J. Am. Chem. Soc.* **135,** 12886–12891 (2013).
18. Chui, S. S. Y., Lo, S. M. F., Charmant, J. P. H., Guy Orpen, A. & Williams, I. D. A chemically functionalizable nanoporous material $[Cu_3(TMA)_2(H_2O)_3]_n$. *Science* **283,** 1148–1150 (1999).
19. Majano, G. & Pérez-Ramírez, J. Scalable room-temperature conversion of copper(II) hydroxide into HKUST-1 ($Cu_3(btc)_2$. *Adv. Mater.* **25,** 1052–1057 (2013).
20. Burtch, N. C., Jasuja, H. & Walton, K. S. Water stability and adsorption in metal-organic frameworks. *Chem. Rev.* **114,** 10575–10612 (2014).
21. Férey, G. *et al.* A chromium terephthalate-based solid with unusually large pore volumes and surface area. *Science* **309,** 2040–2042 (2005).
22. Horcajada, P. *et al.* Synthesis and catalytic properties of MIL-100(Fe), an iron(III) carboxylate with large pores. *Chem. Commun.* 2820–2822 (2007).
23. Márquez, A. G. *et al.* Green microwave synthesis of MIL-100(Al, Cr, Fe) nanoparticles for thin-film elaboration. *Eur. J. Inorg. Chem.* 5165–5174 (2012).
24. Zhang, F. *et al.* Facile synthesis of MIL-100(Fe) under HF-free conditions and its application in the acetalization of aldehydes with diols. *Chem. Eng. J.* **259,** 183–190 (2015).
25. Mao, Y. *et al.* General incorporation of diverse components inside metal-organic framework thin films at room temperature. *Nat. Commun.* **5,** 5532 (2014).
26. Li, W. *et al.* Assembly of MOF microcapsules with size-selective permeability on cell walls. *Angew. Chem. Int. Ed.* **55,** 955–959 (2016).
27. Zhao, J. *et al.* Facile conversion of hydroxy double salts to metal-organic frameworks using metal oxide particles and atomic layer deposition thin-film templates. *J. Am. Chem. Soc.* **137,** 13756–13759 (2015).
28. Wehring, M. *et al.* Self-diffusion studies in CuBTC by PFG NMR and MD simulations. *J. Phys. Chem. C* **114,** 10527–10534 (2010).
29. Mao, Y. *et al.* Specific oriented metal-organic framework membranes and their facet-tuned separation performance. *ACS Appl. Mater. Interfaces* **6,** 15676–15685 (2014).
30. Umemura, A. *et al.* Morphology design of porous coordination polymer crystals by coordination modulation. *J. Am. Chem. Soc.* **133,** 15506–15513 (2011).
31. Li, H. L., Eddaoudi, M., O'Keeffe, M. & Yaghi, O. M. Design and synthesis of an exceptionally stable and highly porous metal organic framework. *Nature* **402,** 276–279 (1999).
32. Tranchemontagne, D. J., Hunt, J. R. & Yaghi, O. M. Room temperature synthesis of metal-organic frameworks: MOF-5, MOF-74, MOF-177, MOF-199, and IRMOF-0. *Tetrahedron* **64,** 8553–8557 (2008).
33. Mori, W. *et al.* Synthesis of new adsorbent copper(II) terephthalate. *Chem. Lett.* **26,** 1219–1220 (1997).
34. Adams, R., Carson, C., Ward, J., Tannenbaum, R. & Koros, W. Metal organic framework mixed matrix membranes for gas separations. *Microporous Mesoporous Mater.* **131,** 13–20 (2010).
35. Carson, C. G. *et al.* Synthesis and structure characterization of copper terephthalate metal-organic frameworks. *Eur. J. Inorg. Chem.* 2140–2145 (2014).
36. Rodenas, T. *et al.* Metal-organic framework nanosheets in polymer composite materials for gas separation. *Nat. Mater.* **14,** 48–55 (2015).

37. Serre, C. *et al.* Very large breathing effect in the first nanoporous chromium(III)-based solids: MIL-53 or Cr-III(OH){O_2C-C_6H_4-CO_2}{HO_2C-C_6H_4-CO_2H}(x)H_2O_y. *J. Am. Chem. Soc.* **124,** 13519–13526 (2002).
38. Horcajada, P. *et al.* Porous metal-organic-framework nanoscale carriers as a potential platform for drug delivery and imaging. *Nat. Mater.* **9,** 172–178 (2009).
39. Llewellyn, P. L. *et al.* Complex adsorption of short linear alkanes in the flexible metal-organic-framework MIL-53(Fe). *J. Am. Chem. Soc.* **131,** 13002–13008 (2009).
40. Gordon, J., Kazemian, H. & Rohani, S. Rapid and efficient crystallization of MIL-53(Fe) by ultrasound and microwave irradiation. *Microporous Mesoporous Mater.* **162,** 36–43 (2012).
41. Robens, E., Rouquerol, F. & Rouquerol, J. *Adsorption by Powders and Porous Solids* (WILEY-VCH Verlag, 1999).
42. Canioni, R. *et al.* Stable polyoxometalate insertion within the mesoporous metal organic framework MIL-100(Fe). *J. Mater. Chem.* **21,** 1226–1233 (2011).
43. Ahmed, I., Jun, J. W., Jung, B. K. & Jhung, S. H. Adsorptive denitrogenation of model fossil fuels with Lewis acid-loaded metal-organic frameworks (MOFs). *Chem. Eng. J.* **255,** 623–629 (2014).
44. Tan, F. *et al.* Facile synthesis of size-controlled MIL-100(Fe) with excellent adsorption capacity for methylene blue. *Chem. Eng. J.* **281,** 260–367 (2015).
45. Li, W. *et al.* Non-activation ZnO array as a buffering layer to fabricate strongly adhesive metal-organic framework/PVDF hollow fiber membranes. *Chem. Commun.* **50,** 9711–9713 (2014).
46. Guo, H., Zhu, G., Hewitt, I. J. & Qiu, S. "Twin copper source" growth of metal-organic framework membrane: $Cu_3(BTC)_2$ with high permeability and selectivity for recycling H_2. *J. Am. Chem. Soc.* **131,** 1646–1647 (2009).
47. Carta, M. *et al.* An efficient polymer molecular sieve for membrane gas separations. *Science* **339,** 303–307 (2013).
48. Liu, Q., Wang, N., Caro, J. & Huang, A. Bio-inspired polydopamine: a versatile and powerful platform for covalent synthesis of molecular sieve membranes. *J. Am. Chem. Soc.* **135,** 17679–17682 (2013).
49. Zhang, F. *et al.* Hydrogen selective NH_2-MIL-53(Al) MOF membranes with high permeability. *Adv. Funct. Mater.* **22,** 3583–3590 (2012).
50. Yoo, Y. & Jeong, H. K. Heteroepitaxial growth of isoreticular metal-organic frameworks and their hybrid films. *Cryst. Growth Des.* **10,** 1283–1288 (2010).

Acknowledgements

This work was supported by the National Natural Science Foundation of China (Grant Nos. 21236008 and 21476206), Fujian Provincial Department of Ocean (Grant No. 2014-06), Zhejiang Provincial Bureau of Science and Technology (Grant No. 2014C33032), Taishan Scholarship Blue Industry Program from Shandong Provincial Government (Grant No. 2014008) and Huzhou Gemking Biotechnology Co. Ltd. G.Z. thanks Fujian Provincial Government for the Minjiang Scholarship. W.L. also thanks Chinese Government for the first-level prize fellowship.

Author contributions

G.Z. and W.L. conceived the research idea. W.L., Y.Z., C.Z., Z.X. and P.S. synthesized the MOF membranes and particles. W.L., Y.Z., Z.X., P.S., C.S., L.Q. and Z.F. carried out related characterizations and measurements. Q.M., Q.L. and Y.Z. contributed to the general methodology and reviewed the manuscript. G.Z. supervised the project, helped design the experiments and co-drafted the manuscript. All authors contributed to the analysis of the manuscript.

Additional information

Competing financial interests: A patent application related to this work has been filed.

21

A modular design of molecular qubits to implement universal quantum gates

Jesús Ferrando-Soria[1], Eufemio Moreno Pineda[1], Alessandro Chiesa[2], Antonio Fernandez[1], Samantha A. Magee[1], Stefano Carretta[2], Paolo Santini[2], Iñigo J. Vitorica-Yrezabal[1], Floriana Tuna[1], Grigore A. Timco[1], Eric J.L. McInnes[1] & Richard E.P. Winpenny[1]

The physical implementation of quantum information processing relies on individual modules—qubits—and operations that modify such modules either individually or in groups—quantum gates. Two examples of gates that entangle pairs of qubits are the controlled NOT-gate (CNOT) gate, which flips the state of one qubit depending on the state of another, and the $\sqrt{\mathrm{iSWAP}}$ gate that brings a two-qubit product state into a superposition involving partially swapping the qubit states. Here we show that through supramolecular chemistry a single simple module, molecular $\{Cr_7Ni\}$ rings, which act as the qubits, can be assembled into structures suitable for either the CNOT or $\sqrt{\mathrm{iSWAP}}$ gate by choice of linker, and we characterize these structures by electron spin resonance spectroscopy. We introduce two schemes for implementing such gates with these supramolecular assemblies and perform detailed simulations, based on the measured parameters including decoherence, to demonstrate how the gates would operate.

[1] School of Chemistry and Photon Science Institute, The University of Manchester, Oxford Road, Manchester M13 9PL, UK. [2] Dipartimento di Fisica e Scienze della Terra, Università di Parma, viale delle Scienze 7/a, Parma 43123, Italy. Correspondence and requests for materials should be addressed to R.E.P.W. (email: richard.winpenny@manchester.ac.uk).

The inability of conventional computers to solve certain problems efficiently, such as the simulation of quantum systems[1,2], is one of the main driving forces for the implementation of quantum computing and quantum information processing (QIP)[3,4], which exploit the laws of quantum mechanics. At the theoretical level, several algorithms have been shown to outperform classical computers in certain computational tasks such as factoring large numbers into primes[5] and searching of unsorted directories[6]. A variety of quantum systems have shown an excellent performance as the basic units for quantum information (qubits)[7–11]. However, they are difficult to link controllably into useful arrays producing the two-qubit entangling quantum gates (QGs) crucial for any quantum algorithm. Among the most important entangling QGs are the controlled NOT-gate (CNOT) and the $\sqrt{iSWAP}$ gate. The $\sqrt{iSWAP}$ gate brings the two-qubit state $|01\rangle$ to the superposition $(|01\rangle + i|10\rangle)/\sqrt{2}$. The CNOT gate flips the state of the target qubit if, and only if, the control qubit is in the $|1\rangle$ state; this implies that each qubit has to respond inequivalently to an external stimulus. All other quantum gates, however complex, can be constructed from these two-qubit gates and single-qubit gates.

Therefore the major challenge for the physical implementation of QIP is bringing together qubits in an organized, scalable and addressable way to make such QGs[3,4]. Here we demonstrate that supramolecular chemistry could have a major impact in addressing this challenge.

Molecular nanomagnets have been proposed as qubits[12–26] and coherence times of individual molecular qubits have been studied and have improved[24,25]. Supramolecular chemistry[27] allows us to bring together, with great control, complex molecules into arrays, and recently this has been exploited to tune two-qubit interactions built from molecular nanomagnets[28]. Hence supramolecular chemistry is a promising tool to build multi-qubit devices; if this could be made scalable this is a competitive route towards the realization of a quantum computer.

Demonstrations of CNOT gates based on molecular nanomagnets have been published[18,19,23]. The interaction between two dissimilar qubits produces a splitting in the low-lying two-qubit states that allows selective addressing of the transition by means of resonant EPR pulses, while keeping the other components of the wave-function frozen. In this way, the excitation of the target depends on the state of the control qubit and a CNOT gate is implemented. The drawback of this approach is the permanent direct coupling between the qubits, which makes these proposals hard to scale as with a permanent coupling the state of the qubits experiences an unwanted many-body spontaneous evolution in time (as in NMR QIP schemes[29]), whose harmful effects increase with the number of qubits. Conversely, a switchable indirect coupling between the two qubits would make the register much more easily scalable as when the switch is in the off state the unwanted spontaneous evolution is suppressed.

We have proposed using $\{Cr_7Ni\}$ heterometallic rings as qubits[14]. They are two-level systems ($S = 1/2$ ground state) that have sufficiently long phase memory times to allow many gate operations before state degradation occurs[25], and assemblies of $\{Cr_7Ni\}$ rings have been made which show a permanent coupling between the spins[26]. Here we report two-qubit assemblies which include switchable links, that allow us to propose a fully modular supramolecular design strategy[27,28] towards quantum computation schemes based on either the CNOT or $\sqrt{iSWAP}$ gate. The strategy uses supramolecular chemistry to tailor the individual components, spatial configuration and hence the properties of the resulting supramolecules. These supramolecules have been studied by electrochemistry, continuous-wave and pulsed EPR spectroscopy to understand their static and dynamic spin properties. These measured parameters have been used to perform detailed simulations of the performance of both the CNOT and $\sqrt{iSWAP}$ gates including the effects of decoherence. The two gates we propose are based on either global or local control of the qubit–qubit interaction. The first proposal exploits uniform magnetic pulses to manipulate two inequivalent Cr_7Ni qubits in an asymmetric supramolecule, which would implement the CNOT gate. The second exploits the local electric control of a redox-active linker, which might be addressed by a tip to reversibly switch on and off the qubit–qubit interaction, to implement the $\sqrt{iSWAP}$ gate. The scalability of these approaches is discussed in the last part of the paper.

Results

Syntheses and structural characterization. We have earlier reported the selective functionalization of $[^nPr_2NH_2][Cr_7NiF_8(O_2C^tBu)_{16}]$ **1** (ref. 30) with *iso*-nicotinate (O_2C-py) to obtain $[^nPr_2NH_2][Cr_7NiF_8(O_2C^tBu)_{15}(O_2C\text{-py})]$ **2** (ref. 26) (Supplementary Fig. 1a). This synthetic method can be extended to produce $[^nPr_2NH_2][Cr_7NiF_8(O_2C^tBu)_{15}(O_2C\text{-terpy})]$ **3** and $[^nPr_2NH_2][Cr_7NiF_8(O_2C^tBu)_{15}(O_2C\text{-Ph-terpy})]$ **4** (Supplementary Fig. 1b,c), from the controlled reaction of **1** with 4-carboxy-2,2′:6′,2″-terpyridine (O_2C-terpy) and 4′-(4-carboxyphenyl)-2,2′:6′,2″-terpyridine (O_2C-Ph-terpy), respectively (full experimental details are given in the Supplementary Methods)[31]. **2**, **3** and **4** are hereafter abbreviated as $\{Cr_7Ni\text{-}O_2C\text{-py}\}$, $\{Cr_7Ni\text{-}O_2C\text{-terpy}\}$ and $\{Cr_7Ni\text{-}O_2C\text{-Ph-terpy}\}$, respectively. They consist of $Cr^{III}{}_7Ni^{II}$ rings, containing an octagon of metal centres, with the inner rim bridged by fluoride ions and the outer rim by carboxylates: the functionalized O_2C-py **2**, O_2C-terpy **3** and O_2C-Ph-terpy **4** ligands sit on a Cr…Ni edge of the octagon. This provides us with three supramolecular qubits with two different denticities, which allow us to assemble different QGs by appropriate choice of the central node (Fig. 1).

Reaction of equimolar quantities of **2** and **3** with cobalt(II) thiocyanate in a mixture of Et_2O/acetone leads to $[\{Cr_7Ni\text{-}O_2C\text{-py}\}\rightarrow Co(SCN)_2\leftarrow\{Cr_7Ni\text{-}O_2C\text{-terpy}\}]$ **5** in good yield, which has been characterized by X-ray single crystal diffraction (Fig. 2a; for all crystallographic information see Supplementary Data 1). Compound **5** contains two inequivalent Cr_7Ni qubits coordinated to a central cobalt(II) ion, which has a six-coordinate CoN_6 octahedral environment with a *cis*-arrangement of the two thiocyanate N atoms. The Co-N bond distances are typical of high spin cobalt(II) ions[32], with the bonds to the thiocyanate ligands shorter (average 2.056(12) Å) than those to terpy or pyridine N-donors (2.137(6) to 2.224(4) Å). Hence the d^7 Co^{II} site ($S_{Co} = {}^3/_2$) has a $^4T_{1g}$ ground term (using O_h symmetry labels)[33], which leads to a well-isolated effective $S_{eff} = {}^1/_2$ ground state at low temperature (see Characterization of assemblies via EPR spectroscopy below). The *cis* coordination geometry at the Co^{II} node means the $\{Cr_7Ni\text{-}O_2C\text{-terpy}\}$- and $\{Cr_7Ni\text{-}O_2C\text{-py}\}$-based qubits are arranged in an almost orthogonal orientation. Therefore the two qubits in **5** are symmetry inequivalent, as required for implementing a CNOT gate (Fig. 2b).

Reaction of two equivalents of **2** with a preformed oxo-centred pivalate-bridged triangular cluster with terminal pyridine groups $[Ru^{III}{}_2Co^{II}O(O_2C^tBu)_6(py)_3]$ **6**, hereafter abbreviated as $[Ru^{III}{}_2Co^{II}]$, in acetone gives $[\{Cr_7Ni\text{-}O_2C\text{-py}\}\rightarrow[Ru^{III}{}_2Co^{II}O(^tBuCO_2)_6(py)]\leftarrow\{Cr_7Ni\text{-}O_2C\text{-py}\}]$ **7** (Fig. 3a), where two of the terminal pyridine ligands of **6** were replaced by the *iso*-nicotinate group of **2**. Reaction of two equivalents of **3** or **4** and either cobalt(II) perchlorate or tetrafluoroborate gives $[\{Cr_7Ni\text{-}O_2C\text{-terpy}\}\rightarrow Co\leftarrow\{Cr_7Ni\text{-}O_2C\text{-terpy}\}][X]_2$ [$X = ClO_4^-$ **8a** or BF_4^- **8b**] and $[\{Cr_7Ni\text{-}O_2C\text{-Ph-terpy}\}\rightarrow Co\leftarrow\{Cr_7Ni\text{-}O_2C\text{-Ph-terpy}\}][X]_2$ [$X = ClO_4^-$ **9a** or BF_4^- **9b**] (Fig. 3b,c) (full experimental details are given in the Supplementary Methods). The

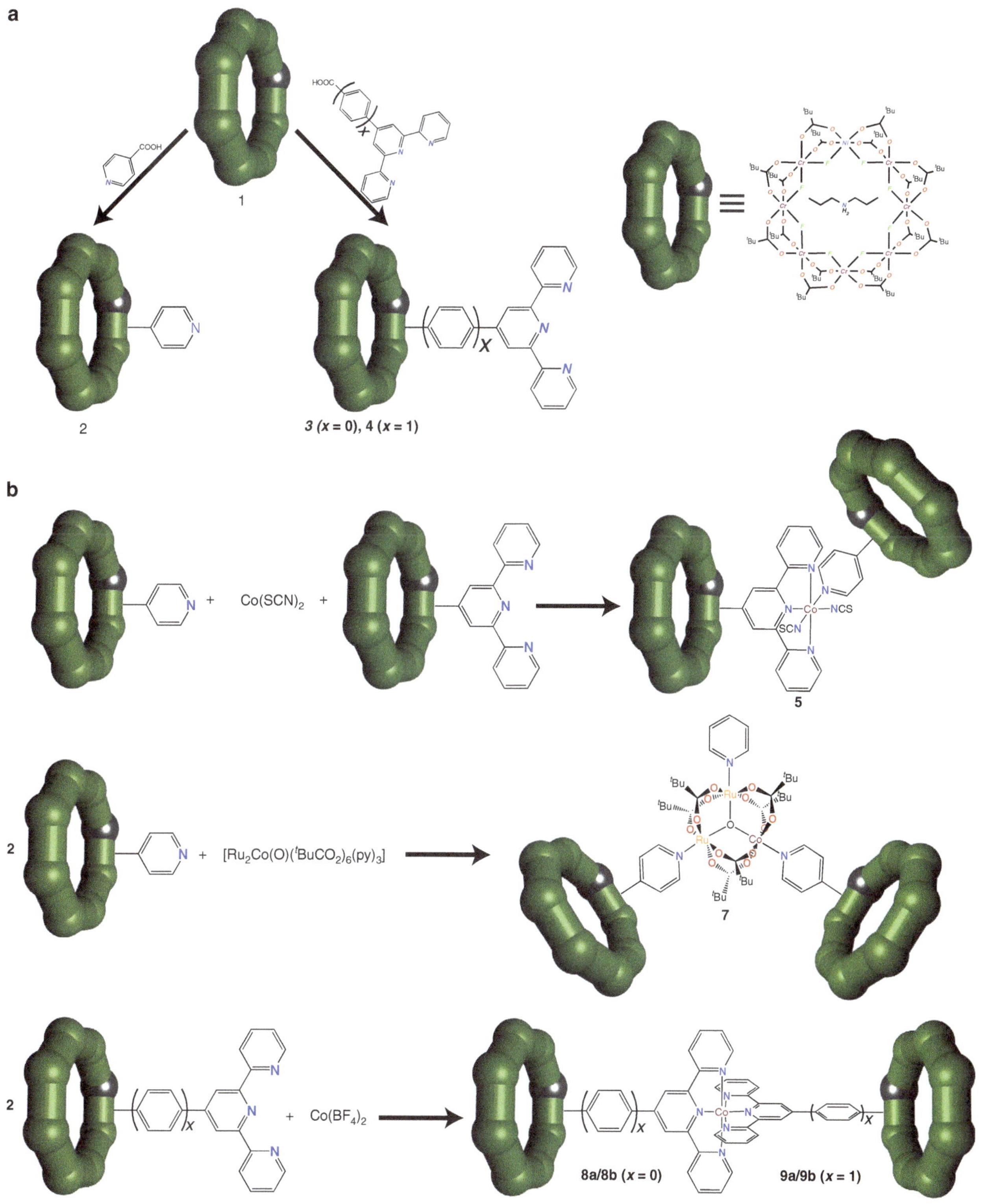

Figure 1 | Supramolecular design strategy for the construction of two-qubits assemblies. (**a**) Synthesis of functionalized {Cr_7Ni} rings (where $x = 0$ and 1). (**b**) Linkage of {Cr_7Ni} coordination cages to different central nodes.

architecture in **7**, **8** and **9** contains two equivalent qubits, separated by a redox-switchable centre[34,35], which makes it suitable for implementation of a $\sqrt{iSWAP}$ gate (Fig. 3d).

The crystal structure of **7** consists of two {Cr_7Ni-O_2C-py} rings linked through the *iso*-nicotinate groups to a [$Ru^{III}{}_2Co^{II}O$ ($^tBuCO_2)_6$(py)] triangle (Fig. 3a)[34,36]. The metal ions are statistically disordered over the three sites within the triangular M_3O unit. They have a six-coordinate, octahedral MO_5N environment formed by four carboxylate oxygen atoms from the bridging pivalate ligands in the equatorial plane, with the central oxide and one nitrogen atom from *iso*-nicotinate or pyridine groups occupying the axial positions. The three metal ions are positioned at the corners of an isosceles triangle with two distinct intermetallic distances of 3.296 and 3.315 Å. The central oxide lies within the M_3 plane of the metal atoms, while the six pivalate bridging ligands lie above and below the M_3O plane and the pyridine N-donors rest perpendicular to this plane.

The crystal structure of **8a** and **9a** are made up of cationic [{Cr_7Ni-O_2C-terpy}Co^{II}{Cr_7Ni-O_2C-terpy}] and [{Cr_7Ni-O_2C-Ph-terpy}Co^{II}{Cr_7Ni-O_2C-Ph-terpy}] units together with counter-balancing ClO_4^- anions (Fig. 3b,c). Both compounds have a central cobalt(II) ion with a six-coordinate CoN_6 octahedral environment, with the six Co-N bonds in the range expected for a low spin cobalt(II) ion[37]. The average value of the two N-donors from the central pyridines of the terpy are shorter (1.923(6) **8a** and 1.920(8) **9a** Å) than the other Co-N contacts (2.118(7) **8a** and 1.986(9) **9a** Å). This feature leads to an axially compressed octahedral environment for the low-spin d^7 Co^{II} ions giving a $S_{Co} = {}^1/_2$ ground state. Compounds **8a** and **9a** differ in the shortest Co…M(ring) contact (8.675(2) **8a** and 10.979(5) **9a** Å) and the staggered **8a** or eclipsed **9a** arrangement of rings in the assemblies.

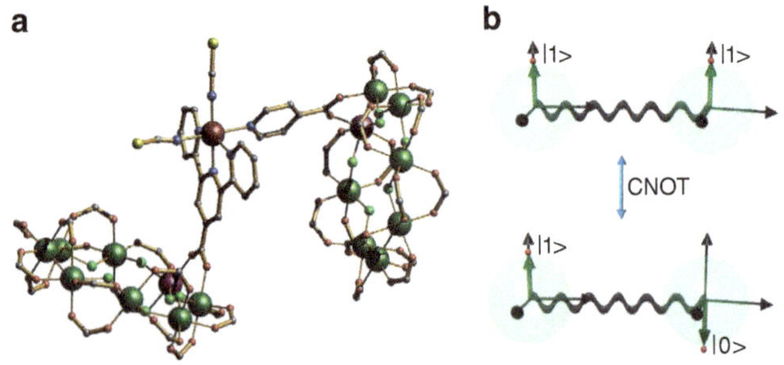

Figure 2 | Molecular structure of the asymmetric two-qubit assembly for proposed implementation of a CNOT gate. (**a**) Molecular structure of [{Cr_7Ni-O_2C-py}→Co$(SCN)_2$←{Cr_7Ni-O_2C-terpy}] **5**. $^nPr_2NH_2^+$ cations are not shown (H atoms and tert-butyl groups are omitted for clarity). Colour code: Co, dark red; Cr, green; Ni, purple; Ru, brown; N, cyan; O, red; S, yellow; C, grey; F, pale green. (**b**) Schematic representation of the effect of the CNOT gate on a pair of qubits, initialized in the computational basis states $|11\rangle$ and $|10\rangle$, respectively. The CNOT flips the target qubit if the control is set to $|1\rangle$.

The two Cr_7Ni qubits in each of **7–9** are linked by a redox-switchable central node[34,38]. Cyclic voltammetry on **7–9**, and on the central nodes in isolation, show a one-electron reversible oxidation in each case (measured in CH_2Cl_2, 0.1 M nBu_4NPF_6). For **6** and **7** the half-wave potential ($E_{1/2}$) = − 0.31 (**6**), − 0.35 (**7**) V versus Fc^+/Fc (Fc = ferrocene)[39], and this oxidation is assigned to the [$Ru^{III}{}_2Co^{II}$] to [$Ru^{III,IV}{}_2Co^{II}$] couple[34]. For **8b**, **9b** and reference complexes [Co$(HO_2C$-terpy$)_2$]$[BF_4]_2$ **10** and [Co$(HO_2C$-Ph-terpy$)_2$]$[BF_4]_2$ **11**, $E_{1/2}$ = − 0.30 (**8b**), − 0.27 (**9b**), − 0.27 (**10**) and − 0.24 (**11**) V, which are assigned to the $Co^{II/III}$ couple. The anodic to cathodic peak separation values are similar to those of ferrocene under the same conditions (see Electrochemistry section in Supplementary Methods, Supplementary Fig. 4 and Supplementary Table 3).

Characterization of assemblies via EPR spectroscopy. To study the interactions between the molecular components in the supramolecular structures **5**, **7**, **8b** and **9b** we performed magnetometry and multi-frequency continuous wave EPR studies. The measured magnetometry data are essentially the sum of the Cr_7Ni qubits and the linking nodes (Supplementary Figs 5–9), and therefore are uninformative other than confirming that the interactions are very weak. EPR spectroscopy is much more sensitive to such weak interactions.

The EPR spectra of a powder of **5** at 5K (Fig. 4a and Supplementary Fig. 10) have a complex multiplet structure because the S_{eff} = 1/2 of the Co(II) node has a very anisotropic effective g-tensor and also gives rise to very anisotropic exchange tensors (J) with the {Cr_7Ni} rings. This structure is best resolved at W-band (94 GHz; Fig. 4a). The spectra can be simulated[40] with the effective three-spin Hamiltonian (1), with two different

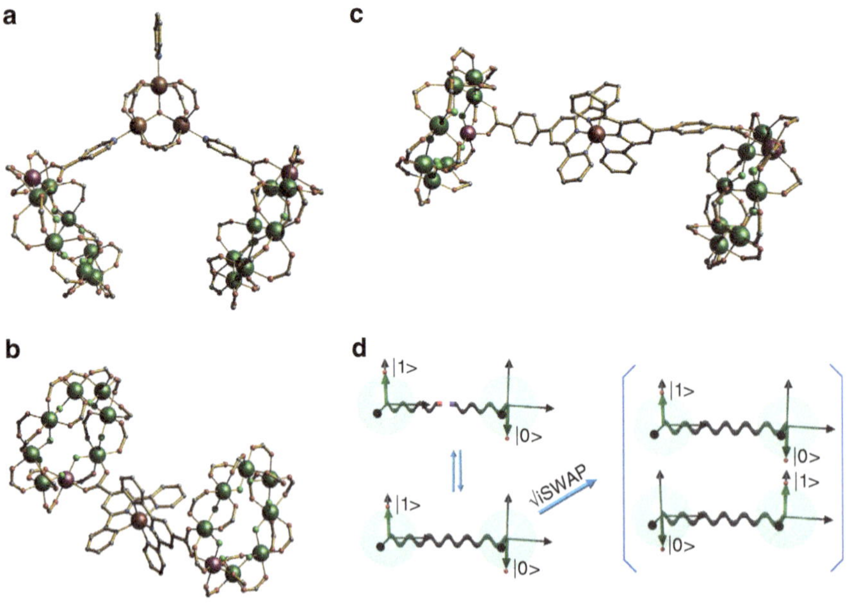

Figure 3 | Molecular structures of the redox-active two-qubit assemblies for proposed implementation of an $\sqrt{i}$SWAP gate. (**a**) [{Cr_7Ni-O_2C-py}→[$Ru^{III}{}_2Co^{II}O(^tBuCO_2)_6(py)$]←{$Cr_7Ni$-$O_2C$-py}] **7**. (**b**) the cation of [{Cr_7Ni-O_2C-terpy}→Co←{Cr_7Ni-O_2C-terpy}]$[ClO_4]_2$ **8a**. (**c**) the cation of [{Cr_7Ni-O_2C-Ph-terpy}→Co←{Cr_7Ni-O_2C-Ph-terpy}]$[ClO_4]_2$ **9a**. $^nPr_2NH_2^+$ cations and ClO_4^- anions are not shown (H atoms and tert-butyl groups are omitted for clarity). Colour codes as Fig. 2. (**d**) Schematic representation of the effect of the $\sqrt{i}$SWAP gate on a pair of qubits, initialized in the computational basis state $|10\rangle$. The gate brings $|10\rangle$ to the equal-weight superposition $(|10\rangle + i|01\rangle)/\sqrt{2}$. In the scheme proposed below, it operates as soon as the inter-qubit interaction is turned on (double arrow).

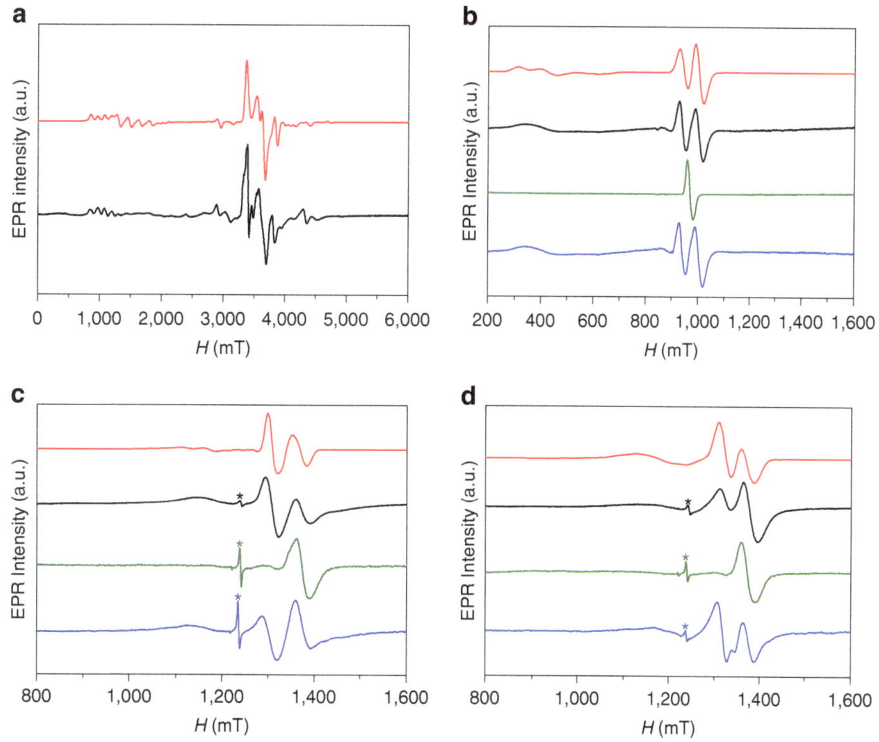

Figure 4 | EPR spectroscopy of QG assemblies. (**a**) Experimental powder W-band (94 GHz) EPR spectrum of **5** at 5K (black) and simulation (red) using Hamiltonian (1) and parameters in Table 1. (**b**) Experimental K-band (24 GHz) EPR spectra of **7** frozen solution (black), and simulation (red) using Hamiltonian (2) and parameters given in the text: experimental spectra after oxidation of **7** to **7b^{ox}** with $[FeCp_2](PF_6)$ (ca. 3 mM, 1:1) (green), and after reducing **7b^{ox}** with cobaltocene (3 mM, 1:1) (blue). (**c**,**d**) Experimental Q-band (34 GHz) EPR spectra of **8b** and **9b**, respectively, in frozen solution (black), and simulations (red) using Hamiltonian (2) and parameters given in the text; experimental spectra after oxidation to **8b^{ox}** and **9b^{ox}** with $AgBF_4$ (ca. 3 mM, 1:1) (green), and after reducing using with cobaltocene (3 mM, 1:1) (blue). The sharp peak marked * is a trace radical impurity.

anisotropic exchange interactions (J_{12} and J_{23}) between the $S_{eff}=½$ of the Co(II) ion (S_2) and the two $S=½$ of {Cr_7Ni-O_2C-terpy} (S_1) and {Cr_7Ni-O_2C-py} (S_3) rings, respectively (Fig. 4a).

$$\hat{H} = -2\left[\left(\hat{S}_1 \cdot \underline{J}_{12} \cdot \hat{S}_2\right) + \left(\hat{S}_2 \cdot \underline{J}_{23} \cdot \hat{S}_3\right)\right] + \mu_B\left(\hat{S}_1 \cdot \underline{g}_1 + \hat{S}_2 \cdot \underline{g}_2 + \hat{S}_3 \cdot \underline{g}_3\right) \cdot \mathbf{B} \quad (1)$$

The multiplets centred at ca. 1, 1.5 and 3.5 T (Fig. 4a) mark the effective g values of the Co(II) ion as $g=6.50$, 4.25 and ca. 2, while the Cr_7Ni rings have well defined $g=1.78$, 1.78 and 1.74 (the unique value being for the orientation perpendicular to the ring)[30]. The large difference (Δg) between the $g=6.5$ and 4.3 orientations of the Co(II) ion and those of the rings, and the weak exchange interactions ($\mu_B \Delta g B \gg |J|$) is useful as it allows us to treat the problem as an *ABX* spin system. The multiplet structure of the $g=6.5$ and 4.3 features is then due to weak exchange with two different $S=1/2$, giving doublets-of-doublets from which we can read the J_y and J_z components of J_{12} and J_{23}. The high field region is more complicated, due to the much smaller difference between the third Co(II) g value and those of the {Cr_7Ni} rings. The remaining parameters (the final components of J_{12}, J_{23} and g_2) were obtained by simulation (Fig. 4a). The parameters in Table 1 are labelled to a common reference frame, that is, taking into account the orthogonal orientation of the two {Cr_7Ni} rings.

7, **8b** and **9b**, contain redox-active linking nodes and their EPR spectra change with the oxidation state. Hence, EPR was performed on both the reduced and oxidized forms at 5K as frozen solutions, allowing cycling of the redox state (Fig. 4b–d). Spectra were measured on the as prepared samples, then solutions were warmed to room temperature and the oxidized forms **7^{ox}**, **8b^{ox}** and **9b^{ox}** generated *in situ* by addition of $[FeCp_2](PF_6)$ or $AgBF_4$ followed by freezing and measurement. To complete the cycle, the solutions were thawed again and **7**, **8b** and **9b** regenerated by reduction with cobaltocene. We have also measured spectra of the isolated nodes in the paramagnetic Co(II) form, viz. complex **6**, and $[Co(HO_2C\text{-terpy})_2](BF_4)_2$ **10** and $[Co(HO_2C\text{-Ph-terpy})_2](BF_4)_2$ **11** (Supplementary Figs 2 and 3). Like **5**, **6** also contains a ^4T high-spin Co(II) ion with an $S_{eff}=1/2$ ground state, giving $g=5.61$, 4.05, 2.77 (Supplementary Fig. 11), while **10** and **11** have low-spin Co(II) hence $S=1/2$, giving $g=2.047$, 2.076, 2.195 (Supplementary Fig. 12) and $g=2.022$, 2.111 and 2.215 (Supplementary Fig. 13), respectively.

Table 1 | Spin-Hamiltonian parameters used for simulation of the W-band EPR spectra of 5.

	g_x	g_y	g_z		J_x (cm^{-1})	J_y (cm^{-1})	J_z (cm^{-1})
{Cr_7Ni-O_2C-terpy}	1.78	1.78	1.74	J_{12}	−0.14	0.34	0.17
{Cr_7Ni-O_2C-py}	1.74	1.78	1.78	J_{23}	−0.07	0.17	0.34
Co(II)	1.78	4.25	6.50				

$\underline{J}_{12}$, $\underline{J}_{23}$ and $\underline{g}=\{g_x g_y g_z\}$ are diagonal tensors.

For the supramolecular structures **7**, **8b** and **9b**, the exchange coupling is nicely resolved in K- and Q-band (24 and 34 GHz) EPR spectra (Fig. 4b–d). In each case there is weak coupling with respect to the difference in Zeeman energies ($\mu_B \Delta g B \gg |J|$)[41]. This, and the equivalence of the two Cr_7Ni rings, gives AB_2 spin systems. The {Cr_7Ni} ring resonances (the '*B*' spins) are split into doublets (Fig. 4b–d), giving a direct measure of J (the exchange splittings of the central node resonances are not resolved). The spectra can be simulated with the simple Hamiltonian (2) using an isotropic exchange interaction between the central nodes (S_2) and the two {Cr_7Ni} rings (S_1 and S_3), even for **7**, which has an effective spin 1/2 centre at the node.

$$\widehat{\mathrm{H}} = -2J\left(\widehat{\mathbf{S}}_1 \cdot \widehat{\mathbf{S}}_2 + \widehat{\mathbf{S}}_2 \cdot \widehat{\mathbf{S}}_3\right) + \mu_B\left(\widehat{\mathbf{S}}_1 \cdot \underline{g}_1 + \widehat{\mathbf{S}}_2 \cdot \underline{g}_2 + \widehat{\mathbf{S}}_3 \cdot \underline{g}_3\right) \cdot \mathbf{B} \quad (2)$$

Simulations with the isotropic exchange interaction J as the only variable gives $J = -0.026$, -0.026 and $-0.024\,\mathrm{cm}^{-1}$ for **7**, **8b** and **9b**, respectively; the g values were fixed to those measured for individual components. The EPR spectra of **7^{ox}**, **8b^{ox}** and **9b^{ox}** are very simple, resembling isolated {Cr_7Ni} rings, because the oxidized forms of the central nodes are diamagnetic. Reduction of **7^{ox}**, **8b^{ox}** and **9b^{ox}** back to **7**, **8b** and **9b** regenerates the original spectrum.

Implementing interesting quantum algorithms requires qubits with long phase memory times, such that they can be manipulated many times without errors. Individual {Cr_7Ni} heterometallic rings and simpler paramagnetic centres have shown long enough phase memory times (T_M)[21,22,24,25] to perform coherent electron spin manipulations before state degradation occurs. To check that this key property is preserved when they are incorporated in the supramolecular assemblies, we have performed pulsed EPR measurements at resonances corresponding to both the central nodes and the {Cr_7Ni} rings (Supplementary Figs 14–26). For the latter we find similar values for all compounds, and in both oxidation states where relevant ($T_M = 683$, 749, 767, 790, 984, 750 and 1,031 ns for **5**, **7**, **7^{ox}**, **8b**, **8b^{ox}**, **9b** and **9b^{ox}**, respectively, measured at Q-band and 3K; Supplementary Table 4), demonstrating that the phase memory times are not strongly influenced by structural or magnetic differences in the linkage of the rings. In addition, values of T_M measured at resonances fields corresponding to the central node (in the paramagnetic state) are also similar for all compounds (500–700 ns at Q-band and 3K; Supplementary Tables 5-12). Hence, the phase memory time of the central node does not represent a limitation (see also the Discussion below) for the applicability of our schemes for implementing a universal set of quantum gates. These results are used in the simulations below, and are extremely promising, as these T_M times are sufficiently long for spin manipulation even when they are integrated in a supramolecular assembly.

CNOT gate with uniform magnetic pulses. In the following, we introduce two quantum computation schemes, based on either local or non-local control of the inter-qubit interaction employing the structures described and the parameters obtained by EPR spectroscopy. First, we show that compound **5** is suitable for a CNOT gate, using uniform magnetic pulses as the only manipulation tool[42]. As we are treating three interacting inequivalent doublets this produces 2^3 energy levels. We define the computational basis within the low-energy subspace where Co^{II} is frozen into its $S_z = -1/2$ state, which corresponds to the four lowest levels shown in red in Fig. 5a. The four levels correspond to arrangements of the spins of the two qubits, S_1 and S_3, having the relative orientations $|\downarrow\downarrow>$, $|\downarrow\uparrow>$, $|\uparrow\downarrow>$ and $|\uparrow\uparrow>$, respectively, which we label as $|00>$, $|01>$, $|10>$ and $|11>$ in Fig. 5a.

In a field of a few Teslas the eigenstates are factorized, that is, the eigenfunctions of the rings and the Co^{II} are not entangled. Hence it is possible to implement high-fidelity single-qubit rotations by EPR pulses resonant with low-energy gaps (see, for example, the shorter arrow in Fig. 5a). The combination of the inequivalent and anisotropic ring-Co exchange interaction and of the perpendicular arrangement of the two rings makes the two qubits significantly inequivalent. In particular, there is a sizeable gap (about 33 μeV in a field of 5 T) between the $|00> \rightarrow |01>$ and $|00> \rightarrow |10>$ transitions, which enables independent single-qubit rotations. We note that a distribution of the spin Hamiltonian parameters arising from a reasonable g or J strain (~1%, see ref. 43) would yield a broadening of the energy levels with s.d. ~5 μeV, thus keeping the two transitions distinguishable in realistic conditions, possibly with the help of narrow and/or composite pulses[44,45].

A controlled phase-shift (C_ϕ) gate is obtained by a pulse resonant with the transition corresponding to the longer arrow in Fig. 5a where the cobalt(II) ion is temporarily in its $S_z = +1/2$ state (blue energy levels), followed by a repetition of the same pulse that would bring the state back with an additional phase ϕ. The value of ϕ is controlled by the phase difference between the first and the second pulse. For a pulse with field along the y direction, the implementation of this gate is fast, owing to the

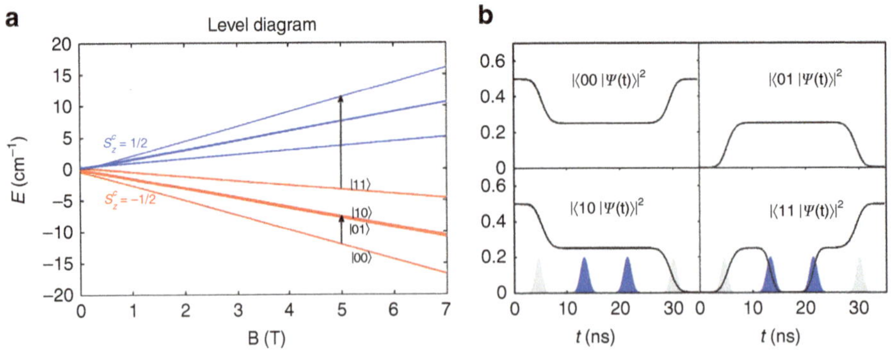

Figure 5 | A CNOT gate based on the structure of 5. (**a**) Field-dependence of the energy levels of **5** resulting from the Hamiltonian (1). The low-energy group of levels (red), where Co is frozen into its $S_z = -1/2$ state, defines the computational basis. The high-energy group of levels, where the Co spin is inverted, is exploited to perform two-qubit gates. (**b**) Simulation of the pulse sequence implementing CNOT as $R_y(\pi/2)C_Z R_y(-\pi/2)$, where $R_y(\alpha)$ is a rotation of the target qubit by an angle α around the y axis and C_Z is a controlled-Z gate. We illustrate the gate by starting at time $t = 0$ with a superposition state $|\psi\rangle = (|0,0\rangle + |1,0\rangle)/\sqrt{2}$, which transforms under a CNOT gate (with the left qubit acting as control) into $(|0,0\rangle + |1,1\rangle)/\sqrt{2}$. The latter state is actually obtained by the pulse sequence implementing $R_y(\pi/2)C_Z R_y(-\pi/2)$, with a fidelity of 99.7%. The envelope of the pulses implementing the two R_y rotations and the C_Z are outlined at the bottom. Note that performing the C_Z gate (two central pulses) requires temporarily leaving the computational subspace. The intensity of the oscillating field at the pulse maximum is 50 G, and we assume a static field of 5 T directed along z.

large value of g_y for Co^{II} ion. Moreover, the relatively large exchange interaction in equation (1) allows us to employ large oscillating fields (50 G), since the desired transition is spectroscopically well resolved from all the others. Consequently, the C_ϕ gate can be performed in only about 12 ns, with fidelities close to 99.99 % (see Computational Details section in Supplementary Methods).

The CNOT gate can then be obtained by the sequence of gates $R_y(\pi/2)\ C_Z\ R_y(-\pi/2)$, where $R_y(\alpha)$ is a rotation of the target Cr_7Ni qubit by an angle α around the y axis (a single-qubit operation) and where C_Z is the phase-shift gate described above with $\phi=\pi$. We have numerically solved the time-dependent Schrödinger equation for the Hamiltonian (1) in presence of this pulse sequence. Results are reported in Fig. 5b and show that this two-qubit gate can be obtained with very high fidelity in only 30 ns. As noted above, the two qubits are significantly inequivalent, even if the g tensor of the rings is nearly isotropic. This makes this complex well suited for the quantum simulation of antisymmetric Hamiltonians (Supplementary Figs 28 and 29 for an illustrative example).

$\sqrt{\text{iSWAP}}$ gate with local electric control. The redox properties of the central node in **7**, **8** and **9** can be exploited to perform the universal $\sqrt{\text{iSWAP}}$ gate on the {Cr_7Ni} qubits, whose effect on the computational basis is given by:

$$|01> \rightarrow (|01>+i\,|10>)/\sqrt{2};$$
$$|10> \rightarrow (|10>+i\,|01>)/\sqrt{2};$$
$$|00> \rightarrow |00> \text{ and } |11> \rightarrow |11>.$$

When the central node is in the diamagnetic state ($[Ru^{III,IV}{}_2\text{-}Co^{II}]$ **7** or Co^{III} **8–9**), the two qubits are decoupled and only single-qubit operations can be performed. Conversely, when the paramagnetic oxidation state, $[Ru^{III}{}_2Co^{II}]$ **7** or Co^{II} **8–9**, is present the system behaves as a trimer described by the Hamiltonian (2). Previously, Lehmann *et al.*[16] have proposed that two spins $S=1/2$ connected by a redox-active unit can be exploited for the implementation of the $\sqrt{\text{SWAP}}$ gate by switching the redox unit with a scanning tunnelling microscope (STM) tip at an appropriate potential. In this way one electron can be added or removed from the redox unit very quickly. It was demonstrated that particular sets of parameters of the trimer Hamiltonian lead to a pure $\sqrt{\text{SWAP}}$ evolution of the two qubits after specific time intervals. However, high fidelity for the gate is guaranteed only for fixed ratios between the qubit–qubit exchange (J_{qq}) and the qubit-redox unit exchange (J_{qr}). For the measured parameters of **7**, **8** and **9** ($J_{qr}\equiv J$ and $J_{qq}=0$ in equation (2)) the fidelity of the $\sqrt{\text{SWAP}}$ gate would be low.

Here we propose a more flexible scheme for the $\sqrt{\text{iSWAP}}$ gate based on our measured parameters, which works if the two qubits have the same Zeeman energy but one different to that of the switch in the 'on' state. This is the case here, where the g values of the qubits and switches are very different, $g_{1z}=g_{3z}=1.74$ for {Cr_7Ni} rings and $g_{2z}=2.77$ (**7**), 2.195 (**8**) and 2.215 (**9**) for the central node. This difference in the g values means that in magnetic fields of a few T, the ring-central node exchange J is small compared with the difference between the Zeeman energies of the ring and the central node. Hence, the spin state of the central node is nearly frozen in the $S_z=-1/2$ state and has only tiny virtual fluctuations that lead to an effective interaction between the two {Cr_7Ni} qubits given by

$$\hat{H}_{qq}=\Gamma\left(\hat{S}_{1x}\hat{S}_{3x}+\hat{S}_{1y}\hat{S}_{3y}\right)+\lambda\left(\hat{S}_{1z}+\hat{S}_{3z}\right) \quad (3)$$

where the field is along z, S_1 and S_3 refer to the spins of the first and second Cr_7Ni qubits and

$$\Gamma=\frac{2J^2}{B\mu_B(g_{1z}-g_{2z})} \qquad \lambda=J+\frac{J^2}{B\mu_B(g_{1z}-g_{2z})}$$

The feasibility of the scheme relies only on the equivalence of the qubits and on the hierarchy of the interactions $[J\ll\mu_B B\,(g_{1z}-g_{2z})]$ (see Supplementary Methods for more details).

Therefore when the central node is paramagnetic, the state of the two qubits evolves according to (3), with negligible entanglement with the central node. For specific times, this evolution coincides with the $\sqrt{\text{iSWAP}}$ gate apart from single-qubit rotations along z due to the second term in (3).

This perturbative picture is confirmed by the results of detailed calculations for compound **9** using the full Hamiltonian (2) (Fig. 6). Starting from the $|10>$ logical state, we report in Fig. 6a the time evolution of the trimer wavefunction that would implement the $\sqrt{\text{iSWAP}}$ gate. In a magnetic field of 3 T, after 4 ns the wavefunction has equal-weight contributions from $|10>$ and $|01>$, which is the $\sqrt{\text{iSWAP}}$ gate, while after 8 ns

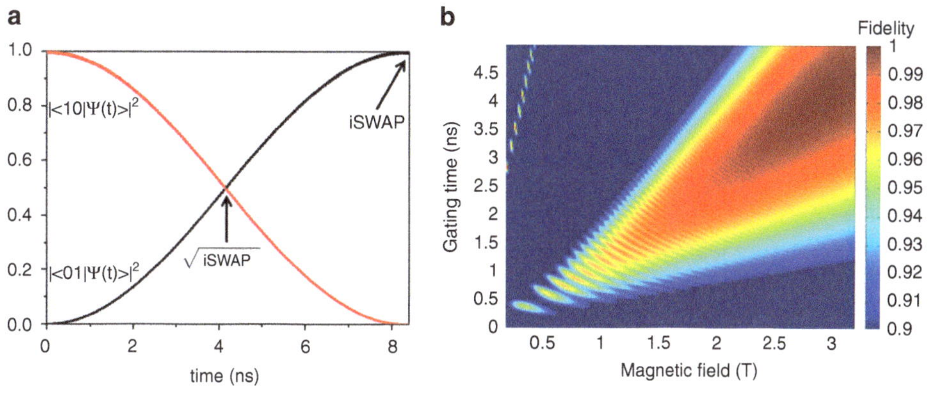

Figure 6 | Simulation of the $\sqrt{\text{iSWAP}}$ gate. (**a**) For **9**, taking $|10>\equiv|1/2,-1/2>\otimes|-1/2>_{Co}$ as the initial state and an applied field of 3 T, we calculate the time-dependence of the oscillation of the trimer wavefunction between $|10>$ and $|01>\equiv|-1/2,1/2>\otimes|-1/2>_{Co}$. Other components are negligible. (**b**) Calculated average fidelity $\bar{F}$ for **9** as a function of the magnetic field B and of the gating time, that is, the time the Co switch is in the on state. The fidelity is defined by $F=|\langle\psi_0|\psi(t_f)\rangle|$, where for a given starting logical state, $|\psi_0>$ is the final state after an ideal gate, whereas $|\psi(t_f)>$ is the actual final state. The average has been made over four random starting states. For each value of the field, the optimal gating time t_f is the one maximizing $\bar{F}$ ca. 4 ns for 3 T as shown in **a**. The oscillations corresponding to the fringes in the picture are associated with fluctuations of the Co spin state. As long as the frequency of these fluctuations is much larger than $1/t_f$ these fluctuations are negligible, that is, the perturbative description of equation (3) is valid.

the states of the two spins are exchanged, that is, we have the $|01>$ state. An extremely good fidelity F (larger than 0.99) for compound **9** is obtained for fields of the order of about 2.5 T or larger, after a suitable gating time t_f of the order of a few nanoseconds (Fig. 6b). For such fields the perturbative picture of equation (3) holds very well, and t_f is proportional to B (Fig. 6b), which is consistent with the form of the effective qubit–qubit coupling $\Gamma \propto 1/B$.

Analogous or better fidelities can be obtained with the parameters derived for compounds **7** and **8**. Indeed, larger values of $|g_{1z} - g_{2z}|$ (as in **7**) increase the validity of the perturbative picture (3), even for smaller values of B. Hence, we can exploit the modular strategy and optimize the performance of the gates by targeted chemical manipulations.

Discussion

Our initial simulations above did not include decoherence. To gain a deeper insight into the performance of the proposed quantum gates, we have performed further simulations that include the effect of both relaxation and pure dephasing, allowing for the finite and measured values of T_1 and T_M, respectively. For the values of T_1 measured by pulsed EPR ($>10\,\mu s$ in all compounds, see the Supplementary Tables 4–12), the effect of relaxation is found to be completely negligible.

Pure dephasing is accounted for by numerically solving the Lindblad master equation for the system density matrix ρ (refs 46,47):

$$\frac{d\rho}{dt} = -i[H,\rho] + \sum_k \frac{1}{2T_M^k}\left(2S_{k,+}S_{k,-}\rho S_{k,+}S_{k,-} - \rho S_{k,+}S_{k,-} - S_{k,+}S_{k,-}\rho\right) \text{ with } \hbar = 1$$

Here the commutator describes the coherent evolution induced by the full system Hamiltonian H, while the second term describes pure dephasing mechanisms. The subscript $k = 1$, 2 and 3 labels the spins of the qubits and of the switch, while $S_{k,+} = |1_k\rangle\langle 0_k|$ and $S_{k,-} = |0_k\rangle\langle 1_k|$ are spin 1/2 raising and lowering operators.

Using the values of T_M of the Cr_7Ni qubits measured by pulsed-EPR spectroscopy (between 700 and 1,000 ns, see above) we still find very high fidelities: 99.3% for the implementation of the CNOT gate on compound **5** and 99.6% for the $\sqrt{iSWAP}$ gate on **9**. As expected, the fidelities are only marginally affected by decoherence because gating times are much shorter than T_M.

It is also worth discussing the role of the decoherence of the switch in the present schemes. As far as the CNOT gate scheme (compound **5**) is concerned, the contribution of the central Co switch to decoherence is similar to that of the rings, because the switch is temporarily excited to the $S_Z = +1/2$ state during the CNOT gates. Conversely, only virtual excitations of the switch are involved in the implementation of the $\sqrt{iSWAP}$ gate exploiting the redox-active linker. Hence, the fidelity remains very high even for dephasing time of the switch of the order of the gating time (some ns). This difference is shown by the two colour maps reported in Fig. 7. They show the dependence of the fidelity of a $\sqrt{iSWAP}$ (left) and of a CNOT gate (right) on the dephasing times T_M of the rings and of the switch. It is evident that the fidelity of the $\sqrt{iSWAP}$ is practically independent on T_M^{Co} whereas its dependence on T_M^{rings} is much more pronounced. Conversely, the implementation of the CNOT on compound **5** leads to a similar dependence of the fidelity on T_M^{Co} and T_M^{rings}.

The chemistry described above can be extended to make one-dimensional (1D) chains incorporating the two-Cr_7Ni supramolecules, linked by either single cobalt sites or oxo-centred metal triangles. Such 1D chains have already been made involving single Cr_7Ni units and [2]- and [3]-rotaxanes and the principles established, especially for 1D chains of rotaxanes[48], should work well. The key steps are to include two functional groups per Cr_7Ni ring, which has already been done for *iso*-nicotinate acid, and by functionalizing the termini of the central organic thread of rotaxanes. Inclusion of two different functionalities is more taxing but entirely feasible (see the detailed schemes in Supplementary Figs 30 and 31). Therefore the chemistry is potentially scalable.

While this challenging chemistry proceeds, it is possible to show that the schemes are theoretically scalable. The extension of a two-qubit QG to a multi-qubit register raises important issues concerning the propagation of errors. In such a register, we can identify two main sources of errors, whose effect increases with the number of qubits: pure dephasing and imperfect operation of the switch arising from a residual inter-qubit interaction still present in the 'off' state.

Errors induced by pure dephasing (decoherence) increase with the overall computational time. However, a finite chain of qubits with interposed switches would allow simultaneous (parallel) manipulation of non-overlapping parts of the register, which drastically reduces the computation time, and hence decoherence-induced errors, with respect to a serial implementation. For instance, a setup based on compound **5** can be manipulated in parallel for interesting classes of quantum simulation algorithms[17]. This requires inclusion of two distinct switches in the structure—$Switch_A$ and $Switch_B$—with different responses to external stimuli, such that the 1D chain has as a repeat pattern -Qubit-$Switch_A$-Qubit-$Switch_B$-.

For compounds **7**, **8** and **9**, the scheme requires a local addressability of individual switches on the chain, which are

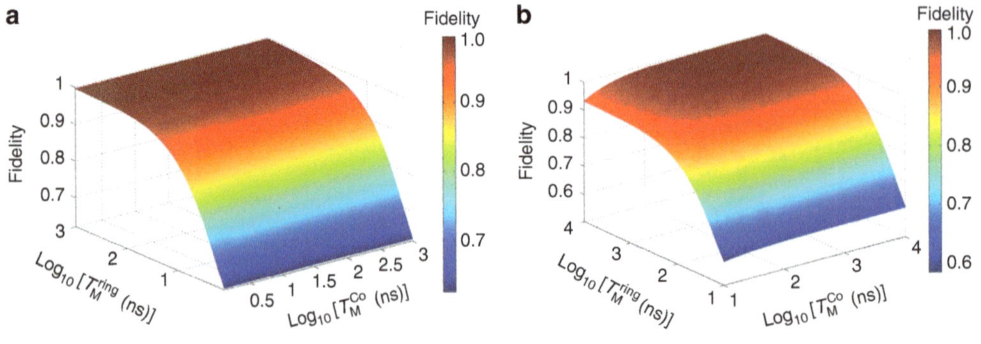

Figure 7 | Effect of decoherence on quantum gates. (**a**) Fidelity of the $\sqrt{iSWAP}$ as a function of the dephasing times of the rings and of the switch. Remarkably, the fidelity is nearly independent on the T_M^{Co}

separated by about 3 nm. For a first proof-of-principle experiment with a single assembly, an STM tip could be used to provide the best control. A parallel implementation of gates would be possible if different Co switches could be addressed individually at the same time. To achieve a scalable structure, a molecular chain of qubits might be layered onto a surface and addressed by means of a cross-bar architecture similar to that proposed in refs 16,49. It is worth noting that Cr_7Ni rings have already been deposited on surfaces, without significant modification of their magnetic properties[50].

We now examine the scaling of the errors with the number of qubits in the register. As a first step, we consider the effect of pure dephasing on a set of non-interacting qubits subject to Lindblad (Markovian) dynamics. It can be shown (see ref. 51 and Supplementary Information) that the decoherence error $\varepsilon = 1 - F^2$ on N qubits scales at most as $\varepsilon_N = N\varepsilon_1 = \frac{N}{2}(1 - e^{-t_f/2T_M}) \approx N\frac{t_f}{4T_M}$, valid for $t_f \ll T_M$ (where t_f is the optimal gating time). In the parallel quantum-simulator implementation considered in ref. 17, manipulating a chain of N qubits takes the same time as manipulating the shortest chain that contains the two distinct switches (qubit-Switch$_A$-qubit-Switch$_B$-qubit). Therefore t_f is limited by the value it assumes for a chain of three qubits, while in a serial scheme t_f increases linearly with N. By chemical engineering of Cr_7Ni qubits, a T_M of 15 μs has been obtained[25]; this should allow the implementation of around twenty 2-qubit gate operations on a chain consisting of 10 qubits, while keeping the error below 0.1.

Finally, we analyse the consequences of an imperfect operation of the switch and their effect on scalability. For an ideal $\sqrt{\text{iSWAP}}$ gate the inter-qubit exchange interaction would be completely turned off when the central node is in its diamagnetic state. Double electron–electron resonance measurements (Supplementary Fig. 27) reveal a very weak, residual interaction between the qubits. The resulting oscillations are on a time scale (ca. 0.2 μs) much longer than our gating times (a few ns) and could be corrected by means of refocusing techniques[29]. In the case of the CNOT scheme (compound **5**), the always 'on' exchange interaction between the rings and the Co ion yields a small residual qubit–qubit coupling (Supplementary Methods). As a consequence, an unwanted slow evolution of the qubits occurs on a timescale T_{UE} of the order of a few hundreds of ns. Although this is longer than the CNOT gate time (about 30 ns), larger values of T_{UE} would be needed for performing sequences of many gates. This can be obtained by modifying the molecule to decrease the Co-ring exchange interaction. For instance, T_{UE} can be increased by a factor of 50 by reducing the Co-ring exchange interaction by a factor of three, which could be achieved chemically by adding an extra phenyl group between the ring and the central node. A similar reduction of the residual qubit–qubit coupling can be obtained by increasing the static magnetic field.

In summary, we have described two different schemes for universal quantum information processing, based on either local or global control of the qubit–qubit interaction. We have demonstrated that the flexibility of molecular $\{Cr_7Ni\}$ qubits makes them suitable for the implementation of each of these two schemes, if properly functionalized and linked by means of a supra-molecular design strategy. The two-qubit units can be controlled either magnetically or electrically, and implement either the CNOT or $\sqrt{\text{iSWAP}}$ perfectly entangling gates. Our realistic simulations, based on experimental parameters and including decoherence, show that these gates can be performed with remarkably high fidelity. The modular strategy proposed here offers a degree of control in terms of the magnitude of the coupling between molecular spin qubits, the spatial orientation of the modules and the possibility to have a switchable interaction that represents a significant step forward with respect to previous achievements on the assembly of qubits.

Since future developments of quantum technology are unpredictable, we emphasize the importance of pursuing both these two parallel roads towards the actual realization of a quantum computer.

References

1. Feynman, R. P. Simulating physics with computers. *Int. J. Theor. Phys.* **21,** 467–488 (1982).
2. Lloyd, S. Universal quantum simulators. *Science* **273,** 1073–1078 (1996).
3. Nielsen, M. A. & Chuang, I. L. *Quantum Computation and Quantum Information* (Cambridge University Press, 2000).
4. Ladd, T. D. *et al.* Quantum computers. *Nature* **464,** 45–53 (2010).
5. Shor, P. W. Polynomial-time algorithms for prime factorization and discrete logarithms on a quantum computer. *SIAM J. Comput.* **26,** 1484–1509 (1997).
6. Grover, L. K. Quantum computers can search arbitrarily large databases by a single query. *Phys. Rev. Lett.* **79,** 4709–4712 (1997).
7. Burkard, G., Loss, D. & DiVincenzo, D. P. Coupled quantum dots as quantum gates. *Phys. Rev. B* **59,** 2070–2078 (1999).
8. Hanson, R. & Awschalom, D. D. Coherent manipulation of single spins in semiconductors. *Nature* **453,** 1043–1049 (2008).
9. Balasubramanian, G. *et al.* Ultralong spin coherence time in isotopically engineered diamond. *Nat. Mater.* **8,** 383–387 (2009).
10. Knowles, H. S., Kara, D. M. & Atatüre, M. Observing bulk diamond spin coherence in high-purity nanodiamonds. *Nat. Mater.* **13,** 21–25 (2014).
11. Georgescu, I. M., Ashhab, S. & Nori, F. Quantum simulation. *Rev. Mod. Phys.* **86,** 153 (2014).
12. Leuenberger, M. & Loss, D. Quantum computing in molecular magnets. *Nature* **410,** 789–793 (2001).
13. Meier, F., Levy, J. & Loss, D. Quantum computing with spin cluster qubits. *Phys. Rev. Lett.* **90,** 47901–47904 (2003).
14. Troiani, F. *et al.* Molecular engineering of antiferromagnetic rings for quantum computation. *Phys. Rev. Lett.* **94,** 207208 (2005).
15. Troiani, F., Affronte, M., Carretta, S., Santini, P. & Amoretti, G. Proposal for quantum gates in permanently coupled antiferromagnetic spin rings without need of local fields. *Phys. Rev. Lett.* **94,** 190501 (2005).
16. Lehmann, J., Gaita-Ariño, A., Coronado, E. & Loss, D. Spin qubits with electrically gated polyoxometalate molecules. *Nat. Nanotechnol.* **2,** 312–317 (2007).
17. Santini, P., Carretta, S., Troiani, F. & Amoretti, G. Molecular nanomagnets as quantum simulators. *Phys. Rev. Lett.* **107,** 230502 (2011).
18. Nakazawa, S. *et al.* A synthetic two-spin quantum bit: *g*-engineered exchange-coupled biradical designed for controlled-NOT gate operations. *Angew. Chem. Int. Ed.* **51,** 9860–9864 (2012).
19. Luis, F. *et al.* Molecular prototypes for spin-based CNOT and SWAP quantum gates. *Phys. Rev. Lett.* **107,** 117203 (2011).
20. Aromí, G., Aguilà, D., Gamez, P., Luis, F. & Roubeau, O. Design of magnetic coordination complexes for quantum computing. *Chem. Soc. Rev.* **41,** 537–546 (2012).
21. Warner, M. *et al.* Potential for spin-based information processing in a thin-film molecular semiconductor. *Nature* **503,** 504–509 (2013).
22. Graham, M. J. *et al.* Influence of electronic spin and spin-orbit coupling on decoherence in mononuclear transition metal complexes. *J. Am. Chem. Soc.* **136,** 7623–7626 (2014).
23. Aguilà, D. *et al.* Heterodimetallic [LnLn'] lanthanide complexes: toward a chemical design of two-qubit molecular spin quantum gates. *J. Am. Chem. Soc.* **136,** 14215–14222 (2014).
24. Bader, K. *et al.* Room temperature quantum coherence in a potential molecular qubit. *Nat. Commun.* **5,** 5304 (2014).
25. Wedge, C. J. *et al.* Chemical engineering of molecular qubits. *Phys. Rev. Lett.* **108,** 107204(1)–107204(5) (2012).
26. Timco, G. A. *et al.* Engineering the coupling between molecular spin qubits by coordination chemistry. *Nat. Nanotechnol.* **4,** 173–178 (2008).
27. Lehn, J.-M. *Supramolecular Chemistry: Concepts and Perspectives* (Wiley-VCH, 1995).
28. Ardavan, A. *et al.* Engineering coherent interactions in molecular nanomagnet dimers. *NPJ Quantum Inf.* **1,** 15012 (2015).
29. Jones, J. A. Quantum computing with NMR. *Prog. Nucl. Magn. Reson. Spectrosc.* **59,** 91 (2011).
30. McInnes, E. J. L., Timco, G. A., Whitehead, G. F. S. & Winpenny, R. E. P. Heterometallic rings as a playground for physics and supramolecular building blocks. *Angew. Chem., Int. Ed.* **54,** 14244–14269 (2015).
31. Constable, E. C. *et al.* Expanded ligands: bis(2,2':6',2"-terpyridine carboxylic acid)ruthenium(II) complexes as metallosupramolecular analogues of dicarboxylic acids. *Dalton Trans.* **38,** 4323–4332 (2007).

32. Hayami, S., Komatsu, Y., Shimizu, T., Kamihata, H. & Hoon Lee, Y. Spin-crossover in cobalt(II) compounds containing terpyridine and its derivatives. *Coord. Chem. Rev.* **255,** 1981–1990 (2011).
33. Lloret, F., Julve, M., Cano, J., Ruiz-García, R. & Pardo, E. Magnetic properties of six-coordinated high-spin cobalt(II) complexes: theoretical background and its application. *Inorg. Chim. Acta* **361,** 3432–3445 (2008).
34. Ohto, A., Sasaki, Y. & Ito, T. Mixed-metal trinuclear complexes containing two ruthenium(III) ions and a divalent metal ion, $[Ru_2M(\mu_3\text{-}O)(\mu\text{-}CH_3COO)_6(L)_3]$ (M = Mg, Mn, Co, Ni, Zn; L = H_2O, Pyridine). *Inorg. Chem.* **33,** 1245–1246 (1994).
35. Schubert, U. S., Hofmeier, H. & Newkome, G. R. *Modern Terpyridine Chemistry* (Wiley-VCH, 2006).
36. Cannon, R. D. & White, R. P. Chemical properties of triangular bridged metal complexes. *Prog. Inorg. Chem.* **36,** 195–298 (1988).
37. Murray, K. S. in *Spin-Crossover Materials: Properties and Applications* (ed. Halcrow, M. A.), Ch. 1, 1–54 (John Wiley and Sons, 2013).
38. Aribia, K. B., Moehl, T., Zakeeruddin, S. M. & Grätzel, M. Tridentate cobalt complexes as alternative redox couples for high-efficiency dye-sensitized solar cells. *Chem. Sci.* **4,** 454–459 (2013).
39. Connelly, N. G. & Geiger, W. E. Chemical redox agents for organometallic chemistry. *Chem. Rev.* **96,** 877–910 (1996).
40. Stoll, S. & Schweiger, A. EasySpin, a comprehensive software package for spectral simulation and analysis in EPR. *J. Magn. Reson.* **178,** 42–55 (2006).
41. Fernandez, A. *et al.* g-engineering in hybrid rotaxanes to create *AB* and AB_2 electron spin systems: EPR studies of weak interactions between dissimilar electron spin qubits. *Angew. Chem. Int. Ed.* **54,** 10858–10861 (2015).
42. Chiesa, A. *et al.* Molecular nanomagnets with switchable coupling for quantum simulation. *Sci. Rep.* **4,** 7423 (2014).
43. Park, K., Novotny, M. A., Dalal, N. S., Hill, S. & Rikvold, P. A. Effects of *D*-strain, *g*-strain, and dipolar interactions on EPR linewidths of the molecular magnets Fe_8 and Mn_{12}. *Phys. Rev. B* **65,** 014426 (2001).
44. Wesenberg, J. & Mølmer, K. Robust quantum gates and a bus architecture for quantum computing with rare-earth-ion-doped crystals. *Phys. Rev. A* **68,** 012320 (2003).
45. Cummins, H. K., Llewellyn, G. & Jones, J. A. Tackling systematic errors in quantum logic gates with composite rotations. *Phys. Rev. A* **67,** 042308 (2003).
46. Tempel, D. G. & Aspuru-Guzik, A. Relaxation and dephasing in open quantum systems time-dependent density functional theory: properties of exact functionals from an exactly-solvable model system. *Chem. Phys.* **391,** 130 (2011).
47. Breuer, H. P. & Petruccione, F. *The Theory of Open Quantum Systems* (Oxford University Press, 2002).
48. Whitehead, G. F. S. *et al.* Rings and threads as linkers in metal-organic frameworks and poly-rotaxanes. *Chem. Commun.* **49,** 7195–7197 (2013).
49. Green, J. E. *et al.* A 160-kilobit molecular electronic memory patterned at 10^{11} bits per square centimeter. *Nature* **445,** 414–417 (2007).
50. Corradini, V. *et al.* Magnetic anisotropy of Cr_7Ni spin clusters on surfaces. *Adv. Funct. Mat* **22,** 3706 (2012).
51. Jing, J. & Hu, X. Scaling of decoherence for a system of uncoupled spin qubits. *Sci. Rep.* **5,** 17013 (2015).

Acknowledgements

This work was supported by the EPSRC (UK), the European Commission (Marie Curie Intra-European Fellowship to J.F.-S. (622659) and A.F. (300402)). E.M.P. thanks the Panamanian agency SENACYT-IFARHU for funding. R.E.P.W. thanks the Royal Society for a Wolfson Merit Award. We also thank EPSRC (UK) for funding an X-ray diffractometer (grant number EP/K039547/1) and the National EPR Facility. We thank Diamond Light Source for access to synchrotron X-ray facilities. A.C., S.C. and P.S. acknowledge financial support from the FIRB Project No. RBFR12RPD1 of the Italian Ministry of Education and Research.

Author contributions

J.F.S., E.M.P. and A.F. synthesized the majority of the compounds discussed; S.A.M. synthesized compound **6**. G.T. was involved in design of the synthetic strategies. E.M.P. and I.J.V.-Y. carried out the X-ray single crystal diffraction studies. E.M.P., J.F.-S. and F.T. performed the EPR spectroscopic studies, and J.F.-S. carried out the electrochemical studies reported here. E.J.L.M. helped interpret the EPR spectra. A.C., S.C. and P.S. designed the schemes for quantum gates and A.C. carried out their detailed simulations. R.E.P.W. oversaw the project and wrote the paper with J.F.S., S.C. and P.S., with significant input from all other authors.

Additional information

Accession codes: The X-ray crystallographic coordinates for structures reported in this Article have been deposited at the Cambridge Crystallographic Data Centre (CCDC), under deposition number CCDC 1029608–1029613 and 1415380–1415383. These data can be obtained free of charge from The Cambridge Crystallographic Data Centre via www.ccdc.cam.ac.uk/data_request/cif.

Competing financial interests: The authors declare no competing financial interests.

22

Structural control of mixed ionic and electronic transport in conducting polymers

Jonathan Rivnay[1,†], Sahika Inal[1], Brian A. Collins[2,3], Michele Sessolo[4], Eleni Stavrinidou[1], Xenofon Strakosas[1], Christopher Tassone[5], Dean M. Delongchamp[2] & George G. Malliaras[1]

Poly(3,4-ethylenedioxythiophene) doped with poly(styrenesulfonate), PEDOT:PSS, has been utilized for over two decades as a stable, solution-processable hole conductor. While its hole transport properties have been the subject of intense investigation, recent work has turned to PEDOT:PSS as a mixed ionic/electronic conductor in applications including bioelectronics, energy storage and management, and soft robotics. Conducting polymers can efficiently transport both holes and ions when sufficiently hydrated, however, little is known about the role of morphology on mixed conduction. Here, we show that bulk ionic and electronic mobilities are simultaneously affected by processing-induced changes in nano- and meso-scale structure in PEDOT:PSS films. We quantify domain composition, and find that domain purification on addition of dispersion co-solvents limits ion mobility, even while electronic conductivity improves. We show that an optimal morphology allows for the balanced ionic and electronic transport that is critical for prototypical mixed conductor devices. These findings may pave the way for the rational design of polymeric materials and processing routes to enhance devices reliant on mixed conduction.

[1] Department of Bioelectronics, Ecole Nationale Supérieure des Mines, CMP-EMSE, MOC, Gardanne 13541, France. [2] Material Science and Engineering Division, National Institute of Standards and Technology (NIST), Gaithersburg, Maryland 20899, USA. [3] Department of Physics and Astronomy, Washington State University, Pullman, Washington 99164, USA. [4] Instituto de Ciencia Molecular, Universidad de Valencia, Paterna 46980, Spain. [5] Stanford Synchrotron Radiation Lightsource (SSRL), SLAC National Accelerator Laboratory, Menlo Park, California 94025, USA. † Present address: PARC, a Xerox Company, 3333 Coyote Hill Road, Palo Alto, California 94304, USA. Correspondence and requests for materials should be addressed to J.R. (email: rivnay@gmail.com).

Mixed electronic/ionic transport in conducting polymers (CPs) is attracting a great deal of attention due to a host of new devices that leverage the coupling of these carriers to enable new modes of operation. One example is the organic electrochemical transistor (OECT), which consists of a CP layer in which ions injected from an electrolyte modify hole conductivity. Owing to their high transconductance[1], these devices have been shown to be efficient transducers of biological activity, with applications ranging from neural interfacing[2,3] to biosensing[4]. The inverse mode of operation, where a CP film injects ions into an electrolyte, has been used for drug delivery with exceptional spatio-temporal control[5,6]. In addition to applications in bioelectronics, OECTs are being developed for printed electronics[7,8], and more recently, neuroinspired electronics[9–11]. Mixed transport in CPs is also leveraged in energy applications including batteries/supercapacitors[12], electrochromic windows[13] and in electromechanical actuators for soft robotics[14,15]. Despite the many opportunities enabled by mixed conduction in CPs, the lack of fundamental understanding of combined ionic/electronic transport and its connection to morphology hinders rational materials and device design.

A prototypical CP used in many mixed conductor devices is poly(3,4-ethylenedioxythiophene) doped with poly(styrene sulfonate); (PEDOT:PSS). This material has attracted interest for its high hole conductivity ($>1{,}000\ \mathrm{S\,cm^{-1}}$), high stability and for its commercial availability as a dispersion for solution processing[16]. PEDOT:PSS dispersions are typically described as being composed of gel-like particles comprising a polyanion (PSS^-)-rich shell which helps to stabilize the PEDOT-enriched particles in aqueous solvents[17,18]. PEDOT oligomers are believed to polymerize onto the PSS template (Fig. 1a), meaning that PEDOT itself is inherently ill-characterized as far as molecular weight and polydispersity. The gel particles (Fig. 1b) deposit in what has been described as a pancake-like morphology[19–21], and show signs of π-stacking[22–25] (Fig. 1c,d). PEDOT:PSS structure has been intensely studied by a variety of techniques including electron microscopy, X-ray scattering, elemental and surface-sensitive scanning probes[17,22–26], largely with the goal of understanding and enhancing hole transport to make better electrodes for optoelectronic devices. Changes in formulation content, processing or post-processing have been shown to lead to enhancements in hole conductivity by inducing morphological re-arrangements, minimizing excess dopant phase (that is, insulating PSS) or both. A thorough review of the methods and suggested mechanisms to enhance conductivity can be found in ref. 27.

Ion transport in PEDOT:PSS (and in CPs, in general) has been explored only on a limited basis. The presence of electronic carriers in the film renders most techniques used to study ion conduction inutile. Nevertheless, the electrochromic nature of PEDOT:PSS, which gives rise to changes in film absorption on electrochemical (de)doping, allows us to follow ion motion in the film[28–30], and to extract ion-drift mobility[31]. Similar to ion-exchange membranes such as Nafion[32], hydration of PEDOT:PSS was found to establish ion-transport pathways, leading to a high ion mobility[31]. The nature of these ion pathways and their connection to film structure and morphology is not well-understood. Moreover, unlike the case of hole transport, attempts to modify the structure of PEDOT:PSS in a systematic manner and to monitor the effect on ion transport are lacking. It is important, for example, to deduce if hole and ion transport can be improved simultaneously, and, if not, whether an optimal morphology exists wherein device performance is maximized. This calls for a comprehensive study where both ion and hole transport are measured in films processed to have different structure and morphology. Thorough quantification of the latter is particularly important, as there have been differing reports on the influence of the same additive on film morphology[24,25]. This is a challenging proposition, as PEDOT and PSS have similar densities and bonding environments (limiting X-ray scattering contrast), and are disordered on multiple length scales.

Here, we investigate the microstructure and morphology of PEDOT:PSS films modified with varying amounts of co-solvent dispersion additive, ethylene glycol (EG), and relate these findings to measurements of electronic and ionic transport. To overcome the low contrast and a relatively disordered film microstructure, we use Synchrotron radiation and resonant soft X-ray scattering (rSoXS) and investigate the multi-scale microstructure of PEDOT:PSS films. The combined use of carbon-edge X-ray scattering and spectroscopy allows the deduction of the absolute nano-scale domain composition of the films. We find that the PEDOT:PSS-rich cores and PSS-rich matrix are surprisingly impure even with a large dispersion content of EG, but that the co-solvent enhances film heterogeneity. We then investigate both ionic and electronic (hole) transport in the same formulations. Ionic transport is probed using a spectroscopic analysis of the electrochromic moving front, revealing an ion motion heavily affected by the meso-scale domain size and composition. We find that ion and hole transport are oppositely affected by the addition of EG—a finding that is corroborated by the structural work herein. We show that this suggests an optimal morphology exists, enabling enhanced ionic mobility and hole conductivity, and leading to improved transconductance in OECTs. Therefore, this

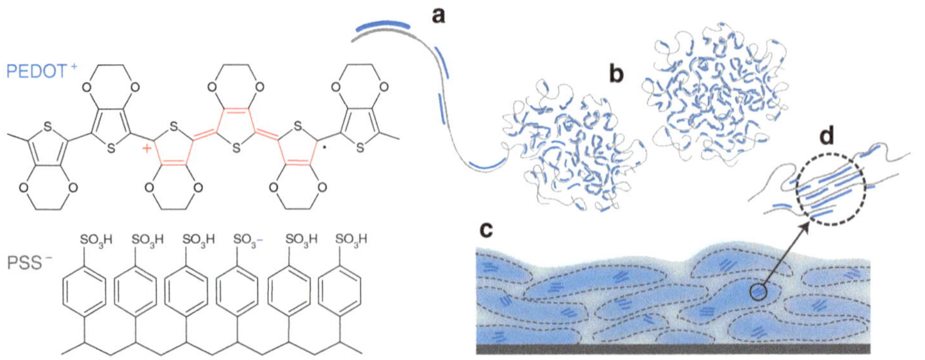

Figure 1 | PEDOT:PSS structure and morphology. The chemical structure of PEDOT:PSS and commonly described microstructure of the CP system (**a**) synthesis onto PSS template, (**b**) formation of colloidal gel particles in dispersion and (**c**) resulting film with PEDOT:PSS-rich (blue) and PSS-rich (grey) phases. (**d**) Aggregates/crystallites support enhanced electronic transport.

work presents a study of the bulk composition and microstructure of an organic electronic system, and its effect on mixed ionic and electronic transport.

Results

Structure in PEDOT:PSS films. The low-density contrast in PEDOT:PSS largely limits the utility of traditional electron or X-ray methods, however, the use of synchrotron radiation, as well as rSoXS allows for multi-scale structure determination. Wide-angle grazing incidence scattering (GIXS) of PEDOT:PSS films allows one to focus on the relative molecular packing arising from aggregation/crystallization[25,33]. Figure 2a and Supplementary Fig. 1 show the changes in wide-angle scattering on addition of EG as co-solvent to the PEDOT:PSS formulation. Dispersion co-solvents should not be present within the film (after film formation, annealing and vacuum treatment)[34,35]. π-stacking scattered intensity at $Q = 1.83\,\text{Å}^{-1}$ is enhanced with higher EG formulation content. This enhancement is accompanied by a narrower peak width, indicating an increase in coherence length from $\cong 1$ to 3 nm (due to larger crystallite/aggregate size, or a higher degree of aggregate ordering)[36]. A tendency for the π-stacked aggregates to slightly favour a face-on texture is observed, similar to previous reports[25,33].

Meso-scale structure of PEDOT:PSS films is investigated using C-edge rSoXS. The optical constants are determined from films of Na:PSS and PEDOT:Cl, which are dominated by the bonding of the polymeric/oligomeric species in the soft X-ray regime. The Na^+ and Cl^- ions contribute a linear background absorbance but do not participate in the energetic transitions at the C-edge, enabling facile removal of their signals. The calculated materials and vacuum contrast functions can be computed[37], and are shown in Supplementary Fig. 2. Two energies are selected for investigation, the pre-edge energy 270 eV, and the 1 s to π^* C=C resonance at 285.1 eV. Figure 2b shows scattering profiles of PEDOT:PSS samples in the transmission geometry. The scattered intensity at $Q < 0.01\,\text{nm}^{-1}$ ($2\pi/Q > 600$ nm) matches well between the two incident energies tested for all samples measured, indicating a predominant contribution from vacuum contrast (that is, film roughness). The scattering feature in the $Q \cong 0.1\,\text{nm}^{-1}$ range is enhanced at resonance owing to materials (PEDOT to PSS) contrast (see Supplementary Fig. 2) and is thus a measure of the bulk meso-scale domain morphology. The upturn at the highest Q-values is primarily attributed to X-ray fluorescence (scaled by Q^2), since it follows the energy dependence of absorbance and approaches a constant $I(Q \rightarrow \infty)$. Some of the scattering at the highest Q-values may also originate from nanocrystals within PEDOT:PSS-rich domains. Previous works suggest the overall film morphology consists of closely packed PEDOT-rich domains in a PSS-rich matrix (Fig. 1c)[17,20,21,26] thus, to quantify how the average meso-scale domain size and purity varies with EG, the features on a Lorentz plot $IQ^2(Q)$ are fit to a log-normal peak with a cubic background (Supplementary Fig. 3). The peak intensity (area) is found to scale linearly by one order of magnitude when EG content ranges from 0 to 50 vol%. This suggests a substantial increase in either domain purity or volume fraction, in concordance with the Porod Invariant formalism (see Supplementary Note 1 for specifics of the calculation)[38]. The feature position (Fig. 2c, top) is indicative of domain spacing (gel-like particle spacing in plane), and indicates the size of particles increases from 16 to 42 nm over the same range of EG concentrations, a size scale range in accordance with others[23,26]. Taken together, these data strongly suggest an increase of film heterogeneity with the addition of EG rather than having a homogenizing/inversion effect.

Compositional analysis using transmission NEXAFS of the same samples (same X-ray beam on the same spot as scattering) reveals only minimal variation in average film composition[39], and corroborates the expectation that EG is effectively absent in the films[34,35]. The linear combination of the background-subtracted PEDOT:Cl and Na:PSS NEXAFS spectra is used to determine film thickness and composition. The PEDOT to PSS content in the film varies around 1:1.5 to 1:1.8 by weight, without a consistent trend with EG concentration, and is slightly lower than the as-received content of 1:2.5. Since there is not a large difference in the scattering population, the increased scattering intensity must arise from increased contrast between the scatterers. These findings confirm the domain purification on addition of the co-solvent, rather than loss of overall PSS content during the casting process.

Combining the NEXAFS and rSoXS analyses, we calculate an estimate of the absolute PEDOT concentration of the PEDOT:PSS-rich phase and surrounding PSS-rich matrix (Fig. 2c, bottom). The composition analysis (detailed in the

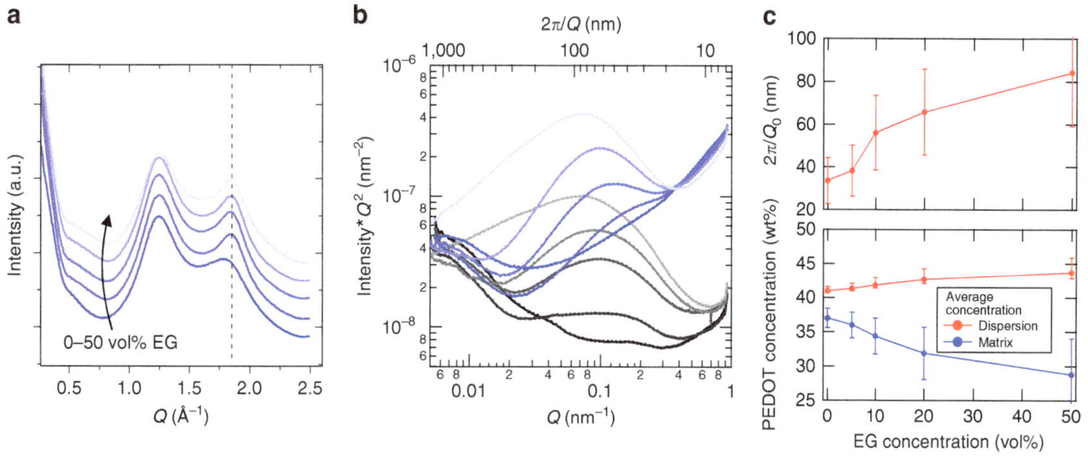

Figure 2 | Microstructure and morphology of PEDOT:PSS from X-ray scattering. (**a**) Linearly offset wide-angle GIXS of PEDOT:PSS films cast from dispersions with EG concentration of 0, 5, 10, 20 and 50 vol% (varying from dark to light colours). The dotted vertical line is the position of the PEDOT:PSS π-stacking scattered intensity (020). (**b**) C-edge rSoXS of PEDOT:PSS films at pre-edge energy (270 eV, black) and at the C 1s to π^* resonance (285.1 eV, blue). (**c**) Domain spacing from peak position ($2\pi/Q_0$) of the features observed at resonance in **b** and quantification of the absolute domain purity (as PEDOT concentration) from peak area. Uncertainties for purity are derived in Supplementary Fig. 3 (mainly from uncertainty in domain volume fraction), while uncertainties for spacing are calculated using the peak width to demonstrate the distribution of domain spacings.

Supplementary Figs 3–5) reveals that neither phase in the film is pure. As EG content in the dispersion is increased, the PEDOT concentration in the film's collapsed gel particles increases mildly from 41 to 44 wt%. Considering our X-ray diffraction evidence of PEDOT chain packing along the π-stacking direction, such a low PEDOT concentration suggests the aggregates that lie within these domains consist of alternating PEDOT and PSS chains along the non-π-stacking direction (similar to that of PEDOT: tosylate)[40], rather than a crystal of pure PEDOT as suggested by others[24]. The PEDOT concentration varies more significantly in the matrix, where it decreases from 37 to 29 wt% with increasing EG co-solvent. These calculations assume a constant 0.75 volume fraction of the PEDOT enriched gel particles—an assumption based on cross-sectional AFMs reported by Nardes *et al.*[23]—which show a skin layer around the gel particles in this approximate volume ratio. The possibility of a changing volume ratio is included in the uncertainty calculations (along with other possible sources—see Supplementary Fig. 3). It is unlikely that the marked increase in scattering intensity, ΔTSI, originates from a change in domain volume fraction since our calculations show that such a ΔTSI would require an unrealistic volume fraction of >0.9. Indeed, with rSoXS results indicating a purification and rarefication of PEDOT: PSS-rich domains with the EG co-solvent, a pure skin layer covering these domains would be expected to pose a significant barrier to hole conductivity. Although our measurements cannot address connectivity directly, the calculated PEDOT concentration in this skin layer above 20 wt% indicates a possible means for percolation of charges between the PEDOT:PSS-rich particles.

To the best of our knowledge, this is the first time an analysis of absolute nano-scale domain composition in an all-polymer blend has been achieved using a combination of NEXAFS and X-ray scattering at a resonant photon energy of a molecule. It should be noted that the resonant scattering experiments described above are performed in transmission, with the incident X-ray beam normal to the plane of the film. Therefore, only the lateral meso-scale structure is quantified. Previous works show the collapsed gel particles (the pancake-like morphology) possess an approximate aspect ratio of 0.1–0.5 (ref. 41), with the out-of-plane direction possessing a *Q*-value out of the range of the current experimental setup. Despite this known anisotropic morphology, the trends in domain composition should not be significantly affected, and are discussed in detail in Supplementary Fig. 4.

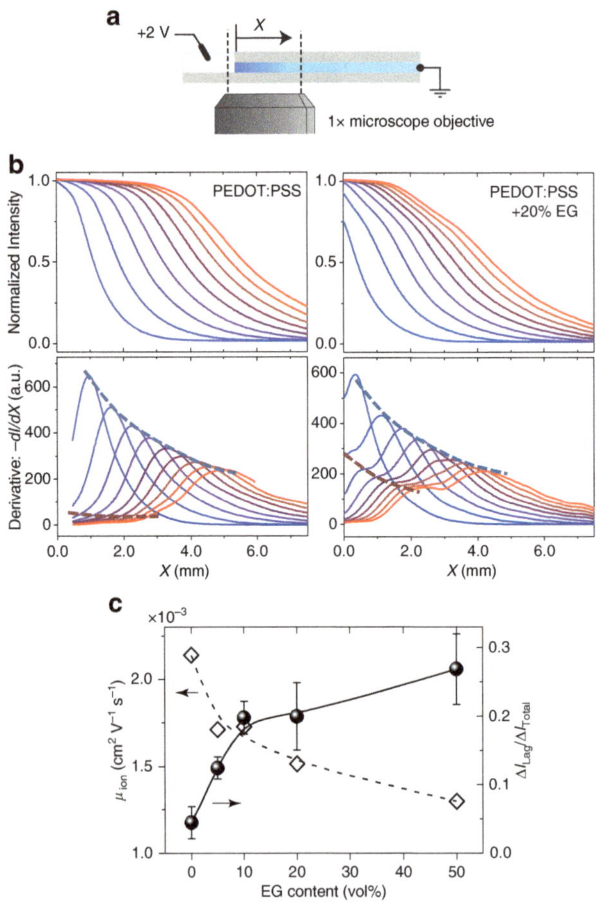

Figure 3 | Co-solvent influences ion transport. (**a**) Temporal characteristics of spatial moving front profiles due to ion transport. (**b**) 1D normalized-intensity profiles as a function of distance, X, from the electrolyte-PEDOT:PSS interface (from $t = 5$ s, blue, to $t = 45$ s, red; $\Delta t = 5$ s) are obtained from time-lapse optical micrographs. Use of EG as a co-solvent affects the shape of the moving front profiles, which is highlighted by investigating the spatial derivative of the intensity profiles, revealing a 'leading' and 'lagging' front (blue and red dotted lines, respectively). (**c**) The effective ion mobility as derived from the moving front data (open symbols/dotted line), and the fractional contribution of the lagging front ($\Delta I_{Lag}/\Delta I_{Total}$) (solid symbols/line), as a function of the EG content in the dispersion. Error bars were determined from the standard deviation of $\Delta I_{Lag}/\Delta I_{Total}$ from 5-10 consecutive spatial derivative profiles in the $t = 5$-50 s range (the inclusion of a specific profile in quantification depended on the ability to fit both leading and lagging contributions within the experimentally restricted spatial range, X).

Ionic transport in PEDOT:PSS. We probed ion transport in PEDOT:PSS films with a one-dimensional version of the 'moving front' experiment[31]. The method involves monitoring visible light transmission through an electrochromic film as it is (de)doped due to lateral injection of ions from a planar junction with an electrolyte (Fig. 3a). The drift mobility of ions is then determined from the time-dependent position of the moving front, by assuming a simple equivalent circuit model[42]. Applying this methodology to our EG-containing PEDOT:PSS formulations allows us to relate the meso-scale structure of films, discussed above, with ion motion therein.

The electrochromic moving front is represented as a 1D normalized-intensity profile at each time point (Fig. 3b, top and Supplementary Fig. 6). The effective mobility of K^+ ions, μ_{ion}, drops consistently and in a reproducible manner from $2.2 \times 10^{-3}\,cm^2\,V^{-1}\,s^{-1}$ to $1.3 \times 10^{-3}\,cm^2\,V^{-1}\,s^{-1}$ for films prepared with 0 and 50 vol% EG, respectively (Fig. 3c). The EG-induced reduction in effective ionic mobility is most likely due to the changes in meso-scale arrangements.

The subtle change in the shape of the moving front profiles on addition of EG reveals that the ion motion can be described by two moving fronts. This behaviour is more clearly observed in the derivative of the front profiles (Fig. 3b, bottom and Supplementary Fig. 6, bottom). The derivative profiles can be used to deconvolute the moving front into a 'leading' front and a 'lagging' front (Supplementary Fig. 7), whereby the fraction of PEDOT:PSS de-doping attributed to each wave of effective ion motion is deduced. With 50 vol% EG, up to 30% of the complete film de-doping is due to the lagging front alone—a contribution which is negligible when no co-solvent is used (Fig. 3c). The evolution of the lagging front is correlated with the position and breadth of the leading front, ruling out the possibility of different (impurity) cations.

To further deduce the origin of different moving fronts observed in EG-modified films, we add an element of absorption spectroscopy to the lateral moving front experiment and monitor

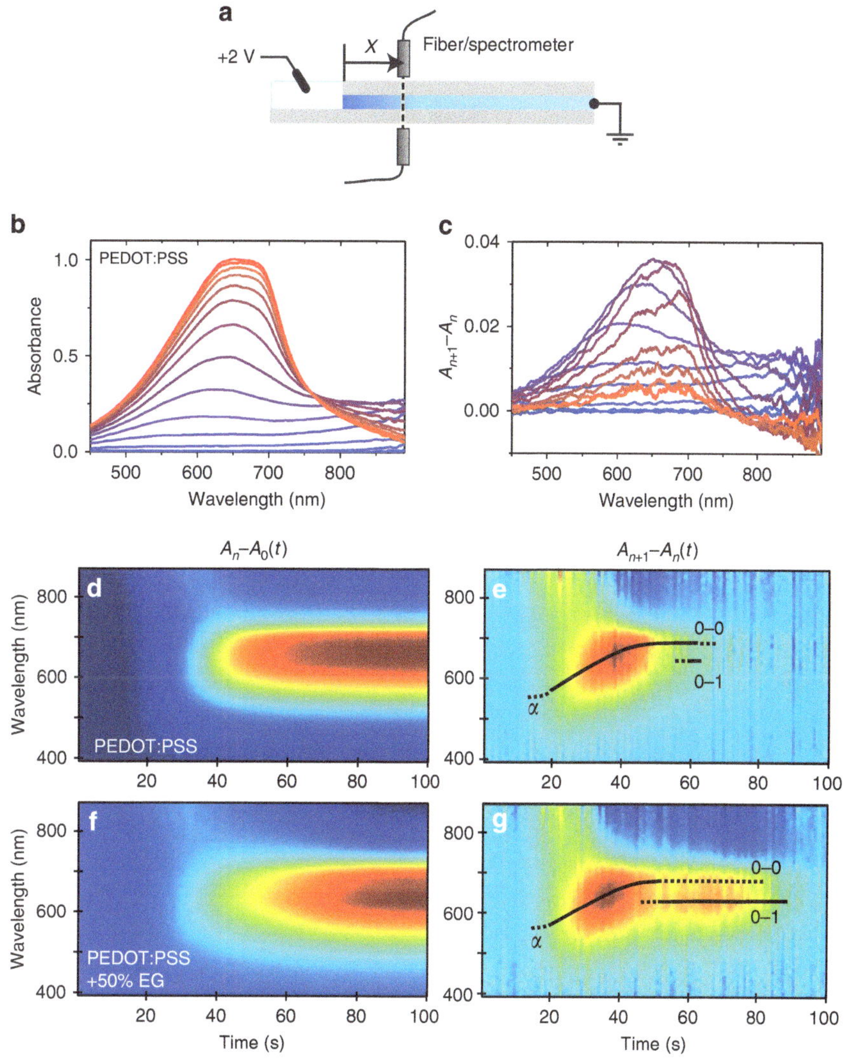

Figure 4 | Spectro-temporal characteristics of ion transport and the effect of EG content. (**a**) Spectro-temporal data are collected with a white-light fibre spectrometer at a fixed position ($X \cong 4$ mm) from the edge of the well. (**b**) Absorbance data ($A_n - A_0$, where the subscripts represent time $t = n$ and $t = 0$ s) of a hydrated PEDOT:PSS film (0 vol% EG) as the de-doping front moves past the position of the optical fibre ($t = 0$ s, blue to $t = 60$ s, red). (**c**) Temporal derivative spectra ($A_{n+1} - A_n$) of the same film revealing the variable electrochromic signatures associated with K^+ transport as the moving front passes. (**d**,**e**) Colour plot representations of the data in **b**,**c**. (**f**,**g**) Colour plot spectra of PEDOT:PSS with 50 vol% EG in the formulation. Note that the time to reach saturation increases with EG formulation content; the added time is due to a lengthy tail of aggregate absorption, as noted by the 0-0 and 0-1 vibronic features (**e**,**g**).

the time-resolved ion motion through the local microstructure. When a cation effectively de-dopes a PEDOT chromophore, the associated change in the visible absorption spectrum of the film can reveal not only the presence of ions, but also the local molecular order of the chromophore in question. The higher energy feature of a polythiophene absorption spectrum is typically associated with amorphous molecular arrangements, as in solution phase, while the low-energy vibronic progression is attributed to weakly interacting H- and J-aggregates[43]. The intensity and ratios of the vibronic peak progression are correlated with the degree of local order of the conjugated backbone[44].

A fibre light source/spectrometer is mounted at a distance of $X \cong 4$ mm from the edge of the CP film/electrolyte interface, in the same configuration as that of the microscopy measurements (Fig. 4a). The ultraviolet–visible spectra are recorded in 1 s intervals as a +2 V bias is applied across the film (Fig. 4b,d,f). To deduce the time course of the instantaneous optical signature and to better understand the local ionic environment, we plot the absorption of individual 1 s intervals (Fig. 4c,e,g). For all samples, we observe first the arrival of the front at $t \cong 10$ s, after which the spectra evolve from a broad peak absorption at ~550–600 nm, to A_{0-1} and A_{0-0} vibronic features at 625 and 685 nm, respectively, over a 40–50 s time span. The sustained addition of vibronic absorption persists in EG-rich samples (Fig. 4g) for an additional 30–40 s (for 50 vol% EG), and is responsible for the delay in reaching saturation in Fig. 4f. The spectroscopy data thus suggest that the rapidly moving leading front is due to ions passing through regions of disordered PEDOT chromophores (most likely those in the PSS-rich matrix), while the lagging front is dictated by ion penetration into aggregate-rich domains such as those in the PEDOT:PSS-densified cores. The persistence of the aggregate absorption recovery within films cast from a higher EG content directly relates to the existence of a lagging front, and thus suggests that aggregates and domain content dictate effective ion mobility.

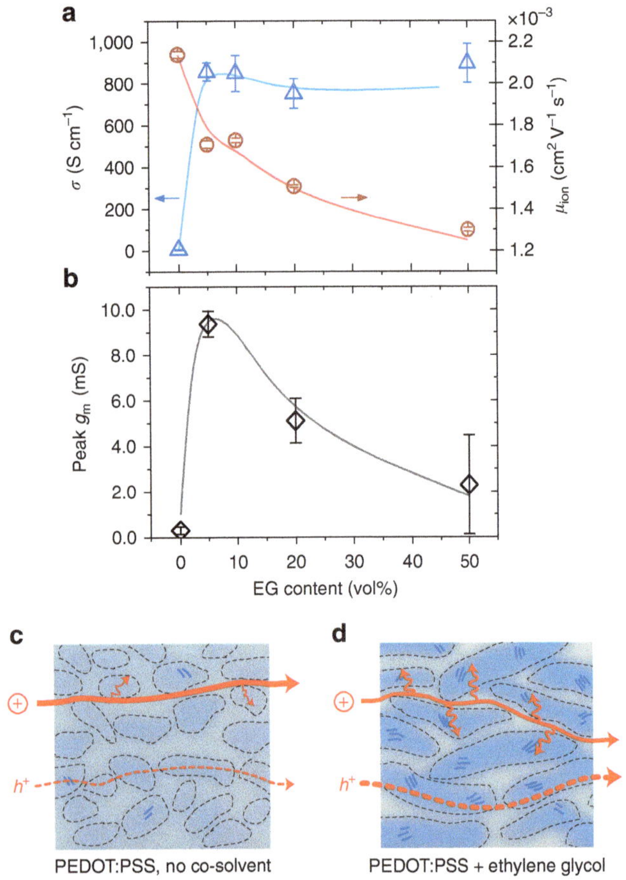

Figure 5 | Effect of mixed conduction in PEDOT:PSS on OECT performance. (**a**) Electronic conductivity (blue), K^+ ion mobility (red) and (**b**) OECT peak transconductance (g_m) at $V_D = -0.6$ V as a function of EG formulation content. Devices have dimensions $W/L = 50\,\mu m/50\,\mu m$ and an equivalent thickness of $d = 200$ nm. Solid lines are guides to the eye. Note that neither efficient electronic nor ionic transport alone is sufficient for achieving the highest g_m. A schematic summary of the morphological changes and associated transport of ions and holes viewed in the cross section of bulk film of PEDOT:PSS (**c**), and PEDOT:PSS with EG (**d**). Addition of EG varies the PEDOT content (shade of blue), size/purity of PEDOT:PSS-rich phases, and amount of PEDOT:PSS aggregates/crystallites. Error bars in **a** represent standard deviations of conductivity values from $n = 8$ measurements, while those in **b** are from OECT measurements of four transistors for each EG condition.

Implications for device performance. Figure 5a shows that the conductivity improves from 6 to $800\,S\,cm^{-1}$ upon addition of just 5 vol% EG, in agreement with previous reports[45,46]. Increasing the EG content > 10 vol% does not lead to a notable change in conductivity of the films. The trends in effective ion mobility in films of the same formulation are shown overlayed in Fig. 5a. The film cast from PEDOT:PSS dispersion without EG displays maximum ionic mobility but poor electrical conductivity, while on addition of 50 vol% EG to the dispersion, the electrical conductivity is maximized at the expense of ionic mobility. Figure 5b shows that neither of these cases is optimal, but it is the 5 vol% EG formulation which yields OECTs with the highest transconductance (g_m) (Supplementary Fig. 8). The formulation with 0 vol% EG leads to an $\cong 30\times$ lower g_m despite showing the best ionic mobility, and the one with 50 vol% EG shows $\cong 4\times$ lower g_m despite displaying the highest electrical conductivity (Fig. 5b).

The difference in effective ionic mobility between samples with low (5 vol%) and high (50 vol%) EG content may seem small in comparison with the observed variation in OECT transconductance. The ion mobilities determined from the moving front displacement are, in fact, dominated by lateral drift of ions, whereas operation of the OECT implies initial through-plane penetration of ions, such that a direct correlation of ionic mobility to device performance is not straightforward. Especially for the films prepared from the dispersions of high EG content, the fractional contribution of the lagging front on the total moving front increases substantially (see Fig. 3c). Furthermore, due to the larger lateral extent of domains, these films may show a higher degree of anisotropy, with more tortuous pathways in the out-of-plane direction through the PSS-rich regions. Taken together, ion mobilities determined from lateral moving front experiments, in particular for the films cast from high-EG-fraction dispersions, are likely overestimations of the through-plane ion transport.

Discussion

Correlating device performance with transport and ultimately film microstructure/morphology is the primary route by which to design new formulations and materials. Addition of the higher boiling point polar solvent, EG, in the casting dispersion leads to changes in the film structure at both the molecular and meso-scale (see Fig. 5c,d). EG leads to an increase in PEDOT aggregation, and slightly tighter π-stacking, in agreement with the work of Palumbiny *et al.*[25,33] At the same time, the domain size of the gel-like particles is known to grow[26,47], increasing by a factor of 3 in the in-plane direction in this work. The growth of domain size with higher EG content is accompanied by a purification of the PEDOT:PSS-rich cores (towards the ultimate limit of 1:1 PEDOT:PSS), and of the PSS matrix (~30% reduction in PEDOT; a lower purity than that observed in drop casting studies)[48].

The coarsened, heterogeneous film morphology of films cast with EG is responsible for the observed dual moving front profiles in Fig. 3b, and leads to lower effective ion mobility. The spectroscopic investigations show that films cast from formulations with higher EG content exhibit a longer, lagging tail of visible absorption recovery (de-doping) attributed to the interaction of ions with aggregated PEDOT chromophores. This aggregate absorption is responsible for the enhanced contribution of the lagging front in the 50 vol% EG film—up to 27% of the total extent of de-doping. We propose, therefore, that the transport of ions takes place through the PSS-rich regions (via the initial leading front), but that the effective ionic mobility is limited by the subsequent diffusion of ions into the densified PEDOT:PSS-rich domains (which slows the total de-doping front).

The observed improvement of hole transport and deterioration of ion transport on the addition of EG thus agree with changes in the multi-scale morphology. At the inter-molecular level, the quality of hopping between molecules is aided by tighter π-stacking and enhanced aggregation[33], largely improving electronic conductivity via an enhancement in hole mobility. At the meso-scale, the growth and purification of PEDOT:PSS-rich domains should aide in electronic transport, however, the percolative nature of the network—the connectivity of the collapsed gel-like particles—cannot be probed with the scattering techniques used. The increase in PEDOT content in the cores, due to potential recruitment of previously inactive PEDOT into the electrically percolating network, may serve to further enhance the film conductivity by increasing hole concentration. Ultimately, the same π–π aggregation and densification of the PEDOT:PSS cores that aides in improving electronic transport, slows down the effective ion motion, leading to opposite trends in conductivity and ion mobility.

The trend in OECT performance, and its similarity to the convoluted trends in electronic and ionic transport give credence to the importance of mixed conduction for device operation. OECT operation relies on efficient lateral bulk electronic transport, and good vertical ion penetration into the film. As can be seen here, these requirements are often contradictory with electron transport being supported by rigid backbones with π-stacked crystallites, while ion transport would benefit from an open, hydrogel-like morphology. Balancing these needs will lead to efficient mixed conduction-based devices. It has been shown that the product of electronic charge carrier mobility (μ), and the volumetric capacitance (C^*) is the material's figure-of-merit for high transconductance devices[3]. Indeed, a high C^* requires that hydrated ions readily access the largest possible percolating conjugated polymer (PEDOT) content, placing greater importance on the lagging front (ions arriving at PEDOT aggregates).

OECT device operation, especially for applications in bioelectronics, necessitates an aqueous environment. Inherently, some of the techniques used here are limited in their ability to probe the films in their fully hydrated state. X-ray scattering and conductivity measurements were performed in the dry state, while the films were in contact with aqueous media during the moving front experiments, the latter most resembling device operation. While film hydration may affect the meso-scale spacings/relative domain sizes, the transport trends observed with EG formulation content are preserved. This verification was performed by extracting the hole mobility of PEDOT:PSS films from OECT operation (see the 'Methods' section) for different EG casting conditions, and comparing the trend with the conductivity/mobility obtained for dry films (Supplementary Fig. 9). Hydrated films were found to have a slightly higher mobility, with larger discrepancy noted for the pristine PEDOT:PSS films. Estimates of hole mobility from Hall effect measurements in this work, for example, $\mu = 1.3\,cm^2\,V^{-1}\,s^{-1}$ for the 20 vol% EG-cast film, are in agreement with previous reports[49], and are on par with OECT-determined hole mobility for the same formulation ($\mu = 2.7\,cm^2\,V^{-1}\,s^{-1}$). This result is in agreement with Wang *et al.*[50] who reported that relative humidity affects the electronic conductivity of PEDOT-based films only when the ionic transport governs the electrical properties. Accordingly, environmental humidity is crucial for poorly electrically conducting films, such as melanin, where conductivity is dominated by proton hopping[51]. Taken together, the agreement of mobility between dry and hydrated PEDOT:PSS films, and the conservation of the general trends observed, confirm that structure–electronic property correlations can be made for the present system independently of the polymer hydration.

By revealing the multi-scale morphological changes triggered by the addition of a commonly used co-solvent, we are able to establish structure-property relations describing mixed conduction in a prototypical CP. We extend the state of understanding of PEDOT:PSS film structure by quantifying domain composition using NEXAFS and resonant scattering techniques. Even with a large volume fraction of EG in the casting dispersion, the phases are seen to trend towards purification, with the PSS phase being far from pure. The observed phase purification enhances electrical conductivity while adversely affecting effective ion mobility as observed in electrochromic moving front studies. Here, the co-solvent EG is used as a means to tune both electronic and ionic transport, and to maximize the performance of OECTs. Finding routes to simultaneously improve ionic and electronic conductivity of CPs and their devices is relevant for many rising materials for applications in bioelectronics, electronics, energy storage and management, and robotics. Enhancing ion transport by approaching hydrogel-like structures without limiting long-range electronic transport remains a materials design hurdle in CP systems. Nevertheless, the structure-mixed conduction property work presented here lays the foundation for future materials design of organic mixed conductors.

Methods

Materials and film formation. Films for structural, electronic and planar junction ion-transport measurements were spun from aqueous dispersion formulation onto glass, Si or Si/SiN window substrates (cleaned by ultra-sonication in dilute soap and acetone/IPA, and/or rinsed in DI water before drying). Substrates are cleaned for 2 min in oxygen plasma before spin casting the dispersions: PEDOT:PSS (Clevios PH-1000 from Heraeus Holding, GmbH) with dodecyl benzene sulfonic acid (DBSA; 0.002 vol%) is mixed with EG (between 0–50% by volume of dispersion as noted in the text/figures) before ultra-sonication (10–15 min) and filtration (through 1.2 µm cellulose acetate filters). Unless otherwise noted, all chemicals are purchased from Sigma Aldrich. Formulations are spun at 1,000–3,000 r.p.m., 1–3 times, soft baked after each spin (90 °C, 60 s), followed by a hard bake (140 °C, 40 min).

X-ray scattering. Grazing incidence wide-angle X-ray scattering was carried out on films spun on native oxide silicon and was performed at the Stanford Synchrotron Radiation Lightsource (SSRL) on beamline 11-3 (2D scattering with an area detector, MAR345 image plate, at grazing incidence), with incident energy of 12.7 keV. The incidence angle was slightly larger than the critical angle, ensuring that we sampled the full film depth, and measurements were performed in a helium-filled chamber to minimize beam damage and reduce air scattering. Data are expressed as a function of the scattering vector $Q = 4\pi \sin(\theta)/\lambda$, where θ is half the scattering angle and λ is the wavelength of the incident radiation. Wide-angle GIXS data were distortion corrected and integrated from $\chi = 0$ to 90°, where χ is defined as the angle between the orientation of a scatterer and the surface normal. Films for rSoXS were solution cast onto 1.5 mm × 1.5 mm, 75 nm thick Si_3N_4 membranes (Norcada, Inc.). Scattering measurements were acquired in transmission geometry onto a CCD at beamline 11.0.1.2 of the Advanced Light Source (ALS, Lawrence Berkeley National Lab) using previously published methods[52].

Near-edge absorption fine structure. NEXAFS spectra for optical constants were measured in transmission mode via a phosphor-coated PMT in a scanning transmission X-ray microscope at beamline 5.3.2.2 of the ALS. The PEDOT:Cl (as in ref. 53) and Na:PSS (from Polymer Souce, spin cast from $1\,mg\,ml^{-1}$ solution) films were cast onto Si_3N_4 membranes. The measurement and data processing followed previously published methods[37]. NEXAFS spectra for average film composition were measured at beamline 11.0.1.2 of the ALS in transmission mode using a photodiode and illuminating the same film spot under investigation for scattering. NEXAFS measurements before and after the scattering experiment were used to confirm no sample damage occurred during the measurement.

Moving front characterization. Planar junctions for electrochromic moving front experiments were prepared as described previously[31]. Briefly, The PEDOT:PSS formulations described above are spun cast on 2-µm thick parylene-C (SCS Labcoater 2) on a 25 × 75 mm glass substrate, before vapour depositing a gold electrode on one side of the film. A ≅ 40-µm-thick SU-8 film was spin-coated, patterned and baked on top of the PEDOT:PSS film to serve as an ion barrier. A polydimethylsiloxane rim was placed on top of the SU-8 well to confine 1.5–2 ml of electrolyte. The length of PEDOT:PSS film between the electrolyte and the Au contact was 3.2 cm. A Ag/AgCl electrode (Warner Instruments) was immersed in the 100 mM KCl in DI water electrolyte. De-doping was achieved by applying a + 2 V bias to the Ag/AgCl electrode; the device was then short-circuited to return to its doped state (this defined one cycle). Measurements were reproducible after 2–3 cycles. Spatio-temporal measurements were conducted on an inverted Carl Zeiss Axio-Observer Z1 in the bright-field mode with a 1 × objective. Voltage was applied, and current measured using a Keithley 2612A source-measure unit connected to a computer running custom LabView software. Grey-level profiles, front shape normalization and deconvolution were calculated using a custom Matlab program. UV-visible spectro-temporal measurements were carried out by mounting the planar junction devices in a 3D-printed holder with fibres for the light source (Hamamatsu) and detector (Ocean Optics USB4000) aligned across from each other at a fixed X distance from the edge of the well. Time-lapse spectra were taken every second using Ocean Optics OceanView software. $A_n - A_0$, $A_{n+1} - A_n$ spectra, normalization and spectral-time colour plots were obtained through custom Matlab software.

Electrical characterization. Conductivity of the polymer thin films were measured with a Van der Pauw Ecopia HMS-3000 Hall Measurement System. PEDOT:PSS formulations with increasing EG concentrations were coated on $10 \times 10\,mm^2$ glass slides with patterned ITO contacts on the corners. IV scans confirmed the ohmic behaviour of the sample contacts. Measurements were performed at room temperature using a current source of 1 mA and a magnetic field strength of 0.56 T.

OECTs were fabricated in a cleanroom and patterned photolithographically, as previously described[1]. They consisted of a PEDOT:PSS channel (formulation described above, with 0.1 wt% of (3-Glycidyloxypropyl)trimethoxysilane to stabilize the film in aqueous) with Au source and drain electrodes and interconnect lines. Parylene-C was used to insulate the Au interconnects from the electrolyte. A Ag/AgCl pellet (Warner Instruments) was used as the gate electrode. Electrical measurements followed previously defined protocols[1]. Device characterization/analysis were carried out with custom LabView and MATLAB software. Hole mobilities from working OECTs were determined using drain current transients at different constant gate currents, which allows for extraction of hole transit times. The approximate electric field at a low drain bias along with the hole velocity allows for mobility estimation, as shown in refs 3 and 54.

References

1. Khodagholy, D. *et al.* High transconductance organic electrochemical transistors. *Nat. Commun.* **4,** 2133 (2013).
2. Khodagholy, D. *et al. In vivo* recordings of brain activity using organic transistors. *Nat. Commun.* **4,** 1575 (2013).
3. Rivnay, J. *et al.* High-performance transistors for bioelectronics through tuning of channel thickness. *Sci. Adv.* **1,** e1400251 (2015).
4. Lin, P., Yan, F., Yu, J., Chan, H. L. W. & Yang, M. The application of organic electrochemical transistors in cell-based biosensors. *Adv. Mater.* **22,** 3655–3660 (2010).
5. Jonsson, A. *et al.* Therapy using implanted organic bioelectronics. *Sci. Adv.* **1,** e1500039 (2015).
6. Williamson, A. *et al.* Controlling epileptiform activity with organic electronic ion pumps. *Adv. Mater.* **27,** 3138–3144 (2015).
7. Kim, S. H. *et al.* Electrolyte-gated transistors for organic and printed electronics. *Adv. Mater.* **25,** 1822–1846 (2013).
8. Kaihovirta, N. *et al.* Printed all-polymer electrochemical transistors on patterned ion conducting membranes. *Org. Electron.* **11,** 1207–1211 (2010).
9. Gkoupidenis, P., Schaefer, N., Garlan, B. & Malliaras, G. G. Neuromorphic functions in PEDOT:PSS organic electrochemical transistors. *Adv. Mater.* **27,** 7176–7180 (2015).
10. Tarabella, G. *et al.* A hybrid living/organic electrochemical transistor based on the *Physarum polycephalum* cell endowed with both sensing and memristive properties. *Chem. Sci.* **6,** 2859–2868 (2015).
11. Josberger, E. E., Deng, Y., Sun, W., Kautz, R. & Rolandi, M. Two-terminal protonic devices with synaptic-like short-term depression and device memory. *Adv. Mater.* **26,** 4986–4990 (2014).
12. Malti, A. *et al.* An organic mixed ion–electron conductor for power electronics. *Adv. Sci.* **3,** 1500305 (2016).
13. Mortimer, R. J., Dyer, A. L. & Reynolds, J. R. Electrochromic organic and polymeric materials for display applications. *Displays* **27,** 2–18 (2006).
14. Haines, C. S. *et al.* Artificial muscles from fishing line and sewing thread. *Science* **343,** 868–872 (2014).
15. Smela, E. Conjugated polymer actuators for biomedical applications. *Adv. Mater.* **15,** 481–494 (2003).
16. Elschner, A., Kirchmeyer, S., Lövenich, W., Merker, U. & Reuter, K. *PEDOT: Pinciples and Applications of an Intrinsically Conductive Polymer* (CRC Press, 2010).
17. Crispin, X. *et al.* The origin of the high conductivity of Poly(3,4-ethylenedioxythiophene) – Poly(styrenesulfonate) (PEDOT – PSS) plastic electrodes. *Chem. Mater.* **18,** 4354–4360 (2006).
18. Greczynski, G., Kugler, T. & Salaneck, W. R. Characterization of the PEDOT-PSS system by means of X-ray and ultraviolet photoelectron spectroscopy. *Thin Solid Films* **354,** 129–135 (1999).
19. Lang, U., Müller, E., Naujoks, N. & Dual, J. Microscopical investigations of PEDOT:PSS thin films. *Adv. Funct. Mater.* **19,** 1215–1220 (2009).
20. Nardes, A. M. *et al.* Microscopic understanding of the anisotropic conductivity of PEDOT:PSS thin films. *Adv. Mater.* **19,** 1196–1200 (2007).
21. Kemerink, M., Timpanaro, S., de Kok, M. M., Meulenkamp, E. A. & Touwslager, F. J. Three-dimensional inhomogeneities in PEDOT:PSS films. *J. Phys. Chem. B* **108,** 18820–18825 (2004).
22. Alemu Mengistie, D., Wang, P.-C. & Chu, C.-W. Effect of molecular weight of additives on the conductivity of PEDOT:PSS and efficiency for ITO-free organic solar cells. *J. Mater. Chem. A* **1,** 9907–9915 (2013).
23. Nardes, A. M., Janssen, R. A. J. & Kemerink, M. A morphological model for the solvent-enhanced conductivity of PEDOT:PSS thin films. *Adv. Funct. Mater.* **18,** 865–871 (2008).
24. Takano, T., Masunaga, H., Fujiwara, A., Okuzaki, H. & Sasaki, T. PEDOT nanocrystal in highly conductive PEDOT:PSS polymer films. *Macromolecules* **45,** 3859–3865 (2012).
25. Palumbiny, C. M. *et al.* Molecular reorientation and structural changes in cosolvent-treated highly conductive PEDOT:PSS electrodes for flexible indium tin oxide-free organic electronics. *J. Phys. Chem. C* **118,** 13598–13606 (2014).
26. Timpanaro, S., Kemerink, M., Touwslager, F. J., De Kok, M. M. & Schrader, S. Morphology and conductivity of PEDOT/PSS films studied by scanning-tunneling microscopy. *Chem. Phys. Lett.* **394,** 339–343 (2004).
27. Shi, H., Liu, C., Jiang, Q. & Xu, J. Effective approaches to improve the electrical conductivity of PEDOT:PSS: a review. *Adv. Electron. Mater.* **1,** 1500017 (2015).
28. Aoki, K., Aramoto, T. & Hoshino, Y. Photographic measurements of propagation speeds of the conducting zone in polyaniline films during electrochemical switching. *J. Electroanal. Chem.* **340,** 127–135 (1992).
29. Johansson, T., Persson, N.-K. & Inganäs, O. Moving redox fronts in conjugated polymers studies from lateral electrochemistry in polythiophenes. *J. Electrochem. Soc.* **151,** E119–E124 (2004).
30. Wang, X., Shapiro, B. & Smela, E. Visualizing ion currents in conjugated polymers. *Adv. Mater.* **16,** 1605–1609 (2004).
31. Stavrinidou, E. *et al.* Direct measurement of ion mobility in a conducting polymer. *Adv. Mater.* **25,** 4488–4493 (2013).
32. Mauritz, K. A. & Moore, R. B. State of understanding of nafion. *Chem. Rev.* **104,** 4535–4585 (2004).
33. Palumbiny, C. M. *et al.* The crystallization of PEDOT:PSS polymeric electrodes probed *in situ* during printing. *Adv. Mater.* **27,** 3391–3397 (2015).
34. Jönsson, S. K. M. *et al.* The effects of solvents on the morphology and sheet resistance in poly(3,4-ethylenedioxythiophene)–polystyrenesulfonic acid (PEDOT–PSS) films. *Synt. Met.* **139,** 1–10 (2003).
35. Huang, J. *et al.* Investigation of the Effects of doping and post-deposition treatments on the conductivity, morphology, and work function of Poly(3,4-ethylenedioxythiophene)/Poly(styrene sulfonate) films. *Adv. Funct. Mater.* **15,** 290–296 (2005).
36. Rivnay, J., Mannsfeld, S. C. B., Miller, C. E., Salleo, A. & Toney, M. F. Quantitative determination of organic semiconductor microstructure from the molecular to device scale. *Chem. Rev.* **112,** 5488–5519 (2012).
37. Yan, H., Wang, C., McCarn, A. R. & Ade, H. Accurate and facile determination of the index of refraction of organic thin films near the carbon 1s absorption edge. *Phys. Rev. Lett.* **110,** 177401 (2013).
38. Roe, R.-J. *Methods of X-Ray and Neutron Scattering in Polymer Science* (Oxford University Press, 2000).
39. Collins, B. A. & Ade, H. Quantitative compositional analysis of organic thin films using transmission NEXAFS spectroscopy in an X-ray microscope. *J. Electron. Spectrosc. Relat. Phenom.* **185,** 119–128 (2012).
40. Aasmundtveit, K. E. *et al.* Structural aspects of electrochemical doping and dedoping of poly(3,4-ethylenedioxythiophene). *Synt. Met.* **113,** 93–97 (2000).
41. Nardes, A. M., Kemerink, M. & Janssen, R. A. J. Anisotropic hopping conduction in spin-coated PEDOT:PSS thin films. *Phys. Rev. B* **76,** 085208 (2007).
42. Stavrinidou, E. *et al.* A simple model for ion injection and transport in conducting polymers. *J. Appl. Phys.* **113,** 244501 (2013).
43. Clark, J., Silva, C., Friend, R. & Spano, F. Role of intermolecular coupling in the photophysics of disordered organic semiconductors: aggregate emission in regioregular polythiophene. *Phys. Rev. Lett.* **98,** 206406 (2007).
44. Spano, F. C. Modeling disorder in polymer aggregates: the optical spectroscopy of regioregular poly(3-hexylthiophene) thin films. *J. Chem. Phys.* **122,** 234701 (2005).
45. Kim, Y. H. *et al.* Highly conductive PEDOT:PSS electrode with optimized solvent and thermal post-treatment for ITO-free organic solar cells. *Adv. Funct. Mater.* **21,** 1076–1081 (2011).
46. Wei, Q., Mukaida, M., Naitoh, Y. & Ishida, T. Morphological change and mobility enhancement in PEDOT:PSS by adding co-solvents. *Adv. Mater.* **25,** 2831–2836 (2013).
47. Thomas, J. P., Zhao, L., McGillivray, D. & Leung, K. T. High-efficiency hybrid solar cells by nanostructural modification in PEDOT:PSS with co-solvent addition. *J. Mater. Chem. A* **2,** 2383–2389 (2014).
48. Ouyang, L., Musumeci, C., Jafari, M. J., Ederth, T. & Inganäs, O. Imaging the phase separation between PEDOT and polyelectrolytes during processing of highly conductive PEDOT:PSS films. *ACS Appl. Mater. Interfaces* **7,** 19764–19773 (2015).
49. Tsai, T.-C., Chang, H.-C., Chen, C.-H. & Whang, W.-T. Widely variable Seebeck coefficient and enhanced thermoelectric power of PEDOT:PSS films by blending thermal decomposable ammonium formate. *Org. Electron.* **12,** 2159–2164 (2011).
50. Wang, H., Ail, U., Gabrielsson, R., Berggren, M. & Crispin, X. Ionic Seebeck effect in conducting polymers. *Adv. Energy Mater.* **5,** 1500044 (2015).
51. Mostert, A. B. *et al.* Role of semiconductivity and ion transport in the electrical conduction of melanin. *Proc. Natl Acad. Sci. USA* **109,** 8943–8947 (2012).
52. Gann, E. *et al.* Soft x-ray scattering facility at the advanced light source with real-time data processing and analysis. *Rev. Sci. Instrum.* **83,** 045110 (2012).
53. Winther-Jensen, B. *et al.* Order-disorder transitions in poly(3,4-ethylenedioxythiophene). *Polymer* **49,** 481–487 (2008).
54. Bernards, D. A. & Malliaras, G. G. Steady-state and transient behavior of organic electrochemical transistors. *Adv. Func. Mater.* **17,** 3538–3544 (2007).

Acknowledgements
J.R. acknowledges support from a Marie Curie post-doctoral fellowship (FP7). M.S. acknowledges the support by the Fundación BBVA, Spain. We acknowledge Prof. B. Winther-Jensen (Monash) for providing vapour-phase polymerized PEDOT:Cl films, Dr E. Gann (Monash) for NEXAFS measurements, Dr M. Ramuz (EMSE) for 3D printing efforts and I. Uguz (EMSE) for film thickness measurements. We additionally thank Prof. C. Silva (Montreal) for fruitful discussions. Portions of this research were carried out at the Stanford Synchrotron Radiation Laboratory (SSRL), a national user facility operated by the Stanford University on behalf of the US Department of Energy, Office of Basic Energy Sciences.

Author contributions
J.R. conceived the research. B.A.C. and J.R. prepared samples and performed resonance scattering experiments. C.T. performed GIXD experiments. B.A.C. and J.R. analysed and interpreted all X-ray scattering data with input from D.M.D. J.R., S.I., and E.S. prepared samples, performed and analysed all electrochromic moving front experiments. M.S., S.I., and X.S. characterized electrical properties including 4-point probe and Hall measurements. J.R., G.G.M., oversaw the research. J.R., S.I. and G.G.M. wrote the manuscript, and all authors provided input in manuscript preparation. Certain commercial equipment or materials are identified in this paper to specify the experimental procedure adequately. Such identification is not intended to imply recommendation or endorsement by the National Institute of Standards and Technology, nor is it intended to imply that the materials or equipment identified are necessarily the best available for the purpose.

Additional information

Competing financial interests: The authors declare no competing financial interests.

23

Microscopic origins of the terahertz carrier relaxation and cooling dynamics in graphene

Momchil T. Mihnev[1,2,*], Faris Kadi[3,*], Charles J. Divin[1,2], Torben Winzer[3], Seunghyun Lee[1,4], Che-Hung Liu[1], Zhaohui Zhong[1], Claire Berger[5,6], Walt A. de Heer[5,7], Ermin Malic[3,8], Andreas Knorr[3] & Theodore B. Norris[1,2]

The ultrafast dynamics of hot carriers in graphene are key to both understanding of fundamental carrier-carrier interactions and carrier-phonon relaxation processes in two-dimensional materials, and understanding of the physics underlying novel high-speed electronic and optoelectronic devices. Many recent experiments on hot carriers using terahertz spectroscopy and related techniques have interpreted the variety of observed signals within phenomenological frameworks, and sometimes invoke extrinsic effects such as disorder. Here, we present an integrated experimental and theoretical programme, using ultrafast time-resolved terahertz spectroscopy combined with microscopic modelling, to systematically investigate the hot-carrier dynamics in a wide array of graphene samples having varying amounts of disorder and with either high or low doping levels. The theory reproduces the observed dynamics quantitatively without the need to invoke any fitting parameters, phenomenological models or extrinsic effects such as disorder. We demonstrate that the dynamics are dominated by the combined effect of efficient carrier-carrier scattering, which maintains a thermalized carrier distribution, and carrier-optical-phonon scattering, which removes energy from the carrier liquid.

[1] Department of Electrical Engineering and Computer Science, University of Michigan, Ann Arbor, Michigan 48109, USA. [2] Center for Ultrafast Optical Science, University of Michigan, Ann Arbor, Michigan 48109, USA. [3] Institut für Theoretische Physik, Nichtlineare Optik und Quantenelektronik, Technische Universität Berlin, D-10623 Berlin, Germany. [4] Department of Electronics and Radio Engineering, Kyung Hee University, Gyeonggi 446-701, South Korea. [5] School of Physics, Georgia Institute of Technology, Atlanta, Georgia 30332, USA. [6] Institut Neel, CNRS UJF-INP, Grenoble 38042, Cedex 6, France. [7] King Abdulaziz University, Jeddah 22254, Saudi Arabia. [8] Department of Applied Physics, Chalmers University of Technology, SE-412 96 Gothenburg, Sweden. * These authors contributed equally to this work. Correspondence and requests for materials should be addressed to T.B.N. (email: tnorris@umich.edu).

In graphene, a linearly polarized ultrafast optical pulse excitation gives rise to an initially anisotropic distribution of carriers at high energies[1,2]. Efficient carrier–carrier and carrier–phonon interactions quickly relax the hot carriers to an isotropic thermal distribution, which is then followed by carrier cooling as energy is transferred from the carrier population to the lattice[3–7]. A schematic illustration of these processes is given in Fig. 1 (see Supplementary Note 1). A variety of dynamical phenomena, such as the appearance of significant carrier multiplication and transient optical gain, have been theoretically predicted[6,8–10] and experimentally demonstrated[6,11,12]. The dynamics observed by a time-domain terahertz (THz) probe pulse, however, exhibit a number of features that have been interpreted in the framework of phenomenological models, but as yet have not been understood quantitatively or qualitatively in terms of fundamental microscopic many-particle processes. For example, the observed photoinduced THz conductivity may be either positive or negative. The positive photoinduced THz conductivity has been viewed in the context of simple Drude models as stemming from enhanced free-carrier intraband absorption upon photoexcitation[13,14], while the negative photoinduced THz conductivity has been attributed variously to stimulated THz emission[15], enhanced carrier scattering with optical phonons, surface optical phonons or charge impurities[16–18] and carrier heating[19–21].

It is generally understood that the initial cooling of hot thermalized carriers proceeds via the emission of high-energy optical phonons ($\hbar\omega_{op} \approx 200$ meV). Once the carrier temperature is sufficiently below the optical phonon energy, the cooling should proceed via the emission of low-energy acoustic phonons ($\hbar\omega_{ac} \approx 4$ meV). For graphene samples with low doping density, the cooling of hot carriers near the Dirac point is expected to be very slow due to the vanishing density of states, the energetic mismatch with the energy of optical phonons, and the weak scattering with acoustic phonons[4,22,23]. The slowest observed cooling times, however, have been on the order of hundreds of picoseconds[14], much shorter than the few nanoseconds expected from carrier–acoustic–phonon scattering in ideal graphene[22,23], leading to proposals that the cooling is dominated by disorder-assisted electron–phonon (supercollision) scattering[24]. The central role of disorder in these models implies that the underlying enhancement of carrier–acoustic–phonon scattering should strongly depend on the quality and the degree of disorder of the particular graphene sample. Hence, both a methodical experimental investigation and a microscopic theoretical treatment are markedly needed to provide a rigorous foundation for understanding the THz dynamics of hot carriers in graphene.

In the following, we present the results of a systematic experimental study of the THz carrier dynamics in a wide variety of graphene samples, using ultrafast time-resolved THz spectroscopy[25], in which we vary the graphene fabrication method, the number of graphene layers and their stacking orientation, the degree of disorder, the Fermi level, the carrier temperature, the substrate temperature and the type of underlying substrate to determine the dominant mechanisms responsible for the hot-carrier relaxation and cooling dynamics for different graphene material parameters and under different experimental conditions. The experimental programme is complemented by the development of a theoretical model[26,27] based on the density-matrix formalism that provides microscopic access to the time- and momentum-resolved dynamics of the carrier occupation, the phonon population for different optical and acoustic phonon modes, and the microscopic polarization determining the optical excitation of a disorder-free graphene system. An essential component of the theory is that it incorporates explicitly the time-dependent response of the system to the THz probe pulse; the macroscopic intraband current density induced by the THz probe, and hence the resulting dynamic THz response, is calculated by microscopically accounting for the full time- and momentum-dependent carrier–carrier and carrier–phonon interactions. We find that this first-principles microscopic

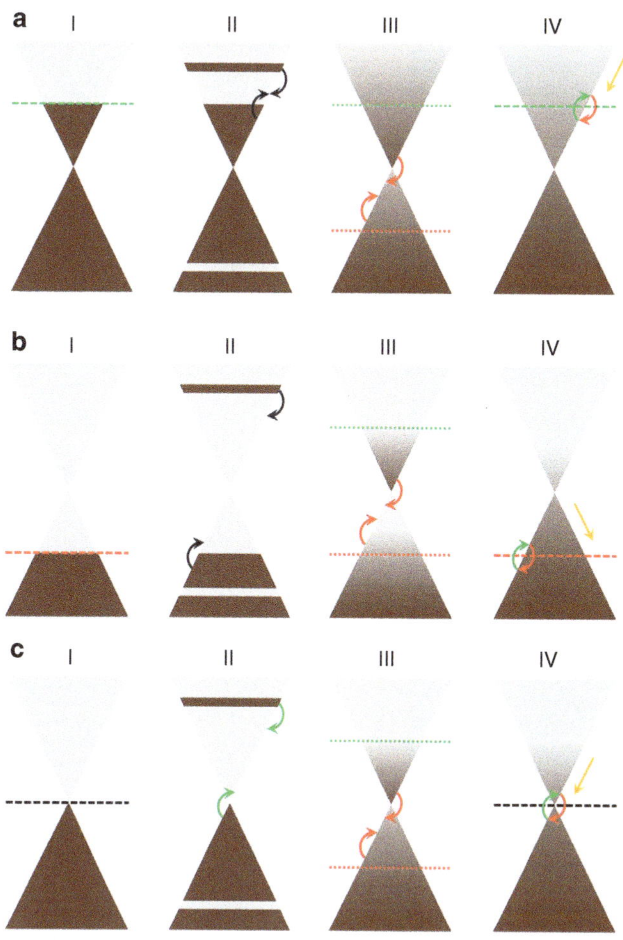

Figure 1 | Hot-carrier relaxation and cooling dynamics in graphene with various doping densities. (**a**,**b**) Hot-carrier relaxation and cooling dynamics in n-type (**a**) and p-type (**b**) highly doped graphene (step I). Initially, the optical pump pulse injects hot non-equilibrium carriers at high energies (step II). The hot electrons and holes thermalize within the conduction and valence bands, respectively, due to very efficient intraband carrier-carrier scattering processes (step III). As the hot carriers relax to lower energies, interband Auger recombination processes become allowed that quickly merge the separate electron and hole quasi-Fermi levels and lead to a single uniform hot-carrier Fermi-Dirac distribution within ~100–200 fs after photoexcitation[3,6,7] (step IV). The hot carriers cool further via optical phonon emission facilitated by very efficient carrier-carrier rethermalization in highly doped graphene. (**c**) Hot-carrier relaxation and cooling dynamics in undoped (very lightly doped) graphene (step I). Initially, the optical pump pulse injects hot non-equilibrium carriers at high energies (step II). In contrast to **a** and **b** interband impact ionization processes are possible and lead for a moderate excitation regime to significant carrier multiplication in the conduction band. Similar to **a** and **b** as the hot carriers relax to lower energies, interband Auger recombination processes become allowed (step III) that lead to a single uniform hot-carrier Fermi-Dirac distribution within ~100–200 fs after photoexcitation[3,6,7] (step IV). The hot carriers cool further via optical phonon emission, which becomes increasingly inefficient at low carrier temperatures due to the small phase space near the Dirac point as the high-energy tail of the hot-carrier distribution diminishes asymptotically. At low carrier temperatures, acoustic phonon emission and/or other slow cooling processes can also have a contribution to hot-carrier cooling in some graphene samples[22–25].

approach explains completely all results without the need for any fitting parameters, phenomenological models or extrinsic effects such as disorder, which strongly suggests that the role of supercollisions in the hot-carrier dynamics has been largely overstated in the literature[24,28,29]. Moreover, the theory allows us to go beyond idealized Drude models for the dynamic THz response, and determine the limitations of such simplified models[13–21]. This work establishes that the hot-carrier dynamics are governed by the coupling of extraordinarily efficient carrier–carrier and carrier–optical–phonon interactions. This is in sharp contrast to electrical transport, which in some graphene samples is dominated by interactions with defects, charge impurities, breaks and ripples (that is, extrinsic effects) since carriers move at the Fermi energy[30–33].

Results

Graphene samples. The specific graphene samples in this study include multilayer epitaxial graphene (MEG), which is grown on the C-face of single-crystal 4H-SiC(000$\bar{1}$) substrates by thermal decomposition of Si atoms[34,35], single-crystal chemical-vapor-deposited (CVD) graphene (sCVDG), which is grown on oxygen-rich copper foil into large individual single crystals exceeding hundreds of micrometres in size[36], and polycrystalline CVD graphene (pCVDG), which is grown on regular copper foil into large continuous layers with domain sizes on the order of hundreds of nanometres[37]. The as-grown CVD graphene samples are transferred to various substrates for the THz measurements. Because the different graphene samples are synthesized using completely different techniques, their doping density and degree of disorder are also very different (see Methods section), which allows us to observe the possible impact of the Fermi level and the degree of disorder on the dynamic THz response.

Ultrafast time-resolved THz spectroscopy experiment. To study the dynamic THz response of graphene, we utilize ultrafast time-resolved THz spectroscopy[25,38,39], which has established itself as a powerful all-optical experimental technique for directly probing the relaxation and cooling dynamics of photoexcited carriers (see Methods section). A 60-fs ultrafast optical pump pulse at 800 nm wavelength injects hot carriers into the graphene sample. The dynamic THz response is monitored by a single-cycle THz probe pulse at a variable time delay after the optical pump. The transmitted THz probe is detected using time-domain electro-optic sampling and frequency-domain THz spectra are obtained via Fourier transformation of the time-domain THz electric field. The THz carrier dynamics are acquired by monitoring the THz transmission only at the peak of the THz probe pulse, while varying the pump–probe delay. The differential THz transmission signal, $\Delta t/t$, is the change in the THz probe transmission through the graphene sample due to photoexcitation by the optical pump, normalized to the THz transmission without photoexcitation.

We begin by summarizing the main features of the data. First, we consider MEG, sCVDG and pCVDG samples having high doping density ($|\varepsilon_F| \sim 100$–400 meV). Figure 2a,b shows representative differential THz transmission signals as a function of pump–probe delay, for variable pump fluence and for variable substrate temperature, respectively, for a MEG sample with three layers. The secondary peak in the $\Delta t/t$ signal at ~ 7 ps is due to a round-trip reflection of the optical pump inside the substrate that photoexcites additional carriers. Similar results are obtained for the other two types of graphene samples (see Supplementary Figs 1–5, Supplementary Table 1 and Supplementary Notes 2–4). Based on extensive measurements on many graphene samples under various experimental conditions, we find that the THz carrier dynamics in all highly doped graphene samples are strikingly similar. The $\Delta t/t$ signal is positive under all experimental conditions, which corresponds to a pump-induced increase of the THz transmission or a decrease of the THz absorption.

Phenomenological fits to the data in Fig. 2a,b (dashed black lines) reveal that the differential THz transmission follows closely a mono-exponential relaxation under all experimental conditions. A summary of the extracted carrier relaxation times as a function of pump fluence and substrate temperature for the highly doped graphene samples is presented in Fig. 3a,b. We note that the relaxation times of all highly doped graphene samples are weakly dependent on the pump fluence and completely independent of the substrate temperature. In addition, they are very similar in value and in the range of ~ 1–3 ps with sample-to-sample variation within ~ 20–30% despite the wide range of disorder present in the array of samples studied. On average, MEG samples exhibit slightly longer relaxation times than sCVDG and pCVDG samples, which can be attributed to a degree of disorder arising from charge impurities, substrate roughness, wrinkling and breaking of the transferred CVDG samples that can provide additional parallel channels for carrier cooling. We note, however, that the relaxation times of some CVDG samples can approach or exceed these of MEG samples, indicating that disorder-assisted electron–phonon (supercollision) cooling[24,28,29] is not the dominant cooling mechanism in our high-quality graphene samples, but generally provides at most only a modest correction.

Figure 2c,d shows the THz carrier dynamics calculated within the microscopic theory for disorder-free highly doped graphene ($|\varepsilon_F| = 300$ meV) under similar experimental conditions (see Methods section) and Fig. 3c,d shows the calculated carrier relaxation times to directly compare with the experiments. We observe that the theory is in excellent agreement with experiment, and reproduces the weak pump fluence dependence and the complete substrate temperature independence. In sharp contrast, the supercollision model[24,28,29] totally fails to capture the substrate temperature independence. The measured THz carrier dynamics can be completely reproduced neglecting carrier–acoustic–phonon scattering in the microscopic theory. Hence, the hot-carrier dynamics are directly the result of an interplay between efficient carrier–carrier and carrier–optical–phonon scattering; optical phonon emission removes energy from the high-energy tail of the hot-carrier distribution, which is maintained by efficient carrier–carrier rethermalization. The physical reason the differential THz transmission of highly doped graphene is independent of the substrate temperature is that the equilibrium THz transmission itself is insensitive to it, when the substrate temperature is far below the Fermi temperature. The slight increase of the relaxation times with increasing pump fluence is due to the re-absorption of hot optical phonons generated during the initial carrier thermalization.

The THz carrier dynamics in graphene samples having very low doping density are very different. We consider specifically MEG samples in which only the first few layers closest to the underlying SiC substrate have high doping density ($|\varepsilon_F| \sim 100$–400 meV) and the large number of top layers have very low doping density ($|\varepsilon_F| \sim 10$ meV). In these MEG samples, the many top lightly doped layers completely dominate the measured THz carrier dynamics. Figure 4a,b shows representative differential THz transmission signals for variable pump fluence and for variable substrate temperature, respectively, for a MEG sample with 63 layers. The secondary dip in the $\Delta t/t$ signal at ~ 7 ps is again due to a round-trip substrate reflection of the optical pump. In sharp contrast to the high doping case, the $\Delta t/t$ signal is negative under all experimental conditions, which corresponds to a pump-induced decrease of the THz transmission or an increase of the THz absorption.

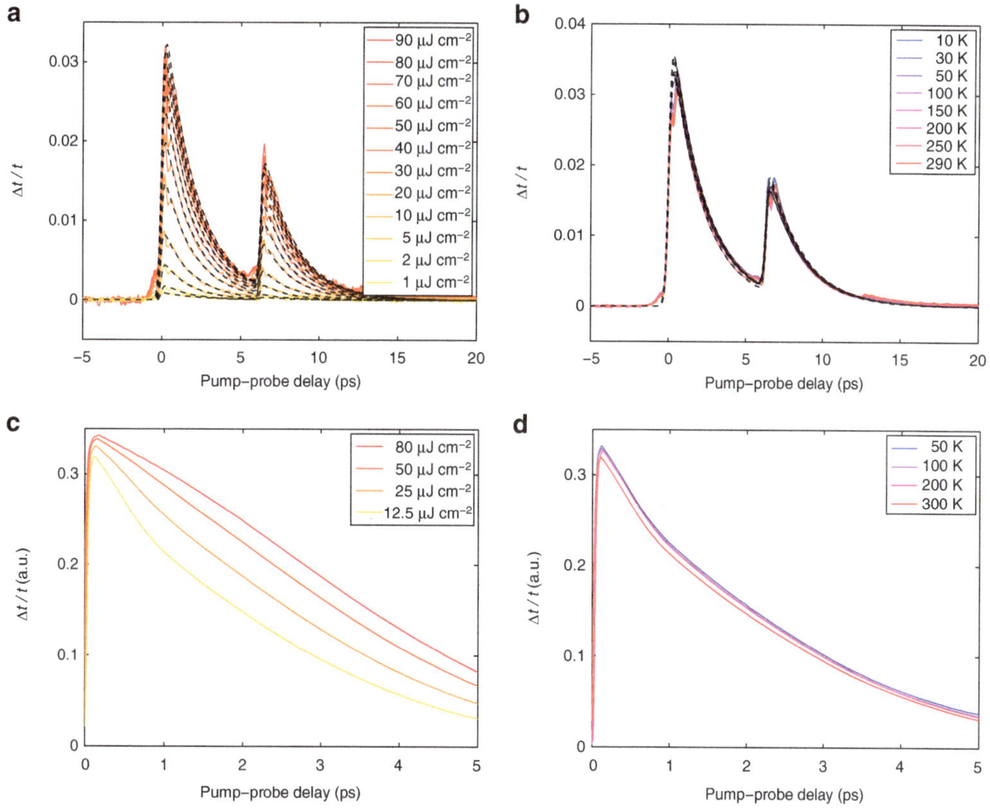

Figure 2 | THz carrier dynamics in graphene with high doping density. (**a**,**b**) Experimental differential THz transmission, $\Delta t/t$, as a function of pump-probe delay recorded at a substrate temperature of 300 K for a few different pump fluences (**a**) and at a pump fluence of 60.0 μJ cm^{-2} for a few different substrate temperatures (**b**) for a highly doped MEG sample with three layers. The THz carrier dynamics follow a fast mono-exponential relaxation at all substrate temperatures and all pump fluences (dashed black lines). (**c**,**d**) Theoretical differential THz transmission, $\Delta t/t$, as a function of pump-probe delay calculated within the microscopic theory at a substrate temperature of 300 K for a few different pump fluences (**c**) and at a pump fluence of 12.5 μJ cm^{-2} for a few different substrate temperatures (**d**) for disorder-free highly doped graphene ($|\varepsilon_F| = 300$ meV). Experiment and theory are in excellent agreement under all conditions.

Phenomenological fits to the data in Fig. 4a,b (dashed black lines) reveal that the differential THz transmission evolves from a faster mono-exponential relaxation at room temperature to a slower bi-exponential relaxation at cryogenic temperatures. A summary of the extracted carrier relaxation times as a function of pump fluence at room temperature for two MEG samples with 35 and 63 layers is presented in Fig. 5a. We note that the relaxation times at room temperature do not depend on the number of layers, which supports the interpretation that the measured THz carrier dynamics are dominated by the many top lightly doped layers. This conclusion is further supported by the fact that the maximum $\Delta t/t$ signal scales roughly linearly with the number of layers (see Supplementary Figs 3–5 and Supplementary Notes 3–4). In addition, the relaxation times of lightly doped graphene samples are in the range of ~4–7 ps, which is longer than the relaxation times of highly doped graphene samples (see Fig. 3a); this is due to a reduced efficiency of carrier–carrier scattering, which scales with carrier density. For cryogenic temperatures, we extract both short and long carrier relaxation times from the bi-exponential fits, which we associate with two distinct cooling mechanisms. As explained below, the fast carrier relaxation component on a timescale of tens of picoseconds is fully accounted for by the combined effect of carrier–carrier and carrier–optical–phonon scattering in the absence of disorder. Figure 5b shows the short relaxation times as a function of substrate temperature for variable pump fluence for the MEG sample with 63 layers. We note that the short relaxation times of all lightly doped graphene samples are weakly dependent on the pump fluence, but strongly dependent on the substrate temperature.

Figure 4c,d shows the THz carrier dynamics calculated within the microscopic theory for disorder-free undoped graphene ($|\varepsilon_F| = 0$ meV) under similar experimental conditions (see Methods section) and Fig. 5c,d shows the calculated carrier relaxation times. The theory captures again all trends observed in the experiments. As in the high-doping case, the hot-carrier relaxation occurs via the interplay between efficient carrier–carrier and carrier–optical–phonon scattering. In contrast to the high-doping case, however, the relaxation times increase significantly with decreasing substrate temperature. This behaviour for undoped graphene is due to the substrate temperature dependence of the equilibrium THz transmission and not the dynamic non-equilibrium part of the differential THz transmission; that is, the initial Fermi surface, where the THz probe pulse acts on the carriers, strongly depends on the substrate temperature, when it is comparable to the Fermi temperature. Similar to the high-doping case, the theory shows a slight increase of the relaxation times with increasing pump fluence that can be traced back to hot phonon effects.

At low substrate temperatures, the THz carrier dynamics in the MEG samples with many lightly doped layers relax on a timescale exceeding hundreds of picoseconds, corresponding to the slow carrier relaxation component in the bi-exponential decay. As the high-energy tail of the hot-carrier distribution diminishes

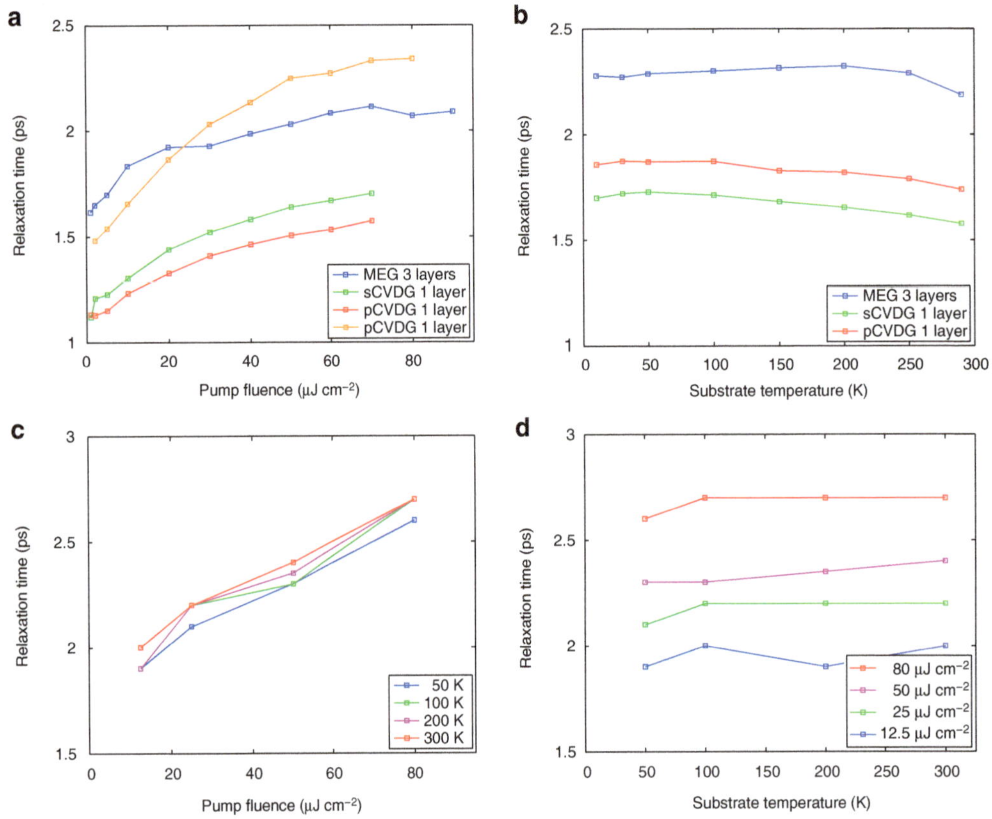

Figure 3 | THz carrier relaxation times in graphene with high doping density. (**a**,**b**) Carrier relaxation times extracted from fits to experimental differential THz transmission, $\Delta t/t$, as a function of pump fluence at a substrate temperature of 300 K (**a**) and as a function of substrate temperature at a pump fluence of 60.0 μJ cm^{-2} (**b**) for highly doped MEG, sCVDG and pCVDG samples. (**c**,**d**) Carrier relaxation times extracted from fits to theoretical differential THz transmission, $\Delta t/t$, as a function of pump fluence for a few different substrate temperatures (**c**) and as a function of substrate temperature for a few different pump fluences (**d**) for disorder-free highly doped graphene ($|\varepsilon_F| = 300$ meV). The theory accurately reproduces the magnitude of the relaxation times and the trends with pump fluence and substrate temperature observed in the experiments.

asymptotically, a different cooling mechanism becomes dominant below ~200 K. We have analysed the physical origin of this mechanism in a separate publication[25]. In particular, the THz carrier dynamics become dependent on the number of layers in the MEG sample, indicating that interlayer thermal coupling effects are important.

Microscopic theory. We now turn to a discussion of the theoretical approach used to calculate the dynamic THz response described above (see Methods section, Supplementary Fig. 6 and Supplementary Note 5). The theory is an extension of the methods developed previously by some of the authors[1,2,5,9,11,12,26,27], and is extended in this work to rigorously include the effect of the time-dependent THz probe electric field. By selectively switching on and off the different scattering processes in the model, we find that the acoustic phonon modes have negligible contribution to the observed THz carrier dynamics on the timescale of tens of picoseconds; the observed dynamics are fully accounted for by the combined effect of carrier–optical–phonon scattering (which transfers energy from the carrier liquid to the lattice) and carrier–carrier scattering (which continuously rethermalizes the carrier population as high-energy carriers lose their energy to the lattice). As seen in the comparisons between experiment and theory above, this first-principles microscopic approach explains completely all experimental results without the need for any fitting parameters, phenomenological models or extrinsic effects such as disorder.

The standard Drude model, which is often employed as a phenomenological basis for the interpretation of graphene transport and optical data, can be obtained by assuming a constant time- and momentum-independent scattering rate in the microscopic theory (see Methods section and Supplementary Note 6). It is thus interesting to consider to what extent the Drude model can account for the observed dynamics. Figure 6a shows the pump-induced temporal evolution of the carrier temperature $T(t)$ and the Fermi level $\varepsilon_F(t)$ obtained by solving the full graphene Bloch equations within this approximation for low (12.5 μJ cm^{-2} (red line)) and high (80 μJ cm^{-2} (blue line)) photoexcitation, for disorder-free highly doped graphene ($|\varepsilon_F| = 300$ meV). For very high carrier temperature, as one has at early time delay and high fluence, the calculated $\Delta t/t$ signal is negative, in contradiction to the experimental observation that the $\Delta t/t$ signal is positive at all time delays for highly doped graphene samples. Only after the carrier temperature drops below ~2,200 K does the $\Delta t/t$ signal become positive. If one makes the further approximation (as is often done in Drude models) that the only effect of the optical pump is to heat the carriers, that is, the Fermi level is assumed to remain constant, than the calculated $\Delta t/t$ signal is negative at all time delays (dashed black line in Fig. 6a). The full microscopic theory including the transient carrier temperature $T(t)$, the transient Fermi level $\varepsilon_F(t)$ and the explicitly time- and momentum-dependent scattering rates

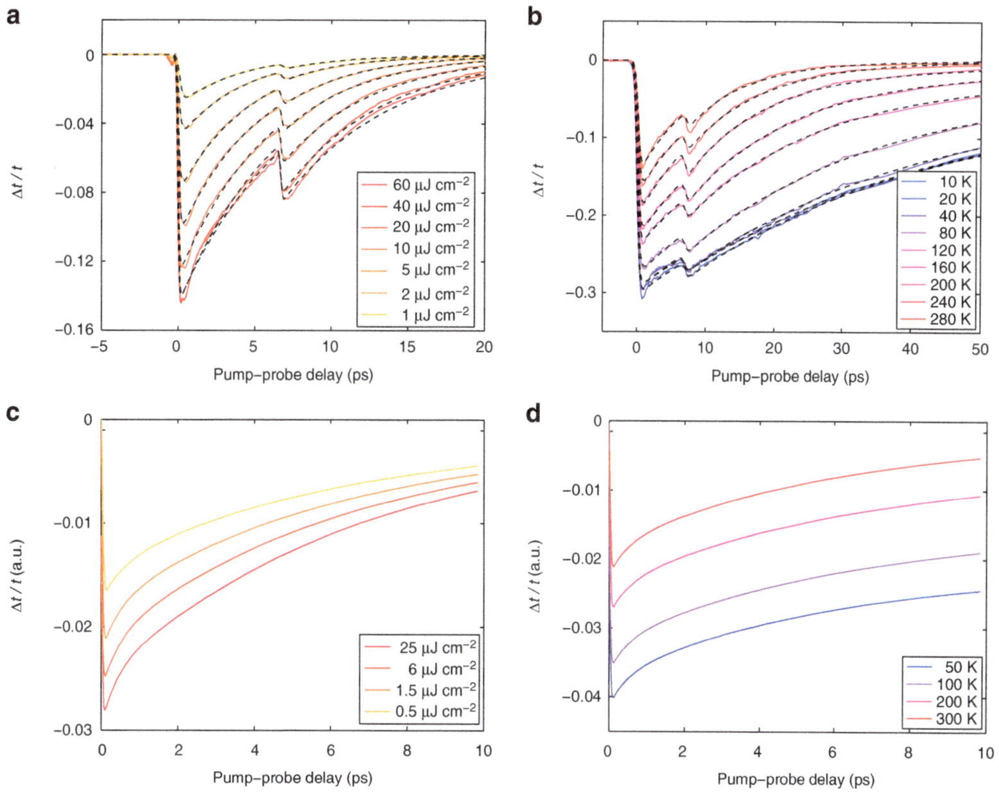

Figure 4 | THz carrier dynamics in graphene with low doping density. (**a**,**b**) Experimental differential THz transmission, $\Delta t/t$, as a function of pump-probe delay recorded at a substrate temperature of 300 K for a few different pump fluences (**a**) and at a pump fluence of 23.4 μJ cm^{-2} for a few different substrate temperatures (**b**) for a lightly doped MEG sample with 63 layers. The THz carrier dynamics evolve from a faster mono-exponential relaxation at room temperature to a slower bi-exponential relaxation at cryogenic temperatures (dashed black lines). (**c**,**d**) Theoretical differential THz transmission, $\Delta t/t$, as a function of pump-probe delay calculated within the microscopic theory at a substrate temperature of 300 K for a few different pump fluences (**c**) and at a pump fluence of 1.5 μJ cm^{-2} for a few different substrate temperatures (**d**) for disorder-free undoped graphene ($|\varepsilon_F| = 0$ meV). Experiment and theory are in excellent agreement under all conditions.

$\Gamma_{\lambda k}^{in/out}(t)$, on the other hand, completely reproduces all the features of the experimental data under all conditions. Drude models are not sufficient to consistently explain all the data, particularly the behaviour at higher pump fluence, for which the short time dynamics would exhibit a negative $\Delta t/t$ signal, unless ad hoc phenomenological parameters such as a carrier heating efficiency[19,20] or a non-monotonic carrier-temperature-dependent Drude weight[21] are added to the model. We conclude that a Drude model approximation to the full microscopic theory can provide only a semi-qualitative framework for interpreting experimental data in highly doped and undoped graphene samples at low photoexcitation; however, the full microscopic theory is required to explain the data consistently for all photoexcitation levels.

The microscopic theory also allows us to address an important question in hot-carrier physics, namely, the energy dependence of the carrier scattering rate (inverse time), defined here as the exponential decay of the carrier occupation (see Methods section). For weak THz probe excitations (as is the case here), the Boltzmann equation reveals a direct relation between the microscopic scattering rates and the exponential decay of the carrier occupation. By fitting the numerically calculated hot-carrier dynamics, we obtain the energy relaxation time of the photoexcited carriers, named the carrier scattering time. Figure 6b shows the calculated carrier scattering time $\tau(\varepsilon)$ as a function of the excess carrier energy ε for disorder-free undoped graphene ($|\varepsilon_F| = 0$ meV) at a substrate temperature of 300 K. We observe that the carrier scattering time is precisely inverse to the carrier energy, $\tau(\varepsilon)=\beta/|\varepsilon|$ (with $\beta \approx 0.9$ eV ps), over a very broad energy range ($|\varepsilon| \sim 0.2$–1.5 eV) in agreement with previous experimental studies on MEG[40] and graphite[41,42]. This behaviour is a direct consequence of the linear density of states and is therefore independent of the substrate temperature. For highly doped graphene, a deviation from the strictly inverse relation can be expected for $\varepsilon \approx \varepsilon_F$. However, our microscopically determined energy dependence of the carrier scattering time is in sharp contrast with the linear relation on the carrier energy, $\tau(\varepsilon)=\alpha|\varepsilon|$ (refs 30,31), that has been inferred from electrical transport measurements in some graphene samples. This is attributed to the fact that in electrical transport measurements carriers have energies close to the Fermi energy, and carrier scattering in these graphene samples is dominated by extrinsic mechanisms such as defects, charge impurities, breaks and ripples[32,33] likely introduced during the multistep synthesis, transfer and fabrication processes. Transferred graphene (for example, on SiO_2) has local spatial charge inhomogeneities under overall charge-neutral conditions due to disorder, charge impurities or surface corrugation[43–45], which obscure the low-energy graphene band structure and prevent one from studying the true graphene physics near the Dirac point. Such effects can be minimized by placing the graphene on ultra-smooth substrates such as hexagonal BN (h-BN)[46] or by suspending it[47]. The lightly doped layers in MEG are naturally protected, which makes them an ideal graphene system for studying the carrier dynamics within a few meV of the Dirac point.

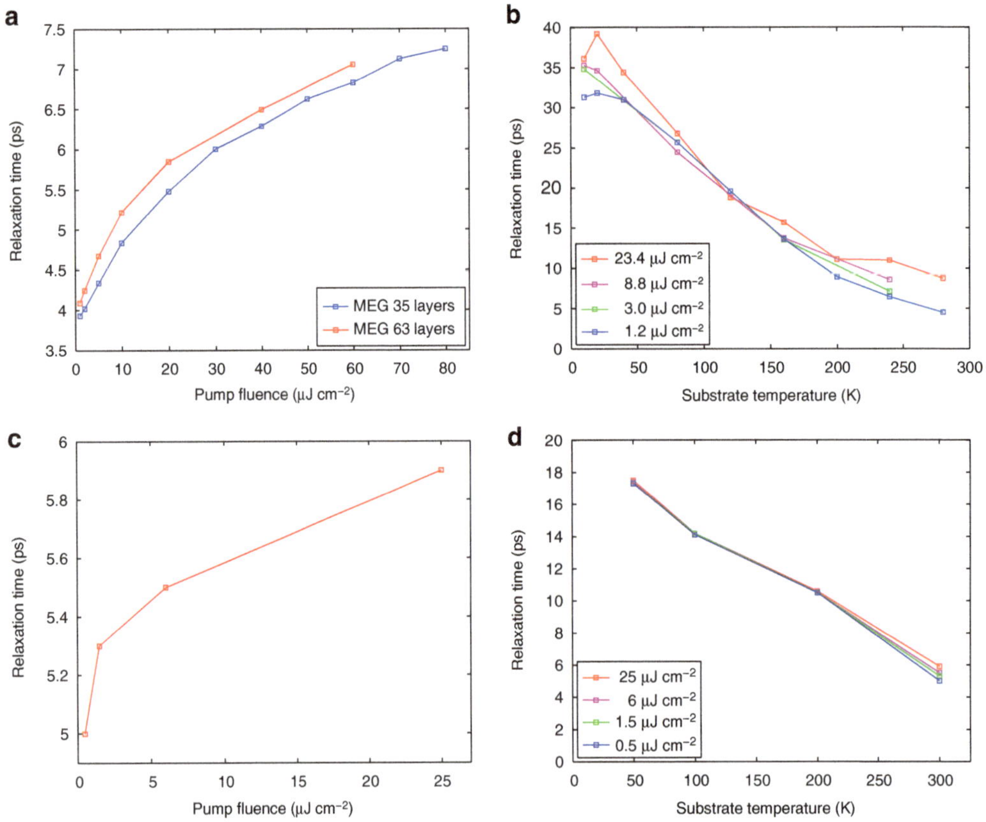

Figure 5 | THz carrier relaxation times in graphene with low doping density. (**a**,**b**) Carrier relaxation times extracted from fits to experimental differential THz transmission, $\Delta t/t$, as a function of pump fluence at a substrate temperature of 300 K (**a**) and as a function of substrate temperature for a few different pump fluences (**b**) for lightly doped MEG samples. (**c**,**d**) Carrier relaxation times extracted from fits to theoretical differential THz transmission, $\Delta t/t$, as a function of pump fluence at a substrate temperature of 300 K (**c**) and as a function of substrate temperature for a few different pump fluences (**d**) for disorder-free undoped graphene ($|\varepsilon_F| = 0$ meV). The theory accurately reproduces the magnitude of the relaxation times and the trends with pump fluence and substrate temperature observed in the experiments.

Finally, we consider the correlation between the differential THz transmission and the carrier temperature dynamics. In the microscopic theory, the dynamic THz response depends on the transient carrier temperature, the transient Fermi level and the time- and momentum-dependent scattering rates (see Methods section). The inset of Fig. 6b shows a direct comparison between the differential THz transmission and the differential carrier temperature (where the substrate temperature is subtracted) calculated within the microscopic theory for disorder-free undoped graphene ($|\varepsilon_F| = 0$ meV) at a substrate temperature of 300 K and a pump fluence of 1.5 μJ cm^{-2}. We see that the differential THz transmission does not exactly follow the carrier temperature dynamics and, in particular, the relaxation times extracted from the decay of the $\Delta t/t$ signals are not exactly equal to the electronic cooling times. In sharp contrast, a simple Drude model, in which only the carrier temperature is assumed to be time-dependent, would predict incorrectly that the two quantities are proportional[13,14,24]. Hence, our microscopic approach clearly reveals that the transient carrier temperature, the Fermi level shifts and the time- and momentum-dependent scattering rates are all essential to capture the dynamic THz response correctly.

Discussion

In summary, we have studied the hot-carrier relaxation and cooling dynamics in highly doped and undoped (very lightly doped) graphene samples synthesized using a wide array of methods, using ultrafast time-resolved THz spectroscopy combined with microscopic modelling. The THz carrier dynamics depend critically on the Fermi level, and are quantitatively explained using a microscopic density-matrix theory of carrier–carrier and carrier–phonon interactions, without the need to invoke any free fitting parameters, phenomenological models or extrinsic effects such as disorder; the theory accounts explicitly for the time-dependent response of the hot carriers to the THz probe field. The hot-carrier dynamics are governed by the interplay of efficient carrier–carrier and carrier–optical–phonon scattering, while carrier–acoustic–phonon scattering is found not to be important on picosecond timescales.

Methods

Graphene samples synthesis, fabrication and characterization. To date a number of techniques have been demonstrated for the synthesis of high-quality graphene including exfoliation, epitaxial and CVD growth methods. We report here comprehensive experiments on graphene synthesized using three different methods. The first type is MEG, which is grown on the C-face of single-crystal 4H-SiC(000$\bar{1}$) substrates by thermal decomposition of Si atoms[34,35]. The MEG grows conformally across atomic terraces on the SiC substrate resulting into large continuous layers with domain sizes exceeding hundreds of micrometres in size. By carefully tuning the chemical recipe, MEG samples having from a few up to a hundred layers can be reliably grown, and the fluctuations in the homogeneity of the MEG samples are estimated not to exceed one layer. The individual layers in MEG are electronically decoupled due to their unique rotational stacking, and each layer exhibits a single graphene layer Dirac cone near the Dirac point, so that MEG behaves in essence as multilayer graphene[48,49]. The second type is sCVDG, which is grown on oxygen-rich copper foil into large individual single crystals exceeding hundreds of micrometres in size[36]. The third type is pCVDG, which is grown on regular copper foil into large continuous layers with domain sizes on the order of hundreds of nanometres[37].

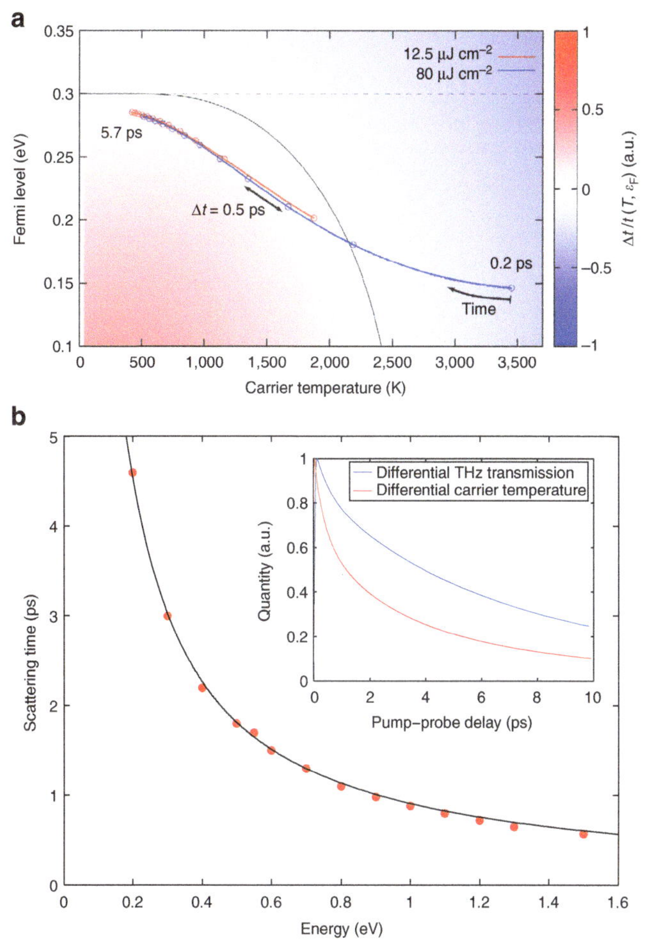

Figure 6 | Microscopic theory calculation of the carrier dynamics and the carrier scattering time in graphene. (**a**) Differential THz transmission, $\Delta t/t$, expected from a Drude model with a constant scattering rate for an initial carrier temperature of 300 K and a Fermi level of 300 meV as a function of the transient carrier temperature $T(t)$ and the transient Fermi level $\varepsilon_F(t)$. The solid black line separates regions of positive and negative $\Delta t/t$ signal. The dashed black line shows the possible differential THz transmission under the assumption that the Fermi level remains constant. The solid red (blue) line shows the path through the $T(t)$-$\varepsilon_F(t)$-map for a pump fluence of 12.5 μJ cm^{-2} (80 μJ cm^{-2}) obtained from the solution of the full graphene Bloch equations for the pump-induced dynamics. The first point reflects the system 0.2 ps after the optical pump pulse, when the carrier distribution can be represented by a single uniform hot-carrier Fermi-Dirac distribution. The time delay between two points is 0.5 ps, respectively. The figure illustrates the fact that the full microscopic theory is required to explain the experimental data consistently for all photo-excitation levels. (**b**) Carrier scattering time $\tau(\varepsilon)$ as a function of excess carrier energy ε calculated within the microscopic theory for disorder-free undoped graphene ($|\varepsilon_F| = 0$ meV) at a substrate temperature of 300 K. The calculated values are precisely inverse to the carrier energy, $\tau(\varepsilon)=\beta/|\varepsilon|$ (with $\beta \approx 0.9$ eV ps). Inset: comparison between the normalized differential THz transmission and the normalized differential carrier temperature dynamics calculated within the microscopic theory for disorder-free undoped graphene ($|\varepsilon_F| = 0$ meV) at a substrate temperature of 300 K and a pump fluence of 1.5 μJ cm^{-2}. The two time-dependent quantities relax on similar, but not exactly equal timescales.

By carefully tuning the chemical recipe, both mono- and bi-layer pCVDG samples can be reliably grown.

The as-grown CVD graphene samples are transferred from the copper foils to various substrates including C-cut single-crystal sapphire (Alfa Aesar) and amorphous polyethylene (TOPAS cyclic olefin copolymer, TOPAS Advanced Polymers) for the ultrafast time-resolved THz spectroscopy measurements. The exact same transfer process is used for both sCVDG and pCVDG samples. One side of the copper samples is spin coated with 950PMMA A2 (MicroChem) photoresist as a protection layer and the other side is exposed to oxygen plasma to etch away the undesired graphene. The samples are left in ammonium persulfate solution (0.025 g ml^{-1}) for 12 h to dissolve the copper foil underneath. Then, the graphene films with PMMA coating are transferred onto the prepared clean substrates and are left to dry for 12 h. The final step is to use acetone for removing the PMMA coating on top and isopropyl alcohol for rinsing the samples.

Because the different graphene samples are synthesized using completely different techniques, their doping density and degree of disorder are also very different, which allows us to study the dynamic THz response for different Fermi levels and different degrees of disorder. The MEG samples have a gradient doping density profile, where the first few layers closest to the underlying SiC substrate are highly n-doped ($|\varepsilon_F| \sim 100$–400 meV) due to electron transfer from the interface, and the Fermi level in subsequent layers decreases exponentially away from the substrate to around $|\varepsilon_F| \sim 10$ meV (refs 34,35,50,51). The CVDG samples are highly p-doped ($|\varepsilon_F| \sim 200$–400 meV) due to water vapour adsorption from the environment[52]. Thus, we can directly compare the dynamic THz response of graphene with Fermi level far above, far below and very close to the Dirac point.

The graphene degree of disorder is in general not straightforward to quantify, but it can be estimated from various characterization techniques including high-resolution angle-resolved photoemission spectroscopy (ARPES), high-resolution scanning tunnelling microscopy, Raman spectroscopy and electrical transport measurements. Raman spectroscopy measurements of all three types of our graphene samples show negligible D peaks suggesting extremely low disorder[36,37,53,54]. The width of the Dirac cone in the graphene band structure directly measured by ARPES can provide a more sensitive measure for the long-range coherence of graphene[55]. Epitaxial graphene exhibits a very sharp, narrow and well-defined Dirac cone indicating very smooth and homogenous graphene films with extremely high quality. On the other hand, exfoliated and especially CVD graphene transferred to an arbitrary substrate exhibits a rather smeared and broad Dirac cone indicating wrinkled and non-homogenous graphene films. From the width of the Dirac cone, we can extract a correlation length, which for epitaxial graphene exceeds ~50 nm, limited only by the instrument resolution, but is expected to be much longer[55]. The correlation length for exfoliated and CVD graphene is ~1–3 nm (ref. 55). A similar disparity in the long-range coherence is inferred also from electrical transport and magneto-optical spectroscopy measurements of the graphene carrier mobility. The carrier mobility of epitaxial graphene has been reported to exceed ~250,000 cm^2 V^{-1} s^{-1} close to the theoretical value for disorder-free graphene[34,35]. On the other hand, the carrier mobilities of exfoliated and CVD graphene transferred to an arbitrary substrate range from a few thousands to tens of thousands dominated by interactions with defects, charge impurities, breaks and ripples. The highest values of up to ~200,000 cm^2 V^{-1} s^{-1} are achieved by minimizing these extrinsic scattering mechanisms including placing the graphene on ultra-smooth substrates such as h-BN[46] or suspending it[47]. Thus, we can also directly investigate the impact of disorder on the dynamic THz response of graphene.

Ultrafast time-resolved THz spectroscopy on graphene. To study the dynamic THz response of graphene, we utilize ultrafast time-resolved THz spectroscopy[25,38,39]. Our laser system consists of a Ti:Sapphire oscillator (Mira 900-F, Coherent) followed by a Ti:Sapphire regenerative amplifier (RegA 9050, Coherent) and produces ultrafast optical pulses with a centre wavelength of 800 nm, a pulse width of ~60 fs and a repetition rate of 250 kHz. A portion of the laser beam is quasi-collimated at the sample position with an intensity spot size diameter of ~1,600 μm, and optically injects hot carriers in the graphene sample. A second portion of the laser beam illuminates a low-temperature-grown GaAs photoconductive emitter (Tera-SED 3/4, Gigaoptics)[56,57] generating a broadband single-cycle THz pulse which is collimated and focused on the graphene sample with an intensity spot size diameter of ~500 μm to probe the dynamic THz response. The transmitted portion of the THz probe is detected by using time-domain electro-optic sampling in a 1-mm-thick ZnTe crystal[58–60] and a pair of balanced Si photodiodes. The electrical signal is modulated by a mechanical chopper, placed in either the optical pump or the THz probe arm, and recorded by using a conventional lock-in amplifier data acquisition technique. The graphene sample is mounted inside a liquid helium continuous flow cryostat (ST-100, Janis) to vary the substrate temperature from 10 to 300 K. The time delays between the optical pump, the THz probe and the sampling pulse are controlled by two motorized stages. All THz optics is surrounded by an enclosure purged with purified nitrogen gas to minimize water vapour absorption. The detection bandwidth of the system is in the range of ~0.2–2.5 THz and the temporal resolution of the measurements is limited by the duration of the THz probe pulse to the sub-picosecond timescale. The experimental error is due primarily to long-term drift of the optomechanical components and the ultrafast Ti:Sapphire laser system, and is estimated not to exceed ~5%.

Microscopic theory calculation of the dynamic THz response of graphene. The dynamic THz response due to an optical excitation can be microscopically addressed by evaluating the graphene Bloch equations, which describe the coupled dynamics of the carrier occupation $\rho_{\mathbf{k}}^{\lambda}$ at the wave vector $\mathbf{k}$ in conduction ($\lambda=c$) and valence band ($\lambda=v$), the microscopic polarization $p_{\mathbf{k}}$, and the phonon population $n_{\mathbf{q}}^{j}$ at the momentum $\mathbf{q}$ for different optical and acoustic phonon modes j (ref. 5):

$$\frac{d}{dt}\rho_{\mathbf{k}}^{\lambda} = -\frac{e_0}{\hbar}\mathbf{E}\cdot\nabla_{\mathbf{k}}\rho_{\mathbf{k}}^{\lambda} + 2\Im\left[\Omega_{\mathbf{k}}^{vc*}p_{\mathbf{k}}\right] + \Gamma_{\lambda\mathbf{k}}^{\mathrm{in}}\left[1-\rho_{\mathbf{k}}^{\lambda}\right] - \Gamma_{\lambda\mathbf{k}}^{\mathrm{out}}\rho_{\mathbf{k}}^{\lambda}, \quad (1)$$

$$\frac{d}{dt}p_{\mathbf{k}} = [i\Delta\omega_{\mathbf{k}} - \gamma_{\mathbf{k}}]p_{\mathbf{k}} - i\Omega_{\mathbf{k}}^{vc}\left[\rho_{\mathbf{k}}^{c} - \rho_{\mathbf{k}}^{v}\right]. \quad (2)$$

An optical excitation of the system is considered via the Rabi-frequency $\Omega_{\mathbf{k}}^{vc}$ accounting for interband transitions, where $\Delta\omega_{\mathbf{k}} = v_F k$ is the transition frequency and v_F is the Fermi velocity. Compared to previous work[5], we include a drift term $\mathbf{E}\cdot\nabla_{\mathbf{k}}\rho_{\mathbf{k}}^{\lambda}$ expressing the light-induced intraband transitions, which are crucial for the THz dynamics driven by the probe pulse. The carrier–carrier and carrier–phonon interactions are taken into account by a Boltzmann-like equation with time- and momentum-dependent scattering rates $\Gamma_{\lambda\mathbf{k}}^{\mathrm{in/out}}$ for the carrier occupation and by the diagonal dephasing $\gamma_{\mathbf{k}} = \frac{1}{2}\sum_{\lambda}\Gamma_{\lambda\mathbf{k}}^{\mathrm{in}} + \Gamma_{\lambda\mathbf{k}}^{\mathrm{out}}$ for the microscopic polarization. The explicit form of the many-particle contributions and the equation for the phonon dynamics can be found elsewhere[5,26,27], for example, in ref. 5 in equations 22–24.

The differential THz transmission is given by:

$$\Delta t/t(t,\omega) \propto \alpha^{(t)}(\omega) - \alpha^{(p,t)}(\omega), \quad (3)$$

with the absorption coefficient $\alpha^{(p,t)}(\omega)$ including both the pump and the probe pulse and $\alpha^{(t)}(\omega)$ including only the probe pulse. The absorption $\alpha(\omega) = \Im[j(\omega)/(\epsilon_0\omega^2 A(\omega))]$ is determined by the macroscopic current density[26,27,61]:

$$\mathbf{j}(\omega) = \frac{4e_0\hbar}{m_0 L^2}\sum_{\mathbf{k}} M_{\mathbf{k}}^{vc}\Im[p_{\mathbf{k}}(\omega)] + \frac{2e_0\hbar}{i m_0 L^2}\sum_{\mathbf{k},\lambda} M_{\mathbf{k}}^{\lambda\lambda}\rho_{\mathbf{k}}^{\lambda}(\omega), \quad (4)$$

where $\mathbf{M}_{\mathbf{k}}^{\lambda\lambda'}$ is the optical matrix element, m_0 is the free electron mass and L^2 is the structure area of the system that cancels out after performing the summation over $\mathbf{k}$. The current contains an interband contribution that is driven by the microscopic polarization $p_{\mathbf{k}}(\omega)$ and an intraband contribution that is determined by the carrier occupation $\rho_{\mathbf{k}}^{\lambda}(\omega)$. While the interband term has the dominant contribution for probe pulses at optical frequencies, we consider in this work the intraband term, which has the dominant contribution for probe pulses at THz frequencies. Assuming a weak THz probe pulse, $\rho_{\mathbf{k}}^{\lambda}(\omega)$ can be treated perturbatively:

$$\rho_{\mathbf{k}}^{\lambda}(t) = \rho_{\mathbf{k}}^{\lambda,0}(t) + \delta\rho_{\mathbf{k}}^{\lambda}(t), \quad (5)$$

where $\rho_{\mathbf{k}}^{\lambda,0}(t)$ is the pump-induced carrier occupation, while $\delta\rho_{\mathbf{k}}^{\lambda}(t)$ describes the weak carrier occupation excited by the THz probe pulse.

To obtain the dynamic THz response from the differential THz transmission spectra, we exploit the fact that ultrafast carrier–carrier scattering in graphene forms a uniform hot-carrier Fermi-Dirac distribution within the first tens of femtoseconds after the excitation[3,6,7] and the subsequent dynamics is fully characterized by the temporal evolution of the transient carrier temperature $T(t)$ and the transient Fermi level $\varepsilon_F(t)$. Thus, by iteratively evaluating $T(t)$ and $\varepsilon_F(t)$ on the basis of the numerically calculated carrier dynamics at each time step, we obtain the pump-induced carrier occupation $\rho_{\mathbf{k}}^{\lambda,0}(t)$. For the dynamics of the probe-induced carrier occupation, we derive from equation (1) and with the ansatz in equation (5) a separate equation of motion yielding:

$$\frac{d}{dt}\delta\rho_{\mathbf{k}}^{\lambda}(t) = -\frac{e_0}{\hbar}\mathbf{E}\cdot\nabla_{\mathbf{k}}\rho_{\mathbf{k}}^{\lambda,0}(t) - \Gamma_{\lambda\mathbf{k}}^{0}(t)\delta\rho_{\mathbf{k}}^{\lambda}(t), \quad (6)$$

where $\Gamma_{\lambda\mathbf{k}}^{0}(t) = \Gamma_{\lambda\mathbf{k}}^{\mathrm{in},0}(t) + \Gamma_{\lambda\mathbf{k}}^{\mathrm{out},0}(t)$ is the diagonal contribution stemming from the Boltzmann-like scattering terms (equation (1)) which are independent of the probe pulse as denoted by the index 0. Non-linear contributions in the probe pulse and non-diagonal terms have been neglected here. Finally, we numerically evaluate the Fourier transform of $\delta\rho_{\mathbf{k}}^{\lambda}(t)$ from equation (6) and, thereby, we account for the fully microscopically determined carrier dynamics in terms of $\rho_{\mathbf{k}}^{\lambda,0}(t)$ and $\Gamma_{\lambda\mathbf{k}}^{0}(t)$. Thus, our microscopic approach provides access to the Coulomb- and phonon-assisted dynamics induced by the THz probe pulse and is used to obtain the results shown in Figs 2–5.

Microscopic theory calculation of the carrier scattering time in graphene. Here, we derive an analytic relation between the microscopic scattering rates and the exponential decay of the carrier occupation. The collision part of the Boltzmann equation, which is given by:

$$\frac{d}{dt}\rho_{\mathbf{k}}^{\lambda} = \Gamma_{\lambda\mathbf{k}}^{\mathrm{in}}\left[1-\rho_{\mathbf{k}}^{\lambda}\right] - \Gamma_{\lambda\mathbf{k}}^{\mathrm{out}}\rho_{\mathbf{k}}^{\lambda}, \quad (7)$$

can be rewritten as:

$$\frac{d}{dt}\rho_{\mathbf{k}}^{\lambda} = -\frac{\rho_{\mathbf{k}}^{\lambda} - \tau_{\lambda\mathbf{k}}\Gamma_{\lambda\mathbf{k}}^{\mathrm{in}}}{\tau_{\lambda\mathbf{k}}}, \quad (8)$$

where we define:

$$\tau_{\lambda\mathbf{k}} := \frac{1}{\Gamma_{\lambda\mathbf{k}}^{\mathrm{in}} + \Gamma_{\lambda\mathbf{k}}^{\mathrm{out}}}. \quad (9)$$

The term $\tau_{\lambda\mathbf{k}}\Gamma_{\lambda\mathbf{k}}^{\mathrm{in}}$ can be written as:

$$\tau_{\lambda\mathbf{k}}\Gamma_{\lambda\mathbf{k}}^{\mathrm{in}} = \frac{\Gamma_{\lambda\mathbf{k}}^{\mathrm{in}}}{\Gamma_{\lambda\mathbf{k}}^{\mathrm{in}} + \Gamma_{\lambda\mathbf{k}}^{\mathrm{out}}} = \frac{1}{1 + \frac{\Gamma_{\lambda\mathbf{k}}^{\mathrm{out}}}{\Gamma_{\lambda\mathbf{k}}^{\mathrm{in}}}}. \quad (10)$$

By accounting only for weak excitations, the scattering rates can be approximated with the equilibrium scattering rates $\Gamma_{\lambda\mathbf{k}}^{\mathrm{in/out},0}$ fulfilling the principle of detailed balance (D.B.):

$$\frac{\Gamma_{\lambda\mathbf{k}}^{\mathrm{out}}}{\Gamma_{\lambda\mathbf{k}}^{\mathrm{in}}} \approx \frac{\Gamma_{\lambda\mathbf{k}}^{\mathrm{out},0}}{\Gamma_{\lambda\mathbf{k}}^{\mathrm{in},0}} \overset{\mathrm{D.B.}}{=} e^{(\varepsilon_k - \varepsilon_F)/k_B T}. \quad (11)$$

Thus, equation (10) represents the initial Fermi distribution $\rho_{\mathbf{k}}^{\lambda,0} \approx \tau_{\lambda\mathbf{k}}\Gamma_{\lambda\mathbf{k}}^{\mathrm{in}}$ and the Boltzmann equation yields the relaxation-time model:

$$\frac{d}{dt}\rho_{\mathbf{k}}^{\lambda} = -\frac{\rho_{\mathbf{k}}^{\lambda} - \rho_{\mathbf{k}}^{\lambda,0}}{\tau_{\lambda\mathbf{k}}}. \quad (12)$$

Within the approximation in equation (11), equation (12) clearly reveals a direct relation between the microscopic scattering rates and the exponential decay of the carrier occupation. However, by fitting the numerical data stemming from the full scattering equation (equation (7)), our method for the determination of the carrier scattering time (see Fig. 6b) goes beyond the relaxation-time approximation.

Drude model as an approximation of the microscopic theory. To obtain a simple Drude model from our microscopic approach (equations (1)–(6)), we approximate the scattering rate as a constant: $\Gamma_{\lambda\mathbf{k}}^{0}(t) = \Gamma$, that is, we neglect the time and the momentum dependence. By assuming a δ-shaped perturbation for the THz probe pulse, the Fourier transform of equation (6) yields:

$$\delta\rho_{\mathbf{k}}^{\lambda}(\omega) = i\omega\mathbf{e}_x\frac{e_0 A_0}{\hbar}\cdot\frac{1}{i\omega - \Gamma}\nabla_{\mathbf{k}}\rho_{\mathbf{k}}^{\lambda,0}. \quad (13)$$

The intraband current density is given by:

$$\mathbf{j}(\omega) = \frac{2e_0\hbar}{i m_0 L^2}\sum_{\mathbf{k},\lambda}\mathbf{M}_{\mathbf{k}}^{\lambda\lambda}\delta\rho_{\mathbf{k}}^{\lambda}(\omega). \quad (14)$$

By assuming a uniform hot-carrier Fermi-Dirac distribution for the carrier occupation, the gradient yields:

$$\nabla_{\mathbf{k}}\rho_{\mathbf{k}}^{\lambda} = -\sigma_{\lambda}\mathbf{e}_{\mathbf{k}}\frac{v_F\hbar}{4k_B T}\mathrm{sech}^2[\sigma_{\lambda}(k v_F\hbar - \varepsilon_F)/2k_B T], \quad (15)$$

where σ_{λ} is $1\ (-1)$ for $\lambda=c$ ($\lambda=v$) and ε_F is the Fermi level. For a constant scattering rate Γ and with the optical intraband matrix element $\mathbf{M}_{\mathbf{k}}^{\lambda\lambda} \approx i\sigma_{\lambda}M\mathbf{e}_{\mathbf{k}}$ ($\sigma_c = 1$ and $\sigma_v = -1$), the intraband current can be evaluated analytically using equations (13) and (14):

$$\begin{aligned} j(\omega) \;=\; & \frac{e_0^2 M A_0 k_B}{m_0 \pi c^2}\left[i\frac{\Gamma}{\Gamma^2+\omega^2} - \frac{\omega}{\Gamma^2+\omega^2}\right]\times \\ & T\left[\ln\left(1+e^{\varepsilon_F/k_B T}\right) + \ln\left(1+e^{-\varepsilon_F/k_B T}\right)\right], \end{aligned} \quad (16)$$

which corresponds to the Drude model. For a constant scattering rate Γ and an arbitrary THz frequency, the resulting Drude-like differential THz transmission (see Fig. 6a) is given by:

$$\Delta t/t(t) \propto j(T_0, \varepsilon_{F,0}) - j(T(t), \varepsilon_F(t)), \quad (17)$$

where T_0 and $\varepsilon_{F,0}$ denote the initial carrier temperature and Fermi level before the arrival of the optical pump pulse at time $t=0$.

We note that in principle a $\mathbf{k}$-dependent $\Gamma_{\mathbf{k}}$ could also be considered in equations (13) and (14) to obtain a more advanced Drude model. However, this would require also an approximate analytical model for $\Gamma_{\mathbf{k}}$. We again emphasize that we go beyond the approximation by using the numerically calculated microscopic scattering rates for the evaluation of the current.

References

1. Malic, E., Winzer, T. & Knorr, A. Efficient orientational carrier relaxation in optically excited graphene. *Appl. Phys. Lett.* **101,** 213110 (2012).
2. Mittendorff, M. *et al.* Anisotropy of excitation and relaxation of photogenerated charge carriers in graphene. *Nano Lett.* **14,** 1504–1507 (2014).
3. Breusing, M. *et al.* Ultrafast nonequilibrium carrier dynamics in a single graphene layer. *Phys. Rev. B* **83,** 153410 (2011).
4. Winnerl, S. *et al.* Carrier relaxation in epitaxial graphene photoexcited near the dirac point. *Phys. Rev. Lett.* **107,** 237401 (2011).
5. Malic, E., Winzer, T., Bobkin, E. & Knorr, A. Microscopic theory of absorption and ultrafast many-particle kinetics in graphene. *Phys. Rev. B* **84,** 205406 (2011).
6. Brida, D. *et al.* Ultrafast collinear scattering and carrier multiplication in graphene. *Nat. Commun.* **4,** 1987 (2013).

7. Gierz, I. *et al.* Snapshots of non-equilibrium Dirac carrier distributions in graphene. *Nat. Mater.* **12,** 1119–1124 (2013).
8. Winzer, T., Knorr, A. & Malic, E. Carrier multiplication in graphene. *Nano Lett.* **10,** 4839–4843 (2010).
9. Winzer, T., Malić, E. & Knorr, A. Microscopic mechanism for transient population inversion and optical gain in graphene. *Phys. Rev. B* **87,** 165413 (2013).
10. Song, J. C. W., Tielrooij, K. J., Koppens, F. H. L. & Levitov, L. S. Photoexcited carrier dynamics and impact-excitation cascade in graphene. *Phys. Rev. B* **87,** 155429 (2013).
11. Li, T. *et al.* Femtosecond population inversion and stimulated emission of dense dirac fermions in graphene. *Phys. Rev. Lett.* **108,** 167401 (2012).
12. Plötzing, T. *et al.* Experimental verification of carrier multiplication in graphene. *Nano Lett.* **14,** 5371–5375 (2014).
13. George, P. A. *et al.* Ultrafast optical-pump terahertz-probe spectroscopy of the carrier relaxation and recombination dynamics in epitaxial graphene. *Nano Lett.* **8,** 4248–4251 (2008).
14. Strait, J. H. *et al.* Very slow cooling dynamics of photoexcited carriers in graphene observed by optical-pump terahertz-probe spectroscopy. *Nano Lett.* **11,** 4902–4906 (2011).
15. Docherty, C. J. *et al.* Extreme sensitivity of graphene photoconductivity to environmental gases. *Nat. Commun.* **3,** 1228 (2012).
16. Jnawali, G., Rao, Y., Yan, H. & Heinz, T. F. Observation of a transient decrease in terahertz conductivity of single-layer graphene induced by ultrafast optical excitation. *Nano Lett.* **13,** 524–530 (2013).
17. Shi, S.-F. *et al.* Controlling graphene ultrafast hot carrier response from metal-like to semiconductor-like by electrostatic gating. *Nano Lett.* **14,** 1578–1582 (2014).
18. Frenzel, A. J. *et al.* Observation of suppressed terahertz absorption in photoexcited graphene. *Appl. Phys. Lett.* **102,** 113111 (2013).
19. Tielrooij, K. J. *et al.* Photoexcitation cascade and multiple hot-carrier generation in graphene. *Nat. Phys.* **9,** 248–252 (2013).
20. Jensen, S. A. *et al.* Competing ultrafast energy relaxation pathways in photoexcited graphene. *Nano Lett.* **14,** 5839–5845 (2014).
21. Frenzel, A. J., Lui, C. H., Shin, Y. C., Kong, J. & Gedik, N. Semiconducting-to-metallic photoconductivity crossover and temperature-dependent Drude weight in graphene. *Phys. Rev. Lett.* **113,** 056602 (2014).
22. Bistritzer, R. & MacDonald, A. H. Electronic cooling in graphene. *Phys. Rev. Lett.* **102,** 206410 (2009).
23. Tse, W.-K. & Das Sarma, S. Energy relaxation of hot Dirac fermions in graphene. *Phys. Rev. B* **79,** 235406 (2009).
24. Song, J. C. W., Reizer, M. Y. & Levitov, L. S. Disorder-assisted electron-phonon scattering and cooling pathways in graphene. *Phys. Rev. Lett.* **109,** 106602 (2012).
25. Mihnev, M. T. *et al.* Electronic cooling via interlayer Coulomb coupling in multilayer epitaxial graphene. *Nat. Commun.* **6,** 8105 (2015).
26. Malic, E., Winnerl, S. & Knorr, A. *Graphene and Carbon Nanotubes: Ultrafast Optics and Relaxation Dynamics* (Wiley-VCH, 2013).
27. Knorr, A., Hughes, S., Stroucken, T. & Koch, S. W. Theory of ultrafast spatio-temporal dynamics in semiconductor heterostructures. *Chem. Phys.* **210,** 27–47 (1996).
28. Graham, M. W., Shi, S.-F., Ralph, D. C., Park, J. & McEuen, P. L. Photocurrent measurements of supercollision cooling in graphene. *Nat. Phys.* **9,** 103–108 (2013).
29. Betz, A. C. *et al.* Supercollision cooling in undoped graphene. *Nat. Phys.* **9,** 109–112 (2013).
30. Ando, T. Screening effect and impurity scattering in monolayer graphene. *J. Phys. Soc. Jpn* **75,** 074716 (2006).
31. Das Sarma, S., Adam, S., Hwang, E. H. & Rossi, E. Electronic transport in two-dimensional graphene. *Rev. Mod. Phys.* **83,** 407 (2011).
32. Tan, Y.-W. *et al.* Measurement of scattering rate and minimum conductivity in graphene. *Phys. Rev. Lett.* **99,** 246803 (2007).
33. Chen, J.-H. *et al.* Charged-impurity scattering in graphene. *Nat. Phys.* **4,** 377–381 (2008).
34. Berger, C. *et al.* Electronic confinement and coherence in patterned epitaxial graphene. *Science* **312,** 1191–1196 (2006).
35. de Heer, W. A. *et al.* Large area and structured epitaxial graphene produced by confinement controlled sublimation of silicon carbide. *Proc. Natl Acad. Sci. USA* **108,** 16900–16905 (2011).
36. Hao, Y. *et al.* The role of surface oxygen in the growth of large single-crystal graphene on copper. *Science* **342,** 720–723 (2013).
37. Lee, S., Lee, K. & Zhong, Z. Wafer scale homogeneous bilayer graphene films by chemical vapor deposition. *Nano Lett.* **10,** 4702–4707 (2010).
38. Schmuttenmaer, C. A. Exploring dynamics in the far-infrared with terahertz spectroscopy. *Chem. Rev.* **104,** 1759–1779 (2004).
39. Baxter, J. B. & Guglietta, G. W. Terahertz spectroscopy. *Anal. Chem.* **83,** 4342–4368 (2011).
40. Orlita, M. *et al.* Carrier scattering from dynamical magnetoconductivity in quasineutral epitaxial graphene. *Phys. Rev. Lett.* **107,** 216603 (2011).
41. Xu, S. *et al.* Energy dependence of electron lifetime in graphite observed with femtosecond photoemission spectroscopy. *Phys. Rev. Lett.* **76,** 483 (1996).
42. González, J., Guinea, F. & Vozmediano, M. A. H. Unconventional quasiparticle lifetime in graphite. *Phys. Rev. Lett.* **77,** 3589 (1996).
43. Martin, J. *et al.* Observations of electron-hole puddles in graphene using a scanning single-electron transistor. *Nat. Phys.* **4,** 144–148 (2008).
44. Zhang, Y. B., Brar, V. W., Girit, C., Zettl, A. & Crommie, M. F. Origin of spatial charge inhomogeneity in graphene. *Nat. Phys.* **5,** 722–726 (2009).
45. Jung, S. Y. *et al.* Evolution of microscopic localization in graphene in a magnetic field from scattering resonances to quantum dots. *Nat. Phys.* **7,** 245–251 (2011).
46. Dean, C. R. *et al.* Boron nitride substrates for high-quality graphene electronics. *Nat. Nanotechnol.* **5,** 722–726 (2010).
47. Du, X., Skachko, I., Barker, A. & Andrei, E. Y. Approaching ballistic transport in suspended graphene. *Nat. Nanotechnol.* **3,** 491–495 (2008).
48. Hass, J. *et al.* Why multilayer graphene on 4H-SiC(0001) behaves like a single sheet of graphene. *Phys. Rev. Lett.* **100,** 125504 (2008).
49. Sprinkle, M. *et al.* First direct observation of a nearly ideal graphene band structure. *Phys. Rev. Lett.* **103,** 226803 (2009).
50. Sun, D. *et al.* Ultrafast relaxation of excited Dirac fermions in epitaxial graphene using optical differential transmission spectroscopy. *Phys. Rev. Lett.* **101,** 157402 (2008).
51. Sun, D. *et al.* Spectroscopic measurements of interlayer screening in multilayer epitaxial graphene. *Phys. Rev. Lett.* **104,** 136802 (2010).
52. Chang, Y.-C., Liu, C.-H., Liu, C.-H., Zhong, Z. & Norris, T. B. Extracting the complex optical conductivity of mono- and bilayer graphene by ellipsometry. *Appl. Phys. Lett.* **104,** 261909 (2014).
53. Faugeras, C. *et al.* Few-layer graphene on SiC, pyrolitic graphite, and graphene: a Raman scattering study. *Appl. Phys. Lett.* **92,** 011914 (2008).
54. Kunc, J., Hu, Y., Palmer, J., Berger, C. & de Heer, W. A. A method to extract pure Raman spectrum of epitaxial graphene on SiC. *Appl. Phys. Lett.* **103,** 201911 (2013).
55. Baringhaus, J. *et al.* Exceptional ballistic transport in epitaxial graphene nanoribbons. *Nature* **506,** 349–354 (2014).
56. Katzenellenbogen, N. & Grischkowsky, D. Efficient generation of 380 fs pulses of THz radiation by ultrafast laser pulse excitation of a biased metal-semiconductor interface. *Appl. Phys. Lett.* **58,** 222–224 (1991).
57. Tani, M., Matsuura, S., Sakai, K. & Nakashima, S. Emission characteristics of photoconductive antennas based on low-temperature-grown GaAs and semi-insulating GaAs. *Appl. Opt.* **36,** 7853–7859 (1997).
58. Wu, Q. & Zhang, X. C. Free-space electro-optic sampling of terahertz beams. *Appl. Phys. Lett.* **67,** 3523–3525 (1995).
59. Wu, Q. & Zhang, X. C. Ultrafast electro-optic field sensors. *Appl. Phys. Lett.* **68,** 1604–1606 (1996).
60. Nahata, A., Weling, A. S. & Heinz, T. F. A wideband coherent terahertz spectroscopy system using optical rectification and electro-optic sampling. *Appl. Phys. Lett.* **69,** 2321–2323 (1996).
61. Kira, M., Hoyer, W. & Koch, S. W. Microscopic theory of the semiconductor terahertz response. *Phys. Stat. Sol. (b)* **238,** 443–450 (2003).

Acknowledgements

We appreciate Xiaohan Wang and Rodney S. Ruoff for providing the single-crystal CVD graphene (sCVDG) samples. M.T.M. and C.J.D. thank Steve Katnik for technical assistance with the ultrafast laser system. F.K., T.W., E.M. and A.K. thank Roland Jago (Research Training Group 1558) for fruitful discussions. The work at the University of Michigan and the Georgia Institute of Technology was supported in part by the National Science Foundation (NSF) Materials Research Science and Engineering Center (MRSEC) under grant DMR-0820382. The work at the University of Michigan was also supported in part by the NSF Center for Photonic and Multiscale Nanomaterials (CPHOM) under grant DMR-1120923 and the NSF CAREER Award (ECCS-1254468). T.W. and A.K. acknowledge partial financial support from the Deutsche Forschungsgemeinschaft (DFG) through SPP-1459 and Sfb 787. E.M. acknowledges partial financial support from the European Commission (EC) under the Graphene Flagship programme (contract No. CNECT-ICT-604391) and the Swedish Research Council (VR). C.B. acknowledges partial financial support from the EC under the Graphene Flagship programme (contract No. CNECT-ICT-604391). This work was performed in part at the Lurie Nanofabrication Facility (LNF), a member of the National Nanotechnology Infrastructure Network (NNIN), which is supported in part by the NSF.

Author contributions

M.T.M. and C.J.D. performed and analysed the ultrafast time-resolved THz spectroscopy experiments. F.K., T.W., E.M. and A.K. developed the microscopic theory. S.L., C.-H.L. and Z.Z. provided the polycrystalline CVD graphene (pCVDG) samples. C.B. and W.A.d.H. provided the multilayer epitaxial graphene (MEG) samples. M.T.M. and T.B.N. with input from other authors wrote the paper. All authors discussed the results.

Additional information

Competing financial interests: The authors declare no competing financial interests.

24

Dispersion and shape engineered plasmonic nanosensors

Hyeon-Ho Jeong[1,2,*], Andrew G. Mark[1,*], Mariana Alarcón-Correa[1,3], Insook Kim[1,3], Peter Oswald[1], Tung-Chun Lee[1,4] & Peer Fischer[1,3]

Biosensors based on the localized surface plasmon resonance (LSPR) of individual metallic nanoparticles promise to deliver modular, low-cost sensing with high-detection thresholds. However, they continue to suffer from relatively low sensitivity and figures of merit (FOMs). Herein we introduce the idea of sensitivity enhancement of LSPR sensors through engineering of the material dispersion function. Employing dispersion and shape engineering of chiral nanoparticles leads to remarkable refractive index sensitivities (1,091 nm RIU^{-1} at $\lambda = 921$ nm) and FOMs (>2,800 RIU^{-1}). A key feature is that the polarization-dependent extinction of the nanoparticles is now characterized by rich spectral features, including bipolar peaks and nulls, suitable for tracking refractive index changes. This sensing modality offers strong optical contrast even in the presence of highly absorbing media, an important consideration for use in complex biological media with limited transmission. The technique is sensitive to surface-specific binding events which we demonstrate through biotin-avidin surface coupling.

[1] Max Planck Institute for Intelligent Systems, Heisenbergstrasse 3, 70569 Stuttgart, Germany. [2] Institute of Materials, École Polytechnique Fédérale de Lausanne (EPFL), CH-1015 Lausanne, Switzerland. [3] Institute for Physical Chemistry, University of Stuttgart, Pfaffenwaldring 55, 70569 Stuttgart, Germany. [4] UCL Institute for Materials Discovery and Department of Chemistry, University College London, Christopher Ingold Building, 20 Gordon Street, London WC1H 0AJ, UK. * These authors contributed equally to this work. Correspondence and requests for materials should be addressed to P.F. (email: fischer@is.mpg.de).

Devices based on surface plasmon resonance (SPR) phenomena detect shifts of the resonance wavelength in response to changes of the refractive index of the medium surrounding the plasmonic material[1]. This may, for instance, be due to biomolecules that bind to the sensor. SPR offers high sensitivities[2], but requires extended smooth surfaces[3]. In many situations, it would be desirable to have a local sensor for use *in situ* or *in vivo* (for instance within a cell) and here the localized SPR (LSPR) supported by nanostructures offers substantial advantages[4–6]. The short penetration depth of plasmon oscillations into the surrounding fluid makes for small, spatially localized sensors which promise to be effective in a range of biomedical applications[7–9]. However, compared with plasmonic biosensors that utilize extended SPR, which serve as the reference standard for optically addressed sensors, LSPR sensors generally have a reduced sensitivity ($S_n < 1{,}000\,\mathrm{nm\,RIU}^{-1}$) and a lower figure of merit (FOM) $< 100\,\mathrm{RIU}^{-1}$ (refs 2,5).

The typical strategy employed to enhance the sensitivity of LSPR nanosensors is to manipulate the aspect ratio of symmetrical nanoparticles, with more elongated particles yielding higher sensitivities[10,11]. This approach has been demonstrated with nanorods and nanoprisms that show improved sensitivites[12–14]. On the other hand, the materials selected for such particles have, quite reasonably, been restricted to those with strong plasmonic properties: for example, Ag (ref. 15), Au (ref. 9), Al (ref. 16). This has been driven by the desire for a high figure of merit (FOM) that comes from the low-interband damping, and sharp resonance of pure metals. However, it means that the material dispersion function, a key factor in the sensitivity of LSPR nanosensors, has been limited to the dielectric functions inherent to those pure metals.

Here, we show how dispersion-engineering introduces a new material-based parameter for improving the sensitivity of LSPR sensors. When combined with shape engineering, this leads to extremely high LSPR sensitivities and FOMs, which we report herein. We introduce an analytical model of chiral plasmonic sensing that illustrates the roles of chirality and materials properties for the important sensing characteristics. Based on this understanding we present colloidal nanostructures that are, to the best of our knowledge, the most sensitive LSPR sensors reported to date[12,17]. Furthermore, we demonstrate the utility of the engineered particles as surface-sensitive probes for biotin–avidin binding. Our scheme is particularly robust as it is immune to changes in the optical density of the background, and can thus be used in complex environments.

Results

Theoretical concept. Plasmon-based LSPR sensors operate by tracking the shift in the resonance peak of the plasmon absorption in response to changes in n, the refractive index of the local medium[5]. The resonance condition is met when $\varepsilon_r^* = \varepsilon_r(\lambda^*) = -\chi n^2$, where λ^* is the peak wavelength, ε_r is the real part of the dielectric function describing the plasmonic material, and the factor χ describes the shape of the particle (for spheres $\chi = 2$). Ideally, the dispersion of ε_r should be such that even very small changes in n cause appreciable changes in the resonance wavelength λ^*, that is, large resonance shifts. The sensitivity of the system is defined as[11]

$$S_n = \frac{d\lambda^*}{dn} = \frac{\frac{d\varepsilon_r^*}{dn}}{\left(\frac{d\varepsilon_r}{d\lambda}\right)_{\lambda^*}} = \frac{-2\chi n}{\left(\frac{d\varepsilon_r}{d\lambda}\right)_{\lambda^*}} \qquad (1)$$

so decreasing the wavelength dependence of the real part of the material's dielectric constant (the denominator) leads to an increase in sensitivity. Here, we engineer the dielectric response of the nanoparticles by alloying a plasmonic material with a weakly dispersive one to yield a negative, but flatter ε_r. This dramatically increases the sensitivity of any plasmonic sensor (of any shape).

In conjunction with the sensitivity, it is also of interest to consider the accuracy with which a spectral feature can be resolved. For a spectral peak, the standard measure is its full width at half maximum (FWHM). Engineering the real part of the particle's dielectric function[18] in equation (1) through alloying is generally accompanied by a concomitant increase in the plasmon damping and broadening of the extinction spectrum[19], and thus suggests that the FOM ($= S_n$/FWHM) is also lowered. This can be addressed by noting that traditional plasmon-based sensors are generally polarization independent, because the particles themselves are highly symmetrical, Fig. 1a. Circular dichroism (CD) offers an alternative to extinction-based sensing techniques[20,21], and CD spectra are typically more feature rich than extinction spectra[22,23], which increases the number of spectral signatures that can be tracked when the local environment changes, Fig. 1b. Crucially, CD spectra are also bipolar, and the crossing points where the signal changes sign are ideal features to track, because of the simplicity of identifying the

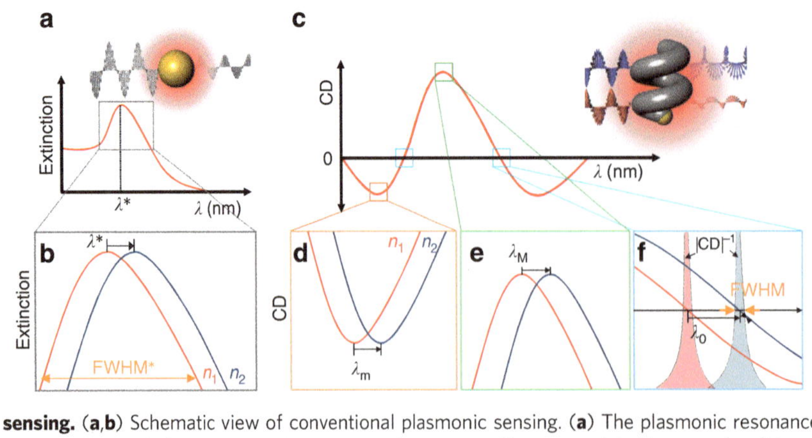

Figure 1 | Chiral plasmonic sensing. (**a**,**b**) Schematic view of conventional plasmonic sensing. (**a**) The plasmonic resonance of the metallic nanoparticle as a function of λ. Inset shows a conventional plasmonic sensing system where a metallic nanoparticle (here a sphere) interacts with light and generates a detectable absorption peak due to plasmon resonance. (**b**) A zoomed-in region near the peak λ^* showing how it shifts in wavelength as the refractive index of the surrounding medium changes $n_1 < n_2$. (**c**-**f**) Schematic view of polarization-dependent chiral plasmonic sensing. (**c**) The CD spectrum of a plasmonic enantiomer as a function of λ. Inset scheme illustrates the interaction of a left-handed nanohelix with circularly polarized light. Three bottom panels indicate the resonance shifts at (**d**) λ_m, (**e**) λ_M and (**f**) λ_0 where the refractive indices of the surrounding media are varied between n_1 and n_2.

null point. Earlier work on CD-based sensing made use of magneto-optical modulation of an achiral nanoantenna LSPR to induce ellipticity in the transmitted beam[24]. In this case, the choice of material was dictated by the requirement to induce a magneto-optical response. However, performing CD-based measurements on chiral particles, that exhibit a natural CD, offers the possibility of tuning the material properties independently of the chirality (shape), and we exploit this approach to maximize the LSPR sensitivity (and FOM).

The exact optical responses for such complex nanoparticles are only calculable using numerical methods[25]. However, we introduce the following chiral form of nanoparticle plasmonic absorption[10,11] that captures the key features of extinction in response to left (−) and right (+) circularly polarized light.

$$E_{\pm} = \frac{2\pi l N V n^3}{\lambda \ln(10)} \left(\frac{\varepsilon_i}{\left(\varepsilon_r + (\bar{\chi} \pm \delta\chi_{D/L})n^2\right)^2 + \varepsilon_i^2} \right). \qquad (2)$$

The shape factor now includes an achiral term $\bar{\chi}$ plus a chiral term $\delta\chi_D = -\delta\chi_L$ specific to right (D) or left (L) handed enantiomers. The CD is the difference between the extinction for ± polarizations, $CD \propto E_- - E_+$. The CD signal is bipolar, and the wavelength at which the extinctions are equal yields a zero-crossing in the spectrum, which serves as a natural point for tracking. Within this formulation the crossover wavelength coincides with the achiral resonance condition $\varepsilon_r^* = -\bar{\chi}n^2$, in the limit of small, slowly-varying ε_i. So the sensitivity of nanohelix sensors follows that of achiral ones. Plotting the reciprocal of the absolute value of the CD (for example, Fig. 2d) gives an intrinsically sharp representation of the crossover point, one whose FWHM is defined by the instrumental resolution σ (ref. 24) and $\frac{dCD}{d\lambda}\big|_{\lambda^*}$, the slope of the CD spectrum at the crossing. This yields a FWHM (see Supplementary Note 1 for details), based on equation (2), of

$$\mathrm{FWHM} = \frac{4\sigma\varepsilon_i^2}{|\delta\chi| n^2 \left|\frac{d\varepsilon_r}{d\lambda}\right|_{\lambda^*}} \qquad (3)$$

and a FOM of

$$\mathrm{FOM} = \frac{S_n}{\mathrm{FWHM}} = \frac{2\bar{\chi}|\delta\chi| n^3}{\varepsilon_i^2 \sigma} \qquad (4)$$

Thus, higher FOM can be achieved by increasing the magnitude of the chiral shape factor $\delta\chi$.

Fabrication and bulk refractive index sensing. We fabricate a series of hybrid nanohelix sensors using a physical vapour deposition technique that allows precise control of the particles' alloy composition and shape $\delta\chi$ (Supplementary Note 2, Supplementary Fig. 1 and Supplementary Table 1)[25,26]. The former can be used to engineer the material dielectric function to maximize sensitivity, while the latter is used to adjust the achiral and chiral shape factors to affect sensitivity and introduce a chiroptical response. A typical particle, a two-turn left-handed 128-nm-tall helix composed of 97% Ag and 3%Ti, is shown in Fig. 2a. The small amount of Ti alloying agent improves the helix fidelity over pure Ag (ref. 27). The particles are tested in the form of a colloid suspension whose stability was confirmed through dynamic light scattering (DLS, more details in Supplementary Note 3 and Supplementary Fig. 2). Figure 2b shows CD spectra of the suspensions in glycerol–water mixtures varying from 0 to 20% concentration (refractive indices between 1.333 and 1.357 (ref. 28)). As expected, the CD spectra exhibit multiple features, shown in Fig. 2c–f: a maximum λ_M, a minimum λ_m and two crossing points λ_{01} and λ_{02}, all of which can be tracked in

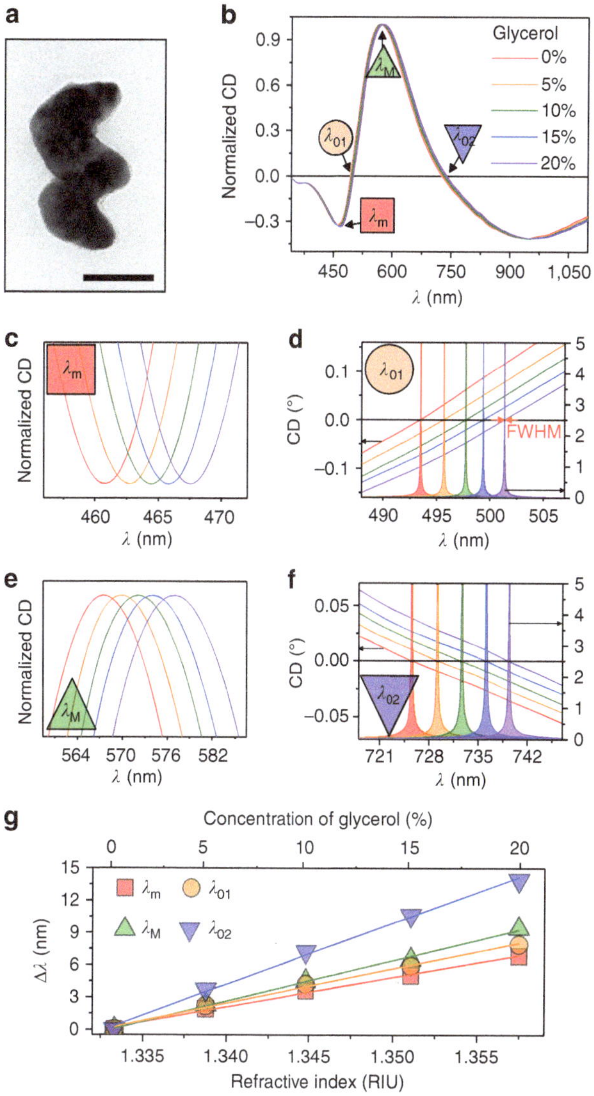

Figure 2 | Bulk refractive index sensing. (**a**) TEM image of a single $Ag_{0.97}Ti_{0.03}$:L_2 nanohelix (Scale bar, 50 nm). (**b**) CD spectra of colloidal $Ag_{0.97}Ti_{0.03}$:L_2 nanohelices in media of five different refractive indices (red: 0%, orange: 5%, green: 10%, blue: 15% and violet: 20% glycerol-water mixtures) over the full spectral range and detailed plots of the resonance shifts at (**c**) λ_m (red square), (**d**) λ_{01} (orange circle), (**e**) λ_M (green top triangle) and (**f**) λ_{02} (blue bottom triangle). For **d** and **f** the filled curves represent $|CD|^{-1}$. (**g**) Wavelength shift, relative to water, of the four spectral features as functions of the glycerol-water concentration (top *x* axis) and its corresponding *n* (bottom *x* axis).

response to changes in the refractive index of the medium. The wavelength shifts, relative to the pure water reference, are shown in Fig. 2g, and indicate sensitivities S_n of 275, 320, 379 and 571 nm RIU^{-1} for λ_m, λ_{01}, λ_M and λ_{02}, respectively.

Dispersion and shape engineering of chiral nanoparticles. These results demonstrate that features, like λ_M, and λ_{02}, found at longer wavelengths exhibit greater sensitivity, consistent with equation (1). Since larger values of the achiral shape factor increase the wavelength of the resonance condition, it is possible to increase the sensitivity by growing particles that are more elongated (Fig. 3a). This has been the approach most often pursued in the search for higher sensitivities from extinction-based LSPR sensors[10,11,12,17]. Here we apply the same principle to

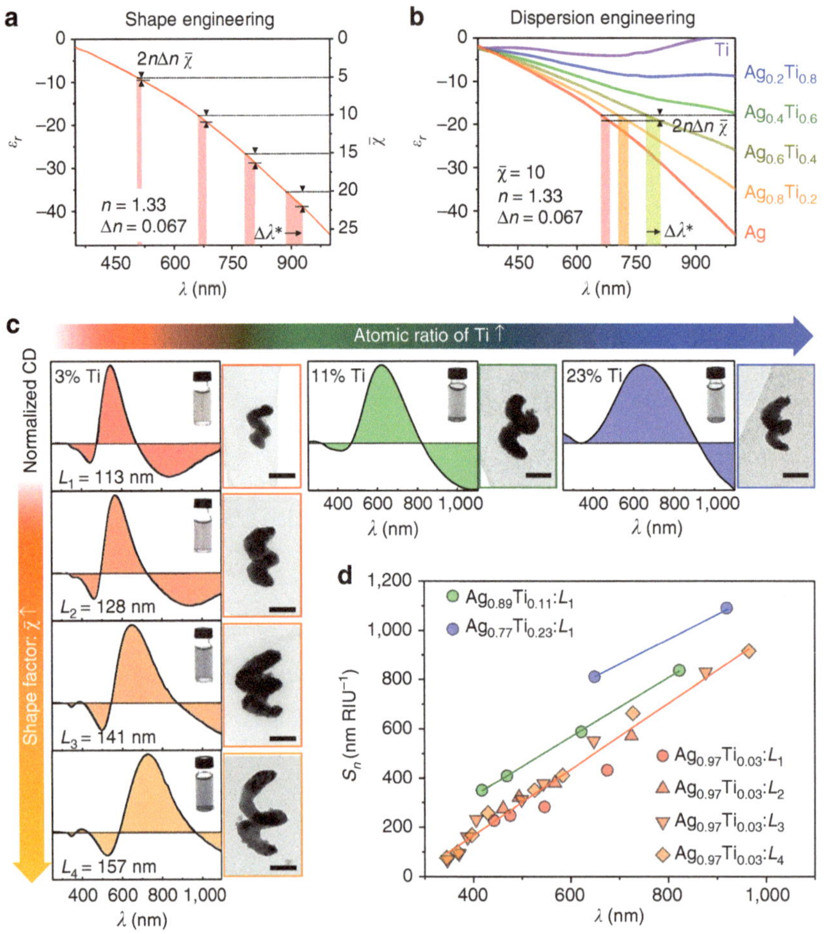

Figure 3 | Shape and dispersion engineering of Ag-Ti nanohelices. (**a**) The effect of the achiral shape factor $\bar{\chi}$ on the wavelength shift $\Delta\lambda$ observed for a given change in refractive index Δn ($=0.067$). (**b**) The effect of ε_r on the wavelength shift based on calculated dielectric constant of Ag-Ti alloys with varying Ti composition. As the proportion of Ti increases the real part of the dielectric function becomes flatter, and the resulting wavelength shift increases. (**c**) CD spectra (panel left), TEM images (panel right) and colloidal solutions (each inset) of the grown Ag-Ti nanohelices. The achiral shape factor $\bar{\chi}$ increases moving down the rows, and the atomic ratio of Ti increases moving to the right. (**d**) The measured S_n of the Ag-Ti nanohelices as a function of λ at 1.333 RIU (the symbols and colours are different for the $\bar{\chi}$ and the atomic ratio of Ti, respectively).

a chiral sensor, by growing a series of nanohelices with a range of heights, labelled L_1 (113 nm) through L_4 (157 nm). The results, shown in Fig. 3c, confirm that the peak and crossing features of longer structures are red-shifted relative to the smaller ones and that the sensitivity of each feature is directly proportional to its wavelength (Fig. 3d, red points).

However, equation (1) also suggests that the sensitivity can be improved by reducing the wavelength dependence of ε_r, the real part of the material dielectric function. We do this by exercising control over the structures' Ag-Ti stoichiometry during their growth. Figure 3b shows the effective ε_r of Ag-Ti alloys of varying compositions (more details in Supplementary Note 4). As the alloy becomes more Ti-rich, the wavelength shift $\Delta\lambda$ increases for a given change in the medium's refractive index, due to a progressive flattening of the material's dielectric function (see also Supplementary Fig. 3 and Supplementary Table 2). The effect of dispersion engineering in practice is illustrated in Fig. 3c for nanohelices having fixed size, but composed of alloys containing 3, 11 and 23% Ti. Higher Ti composition red-shifts the features, and as shown in Fig. 3d, and increases the sensitivity. Indeed, the sensitivity trends show a systematic enhancement for the 11% (green) and 23% (blue) alloy relative to the minimal Ti samples (red; see Supplementary Note 5 and Supplementary Figs 4–9 for detailed plots).

The results are summarized in Fig. 4. The left column in the plots shows the effect of increased nanohelix length on the crossing point wavelength, sensitivity, FWHM and FOM. The null wavelength and sensitivity, both increase as the nanohelices lengthen. However, the FWHM remains unchanged, so that the FOM increases dramatically. The L_4 helix exhibits a FOM = 2,859 RIU^{-1}, which is larger than what has been previously reported for LSPR-based sensors[13,14,29] including those based on a magnetochiral response[24]. Relative to the latter, the improvement comes from a combination of improved sensitivity ($\sim 4\times$) and decreased FWHM ($\sim 5\times$) thanks to a steeper crossing of the CD signal at the null point (Supplementary Note 6 and Supplementary Fig. 10). The right hand column of Fig. 4 illustrates the effect of engineering the dispersion function through the addition of Ti. In this case the increase in sensitivity that comes from flattening the material dispersion function also leads to an increase in the FWHM due to the ε_i contribution in equation (3). The net effect is a decrease in the FOM relative to the low-alloy sample. Nevertheless, the flatter dispersion curve does have the effect of increasing the sensitivity beyond what is achievable in the pure metal or the low-alloy sample. Here we observe a sensitivity $>1{,}000\ nm\ RIU^{-1}$ at a wavelength of 921 nm. For applications where sensitivity is the primary concern, dispersion engineering provides a powerful technique

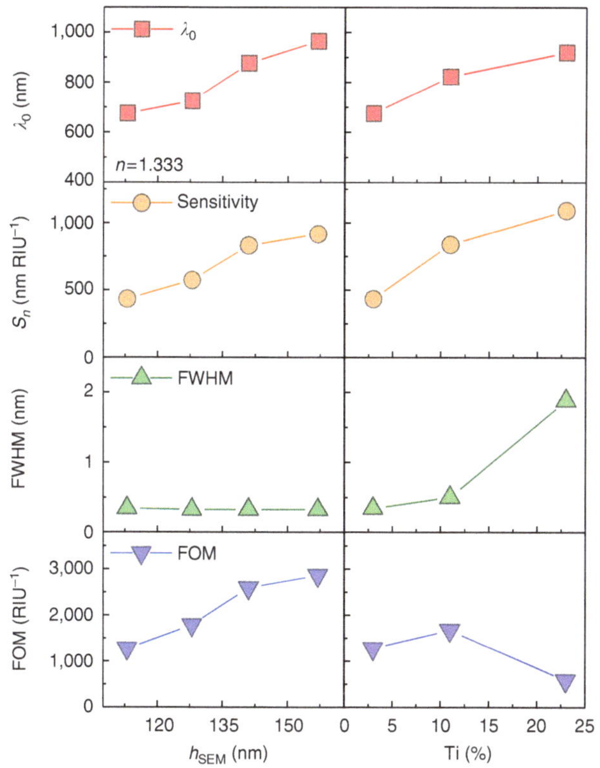

Figure 4 | Summary of shape and dispersion engineering of nanohelices. Left panels show the effect of increasing nanohelix length and the right panels show the effect of increasing Ti composition (red square: zero-crossing wavelengths at $n=1.333$, orange circle: refractive index sensitivities, green top triangles: FWHMs and blue bottom triangles: FOMs).

for reaching the highest possible refractive index sensitivities (Supplementary Note 7 and Supplementary Fig. 11).

Strong optical response in the presence of absorbers. One appealing characteristic of CD is that it is insensitive to achiral absorbers or scatterers[30]. So unlike traditional extinction measurements, high signal-to-noise CD measurements of a chiral analyte are possible even in the presence of strong achiral absorbers. Figure 5a illustrates such a measurement, where we use a colloidal suspension of $Ag_{0.97}Ti_{0.03}$:L_3 nanohelices in water with the addition of absorbers in the form of blue and yellow filters in the optical path. In both cases the maximum absorbance exceeds an optical density of 2 (<0.1% transmission, right panels of Fig. 5a). This introduces small distortions into the CD signal, however, as we have already seen, CD offers several spectral features for sensing, the majority of which are here unperturbed. Notably, since CD-based sensing offers a clearly distinguishable signal even in the presence of absorbance that would be strong enough to obscure optical extinction-based sensing methodologies, Fig. 5b. The same experiment demonstrated with molecular absorbers rhodamine and indigo yield similar results (Supplementary Note 8 and Supplementary Fig. 12).

Surface-sensitive sensing. Since the evanescent field of the excited plasmon decays rapidly into the medium, LSPR sensors are ideally suited to making surface-sensitive measurements[31]. Here we use our chiral plasmonic nanohelices to sense avidin-binding events. First, the surfaces of our nanohelices ($Ag_{0.97}Ti_{0.03}$:L_1 and $Ag_{0.89}Ti_{0.11}$:L_1) are functionalized with biotin (see Methods section for details), which acts as a complementary and specific-binding factor for avidin in solution (Fig. 5c). The wavelength of the λ_{02} crossing in the CD spectrum (Fig. 5d, lower), and the change in the CD amplitude at the initial crossing wavelength (Fig. 5d, upper) were measured at 1 min intervals. Before the addition of avidin both signals are stable, but they experience a prompt shift on the addition of 1 μg ml^{-1} avidin. For the $Ag_{0.97}Ti_{0.03}$:L_1 nanohelices the crossing point $\Delta\lambda_{02}$ redshifts by ~3 nm (blue), and the CD intensity at the initial crossing wavelength increases by 6 mdeg (green). The rise time is ~5 min. The $Ag_{0.89}Ti_{0.11}$:L_1 nanohelices show an even stronger response with shifts of 5 nm (red) and 10 mdeg (yellow) respectively, thanks to the increased sensitivity that comes from the reduced dispersion. Control experiments (grey) where the nanohelices lack biotin functionalization show no distinct shift on the addition of avidin, confirming that the signal is truly surface sensitive and arises from specific binding (see Supplementary Note 9 and Supplementary Figs 13 and 14 for additional details). The two measurement modes suggest alternative sensing schemes for real-time monitoring of binding events: one based on tracking the crossing point wavelength (Fig. 5e, blue and red), and another based on identifying the change in CD intensity at fixed wavelength (Fig. 5e, green and yellow). Both offer a highly sensitive and rapid measurement of the specific binding at surfaces.

Discussion

We have shown that flattening the dielectric function of plasmonic nanomaterials by alloying results in large LSPR shifts in response to changes in their local environment. If these nanostructures are also chiral, then their polarization-dependent spectra offer additional and sharper spectral features relative to optical extinction-based measurement of achiral particles. These characteristics make chiral plasmonic nanohelices especially suited for sensing applications. Furthermore, the bipolar nature of CD spectra of chiral structures leads to points of zero-crossing, which have very small effective line-widths and thus result in high FOM. Both bulk refractive index changes as well as surface-specific binding events are tracked by the engineered nanocolloidal LSPR sensors with sensitivity (~1,100 nm RIU^{-1}) and FOM (~2,900 RIU^{-1}) higher than any previously reported for plasmonic sensors.

Aside from extremely high-detection sensitivities, these chiral sensors offer unique advantages over existing plasmonic devices. Since they are freely suspended colloids, they naturally suggest themselves as sensors for *in situ* or *in vivo* biological applications, particularly, because of their small size. Since the CD signal is robust against background interference, especially in absorbing media, the sensing scheme reported herein is particularly suited to real biological media, which are typically optically dense. For instance, in the presence of whole blood, our scheme has the potential for real-time measurement of protein corona formation, a vitally important factor for the function of nanoparticles in biological media, but one that remains poorly understood.

Methods

Nano glancing angle deposition (nanoGLAD). The shadow growth technique[32] was used to grow the three-dimensional nanocolloids in the manner previously reported[25]. First, a hexagonal array of 12 nm Au nanoparticles was prepared by block-copolymer micelle nanolithography[33]. Next, Ag–Ti nanohelices were grown on the array of the Au nanoparticles in a GLAD system based on co-deposition from dual electron-beam evaporators at $T=90$ K with a base pressure of 1×10^{-6} mbar. Their alloy stoichiometry was controlled by the deposition rates measured by quartz crystal microbalance (QCM) for each evaporator independently. Particles were grown with lengths (measured by scanning electron microscopy (SEM)) of $L_1=113$ nm, $L_2=128$ nm, $L_3=141$ nm, $L_4=157$ nm and Ti contents of 3, 11 and 23%. Finally, the grown Ag–Ti nanohelices were lifted off from the wafer by sonicating a piece of sample wafer (0.5–1 cm^2) in an aqueous solution of 0.15 mM polyvinylpyrrolidone for ~2 min to prepare a stock solution.

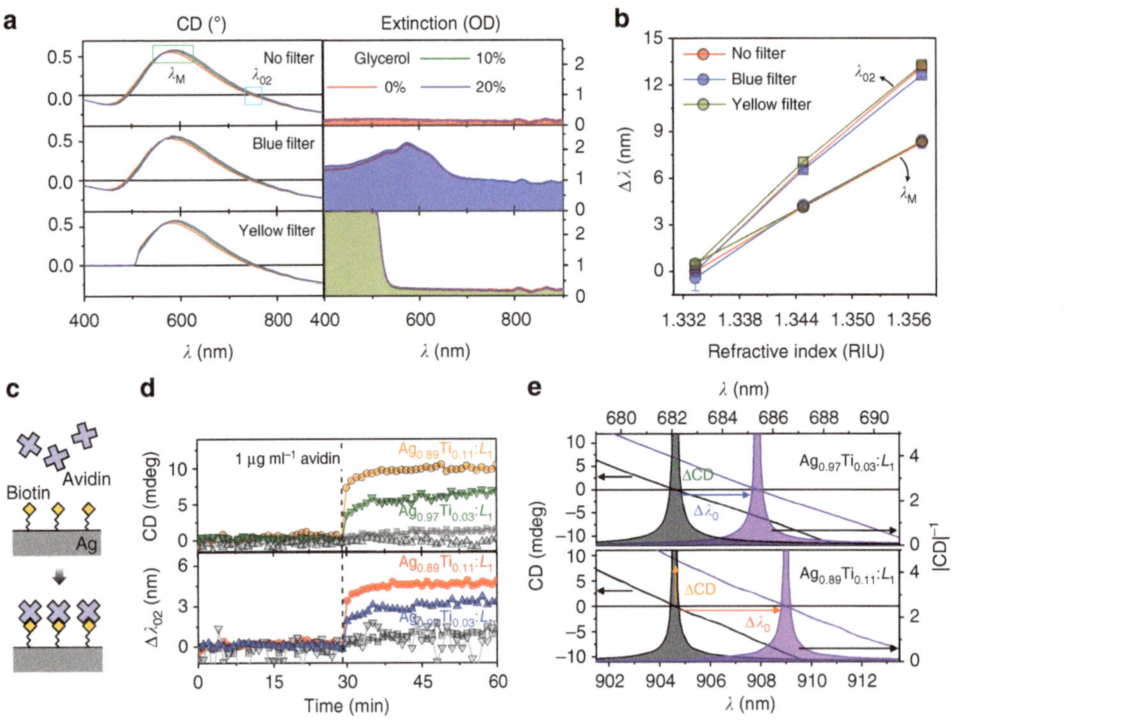

Figure 5 | Sensing in absorbing media and Sensing of specific binding events. CD spectra (left panels) and extinction spectra (right panels) of the colloidal $Ag_{0.97}Ti_{0.03}$:L_3 nanohelices in the presence of complex absorbing environments (top: no filter, middle: a blue filter, bottom: yellow filter), (**b**) their corresponding wavelength shifts, relative to water, of the two spectral features (λ_M and λ_{02}) as functions of the *n* (error bar: s.d.). (**c**) Schematic view of biotin-avidin interaction on the surface of a Ag-Ti nanohelix. (**d**) *In situ* measurements of the biotin-avidin interaction by monitoring the change of CD at the initial λ_{02} (upper plot) and the wavelength shift of $\Delta\lambda_{02}$ (lower plot) with 1 min intervals. The coloured plots indicate the response of specific binding of biotin-avidin ($Ag_{0.97}Ti_{0.03}$:L_1 blue, green; $Ag_{0.89}Ti_{0.11}$:L_1 red, yellow) and the grey plots indicate non-specific binding of avidin with Ag-Ti nanohelices (without biotin) in the control group. (**e**) Close-up view of the CD spectra for the two biotinylated nanoparticle systems showing the wavelength shift and CD amplitude increase after avidin introduction.

To minimize the effect of possible variations in structure and concentration, samples for each set of tests were always drawn from the same stock solution.

SEM and TEM analysis. The Ag–Ti helices were imaged in SEM (Ultra 55, Zeiss) and transmission electron microscopy (TEM) (CM200, Philips) under the accelerating voltages of 10 and 200 kV, respectively.

Circular dichroism analysis. CD spectra were obtained with a Jasco J-810 CD spectrometer. All the spectra were measured with 500 nm min^{-1} scan rate in the wavelength range of 300–1,100 nm with 0.1 nm intervals. For selected regions of interest smaller 0.025 nm intervals were used.

DLS analysis. Colloidal solution (200 μl) of nanohelices was measured using a zeta potential analyser (Zetasizer Nano ZS, Malvern) repeated ~10 times for 20 min with 2 min intervals. The material property of Ag–Ti nanohelices was fixed to Ag (RI: 0.135 and Absorption: 3.990) and the environmental parameter was matched to literature values for the viscosity, and refractive index of the solution based on the temperature and concentration[28].

ICP-OES analysis. Approximately 1 cm^2 as-grown nanohelices supported on a Si substrate was dissolved into HNO_3/HF etchant and this solution was analysed by the inductively coupled plasma atomic emission spectroscopy (ICP-OES) (Ciros CCD, Spectro). The material composition of the nanohelices was evaluated by repeating the analysis three times with samples cleaved from different areas of the growth wafer.

Absorbers. Blue and yellow filters were separately inserted into the optical patch either before or after the sample cuvette. Similarly, 10 μM rhodamine 6G and 100 μM indigo were added to separate samples after acquiring baseline measurements.

Biotin-avidin binding procedure. The mixture of biotin-Polyethylene glycol (PEG)-SH (25 mg ml^{-1} in 3-(N-morpholino)propanesulfonic acid (MOPS)-DNA buffer of pH7.5) and CH_3O-PEG-SH (250 mg ml^{-1} in MOPS-DNA buffer of pH7.5) in the ratio of 1:1 (v v^{-1}) was drop-cast onto ~0.5 cm^2 of as-grown nanohelix substrate and kept under humid condition overnight. For the control group, the nanohelices were exposed to only CH_3O-PEG-SH. Next, each substrate was immersed in 1 ml of 0.15 mM polyvinylpyrrolidone solution and sonicated to remove the particles from the wafer and suspend them in solution. Finally, for the biotin–avidin interaction, 1 μg ml^{-1} of avidin was injected to the colloidal solution and mixed by pipetting.

References

1. Homola, J. Present and future of surface plasmon resonance biosensors. *Anal. Bioanal. Chem.* **377,** 528–539 (2003).
2. Haes, A. J. & Duyne, R. P. V. Preliminary studies and potential applications of localized surface plasmon resonance spectroscopy in medical diagnostics. *Expert Rev. Mol. Diagn.* **4,** 527–537 (2004).
3. Nagpal, P., Lindquist, N. C., Oh, S.-H. & Norris, D. J. Ultrasmooth patterned metals for plasmonics and metamaterials. *Science* **325,** 594–597 (2009).
4. Brolo, A. G. Plasmonics for future biosensors. *Nat. Photon.* **6,** 709–713 (2012).
5. Mayer, K. M. & Hafner, J. H. Localized surface plasmon resonance sensors. *Chem. Rev.* **111,** 3828–3857 (2011).
6. Stockman, M. I. Nanoplasmonic sensing and detection. *Science* **348,** 287–288 (2015).
7. Anker, J. N. *et al.* Biosensing with plasmonic nanosensors. *Nat. Mater.* **7,** 442–453 (2008).
8. Howes, P. D., Rana, S. & Stevens, M. M. Plasmonic nanomaterials for biodiagnostics. *Chem. Soc. Rev.* **43,** 3835–3853 (2014).
9. Yang, X., Yang, M., Pang, B., Vara, M. & Xia, Y. Gold nanomaterials at work in biomedicine. *Chem. Rev.* **115,** 10410–10488 (2015).
10. Lee, K.-S. & El-Sayed, M. A. Gold and silver nanoparticles in sensing and imaging: sensitivity of plasmon response to size, shape, and metal composition. *J. Phys. Chem. B* **110,** 19220–19225 (2006).
11. Miller, M. M. & Lazarides, A. A. Sensitivity of metal nanoparticle surface plasmon resonance to the dielectric environment. *J. Phys. Chem. B* **109,** 21556–21565 (2005).

12. Charles, D. E. *et al.* Versatile solution phase triangular silver nanoplates for highly sensitive plasmon resonance sensing. *ACS Nano* **4,** 55–64 (2010).
13. Shen, Y. *et al.* Plasmonic gold mushroom arrays with refractive index sensing figures of merit approaching the theoretical limit. *Nat. Commun.* **4,** 2381 (2013).
14. Gartia, M. R. *et al.* Colorimetric plasmon resonance imaging using nano lycurgus cup arrays. *Adv. Opt. Mater.* **1,** 68–76 (2013).
15. Jin, R. *et al.* Photoinduced conversion of silver nanospheres to nanoprisms. *Science* **294,** 1901–1903 (2001).
16. Knight, M. W. *et al.* Aluminum for plasmonics. *ACS Nano* **8,** 834–840 (2014).
17. Chen, H., Kou, X., Yang, Z., Ni, W. & Wang, J. Shape- and size-dependent refractive index sensitivity of gold nanoparticles. *Langmuir* **24,** 5233–5237 (2008).
18. Boltasseva, A. & Atwater, H. A. Low-loss plasmonic metamaterials. *Science* **331,** 290–291 (2011).
19. Chang, W.-S. *et al.* Tuning the acoustic frequency of a gold nanodisk through its adhesion layer. *Nat. Commun.* **6,** 7022 (2015).
20. Hendry, E. *et al.* Ultrasensitive detection and characterization of biomolecules using superchiral fields. *Nat. Nanotechnol.* **5,** 783–787 (2010).
21. Ma, W. *et al.* Attomolar DNA detection with chiral nanorod assemblies. *Nat. Commun.* **4,** 2689 (2013).
22. Valev, V. K., Baumberg, J. J., Sibilia, C. & Verbiest, T. Chirality and chiroptical effects in plasmonic nanostructures: fundamentals, recent progress, and outlook. *Adv. Mater.* **25,** 2517–2534 (2013).
23. Zhang, S. *et al.* Photoinduced handedness switching in terahertz chiral metamolecules. *Nat. Commun.* **3,** 942 (2012).
24. Maccaferri, N. *et al.* Ultrasensitive and label-free molecular-level detection enabled by light phase control in magnetoplasmonic nanoantennas. *Nat. Commun.* **6,** 6150 (2015).
25. Mark, A. G., Gibbs, J. G., Lee, T.-C. & Fischer, P. Hybrid nanocolloids with programmed three-dimensional shape and material composition. *Nat. Mater.* **12,** 802–807 (2013).
26. Gibbs, J. G., Mark, A. G., Eslami, S. & Fischer, P. Plasmonic nanohelix metamaterials with tailorable giant circular dichroism. *Appl. Phys. Lett.* **103,** 213101 (2013).
27. Larsen, G. K., He, Y., Wang, J. & Zhao, Y. Scalable fabrication of composite Ti/Ag plasmonic helices: controlling morphology and optical activity by tailoring material properties. *Adv. Opt. Mater.* **2,** 245–249 (2014).
28. Cheng, N.-S. Formula for the viscosity of a glycerol – water mixture. *Ind. Eng. Chem. Res.* **47,** 3285–3288 (2008).
29. Kabashin, A. V. *et al.* Plasmonic nanorod metamaterials for biosensing. *Nat. Mater.* **8,** 867–871 (2009).
30. Claborn, K., Isborn, C., Kaminsky, W. & Kahr, B. Optical rotation of achiral compounds. *Angew. Chem. Int. Ed.* **47,** 5706–5717 (2008).
31. Liu, N., Tang, M. L., Hentschel, M., Giessen, H. & Alivisatos, A. P. Nanoantenna-enhanced gas sensing in a single tailored nanofocus. *Nat. Mater.* **10,** 631–636 (2011).
32. Robbie, K., Sit, J. C. & Brett, M. J. Advanced techniques for glancing angle deposition. *J. Vac. Sci. Technol. B* **16,** 1115–1122 (1998).
33. Glass, R., Möller, M. & Spatz, J. P. Block copolymer micelle nanolithography. *Nanotechnology* **14,** 1153–1160 (2003).

Acknowledgements

We are grateful to C. Miksch and J.P. Spatz for providing us with micellar nanolithographically patterned substrates and for SEM access. We thank the ZWE Analytical Chemistry for the ICP-OES analysis and the Stuttgart Centre for Electron Microscopy for technical support with the TEM imaging. This work was supported by the Max Planck-EPFL centre for molecular nanoscience and technology, and the European Research Council under the ERC Grant agreement 278213.

Author contributions

P.F. proposed the idea and H.-H.J., A.G.M., T.-C.L and P.F. designed the experiments. H.-H.J. performed the experiments except for TEM, ellipsometry and ICP-OES. A.G.M. developed the analytical theory. M.A.-C. carried out the TEM imaging and the DLS. I.K. conducted thin-film analysis theoretically and experimentally. P.O. designed bio-experiment. H.-H.J. and A.G.M. analysed the data and H.-H.J., A.G.M. and P.F. wrote the paper.

Additional information

Competing financial interests: The authors declare no competing financial interests.

25

Automated circuit fabrication and direct characterization of carbon nanotube vibrations

G. Zeevi[1,*], M. Shlafman[1,*], T. Tabachnik[1], Z. Rogachevsky[1], S. Rechnitz[1], I. Goldshtein[1], S. Shlafman[1], N. Gordon[1], G. Alchanati[1], M. Itzhak[1], Y. Moshe[1], E.M. Hajaj[1], H. Nir[1], Y. Milyutin[1], T.Y. Izraeli[1], A. Razin[1], O. Shtempluck[1], V. Kotchtakov[1] & Y.E. Yaish[1]

Since their discovery, carbon nanotubes have fascinated many researchers due to their unprecedented properties. However, a major drawback in utilizing carbon nanotubes for practical applications is the difficulty in positioning or growing them at specific locations. Here we present a simple, rapid, non-invasive and scalable technique that enables optical imaging of carbon nanotubes. The carbon nanotube scaffold serves as a seed for nucleation and growth of small size, optically visible nanocrystals. After imaging the molecules can be removed completely, leaving the surface intact, and thus the carbon nanotube electrical and mechanical properties are preserved. The successful and robust optical imaging allowed us to develop a dedicated image processing algorithm through which we are able to demonstrate a fully automated circuit design resulting in field effect transistors and inverters. Moreover, we demonstrate that this imaging method allows not only to locate carbon nanotubes but also, as in the case of suspended ones, to study their dynamic mechanical motion.

[1] Andrew and Erna Viterbi Faculty of Electrical Engineering, Technion, Haifa 32000, Israel. * These authors contributed equally to this work. Correspondence and requests for materials should be addressed to Y.E.Y. (email: yuvaly@ee.technion.ac.il).

Carbon nanotubes (CNTs) are considered an attractive candidate for future electronics[1-5]. Several approaches were developed to harness their unique properties for different applications. Some studies devote their efforts to grow CNTs between predesigned locations[6-9] or to accurately deposit them from solution[10]. Alternatively, circuit design can be performed post deposition of the CNTs assuming their position and alignment are known. Currently, the two most common imaging techniques are based on scanning electron microscopy (SEM) or atomic force microscopy (AFM). Both methods are invasive, slow and difficult to scale up.

Alternatively, optical imaging techniques for CNTs are faster, less invasive and provide structural information on the tubes, that is, chirality and type. These methods include Raman spectroscopy, Rayleigh scattering, modulation absorption, polarization absorption, fluorescent spectroscopy and conductance imaging by either photocurrent or photothermal microscopy[11-19]. However, most of these methods are based on weak interaction of polarized laser light with the CNT itself, and as a result, the obtained signals are very small in comparison with the background signals. Improving signal to noise ratio requires either invasive sample preparation, specific substrates, aligned CNTs, metallic contacts or specific CNT type, that is, semiconducting CNTs. It is hard to see how these techniques image dense CNT networks with CNTs that cross each other, or CNTs that are bended. Moreover, the requirement that the polarization of the incident light will be parallel to the tube axis imposes a severe constraint for large scale imaging with these techniques. For example, for a given light polarization, the faster mentioned methods can image an area of $100 \times 100\,\mu m^2$ in a few tens of seconds.

Other imaging techniques are based on anchoring different objects to the CNT surface such as fluorescent molecules, metallic nanoparticles and some other adsorbents[20-24]. The common denominator of such markers is that they bind strongly to the CNT surface, either through covalent bonds or π stacking, and as a result, it is hard to remove them after imaging, thus, altering or damaging the properties of the pristine tubes.

In almost all practical scenarios the CNTs would be used on an otherwise flat surface. This motivated us to study the potential of the CNT to act as a nucleation centre for the growth of optically visible nanocrystals (NCs). Namely, we dress the CNT with large enough NCs (few tens of nanometres) to make it visible (like pearls on a chain).

In the following, we present rapid and non-invasive dark field optical imaging of CNTs using organic NCs which preferentially nucleate along the CNT sidewalls. After imaging, the NCs completely remove without leaving any residue on the CNTs and their electrical behaviour remains identical to that before adsorbing the NCs. Homemade image processing software uses the optical image and automatically finds the CNT locations and designs the optimal electrical layout according to predefined rules. Examples of p- and n-type field effect transistors (FETs) and inverters are demonstrated. The optical imaging technique is effective for imaging suspended CNTs as well. Herein, we actuate piezoelectrically or electrostatically the mechanical motion of the suspended CNTs and optically observe their vibrational frequencies, both in the linear and non-linear regimes (Duffing oscillator), and reveal transition from hardening to softening behaviour as the chamber pressure decreases. All these results are discussed within the framework of a finite element model (FEM) specifically developed by us for explaining the experimental results.

Results

Optical imaging of CNTs. A candidate molecule that could nucleate and grow preferentially over the CNT is *p*-nitrobenzoic acid (*p*NBA) which is drawn schematically at the top part of Fig. 1a. This molecule is small so that it should sublime at relatively low temperatures. The crystal properties of *p*NBA

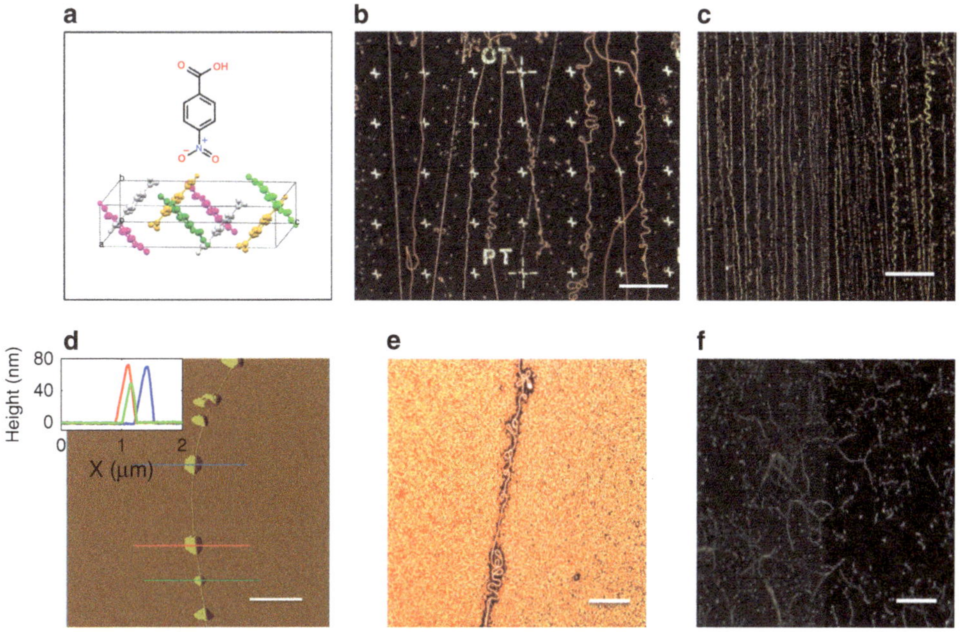

Figure 1 | Preferentially adsorption of *p*-nitrobenzoic acid on carbon nanotubes. (**a**) Top: Chemical structure of *p*-nitrobenzoic acid (*p*NBA). Bottom: Schematic illustration of the monoclinic unit cell of *p*NBA powder as extracted from X-ray diffraction analysis. (**b**,**c**) Dark field optical microscopy images of *p*NBA nanocrystals adsorbed along CVD grown carbon nanotubes (CNTs). Scale bar, 50 and 20 μm, respectively. (**d**) Amplitude image of AFM of a single CNT with a few *p*NBA nanocrystals along. Scale bar, 1 μm. Inset: height cross sections along the marked lines of the main figure. (**e**) Dark field optical microscopy image of *p*NBA nanocrystals after intensive deposition. Note the black voids along the CNT. Scale bar, 20 μm. (**f**) Dark field optical microscopy image of *p*NBA nanocrystals adsorb onto commercial dispersed CNTs. Scale bar, 20 μm.

powder were verified using X-ray diffraction (see details in Supplementary Note 1, and Supplementary Fig. 1) and it was found to have a monoclinic unit cell with cell parameters of $a = 2.453$ nm, $b = 0.505$ nm, $c = 1.291$ nm and $\beta = 93.150002°$ (bottom part of Fig. 1a). To study the crystal formation on the CNT seed we used several deposition methods which are detailed in Supplementary Note 1. Briefly, in most cases, we first used a silicon wafer covered by thin oxide layer to pattern catalyst pads from which CNTs were grown using chemical vapour deposition (CVD) method[6,8,25,26]. Next, the deposition process of *p*NBA, at 150 °C for 90 s, was carried out on the wafers carrying CNTs. Figure 1b,c present dark field optical images of typical results for two substrates with different densities of CNTs. As shown, the molecules nucleated preferentially along the CNTs, forming an organic NCs chain. The arrangement of the NCs around the CNT is revealed by the AFM image and the corresponding cross section of such NCs along the CNT, Fig. 1d. As Fig. 1d shows, the NCs are formed along the CNT in a pearl chain configuration having few tens of nanometres height and one or two hundreds of nanometres width. The height and linear density along the CNT was found to depend on the sublimation temperature and process time of the *p*NBA deposition. A real-time video that captured the deposition process along the CNTs is found in Supplementary Movie 1. In case of a very long process and overexposure of the wafer the entire surface would be covered by *p*NBA crystals. Figure 1e presents dark-field image of a chip following a deposition process having temperature and duration of 180 °C and 300 s, respectively. In this case the CNT is made visible mainly due to it being a preferential nucleation scaffold, which draws the molecules towards it thus leaving two thin empty lines surrounding the CNT.

The results in Fig. 1 show that while one needs to ensure that all the CNTs are marked with the organic NCs, overexposure should be avoided. This is especially true since we aim not only to statically image the CNT but also to follow its mechanical vibrations. Namely, the minimum height for these NCs that can still be imaged by standard dark field optical microscopy is an important parameter. In this context, it is also important to ensure that the optical imaging does capture all the CNTs. Supplementary Fig. 2a–d compare SEM and optical microscope images (see details in Supplementary Note 2). It is clearly observed that all the CNTs have been detected by the optical method, including when the tubes cross each other, or are in great proximity to each other. In addition, CNTs which appear faint in the SEM image (Supplementary Fig. 2d) are clearly visible in the optical image (Supplementary Fig. 2c). Analysis of the minimum height, h_{min}, for optical detection of a single NC located along a CNT reveals $h_{min} \leq 25$ nm, and the lateral resolution between two adjacent CNTs is ≈ 250 nm, as shown in Supplementary Figs 3 and 4, respectively. Since the interaction between the light and the decorated CNT is much stronger than with the bare tube, the required integration time for each image should be much shorter with respect to the other optical techniques. Indeed, we found that the minimum integration time required for the new method is below 1 ms for an area of $120 \times 120\,\mu m^2$ (Supplementary Fig. 5), and surprisingly, it is not restricted to SiO_2 substrates only, but it is applicable to different substrates as well, including high-k dielectrics, insulators and transparent substrates. Several examples are shown in Supplementary Fig. 6.

The last step in testing this nucleation and growth as an imaging tool is to verify that it is not limited to CVD grown CNTs. We obtained commercially available CNTs, dissolved them in organic solvent, and dispersed them on a silicon wafer (see Supplementary Note 1 for more details). The optical image obtained following the *p*NBA deposition process is shown in Fig. 1f, clearly demonstrating the generality of the method.

A key question to address is why the NCs are mainly crystalized along the CNT surface? This question brings us back to non-covalent functionalization of CNT for chemical and biological sensing[21,27]. It was suggested that the electronic properties of CNTs coupled with the specific recognition properties of immobilized biomolecule can serve as ideal miniaturized nanosensor. However, covalent immobilization deteriorates the CNT electrical performance, and thus non-covalent methods were developed. Most of these methods were based on aromatic part of the immobilized molecule, which is known to interact with the basal plane of graphite via π stacking[28]. We believe that, here as well, the aromatic part of the *p*NBA interacts with the sidewalls of the CNTs that act as a nucleation seed for the organic crystal growth. After immobilization of *p*NBA molecules on the CNT sidewall, other molecules, which adsorb onto the oxide surface, diffuse (or sublime and readsorb) until they crystalize on the immobilized molecules and ionic crystal growth occurs.

A natural question that arises is the possibility for the existence of similar molecules to *p*NBA, which can decorate the CNT sidewalls as *p*NBA does. These additional molecules can shed light on the adsorbing mechanism of *p*NBA molecules to the CNTs. Three criteria should be fulfilled by the candidate molecules: (i) have low mass weight, (ii) are in a solid phase at ambient conditions and (iii) have a benzene ring. Indeed, we found few additional candidates for marking the CNTs with organic NCs, yet, the quality of the obtained results are reduced in comparison with *p*NBA molecules. Supplementary Figs 8–12 and discussion within Supplementary Note 3 summarize our results.

π stacking is not restricted to CNTs only, but is very relevant to other nanostructures, such as single layer graphene or few layers graphene. Enabling imaging of graphene layers by optical means is of great importance. One common method is based on constructive interference of the reflected rays collected from single layer graphene or few layers graphene on top of insulating materials[29]. However, this method is restricted to specific insulating thicknesses, which are not always optimal with device performance. Moreover, fabrication of heterostructures based on graphene and hexagonal boron nitride imposes many challenges due to the lack of optical imaging of these new structures on different insulating layers[30]. Therefore, we tried to mark graphene samples with *p*NBA NCs and make them optically visible no matter on which substrates they lie. Successful results of such deposition and marking procedure are summarized in Supplementary Note 4 and Supplementary Figs 13–17. Briefly, *p*NBA NCs preferentially adsorbed on single and few layers exfoliated and CVD graphene, as well as graphene nanoribbons, on top of different insulating thicknesses, which do and do not support constructive interference according to Blake *et al*[29].

Non-invasive imaging. Having established the new mechanism, of nucleation and growth, as an efficient marking method one should verify that following such a step the CNTs maintain their electrical and mechanical properties. The first indication would be if the *p*NBA NCs would completely desorb off the CNTs leaving no residue behind. Figure 2 describes the sublimation process under ambient conditions for these *p*NBA NCs, that is, after they were first grown and nucleated along the CNTs. Figure 2a,b depict the temporal dependence of the AFM and optical images of two marked CNTs with similar *p*NBA NCs. We note in Fig. 2b that the CNTs remain optically visible for about an hour, where, Fig. 2a shows that the complete desorption would require an additional hour. As in any desorption study, the process can be accelerated by elevating the temperature or by introducing the

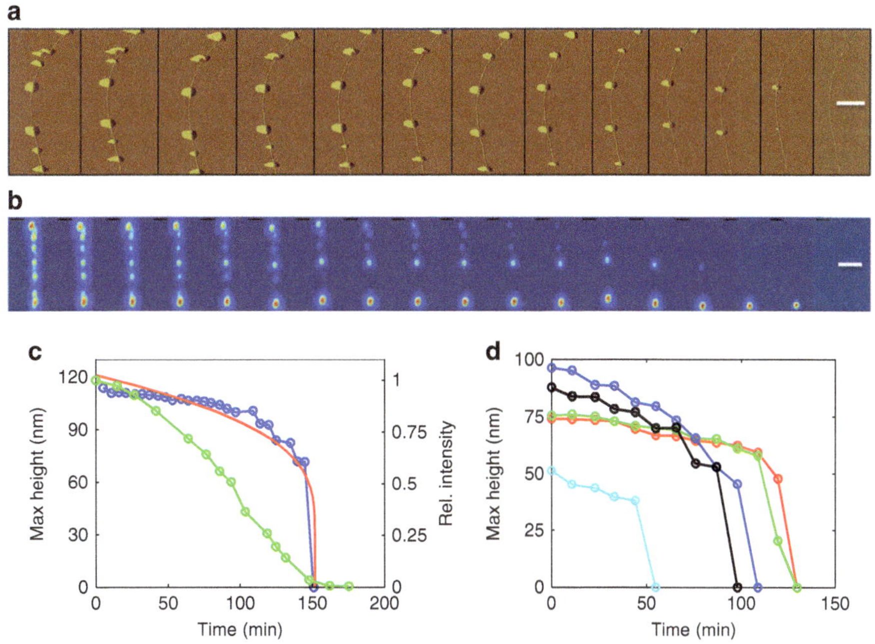

Figure 2 | Sublimation rate of *p*-nitrobenzoic acid (*p*NBA) nanocrystals off carbon nanotubes. (**a**) A temporal set of AFM amplitude images of *p*NBA nanocrystals (NCs) along a single carbon nanotube (CNT). The time interval between each image is 11 min. Scale bar, 1 μm. (**b**) A temporal set of dark field optical microscopy intensity of *p*NBA NCs along a single CNT. The time interval between each image is 6.5 min. Scale bar is 2 μm. (**c**) Left axis: NC height as function of time (blue circles) and theoretical fit according to Supplementary Eq. 7 (red line). Right axis: Relative dark field optical microscopy intensity of *p*NBA NC along CNT as a function of time (green circles). (**d**) Temporal dependence of the maximum height of different NCs along different CNTs.

sample into vacuum. To ensure that the sublimation process off the CNT follows a classical model such that full desorption could be predicted we present in Fig. 2c the temporal evolution of the NCs size as represented by the maximum height, h_{NC} (left axis, blue circles), and the relative optical intensity ($\delta I/I_0=(I_{NC}-I_{SiO_2})/I_{SiO_2}$; right axis, green circles). The red line in Fig. 2c is best fit to a classical sublimation model and the detailed analysis is presented in Supplementary Note 5. Height evolution for a range of marked CNTs is depicted in Fig. 2d. One can notice that the general process is similar, starting with a linear decrease until a critical height is obtained where the process accelerates. In fact, it is quite striking that although the deposited NCs were enormously large in comparison with the pristine CNT diameter, the final height shows no indication for the presence of *p*NBA molecules. To support this conclusions transmission electron microscopy (TEM) imaging was conducted (see Supplementary Note 2 for more details). Briefly, CNTs were grown on TEM grids and imaged before intensive *p*NBA molecules deposition and after they sublimed off the CNT. Supplementary Fig. 7a–d presents the results of this analysis, and confirm that no residues are left on the CNT surface indicating that complete desorption is achieved. Supplementary Fig. 7e,f shows typical optical images of the decorated TEM grid immediately after *p*NBA deposition, ensuring that indeed all the CNTs within the holes were marked with the NCs.

To complete the analysis and verify that not only do the NCs desorb but also that the CNT properties remain intact we repeated the procedure on CNT field effect transistors (CNTFETs) where the CNT are contacted by metallic electrodes on both sides. Figure 3a–d presents AFM images and the corresponding cross sections before *p*NBA deposition, immediately after and after sufficiently long time ($\approx$1 day). The electrodes can be seen at the top and bottom of the sub figures. Two black square areas marked 1 and 2 in the main panel of Fig. 3b depict 1 μm^2 areas where we zoomed in before and after *p*NBA deposition. The zoomed images are presented as insets to Fig. 3a,c and are marked with 1 and 2 according to Fig. 3b. The cross sections of these images were taken along the dotted black lines and are depicted in Fig. 3d. At the top part of Fig. 3d, the cross sections of area 1 (left, blue) and area 2 (right, red) immediately after deposition are plotted. It is evident that the height of these two NCs is above 150 nm. At the bottom of Fig. 3d, the cross sections of area 1\2 (left\right) before (blue) and sufficiently long time after (red) deposition are plotted. The cross sections look similar, and the average height before and after depositions are practically identical ($h_{CNT}=3.25\pm0.25$ nm). These results are in agreement with those reported in Fig. 2 indicating that the presence of the electrodes and the transistor fabrication process do not affect the desorption process. The critical test of the electrical performance of the CNTFET at the stages corresponding to Fig. 3a–c is shown in Fig. 3e,f for two different devices. Figure 3e,f depicts typical transfer characteristic of semiconducting (Fig. 3e) and small band gap (Fig. 3f) CNTFETs before molecule adsorption (blue), immediately after (green), and long after complete desorption (red). Interestingly, immediately after deposition a small shift in the threshold voltage is found, but after cleaning through the desorption process, the electrical response returns to its original values and is identical to the one measured before deposition of the *p*NBA NCs. These results were confirmed using many CNTFETs (above 20) so that we can safely conclude that indeed the process of marking the CNTs, which includes adsorbing of *p*NBA on the CNT surface, organic crystal growth and later on, complete desorption, does not leave any residue, and does not affect the overall electrical performance of the complete device.

On the basis of the above results one can envisage CNT circuit fabrication procedure that consists of growing the CNTs, marking the CNTs on the given substrate with *p*NBA molecules, acquiring the image using an optical microscope, and designing the electrical circuit accordingly. Figure 4a presents an overlay of the microscope image of marked CNT with the electrical circuit design that was made accordingly. The optical image of the completed circuit, as fabricated, is shown in Fig. 4b and its SEM image is shown in the inset of Fig. 4c. The different colours of the source and drain electrodes, Fig. 4b, are attributed to two different metals used for the electrodes (Gold and Aluminium). The blue

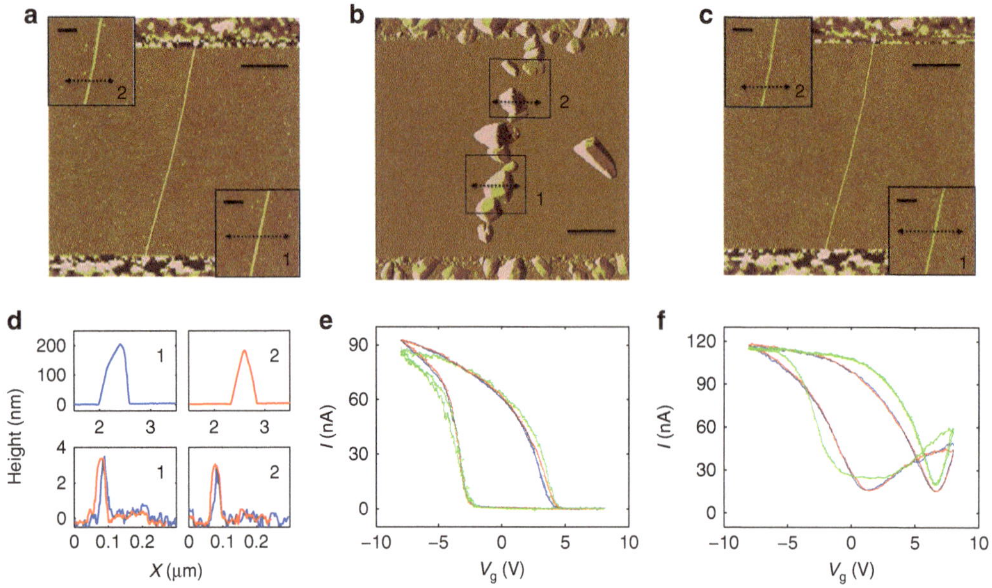

Figure 3 | Electrical characterization of marked and non-marked carbon nanotubes. (**a**-**c**) AFM images of complete carbon nanotube (CNT) device with metallic source and drain electrodes, before *p*-nitrobenzoic acid (*p*NBA) deposition (**a**) immediately after (**b**) and after one day (**c**). Zoom in on the black square areas marked 1 and 2 of the main panel of **b** are shown as four insets before deposition (**a**) and after one day (**c**). The dotted lines mark the locations of the cross sections which are plotted in **d**. The scale bar of the main panels, (**a**-**c**) is 1 μm, and for the insets is 200 nm. (**d**) Cross sections along the black dotted lines of panels (**a**-**c**) of area 1 (left) and area 2 (right). Top panels depict the nanocrystals (NCs) height immediately after deposition (blue for area 1, and red for area 2). Bottom panels plot the CNT diameter before (blue) and long after (red) within area 1 (left) and area 2 (right). (**e**,**f**) Transfer characteristic of two CNT devices before deposition (blue), immediately after deposition along the CNT (green), and after cleaning process (red).

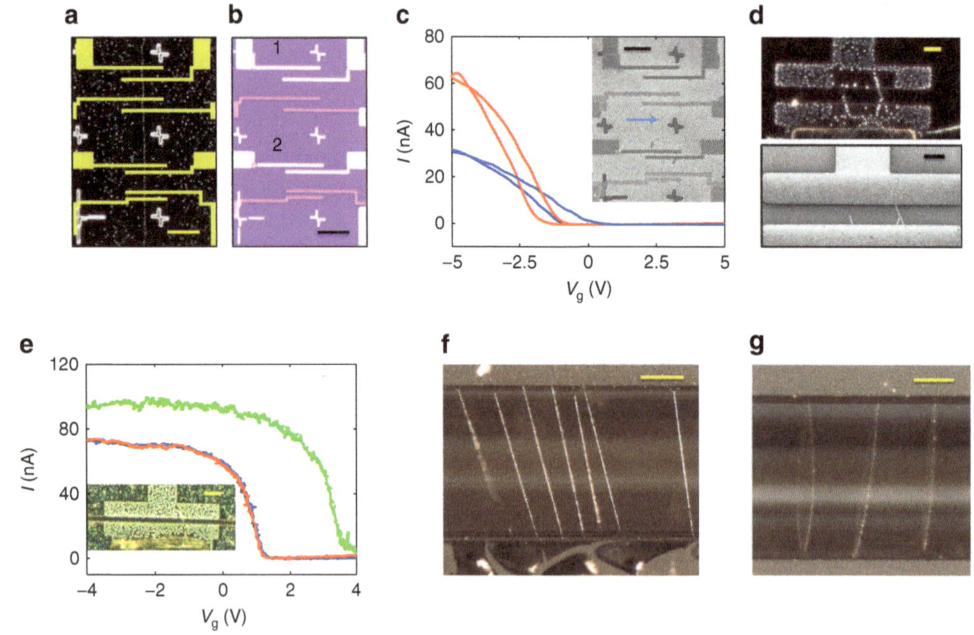

Figure 4 | Optically imaging of on-surface and suspended carbon nanotube devices. (**a**) Dark-field optical microscopy image of marked carbon nanotube (CNT) with *p*-nitrobenzoic acid (*p*NBA) nanocrystals (NCs), and overlay of the lithography circuit design. Scale bar is 10 μm. (**b**) Optical image of the complete fabricated device. The two different colour electrodes are two different metallic contacts. Scale bar is 10 μm. (**c**) Transfer characteristic of the two CNTs devices 1 (blue), and 2 (red). Inset: SEM image of the completed circuit. Blue arrow points towards the CNT that was marked in (**a**). Scale bar, 10 μm. (**d**) Top: Dark field optical microscopy image of marked junction. Bottom: SEM image of the same junction. Scale bar in both images is 3 μm. (**e**) Main panel: Transfer characteristic of suspended CNT before *p*NBA deposition (blue), immediately after deposition (green), and after cleaning process (red). Inset: Dark field optical microscopy image of the measured junction marked with *p*NBA NCs. Scale bar, 3 μm (**f**,**g**) Dark field optical images of long suspended CNTs decorated with *p*NBA NCs. Scale bar, 50 μm.

arrow, Fig. 4c inset, points to a faint line that corresponds to the CNT, see also Fig. 4a. Figure 4c shows the measured transfer characteristic of devices 1 (blue line), and 2 (red line), as depicted in Fig. 4b.

This result confirms that indeed the method presented here is successful and efficient for the fabrication of CNT based devices. We have used this method and fabricated more than 100 CNT-based devices, with more than 95% success rate (see details in Supplementary Note 6). The high yield is attributed to the exact location of the complete CNT network and the possibility to place our electrodes at optimal locations, where the tubes are straight, far from other tubes and do not split to additional tubes. Supplementary Figure 18a, b depicts two histograms for the total resistance and resistance per micron length of the fabricated CNT devices produced by the described method. These results are extremely good and are well within the high-end group of existing devices.

To contrast our method with the case where the CNTs position is not known, and hence is not used as an input to the electrode design, we blindly fabricated devices on CNTs network. The marking by *p*NBA and optical imaging were carried out only after the devices were completed and the images are shown in Fig. 4d,e. Surprisingly, the images show combinations of CNTs bridging narrow junctions as small as 1 μm and suspended tube. The top part of Fig. 4d shows a dark-field optical image of a 3 μm gap junction and the bottom part shows a SEM image of the same junction. Comparing the two images, it is clear that the *p*NBA marking procedure accurately stains all the CNTs within the electrode area and that luckily only a single tube is bridging the two electrodes. To show that, as far as the imaging is concerned, post-imaging is robust we present in Fig. 3e the transfer characteristics of this specific junction before, immediately after, and long period of time after *p*NBA deposition. As before, the electrical characteristics fully recover once the *p*NBA is given time to fully desorb. These results suggest that such imaging procedures could not only be used for circuit design but also in failure analysis of CNT circuits thus enabling the full fabrication chain.

Mechanical vibrations of suspended CNTs. Until now we have mainly discussed the marking procedure of CNTs on surfaces. However, suspended CNTs play a major role in the research and applications of nano-electromechanical systems (NEMS). Due to their high Young's modulus, low mass and presumably low defect density, one can anticipate to obtain extremely high resonance frequencies with high-quality factors[31–34]. The fabrication challenges for such applications are somewhat similar and with respect to imaging and ensuring a given number of tubes in the gap the challenges are almost identical. Currently, identifying the number of tubes bridging the junction is performed either by SEM or by AFM. SEM usually deteriorates the mechanical properties of the CNT resonator, and shifts their resonance frequencies due to deposited materials while imaging. AFM besides being extremely slow, is very detrimental and often ends with device deterioration or even disconnection. Clearly, non-invasive and rapid imaging procedure is highly required here as well. To demonstrate the feasibility of our method, we have fabricated suspended devices and deposited *p*NBA molecules as described before. The results are presented in Fig. 4e–g. The inset of Fig. 4e shows that for narrow trenches, as small as 1 μm, a clear image of the suspended CNT bridging the two metallic contacts is possible. For narrower trenches, the strong light scattering by the metallic electrodes may obscure the *p*NBA NCs along the suspended tube. In such circumstances, the tube can still be identified through its portions that extend beyond the electrode area and one could even extrapolate the location of the suspended part of the tube from its marked portion on the surface.

As before, the validity of the imaging method is tested by comparing the electrical properties of the CNTs at different stages of the process. The main panel of Fig. 4e presents the transfer characteristic of a suspended CNT, similar to that seen in the inset of Fig. 4e, before (blue), immediately after (green) and long time after (red) deposition of *p*NBA molecules. It is clear that although immediately after deposition the electrical measurements are significantly different, due to the presence of *p*NBA NCs, the two measurements that were conducted before and long time after deposition, coincide completely with each other, as was found previously for on-surface CNTs. Working with suspended CNTs and especially with longer ones it may be possible to image their mechanical motion and examine their dynamics in terms of vibrational modes. For that purpose, we fabricated deep trenches, 80 μm depth, 200 μm wide and grew long and straight CNTs across. Later, *p*NBA molecules were deposited and dark-field

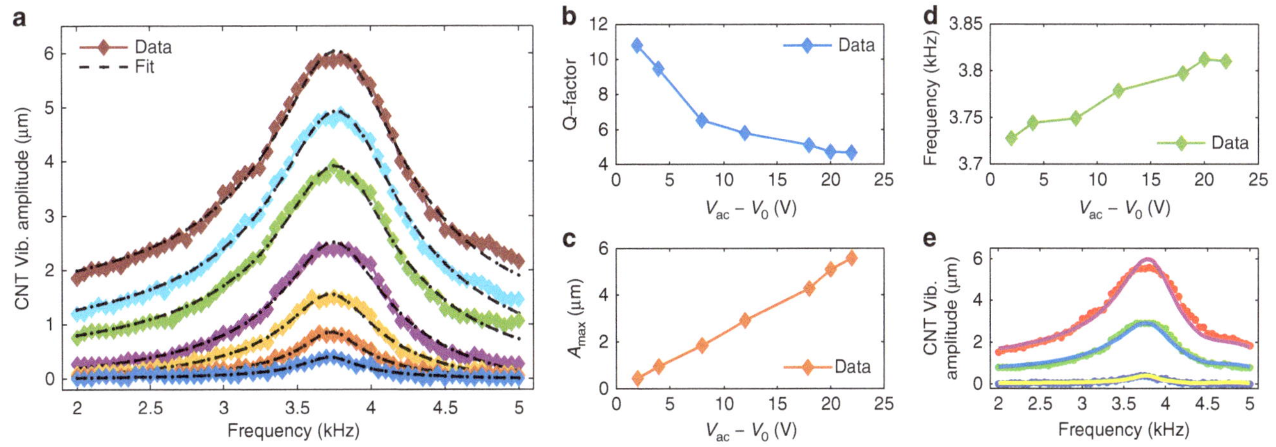

Figure 5 | Optically measured mechanical resonances of marked carbon nanotubes. (**a**) Amplitude of vibration of carbon nanotube (CNT) versus excitation frequency of a driven piezoelectric actuator, V_{ac}. The different coloured curves are for different applied voltages of the piezo as plotted in **c** (2, 4, 8, 12, 18, 20 and 22 V from bottom to top). The different black dashed lines are best fit to Lorentzian curves. (**b**) Quality factors as extracted from the Lorentzian fit. (**c**) Maximum amplitude versus the applied ac piezo voltages (subtracted offset bias originated from the electrical setup, V_0). (**d**) Frequencies of the maximal response for different excitation voltages. The chronological time of the measurements were from right to left. (**e**) The same data as in **a** for three different V_{ac}'s (dotted coloured lines) and resulted fit (colored lines) according to our finite element model (FEM) discussed in Supplementary Note 7. Vib., Vibration.

optical images of the suspended CNTs were collected. One typical image is shown in Fig. 4f. The suspended tubes are decorated with *p*NBA NCs and are clearly seen by the optical image. SEM image of similar NCs is shown in Supplementary Fig. 2e. They crystalize on the CNT surface and after a short period in high vacuum (required for SEM operation) they form discontinuous segments along the tube. Applying mechanical excitation to long suspended CNTs would typically cause such tubes to oscillate. Figure 4g depicts such dynamical behaviour, where the mechanical excitation was provided by gentle air flow. A real-time video of such oscillations is presented in Supplementary Movie 2. From these images and others, one can optically visualize the vibrational modes of CNTs, including higher modes beside the fundamental one, and obtain important information about their slack, dynamics, non-linear behaviour, temporal and spatial correlations, as well as dissipation processes.

Two types of quantitative vibrational experiments were performed, one utilized piezoelectric actuation and the second electrostatic actuation. For both methods, the vibrational modes were detected optically (Supplementary Note 7 and Supplementary Fig. 19). As before, the devices comprised suspended CNTs of different lengths ($L \approx 30$, 100 and 200 μm) with various levels of *p*NBA NCs decoration along their sidewalls. Figure 5a depicts the frequency response of such CNT for different excitations of the piezoelectric actuator. The data nicely fit to Lorentzian with a single peak, as expected for suspended doubly clamped beam. For the entire excitation range, the behaviour is linear and the quality factors and peak heights are plotted in Fig. 5b,c. However, for different deposition condition, the frequency response contains more than a single peak, as presented in Supplementary Fig. 20a. Naively, for CNTs without slack, s, and without built-in tension, T_0, the first vibrational mode is given by solving Euler-Bernoulli equation[35] to yield $f_1 = \beta_1^2 \sqrt{EI/\rho_v S}/(2\pi L^2)$ where $\beta_1 = 1.8751$, E is the beam Young's modulus, I is the moment of inertia, ρ_v is the mass density, and S is the beam cross section area. However, when the tube is decorated with the *p*NBA NCs two opposite effects modify the resonance frequencies. The first effect, which scales as r^2, where r is the radius of the organic circumference, originates from the increase of the tube total mass, and hence causes reduction of the resonance frequencies. The second effect originates from the bending rigidity attributed to the *p*NBA NCs, which involves their Young's modulus, E_{pNBA} as well as their moment of inertia, I_{pNBA}, which scales as r^4. For homogenous coverage the effective bending rigidity is given by the core shell model[36]

$$E_{\text{eff}} I_{\text{eff}} = E_{\text{CNT}} I_{\text{CNT}} + E_{pNBA} I_{pNBA} \tag{1}$$

where $E_{\text{CNT}} \approx 1\,\text{TPa}$, $I_{\text{CNT}} \approx (1\,\text{nm})^4$, $E_{pNBA} \approx 500\,\text{MPa}$, and $I_{pNBA} \approx r^4$. Although $E_{pNBA} \ll E_{\text{CNT}}$, for outer radii which are significantly larger than the typical pristine tube radius, the bending rigidity of the shell will play the dominant role in affecting the resonance frequencies. As the shell becomes wider, the total mass of the tube increases as well. However, the mass per length increases as r^2, where the bending rigidity increases as r^4, hence, overall the resonance frequencies will increase in comparison to a bare tube with the same length, as can be observed in Fig. 5a and Supplementary Fig. 20a. Moreover, while measuring the vibrational modes, *p*NBA molecules sublime and leave the CNT, and the resonance modes are expected to decrease, as is observed in Fig. 5d.

In real experiments, the *p*NBA NCs decorate the CNT sidewalls unevenly, thus, a more rigorous analysis is required. For that purpose, we developed a homemade FEM based on numerical simulation that takes into account non homogenous coverage of the CNT sidewalls with *p*NBA NCs (see details in Supplementary Note 7, and Supplementary Fig. 25). The algorithm includes slack as well, which is typically introduced as a consequence of thermal contraction of the fabricated trench on cooling from the hot CNT growth temperature (~900 °C) to room temperature. Slack plays an important role in affecting the tube resonance modes, dissipation processes[37], and lifting the degeneracy of the in-plane and out of plane resonance frequencies[38]. Usually, this parameter is deduced from three-point bending experiments[39], or as a fitting parameter for the obtained vibrational data. Here, thanks to the optical imaging techniques we could directly measure the slack of our fabricated

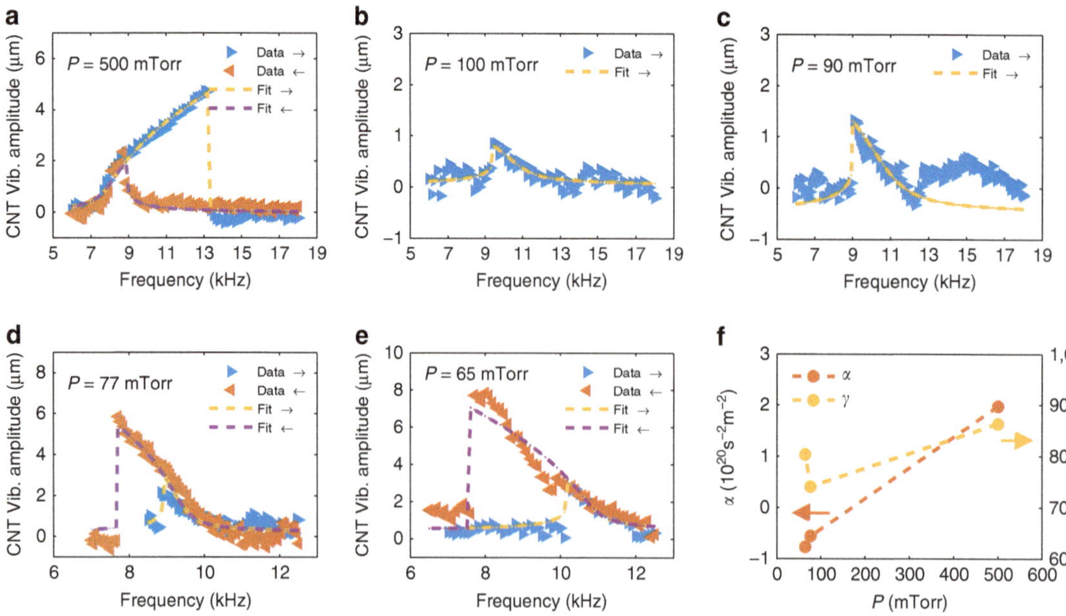

Figure 6 | Hardening to softening behaviour. (**a**-**e**) Amplitude of vibration of carbon nanotube (CNT) versus excitation frequency for different pressures, as marked inside each panel. Blue triangles are data for up sweep and red triangles for down sweep. The yellow and purple dashed lines are the theoretical solutions according to Supplementary Equations 33 and 34, for up and down sweep, respectively. (**f**) Left axis: Extracted non-linear spring constant, α, as function of pressure. Right axis: Extracted coefficient of linear damping, γ, versus chamber pressure. Vib., Vibration.

devices, and use it in our analysis for the vibrational modes. Supplementary Fig. 21a shows how we measure the slack of specific CNT by DF microscopy, and Supplementary Fig. 21b presents the distribution of slack in our devices. Knowing the slack enables us to fit the measured vibrational modes to the theoretical predictions according to our FEM. Two successful examples for two different CNTs are presented in Fig. 5e and Supplementary Fig. 20c.

Optically observed Duffing oscillator of suspended CNTs. All the previous measurements were preformed at ambient conditions. However, it is also possible to obtain the frequency response also in vacuum with the same optical detection technique. This time, the actuation was based on electric force applied between the tube electrodes and an external probe located near the suspended tube. Surprisingly, for this set of measurements strong non-linear behaviour and hysteresis were observed. Supplementary Fig. 22 shows the vibrational behaviour for different excitation powers. Supplementary Fig. 23 presents the frequency response for different pressures. Figure 6 depicts transition from hardening to softening behaviour as the chamber pressure decreases. These set of phenomena can be understood within the framework of Duffing oscillator[40]. In such lumped model the beam displacement, u, satisfies the following differential equation

$$\frac{d^2u}{dt^2} + (\gamma + \gamma_3 u^2)\frac{du}{dt} + \omega_0^2(1 + Ku^2)u = \frac{1}{2M_T}\frac{dC_g}{dz}\left(V_g(dc) + V_g(ac)\right)^2, \quad (2)$$

where K is responsible for hardening/softening and γ_3 for non-linear damping[41], M_T is the total mass of the oscillator, C_g is the tube-external gate mutual capacitance, and V_g(dc/ac) is the dc/ac external gate voltage applied between the CNT and the metallic probe. Under external dc bias the tube is being stretched and the fundamental resonance frequency increases accordingly. Supplementary Fig. 24a presents such increase under the applied gate voltage and our FEM agrees very well with the experimental data. It should be noted that several C_g were tested, all described in Supplementary Note 7 and Supplementary Fig. 26, but the resulted fitting parameters are very similar (within 10%). Then, using these parameters, we could fit the Duffing type behaviour, as presented in Supplementary Figs 22, 23 and Fig. 6, according to the rotation frame analysis[40] discussed in Supplementary Note 7. The model describes very well the experimental data, and even the transition between hardening and softening can be understood from equation 2, as seen in Fig. 6. In that case the force term at the right hand side of equation 2 can be expanded in powers of u, to yield terms which are linear and cubic in u. The linear term may modify the resonance frequency, where the cubic term may affect the non-linearity term, K. As the pressure decreases, and more molecules sublime off the CNT sidewalls, the right hand side becomes more dominant, and the non-linear term is given by

$$\alpha = (E\pi^4)/(2\rho_v L^4) - F_3 \quad (3)$$

according to Supplementary Equation 31, becomes negative. In such circumstances where the total mass is low, and V_g(dc) is high, α changes sign and the expected hardening behaviour becomes softening, as observed in Fig. 6. Detailed analysis of the non-linear term within the Duffing lumped model (Supplementary Note 7) reveals good agreement between the estimated $\alpha = -0.45 \pm 0.1\ 10^{20}\,s^{-2}\,m^{-2}$ to the one found at low pressure from the fitted data in Fig. 6d, that is, $\alpha = -0.56 \pm 0.07\ 10^{20}\,s^{-2}\,m^{-2}$.

Automatic circuit design and fabrication. As a last example of the power in the method presented here we address the potential of CNTs to compete or integrate with silicon fabrication, an issue considered to be a dream by more than a few. A major milestone for realizing such a task ('dream') is the high level of device integration used in the silicon industry. Following the design rules of silicon and the idea that all patterns are predetermined before the fabrication process indeed seem impossible in the context of CNTs. Alternatively, one can choose a path where the functionality is indeed fully predetermined but the patterns are designed and fabricated on the fly as part of the process line. On the basis of the results shown above one may suggest that rapid imaging of the CNTs network, together with image processing capabilities and software design tools, will enable automatic fabrication of CNTs based circuits. To prove the feasibility of such a path we developed an image processing software and a pattern design tool that could automatically design the fabrication pattern of simple components.

The full fabrication process starts with growing CNTs from pre-patterned catalyst islands on top of silicon substrate with 300 nm SiO_2. Following the CNTs growth the sample was marked with *p*NBA molecules, and dark-field optical images were collected. The images were fed to our homemade image processing software and the result of finding all the CNTs and their layout is depicted in Fig. 7a. The red lines are the software output which follow exactly the marked CNTs. Details of the image processing algorithm can be found in Supplementary Note 8. Next, using only the software locations of the CNTs, p- and n-type CNTFETs, as well as inverter circuits based on this network were automatically designed according to pre-design rules. An example is shown in Supplementary Fig. 27 and the

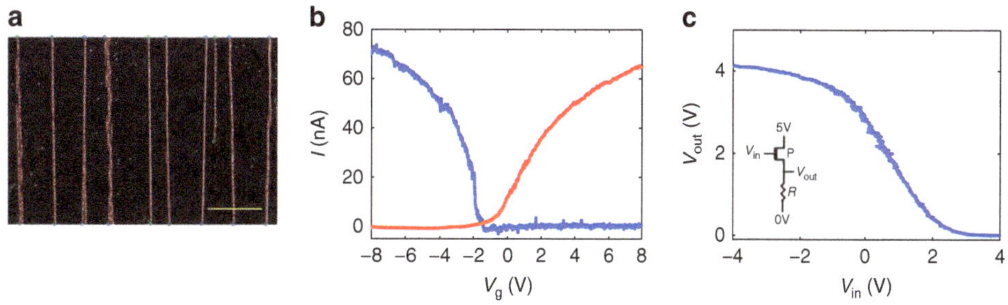

Figure 7 | Automatic design and fabrication of carbon nanotubes based devices. (**a**) Image processing results. Dark field optical microscopy image of marked carbon nanotubes (CNTs) (white dotted lines) superimposed with the image processing analysis results (red lines). Scale bar, 30 μm. (**b**) Transfer characteristics of automatically designed p- (blue line) and n- (red line) type CNT field effect transistors (FETs). (**c**) Automatically designed inverter based on p-type CNTFET and external resistor. Inset: schematic diagram for the inverter.

details of the fabrication process for the p- and n-type FETs are discussed in Supplementary Note 8 as well. The electrical characteristics of the complete devices are plotted in Fig. 7b,c. Two types of inverters were utilized. The first (type I) includes two steps of fabrication, one for the p-type behaviour, and the second for n-type. The second type (type II) was based on p-type CNTFETs and regular resistors. The two approaches were successful, and are plotted in Supplementary Fig. 27c (type I) and Fig. 7c (type II). These p- and n-type CNTFETs, and two kinds of inverters which were automatically fabricated are successful examples for this promising approach.

Summary. In conclusion, we presented a novel and rapid method based on *p*NBA NCs to optically visualize CNTs. The method is based on our finding that CNT surface constitutes a nucleation seed for organic crystal growth that can then be imaged by dark-field optical microscopy. This non-invasive method was found not to leave any residue on the tube sidewall and consequently to have no effect on the CNTs electrical or mechanical performance. The method was demonstrated on both on-surface and suspended devices down to gap size of 1 μm.

Using relatively low level of NCs coverage we were able to not only statically image the CNTs but also to study the dynamic behaviour of suspended CNTs, including oscillations, vibrational modes, non-linear characteristics, and transition from hardening to softening behaviour. The marking and imaging procedure is scalable and enables automatic design and fabrication of electrical circuits based on CNTs. We demonstrated automatic fabrication of both p and n-type CNTFETs as well as inverters and have shown that the CNTs retain their electrical properties throughout the process. We believe that the suggested method can provide a real platform for the integration of nanoelectronics with silicon technology.

Methods

Device fabrication and molecule deposition. CNTs were grown using CVD at 900 °C with 0.5/0.5 SLM flow of H_2/CH_2. The catalyst particles were deposited from ferritin solution onto predefined catalyst pads. Electrical contacts were deposited either before or after CNTs growth. For p-type devices Cr/Au 5/120 nm or Cr/Pt 5/40 nm were deposited[6,8,25,26]. For n-type devices four different processes were used. The first was based on 50 nm of Al deposition capped by 50 nm of Au[42]. The second approach included Ca/Al 30/120 nm deposition[43], and the third consisted of Sc 50 nm metal deposition[44]. The last approach, which was found to be the best, included Cr/Pt 5/40 nm electrical contacts and atomic layer deposition (ALD) of HfO_2 on top of the CNT[45]. The deposition temperature and pressure were 270 °C and 620 mTorr, and the HfO_2 thickness was ≈ 30 nm.

Powder CNTs were purchased from SWeNT and CoMoCAT and were dissolved in chloroform. After tip sonication for 20 min in pulse mode, they were dispersed on silicon wafer. *p*NBA powder was purchased from Fluka. The deposition on the CNTs uses a hotplate. A small metallic tray ~1 cm^2 in size is filled with the *p*NBA powder and placed near the edge of a 10 cm petri dish. Only this edge is sited on a hotplate, and a glass lid covers the whole petri dish. The hotplate is heated up to 150 °C for 15–30 min until *p*NBA molecules start to cover the lid. Then, a silicon chip with the CNTs is placed on the cold side of the petri dish, and the molecules condense on the chip. Typically, the condensation period lasts for 90 s, but shorter or longer times result with diluted or denser *p*NBA coverage.

TEM analysis. Commercial TEM grids with thin silicon nitride mesh film 200 nm thick, and 2 μm diameter holes were used. Catalyst solution was deposited on the circumference of the TEM grid and CNTs were grown using CVD tool. Imaging was performed using 120 kV, LaB6 emitter equipped, FEI T12 G2 TEM, before and sufficiently long time after (2 weeks) *p*NBA deposition.

Electrical measurements. The electrical measurements were done using either DC or AC currents at zero or few Hertz, respectively. The voltage biases were few mV with few hundreds msec time constants.

Vibrational data. The suspended tubes were actuated by piezoelectric membrane or electrostatic electrode connected to RF generator. The detection performed optically by automatically analysing the vibrational amplitude image for each frequency excitation within the frequency range.

References

1. McEuen, P. L., Fuhrer, M. S. & Park, H. K. Single-walled carbon nanotube electronics. *IEEE Trans. Nanotechnol.* **1,** 78–85 (2002).
2. Dai, H. J. Carbon nanotubes: Synthesis, integration, and properties. *Acc. Chem. Res.* **35,** 1035–1044 (2002).
3. Avouris, P., Chen, Z. & Perebeinos, V. Carbon-based electronics. *Nat. Nanotechnol.* **2,** 605–615 (2007).
4. Cao, Q. & Rogers, J. A. Ultrathin films of single-walled carbon nanotubes for electronics and sensors: a review of fundamental and applied aspects. *Adv. Mater.* **21,** 29–53 (2009).
5. Shulaker, M. M. *et al.* Carbon nanotube computer. *Nature* **501,** 526–530 (2013).
6. Kong, J., Soh, H. T., Cassell, A. M., Quate, C. F. & Dai, H. J. Synthesis of individual single-walled carbon nanotubes on patterned silicon wafers. *Nature* **395,** 878–881 (1998).
7. Huang, S. M., Maynor, B., Cai, X. Y. & Liu, J. Ultralong, well-aligned single-walled carbon nanotube architectures on surfaces. *Adv. Mater.* **15,** 1651–1655 (2003).
8. Ismach, A., Segev, L., Wachtel, E. & Joselevich, E. Atomic-step-templated formation of single wall carbon nanotube patterns. *Angew. Chem. Int. Edit.* **43,** 6140–6143 (2004).
9. Kocabas, C. *et al.* Guided growth of large-scale, horizontally aligned arrays of single-walled carbon nanotubes and their use in thin-film transistors. *Small* **1,** 1110–1116 (2005).
10. Vijayaraghavan, A. *et al.* Ultra-large-scale directed assembly of single-walled carbon nanotube devices. *Nano Lett.* **7,** 1556–1560 (2007).
11. Dresselhaus, M. S., Dresselhaus, G., Jorio, A., Souza, A. G. & Saito, R. Raman spectroscopy on isolated single wall carbon nanotubes. *Carbon* **40,** 2043–2061 (2002).
12. Sfeir, M. Y. *et al.* Probing electronic transitions in individual carbon nanotubes by rayleigh scattering. *Science* **306,** 1540–1543 (2004).
13. Joh, D. Y. *et al.* On-chip rayleigh imaging and spectroscopy of carbon nanotubes. *Nano Lett.* **11,** 1–7 (2011).
14. Berciaud, S., Cognet, L., Poulin, P., Weisman, R. B. & Lounis, B. Absorption spectroscopy of individual single-walled carbon nanotubes. *Nano. Lett.* **7,** 1203–1207 (2007).
15. Liu, K. *et al.* High-throughput optical imaging and spectroscopy of individual carbon nanotubes in devices. *Nat. Nanotechnol.* **8,** 917–922 (2013).
16. Bachilo, S. M. *et al.* Structure-assigned optical spectra of single-walled carbon nanotubes. *Science* **298,** 2361–2366 (2002).
17. DeBorde, T., Aspitarte, L., Sharf, T., Kevek, J. W. & Minot, E. D. Determining the chiral index of semiconducting carbon nanotubes using photoconductivity resonances. *J. Phys. Chem. C* **118,** 9946–9950 (2014).
18. Tsen, A. W., Donev, L. A. K., Kurt, H., Herman, L. H. & Park, J. Imaging the electrical conductance of individual carbon nanotubes with photothermal current microscopy. *Nat. Nanotechnol.* **4,** 108–113 (2009).
19. Havener, R. W., Tsen, A. W., Choi, H. C. & Park, J. Laser-based imaging of individual carbon nanostructures. *NPG Asia Mater.* **3,** 91–99 (2011).
20. Choi, H. C., Shim, M., Bangsaruntip, S. & Dai, H. J. Spontaneous reduction of metal ions on the sidewalls of carbon nanotubes. *J. Am. Chem. Soc.* **124,** 9058–9059 (2002).
21. Chen, R. J. *et al.* Noncovalent functionalization of carbon nanotubes for highly specific electronic biosensors. *Proc. Natl Acad. Sci. USA* **100,** 4984–4989 (2003).
22. Welsher, K., Liu, Z., Daranciang, D. & Dai, H. Selective probing and imaging of cells with single walled carbon nanotubes as near-infrared fluorescent molecules. *Nano Lett.* **8,** 586–590 (2008).
23. Meng, L., Fu, C. & Lu, Q. Advanced technology for functionalization of carbon nanotubes. *Prog. Nat. Sci.* **19,** 801–810 (2009).
24. Yoshimura, S. H., Khan, S., Maruyama, H., Nakayama, Y. & Takeyasu, K. Fluorescence labeling of carbon nanotubes and visualization of a nanotube-protein hybrid under fluorescence microscope. *Biomacromolecules* **12,** 1200–1204 (2011).
25. Cao, J., Wang, Q. & Dai, H. Electron transport in very clean, as-grown suspended carbon nanotubes. *Nat. Mater.* **4,** 745–749 (2005).
26. Pascal-Levy, Y. *et al.* Water-assisted mobile charge induced screening and origin of hysteresis in carbon nanotube field-effect transistors. *Phy. Rev. B* **86,** 1–9 (2012).
27. Chin, C.-J. M., Shih, M.-W. & Tsai, H.-J. Adsorption of nonpolar benzene derivatives on single-walled carbon nanotubes. *Appl. Surf. Sci.* **256,** 6035–6039; http://www.sciencedirect.com/science/article/pii/S016943321000406X (2010).

28. Chen, R. J., Zhang, Y. G., Wang, D. W. & Dai, H. J. Noncovalent sidewall functionalization of single-walled carbon nanotubes for protein immobilization. *J. Am. Chem. Soc.* **123,** 3838–3839 (2001).
29. Blake, P. *et al.* Making graphene visible. *Appl. Phys. Lett.* **91,** 063124 (2007).
30. Dean, C. R. *et al.* Boron nitride substrates for high-quality graphene electronics. *Nat. Nanotechnol.* **5,** 722–726 (2010).
31. Sazonova, V. *et al.* A tunable carbon nanotube electromechanical oscillator. *Nature* **431,** 284–287 (2004).
32. Huttel, A. K. *et al.* Carbon nanotubes as ultrahigh quality factor mechanical resonators. *Nano Lett.* **9,** 2547–2552 (2009).
33. Laird, E., Pei, F., Tang, W., Steele, G. & Kouwenhoven, L. A high quality factor carbon nanotube mechanical resonator at 39 ghz. *Nano Lett.* **12,** 193–197 (2012).
34. Chaste, J. *et al.* A nanomechanical mass sensor with yoctogram resolution. *Nat. Nanotechnol.* **7,** 300–303 (2012).
35. Landau, L. D. & Lifshitz, E. M. *Elasticity Theory* (Pergamon, 1986).
36. Chen, C. Q., Shi, Y., Zhang, Y. S., Zhu, J. & Yan, Y. J. Size dependence of young's modulus in zno nanowires. *Phys. Rev. Lett.* **96,** 075505 (2006).
37. Barnard, A. W., Sazonova, V., van der Zande, A. M. & McEuen, P. L. Fluctuation broadening in carbon nanotube resonators. *Proc. Natl Acad. Sci. USA* **109,** 19093–19096 (2012).
38. Ustunel, H., Roundy, D. & Arias, T. A. Modeling a suspended nanotube oscillator. *Nano. Lett.* **5,** 523–526 (2005).
39. Minot, E. D. *et al.* Tuning carbon nanotube band gaps with strain. *Phys. Rev. Lett.* **90,** 156401 (2003).
40. Nayfeh, A. *The Method of Normal Forms* (Wiley-VCH Verlag GmbH & Co. KGaA, 2011).
41. Eichler, A. *et al.* Nonlinear damping in mechanical resonators made from carbon nanotubes and graphene. *Nat. Nanotechnol.* **6,** 339–342 (2011).
42. Javey, A., Wang, Q., Kim, W. & Dai, H. Advancements in complementary carbon nanotube field-effect transistors. *Int. Electron Devices Meeting Tech. Dig.* 741–744 (2003).
43. Nosho, Y., Ohno, Y., Kishimoto, S. & Mizutani, T. n-type carbon nanotube field-effect transistors fabricated by using ca contact electrodes. *Appl. Phys. Lett.* **86,** 073105 (2005).
44. Zhang, Z. *et al.* Doping-free fabrication of carbon nanotube based ballistic cmos devices and circuits. *Nano Lett.* **7,** 3603–3607 (2007).
45. Moriyama, N., Ohno, Y., Kitamura, T., Kishimoto, S. & Mizutani, T. Change in carrier type in high-k gate carbon nanotube field-effect transistors by interface fixed charges. *Nanotechnology* **21,** 165201 (2010).

Acknowledgements

We acknowledge the support of the Russell Berrie Nanotechnology Institute, the Micro Nano Fabrication Unit at the Technion, and the Ministry of Industry, Trade and Labor via the MAGNET program (PRINTEL consortium).

Author contributions

G.Z. led and was involved in all aspects of the projects beside the vibrational part. M.S. was the leader for the vibrational measurements and analysis. M.S., T.T., and Z.R. were responsible for device fabrication, electrical measurements, and marked images. S.R. contributed to measurements and analysis. I.G. contributed to the vibrational experimental setup and measurements. S.S developed the finite element model for the vibrational analysis. N.G. performed the TEM analysis. G.A., M.I., and Y.M. performed the image processing analysis. E.M.H. assisted in the developing of the fabrication process and the theoretical analysis. H.N., Y.M., and T.Y.I. assisted in the developing of the fabrication process and measurement techniques. A.R., O.S., and V.K. designed and built all the experimental tools for the project. Y.E.Y. was in charge and advised on all parts of the project.

Additional information

Competing financial interests: The authors declare no competing financial interests.

Scalable salt-templated synthesis of two-dimensional transition metal oxides

Xu Xiao[1,*], Huaibing Song[1,*], Shizhe Lin[1], Ying Zhou[1], Xiaojun Zhan[1], Zhimi Hu[1], Qi Zhang[2], Jiyu Sun[1], Bo Yang[1], Tianqi Li[1], Liying Jiao[2], Jun Zhou[1], Jiang Tang[1] & Yury Gogotsi[3]

Two-dimensional atomic crystals, such as two-dimensional oxides, have attracted much attention in energy storage because nearly all of the atoms can be exposed to the electrolyte and involved in redox reactions. However, current strategies are largely limited to intrinsically layered compounds. Here we report a general strategy that uses the surfaces of water-soluble salt crystals as growth templates and is applicable to not only layered compounds but also various transition metal oxides, such as hexagonal-MoO_3, MoO_2, MnO and hexagonal-WO_3. The planar growth is hypothesized to occur *via* a match between the crystal lattices of the salt and the growing oxide. Restacked two-dimensional hexagonal-MoO_3 exhibits high pseudocapacitive performances (for example, 300 F cm^{-3} in an $Al_2(SO_4)_3$ electrolyte). The synthesis of various two-dimensional transition metal oxides and the demonstration of high capacitance are expected to enable fundamental studies of dimensionality effects on their properties and facilitate their use in energy storage and other applications.

[1] Wuhan National Laboratory for Optoelectronics and School of Optical and Electronic Information, Huazhong University of Science and Technology, Wuhan, Hubei 430074, China. [2] Key Lab of Organic Optoelectronics & Molecular Engineering, Department of Chemistry, Tsinghua University, Beijing 100084, China. [3] Department of Materials Science and Engineering and A.J. Drexel Nanomaterials Institute, Drexel University, Philadelphia, Pennsylvania 19104, USA.
* These authors contributed equally to this work. Correspondence and requests for materials should be addressed to J.Z. (email: jun.zhou@mail.hust.edu.cn) or to J.T. (email: jtang@mail.hust.edu.cn) or to Y.G. (email: gogotsi@drexel.edu).

Pseudocapacitors are of great interest for energy storage owing to their fast charging/discharging, high power density and excellent cycling stability, which are beneficial to many potential applications ranging from ubiquitous portable electronics to grid energy storage[1–7]. Recently, the synthesis of two-dimensional (2D) nanosheets followed by restacking of these nanosheets to form electrodes has attracted much attention[8–11]. Because of their nanometre/sub-nanometre thickness, nearly all of the metal atoms can be exposed to the electrolyte and potentially involved in redox reactions, resulting in pseudo-capacitances that approach the theoretical values[9]. Prominent examples of 2D nanosheets for pseudocapacitor electrodes include MXenes, which have exhibited a volumetric capacitance as high as 900 F cm^{-3} owing to their high conductivity and hydrophilicity, and 1 T phase MoS_2 nanosheets, which have exhibited a volumetric capacitance of 650 F cm^{-1} (refs 3,9,11).

Transition metal oxides, such as MnO_2 and RuO_2 are classic pseudocapacitor materials that possess a large theoretical capacity owing to their large number of oxidation states[12–14]. Moreover, oxides have a higher chemical stability than sulfides, carbides and nitrides, such as MoS_2 (ref. 11) or MXenes[9–10], especially when processed into 2D flakes, which may provide a better durability and enable use in positive electrodes. In addition, many oxides contain tunnels or pores within the crystal structure, which is also beneficial for energy storage applications[2,3]. Despite the recent demonstration of promising energy application of nanometre-thick liquid-exfoliated α-MoO_3 (ref. 15) and MnO_2 (refs 12,14), the investigation of many other oxides for pseudocapacitor applications is hindered by their lack of a layered structure. Therefore, production of a 2D morphology is challenging.

For pseudocapacitor applications, nanosheets are highly preferable because ion diffusion within the oxide lattice is minimized, enabling the simultaneous achievement of a large volumetric capacitance (high energy density) at a high rate (high power density). This behaviour is typically very challenging to achieve for both bulk and one-dimensional materials. Ideally, the synthesized 2D oxide should be free of contamination from intercalating chemicals, dispersing agents and surfactants that could compromise its electrochemical performance owing to possible blockage of active sites and the added weight/volume of electrochemically inactive contaminants. In addition, the synthesis should be scalable for industrial manufacturing and production of large-size nanosheets with tunable compositions (for example, oxygen vacancies) to compensate for the poor conductivity often encountered in oxides. The current strategies for the production of 2D materials include gas-phase and solution syntheses. Gas-phase synthesis including chemical vapour deposition[16–18] can produce large-size and high-quality 2D nanosheets on a substrate, but is hindered by a low yield. Solution-based processes, such as liquid exfoliation[12,15,19–21] or chemical synthesis[22–26], are simple and scalable. However, these processes lead to small-size particles and have a higher probability of contamination from the synthetic process. One additional limit for exfoliation is its restriction to layered compounds. Only materials bonded by van der Waals or other weak forces in one direction can be easily cleaved by the intercalants. Therefore, the development of a general and scalable strategy for the synthesis of high-quality 2D oxides, including those with non-layered structures and investigation of their potential for pseudocapacitor applications are important.

Herein, we report a general strategy that combines many of the advantages of vacuum-based and solution-based synthetic methods that can be applied to various oxides, including non-layered compounds. Synthesized 2D hexagonal-MoO_3 (h-MoO_3) is shown to be 300 F cm^{-3} at 5 mV s^{-1} in $Al_2(SO_4)_3$ electrolyte and approaches the theoretical capacitance of the material at a charge/discharge rate of 2 mV s^{-1} in organic electrolyte. For the 2D oxide synthesis, a molecular precursor solution was first prepared and mixed with a large volume of an inorganic salt (Fig. 1a,b, and Supplementary Fig. 1 shows scanning electron microscopy (SEM) images of precursor@salt templates). Next, the solution was dried and the mixture was annealed at an elevated temperature to produce 2D oxides, through an intermediate hydroxide phase in some cases (Supplementary Fig. 2). The salts were dissolved and the nanosheets were assembled into a restacked film after rinsing and filtration (Fig. 1c,d). In this process, the salt microcrystals served as the substrate/template to guide the oxide growth at elevated temperatures. The as-produced 2D oxides exhibited a large lateral size (up to 100 µm), which is an advantage of a vacuum-based synthesis approach. The process can be scaled up by increasing the amount of inexpensive salt used, as evidenced by the large amount of precursors and final products shown in Fig. 1e and Supplementary Fig. 3. The salt can be dissolved in water, leaving behind clean products that can be further directly filtered into binder- and additive-free films for pseudocapacitor electrodes and other applications (Fig. 1d,f). When integrated into an electrochemical device, ions can easily diffuse from the electrolyte into the restacked film to react with the surface active sites, which are fully exposed in these atomically thin 2D oxides. Under these circumstances, the rate-limiting step of ion intercalation and diffusion in bulk materials is largely circumvented, and all of the active material participates in redox reactions, which contribute to electrochemical energy storage.

Results

Synthesis of 2D oxides. Using this strategy, several 2D oxides, including h-MoO_3, h-WO_3, MoO_2, and MnO were successfully synthesized as shown in Fig. 2. On the basis of the SEM and optical images (Supplementary Fig. 4), all of the samples possess 2D morphology with the size of some flakes exceeding 400 μm^2, which is much larger than that of liquid-exfoliated or chemically synthesized samples in which sizes are typically limited to a few micrometres[27]. The dimensions of the salt crystals are tens of micrometres and guide the growth of the 2D oxide to a similar size. The transmission electron microscopy analysis of these oxides (Fig. 2e–h; Supplementary Fig. 5) further confirmed the planar 2D morphology, and the high transparency and observed wrinkles suggest that the as-synthesized oxides are nanosheets. All of the oxide crystals had a thickness of <2 nm, which was determined using atomic force microscopy (AFM) (Fig. 2i–l; Supplementary Fig. 6). Selected area electron diffraction, X-ray diffraction (Supplementary Fig. 7) and X-ray photoelectron spectroscopy (XPS; Supplementary Figs 8 and 9) confirmed the formation of h-MoO_3, h-WO_3, monoclinic MoO_2 and cubic MnO. Moreover, X-ray diffraction characterization also revealed broadened diffraction peaks corresponding to 2D MoO_2 that were shifted to small angles compared with those of bulk MoO_2, suggesting lattice relaxation in the 2D MoO_2 (Supplementary Fig. 10). On the basis of the naturally non-layered crystal structures of these oxides (Supplementary Fig. 11), the suitability of salt-assisted templating as a general method for the synthesis of 2D oxides has been convincingly demonstrated. Despite previous studies using NaCl to synthesize oxide nanocubes[28–30], to the best of our knowledge, the use of inorganic salts to synthesize 2D oxides has not been previously reported.

Growth mechanism. Next, we will discuss the synthesis of MnO, which has a non-layered cubic structure (Supplementary Fig. 11d). First, $Mn(CH_3COO)_2$, which was dissolved in ethanol, was used as the molecular precursor to avoid nucleation

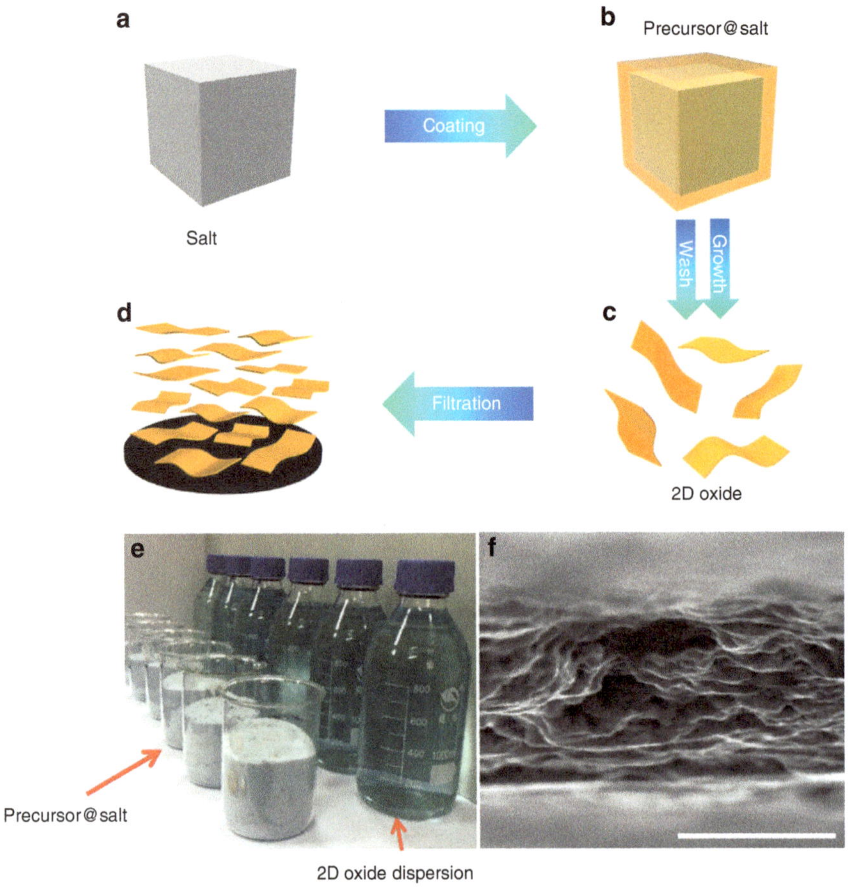

Figure 1 | Schematic representation of 2D *h*-MoO_3 synthesis and electrode fabrication for energy storage. (**a**,**b**) Schematic representation of the fabrication process. The precursor solution was coated on the salt by mixing the solution with a large amount of salt microcrystals. (**c**) After thermal annealing and rinsing in water, 2D oxide flakes were obtained. (**d**) Pseudocapacitor electrodes were fabricated by filtering the 2D oxide/ethanol dispersion using a Celgard separator. (**e**) Optical image of the precursors and 2D *h*-MoO_3-ethonal dispersions. The front row shows six beakers containing the precursor, and the back row shows six bottles containing the 2D *h*-MoO_3-ethanol dispersions. (**f**) Multilayered 2D oxide is shown in the cross-sectional SEM image. Scale bar, 1 μm.

of oxide nanocrystals in the solution. Second, the thickness of the 2D oxide was controlled by mixing a limited volume of the dilute oxide precursor solution with a large quantity of salt microcrystals. Using this approach, the 2D oxide is believed to be confined to a monolayer thickness, and the formation of structurally stable nanocubes[31–34] was avoided because of an insufficient amount of precursor and large-size salt crystals. We measured the size of the salt microcrystals and performed calculations of the precursor-to-salt ratio to meet this requirement (Supplementary Fig. 12; Supplementary Note 1). The AFM analysis indicated that the as-produced 2D MnO crystals were ~1.3 nm thick (Supplementary Fig. 6d). Third, the crystal geometry of the substrate (salt) should be identical to the target oxide to promote Frank–van der Merwe mode film growth and possible heteroepitaxy[35,36]. MnO was cubic with a lattice constant (a_1) of 0.4442 nm. Therefore, KCl was chosen as the template because of its similar cubic structure with a lattice constant (a_2) of 0.3138 nm. Because the crystal symmetry was identical and $\sqrt{2}a_2$ was 0.4437 nm, overlap of the atomic configuration could be achieved by rotating the crystal plane of MnO by 45°. Figure 3a shows the (001) crystal plane of 2D MnO on the (001) plane of KCl. The mismatch between KCl and MnO is only 0.11% (Supplementary Fig. 13). The X-ray diffraction pattern of the mixture of KCl salt and 2D MnO (after annealing but before rinsing) indicated close pairing of the (200) and ($2\bar{2}0$) diffraction peaks of MnO with the ($1\bar{1}0$) and (200) peaks of KCl, confirming our hypothesis (Fig. 3b). The importance of lattice matching is supported by two facts. First, when Na_2SO_4 was chosen as the template salt, no 2D MnO was produced using the same preparation procedure because of the large lattice mismatch between MnO and Na_2SO_4 (Supplementary Figs 13e and 14). Second, when we directly dropped the precursor solution onto the ceramic boat (no template) and annealed it under the same conditions, particles were produced rather than 2D structures (Supplementary Fig. 15a,b).

The same approach was used to synthesize 2D *h*-MoO_3. Mo was dissolved in H_2O_2 to produce the molecular peroxomolybdate precursor solution. A precisely calculated amount of precursor was loaded to restrict the growth of thick nanosheets, and NaCl microcrystals were used as lattice-matched templates. Similar close pairing of the diffraction peaks between *h*-MoO_3 (($1\bar{4}0$) peak) and the NaCl templates (($2\bar{2}0$) peak; cubic $a = 5.62$ Å) was also observed in the X-ray diffraction patterns (Fig. 3d), and the MoO_3 crystal growth on NaCl is shown in the inset and in Fig. 3c. As previously suggested in the literature[37], the formation of *h*-MoO_3 ($a = 10.584$ Å, $c = 3.7278$ Å) rather than α-MoO_3 is most likely due to Na^+ serving as a mineralizer and stabilizer of the hexagonal tunnels in *h*-MoO_3 in addition to the lattice match-induced *h*-MoO_3 nucleation. It is important to note that amorphous MoO_3 particles were obtained when no template

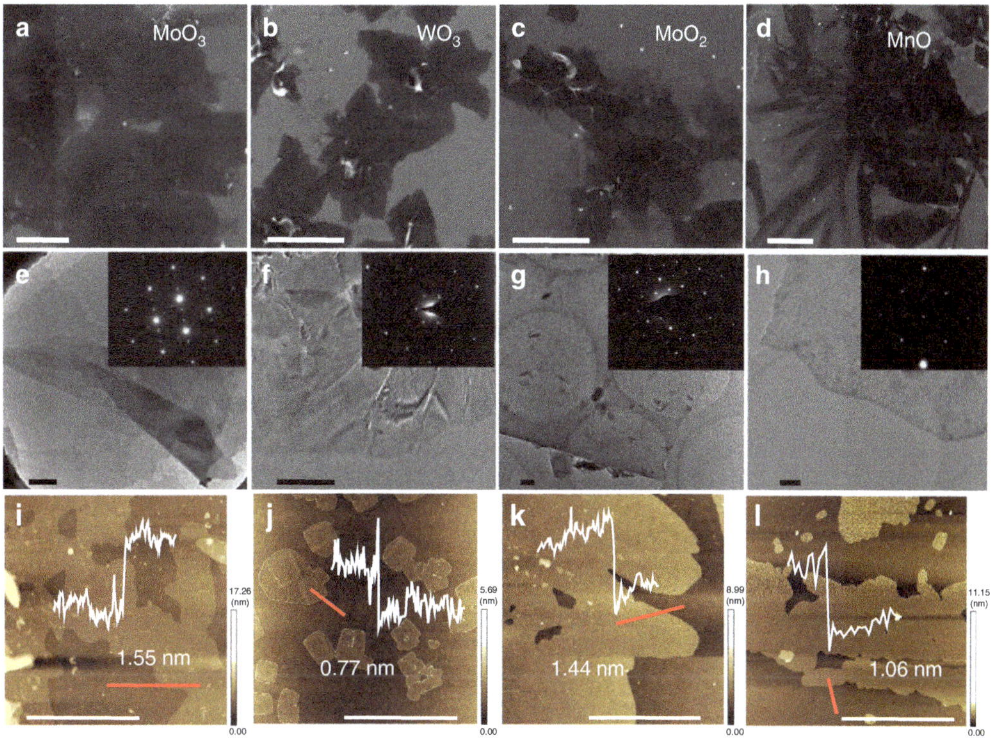

Figure 2 | High-quality 2D oxide synthesized using the salt-assisted templating method. SEM images of *h*-MoO3 (**a**) *h*-WO3 (**b**) MoO2 (**c**) and MnO (**d**). Scale bar, 20 μm (**a**,**b**); 100 μm (**c**); 50 μm (**d**). Low-resolution TEM images of *h*-MoO3 (**e**), *h*-WO3 (**f**), MoO2 (**g**) and MnO (**h**). Scale bar, 500 nm (for all of the TEM images). The insets show the corresponding selected area electron diffraction patterns. AFM images of *h*-MoO3 (**i**), *h*-WO3 (**j**), MoO2 (**k**) and MnO (**l**). The insets show the corresponding thickness of the 2D oxide. Scan lines are shown in red. Scale bar, 1 μm (**i**,**k**); 5 μm (**j**,**l**).

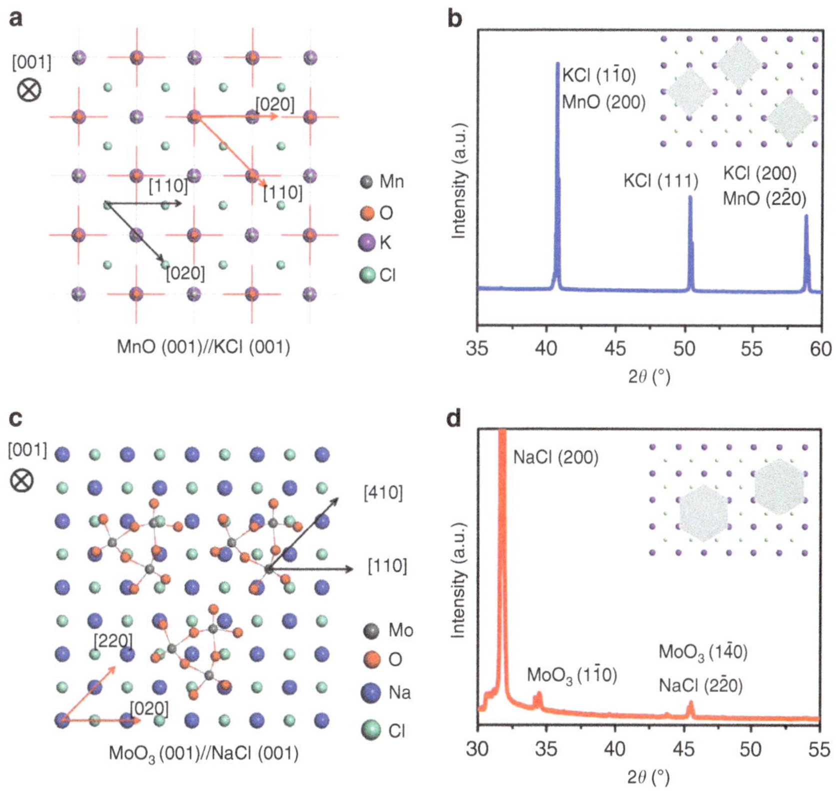

Figure 3 | Proposed mechanism of the salt-assisted templated growth of 2D MnO and *h*-MoO_3. (**a**) Top view of the crystal modes of the MnO (001) plane sitting on the KCl (001) plane. (**b**) X-ray diffraction pattern of 2D MnO on KCl. Please note that the faint diffraction peaks of MnO reside at the left side of KCl diffraction peaks. The inset is a schematic representation of cubic MnO on KCl. (**c**) Top view of the crystal modes of the *h*-MoO_3 (001) plane sitting on the NaCl (001) plane. (**d**) X-ray diffraction pattern of 2D *h*-MoO_3 on NaCl. The inset is a schematic representation of *h*-MoO_3 on NaCl.

was used, further demonstrating the importance of the presence of NaCl (Supplementary Fig. 15c,d). Therefore, we hypothesize that the lateral growth of oxides on the salt surfaces is guided by the salt crystal geometry and promoted by lattice matching, and the growth perpendicular to the salt surface is limited by precursor depletion, which leads to the synthesis of oxide nanosheets with nanometre thickness even for materials with a non-layered crystal structure. In addition, the thickness can be easily controlled by tuning the precursor-to-salt ratio. The thickness increased from 1.6 to ~12 nm when the precursor-to-salt ratio was increased by a factor of 8 (Supplementary Fig. 16). Nonetheless, more characterizations are required to completely elucidate the mechanism of nanosheet growth on salt templates.

Electrochemical performance of 2D *h*-MoO$_3$. To explore the electrochemical performance of a 2D oxide, 2D *h*-MoO$_3$ was selected because MoO$_3$ possesses a high theoretical capacity (1.5 Li/Mo)[38]. Notably, this high theoretical capacity is only achievable at a very low current density (0.01 A g^{-1}) and a slow charging/discharging rate, which is not suitable for electrochemical capacitors[2]. When processed into 2D nanosheets, we expected that the confinement by slow ion diffusion in bulk MoO$_3$ would be relaxed, which would enable fast and reversible surface reactions to achieve high capacitance at high rates to allow for the use of MoO$_3$ in pseudocapacitor applications[39]. Using a three-electrode configuration, six different aqueous electrolytes were explored with different cation sizes (Supplementary Fig. 17a,b), and the results are summarized in Fig. 4a. The highest volumetric capacitance of 600 F cm^{-3} was achieved in H$_2$SO$_4$ at 5 mV s^{-1}, which is larger than the value for the Ti$_3$C$_2$T$_x$ paper in KOH (~400 F cm^{-3})[10] and chemically converted graphene in H$_2$SO$_4$ (256 F cm^{-3})[8]. All of the results indicated superior rate capability, which is crucial for pseudocapacitors. Surprisingly, the performance in the Al$_2$(SO$_4$)$_3$ electrolyte was 300 F cm^{-3} at 5 mV s^{-1}, and to the best of our knowledge, this high capacitance in an Al salt electrolyte has not been previously reported for any material, which opens the door for the use of 2D oxides in Al-ion batteries[40]. Furthermore, the electrochemical impedance spectroscopy results shown in Fig. 4b indicate that the 2D *h*-MoO$_3$ electrode exhibited good capacitive behaviour with no apparent charge transfer resistance, which indicates excellent ionic and electronic conductivity. The enhanced electron conductivity arises from the large dimension (fewer and better contact for each flake in the restacked electrode) and abundant oxygen vacancies present in our 2D *h*-MoO$_3$ (XPS, Raman spectroscopy, photoluminescence characterizations and conductivity measurements; Supplementary Figs 8a, 18, 19 and 20). These properties are intrinsic advantages of the proposed synthesis strategy where the large size is inherited from the NaCl microcrystal templates and the oxygen vacancies originated from thermal annealing at elevated temperatures in an oxygen-free atmosphere. In addition, a control experiment using ~12 nm *h*-MoO$_3$ nanosheets (Supplementary Fig. 17c,d) resulted in a significantly decreased volumetric capacitance compared with that of the 1.6 nm *h*-MoO$_3$ at the same sweep rate, further confirming the effectiveness of nanometre-thin 2D *h*-MoO$_3$ to minimize the restriction of ion diffusion[41]. In addition, the electrochemical performance of the MoO$_3$ particles that were obtained without templates (Supplementary Fig. 21) was investigated, and a capacitance of <50 F cm^{-3} was achieved, which is significantly lower than that for 2D *h*-MoO$_3$ and demonstrates the advantage of the 2D morphology.

Although aqueous electrolytes are widely used, they are inferior to organic electrolytes in terms of the energy density owing to the narrow potential window, which is limited by water splitting. To maximize the energy stored in 2D *h*-MoO$_3$, 2D *h*-MoO$_3$ was tested in a 1 M LiClO$_4$ solution with a 1:1 mixture of ethylene carbonate and dimethyl carbonate. The contact angle measurements indicate that the organic electrolyte provides good wetting of the restacked 2D *h*-MoO$_3$ electrode film (Supplementary Fig. 22). According to the cyclic voltammetry (CV) curves at different sweep rates (Fig. 5a), a pair of wide peaks arising from reduction/oxidation reactions contributed to the pseudocapacitance[41]. The charge–discharge curves at different current densities also indicate pseudocapacitive behaviour (Supplementary Fig. 23a). The volumetric and gravimetric capacities are shown in Fig. 5b within a charging/discharging time that ranged from 10 to 1,000 s. Intriguingly, a gravimetric capacity of 996 C g^{-1} (277 mAh g^{-1}) at 2 mV s^{-1} was much larger than that previously reported for mesoporous α-MoO$_3$ in 1 M LiClO$_4$ with propylene carbonate (~580 C g^{-1}, 161 mAh g^{-1})[41]. When the charging/discharging time was limited to 10 s, the capacity was still 400 C g^{-1}, suggesting that the 2D *h*-MoO$_3$ electrode can provide excellent power handling that is suitable for quick charging of portable or wearable electronics.

The redox reaction can be expressed as follows:

$$MoO_3 + xLi^+ + xe^- \longleftrightarrow Li_xMoO_3 \quad (1)$$

Because the charge of an electron is identical to a lithium ion, the number of Li$^+$ involved in the redox reaction can be determined as follows: $x = QM/mF$, where Q is the stored charge, M is the molecular weight, m is the mass and F is the Faraday constant. As

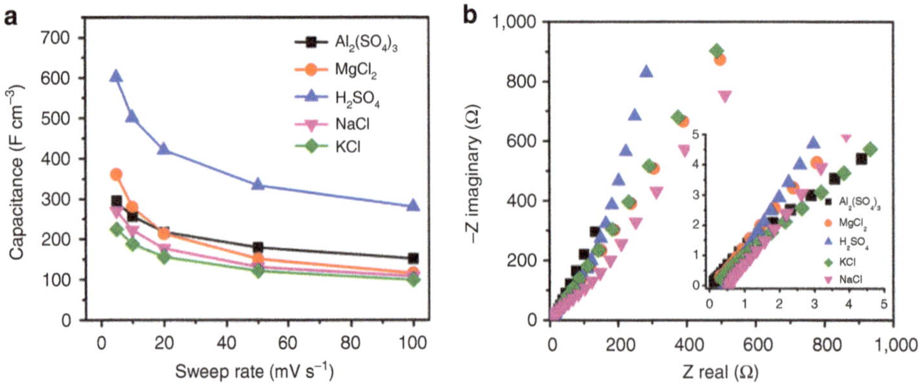

Figure 4 | Electrochemical performance of 2D *h*-MoO$_3$ in aqueous electrolytes. (**a**) Volumetric capacitance as a function of the sweep rate in different aqueous electrolytes. (**b**) Nyquist plots obtained in H$_2$SO$_4$, MgCl$_2$, NaCl, KCl and Al$_2$(SO$_4$)$_3$ electrolytes.

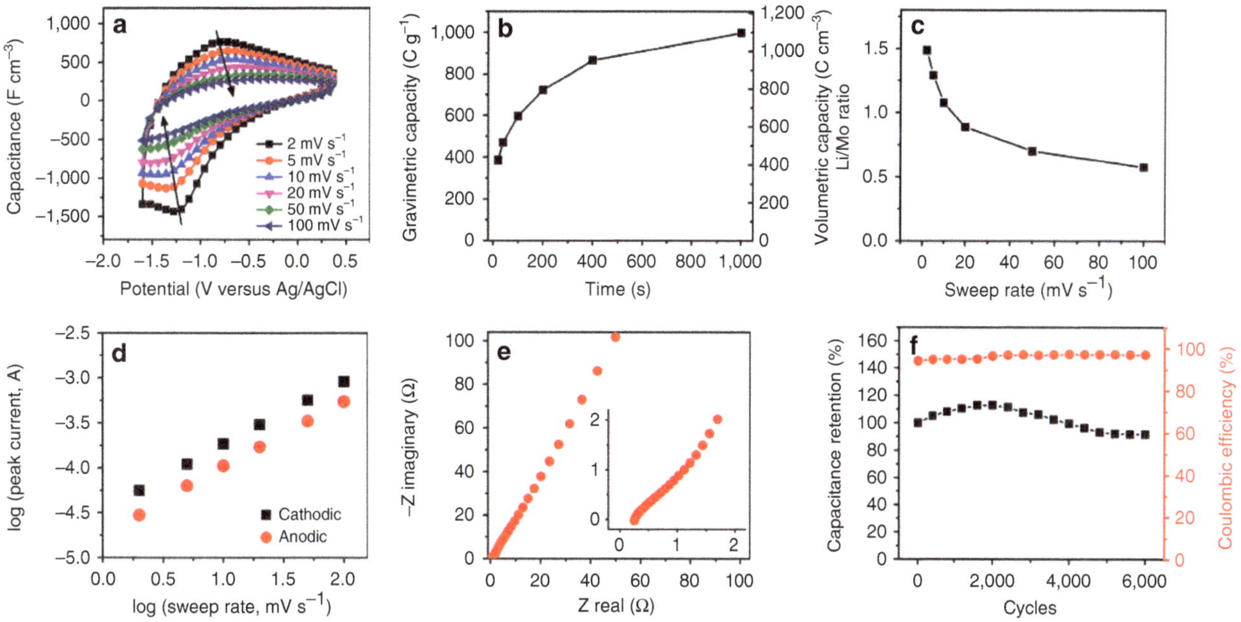

Figure 5 | Electrochemical performance of 2D *h*-MoO$_3$ in a Li-ion-containing organic electrolyte. (**a**) CV curves at sweep rates ranging from 2 to 100 mV s^{-1} indicating the small peak shifts following the arrows. (**b**) Gravimetric and volumetric capacity as a function of charging/discharging time. (**c**) Li/Mo ratio at different sweep rates using $x = QM/mF$, where Q is the stored charge, M is the molecular weight, m is the mass and F is the Faraday constant. After 1,000 s, 2D *h*-MoO$_3$ reaches 1.49 Li/Mo, approaching the theoretical value for bulk α-MoO$_3$ (1.5). (**d**) Plot for calculating *b* values of the cathodic and anodic peaks (both = 0.9). (**e**) Nyquist plots. (**f**) Cycling stability and Coulombic efficiency over 6,000 cycles.

shown in Fig. 5c, 1.49 Li^+ per MoO$_3$ were inserted at 2 mV s^{-1} and 0.58 Li/Mo were inserted at 100 mV s^{-1}. α-MoO$_3$ can accommodate up to 1.5 Li/Mo at a very low current density (0.01 A g^{-1}), and this amount is largely limited to the surface charge storage at high rates. For bulk and even nanostructured materials with thicknesses greater than the atomic level, slow ion diffusion in solids limits the storage capacity and device power[1]. Restacked 2D *h*-MoO$_3$ exhibited a different behaviour because the entire active material was accessible to the electrolyte, allowing for easy Li^+ transport to the surface that facilitated surface redox reactions with the exposed transition metal atoms and led to a high capacitance even at high sweep rates, as shown in Fig. 1e. Therefore, this high capacitance approaches the theoretical capacitance at 2 mV s^{-1}, confirming the advantage of 2D oxides for use in pseudocapacitors.

To gain additional insight into the storage mechanism of the 2D *h*-MoO$_3$ electrode, a peak shift of around 0.1 V was studied at sweep rates ranging from 2 to 100 mV s^{-1} (Fig. 5a). This peak shift indicated that the 2D *h*-MoO$_3$ was involved in the surface-controlled capacitive behaviour[42]. In addition, the relationship between the sweep rate and peak current was also analysed based on the following equation:

$$i = av^b, \tag{2}$$

where i is the peak current (A), v is the sweep rate (mV s^{-1}) and a and b are coefficients[39]. The process is diffusion controlled (battery behaviour) when $b = 0.5$ and is capacitive when $b = 1$. The b values for both the cathodic and anodic peaks were calculated to be 0.9, further supporting the pseudocapacitive storage mechanism (Fig. 5d) that results from the fast and reversible redox reactions of 2D *h*-MoO$_3$ with minimal restriction of the ion diffusion. Therefore, a high capacity of 996 C g^{-1} was obtained at 2 mV s^{-1}. Electrochemical impedance spectroscopy was also performed, and the results are shown in Fig. 5e. No apparent charge transfer resistance was observed, which indicates a combination of excellent ionic and electronic conductivities in the restacked 2D *h*-MoO$_3$ electrodes. The cycling stability of the pseudocapacitor was satisfactory. The initial capacitance increased because of the infiltration of the electrolyte, and 92% of the volumetric capacitance at a sweep rate of 10 mV s^{-1} was maintained after 6,000 cycles (Fig. 5f), which can be further improved by optimizing the electrode architecture. On the basis of the pseudocapacitive mechanism (no phase transformation), these materials are expected to exhibit good cycling stability that exceeds that of Li-ion batteries and simultaneously achieve high power-handling properties and fast charging time.

Discussion

The most significant result of this work is the demonstrated ability to generate high-quality 2D oxides with various compositions. The successful synthesis of various 2D binary transition metal oxides, combined with their outstanding performance in pseudocapacitor applications, will promote the exploration of 2D oxides for energy storage applications and beyond. In addition, achieving a combination of high gravimetric and volumetric capacitances is important for many applications, such as personal electronics, where the volume is limited. In this study, the volumetric capacity of 2D *h*-MoO$_3$ was as high as 1,100 C cm^{-3} at 2 mV s^{-1}, which further indicates the advantage of the 2D morphology. Moreover, in order to compare our results with published data for other materials, the volumetric capacitance and energy/power density of the 2D *h*-MoO$_3$ in a Li-ion electrolyte are plotted in Supplementary Fig. 23b,c. The highest capacitance value was 550 F cm^{-3}, exceeding that for MoS$_2$ in 0.5 M 1-ethyl-3-methylimidazolium tetrafluoroborate (EMIM BF$_4$)/MeCN (~250 F cm^{-3})[11]. In addition, the highest volumetric energy density was 270 mWh cm^{-3} with a power density of 1 W cm^{-3}.

In conclusion, four transition metal oxides with non-layered structures were successfully synthesized with 2D morphologies using salt-assisted templating. On the basis of our experimental observations, we propose that the lateral growth of the 2D oxides was guided by the salt crystal geometry and promoted by lattice matching, and the thickness was restrained by the raw material supply (controlled amount of the dilute solution). The outstanding pseudocapacitive performance of 2D h-MoO_3 indicates the importance of the 2D morphology in energy storage. This material achieved $300\,F\,cm^{-3}$ at $5\,mV\,s^{-1}$ in the $Al_2(SO_4)_3$ electrolyte, which is favourable for the potential use of 2D oxides in Al-ion batteries. For organic electrolytes, capacities as high as $996\,C\,g^{-1}$ (1.49 Li/Mo) and $1,100\,C\,cm^{-3}$ were obtained at a sweep rate of $2\,mV\,s^{-1}$ owing to fast and reversible redox storage, resulting from fast ion diffusion and electron transport in the restacked 2D h-MoO_3 electrodes. Furthermore, the as-synthesized 2D oxides may find applications beyond energy storage, and this synthetic method could potentially be extended to materials other than oxides.

Methods

Synthesis of 2D h-MoO_3 and MoO_2. In total, 0.1 g of Mo powder was dispersed in 20 ml of ethanol with vigorous stirring for 30 min. Then, 0.35 ml of a H_2O_2 (30%) solution was slowly added to the Mo powder suspension, which resulted in a blue solution after stirring for 18 h. Then, the precursor solution was mixed with 320 g of NaCl powder followed by drying at 65 °C with stirring. For h-MoO_3, the as-obtained mixture was further annealed at 280 °C for 1 h at a ramp rate of $2\,°C\,min^{-1}$ under an Ar atmosphere. After being cooled, the product was washed with deionized water to remove the NaCl templates and dispersed in ethanol for further characterization. For MoO_2, the same procedure was applied, but the annealing temperature was 550 °C.

Synthesis of 2D MnO. In total, 0.5 g $Mn(CH_3COO)_2 \cdot 4H_2O$ was dispersed in 25 ml of ethanol with stirring for 5 min to obtain the MnO precursor solution. Then, the precursor solution was mixed with 400 g of potassium chloride power and dried at 70 °C with stirring. The as-obtained mixture was further annealed at 400 °C for 1 h at a ramp rate of $2\,°C\,min^{-1}$ under an Ar atmosphere. After being cooled, the product was washed with deionized water to remove the KCl templates and dispersed in ethanol for further characterization.

Synthesis of 2D h-WO_3. In total, 0.1 g of tungsten powder was dispersed in 20 ml of ethanol with vigorous stirring for 30 min. Then, 1 ml of a H_2O_2 (30%) solution was slowly added to the W powder suspension, which produced a white solution after stirring for 36 h. Then, the precursor solution was mixed with 400 g of KCl powder followed by drying at 70 °C with stirring. The as-obtained mixture was further annealed at 500 °C for 1 h using a ramp rate of $2\,°C\,min^{-1}$ under an Ar atmosphere. After being cooled, the product was washed with deionized water to remove KCl and dispersed in ethanol for further characterization.

Preparation of electrodes. The vacuum-assisted filtration method was used to fabricate the multilayer electrodes. The as-produced 2D oxide/ethanol dispersion was filtered through a membrane (Celgard, USA). The obtained film was dried at room temperature and then punched into circular electrodes. The thickness of the electrodes was 1 μm. The density of each electrode was $\sim 1.1\,g\,cm^{-3}$.

Characterization. The morphology and structure of the samples were studied using field-emission SEM (FE-SEM, FEI Nova 450 Nano), optical microscopy (Nikon), high-resolution transmission electron microscopy (HRTEM, TECNAI), AFM (Shimadzu), X-ray diffraction (X'Pert Pro, PANanalytical), XPS (ESCA-Lab250), Raman spectroscopy and photoluminescence (LabRAM HR800). For the optical microscopy and AFM measurements, the samples were dropped on SiO_2/Si and dried at room temperature. The electron conductivity was measured using a Keithley 4,200. All of the electrochemical tests were carried out using ECLab and CHI660E. For the typical three-electrode set-up, Ag/AgCl (CHI, USA) was used as the reference electrode, YP-50 (Kuraray, Japan) was used as the counter electrode and a Celgard film served as the separator (Celgard, USA). All of the tests were conducted in Swagelok cells (Swagelok, USA). The electrochemical impedance was measured from 1 to 1 MHz with a potential amplitude of 10 mV. The organic electrolyte was fabricated by dissolving $LiClO_4$ in ethylene carbonate/dimethyl carbonate at a concentration of 1 M.

References

1. Simon, P., Gogotsi, Y. & Dunn, B. Where do batteries end and supercapacitors begin? *Science* **343,** 1210–1211 (2014).
2. Augustyn, V., Simon, P. & Dunn, B. Pseudocapacitive oxide materials for high-rate electrochemical energy storage. *Energy Environ. Sci.* **7,** 1597–1614 (2014).
3. Sun, W. M. *et al.* High surface area tunnels in hexagonal WO_3. *Nano Lett.* **15,** 4834–4838 (2015).
4. Mai, L. Q. *et al.* Synergistic interaction between redox-active electrolyte and binder-free functionalized carbon for ultrahigh supercapacitor performance. *Nat. Commun.* **4,** 2923 (2013).
5. Lu, X. *et al.* Improving the cycling stability of metal-nitride supercapacitor electrodes with a thin carbon shell. *Adv. Energy Mater.* **4,** 1300994–1300999 (2013).
6. Béguin, F., Presser, V., Balducci, A. & Frackowiak, E. Carbons and electrolytes for advanced supercapacitors. *Adv. Mater.* **26,** 2219–2251 (2014).
7. Sun, G. *et al.* Hybrid fibers made of molybdenum disulfide, reduced graphene oxide, and multi-walled carbon nanotubes for solid-state, flexible, asymmetric supercapacitors. *Angew. Chem. Int. Ed.* **54,** 4651–4656 (2015).
8. Yang, X., Cheng, C., Wang, Y., Qiu, L. & Li, D. Liquid-mediated dense integration of graphene materials for compact capacitive energy storage. *Science* **341,** 534–537 (2013).
9. Ghidiu, M., Lukatskaya, M. R., Zhao, M. Q., Gogotsi, Y. & Barsoum, M. W. Conductive two-dimensional titanium carbide 'clay' with high volumetric capacitance. *Nature* **516,** 78–81 (2014).
10. Lukatskaya, M. R. *et al.* Cation intercalation and high volumetric capacitance of two-dimensional titanium carbide. *Science* **341,** 1502–1505 (2013).
11. Acerce, M., Voiry, D. & Chhowalla, M. Metallic 1T phase MoS_2 nanosheets as supercapacitor electrode materials. *Nat. Nanotechnol.* **10,** 313–318 (2015).
12. Nicolosi, V., Chhowalla, M., Kanatzidis, M. G., Strano, M. S. & Coleman, J. N. Liquid exfoliation of layered materials. *Science* **340,** 1420–1438 (2013).
13. Conway, B. E., Birss, V. & Wojtowicz, J. The role and utilization of pseudocapacitance for energy storage by supercapacitors. *J. Power Sources* **66,** 1–14 (1997).
14. Liu, Z., Xu, K., Sun, H. & Yin, S. One-step synthesis of single-layer MnO_2 nanosheets with multi-role sodium dodecyl sulfate for high-performance pseudocapacitors. *Small* **11,** 2182–2191 (2015).
15. Hanlon, D. *et al.* Production of molybdenum trioxide nanosheets by liquid exfoliation and their application in high-performance supercapacitors. *Chem. Mater.* **26,** 1751–1763 (2014).
16. Lee, Y. H. *et al.* Synthesis of large-area MoS_2 atomic layers with chemical vapor deposition. *Adv. Mater.* **24,** 2320–2325 (2012).
17. Li, X. *et al.* Large-area graphene single crystals grown by low-pressure chemical vapor deposition of methane on copper. *J. Am. Chem. Soc.* **133,** 2816–2819 (2011).
18. van der Zande, A. M. *et al.* Grains and grain boundaries in highly crystalline monolayer molybdenum disulphide. *Nat. Mater.* **12,** 554–561 (2013).
19. Coleman, J. N. *et al.* Two-dimensional nanosheets produced by liquid exfoliation of layered materials. *Science* **331,** 568–571 (2011).
20. Paton, K. R. *et al.* Scalable production of large quantities of defect-free few-layer graphene by shear exfoliation in liquids. *Nat. Mater.* **13,** 624–630 (2014).
21. Zheng, J. *et al.* High yield exfoliation of two-dimensional chalcogenides using sodium naphthalenide. *Nat. Commun.* **5,** 2995 (2014).
22. Huang, X. *et al.* Freestanding palladium nanosheets with plasmonic and catalytic properties. *Nat. Nanotechnol.* **6,** 28–32 (2011).
23. Du, Y. *et al.* A general method for the large-scale synthesis of uniform ultrathin metal sulphide nanocrystals. *Nat. Commun.* **3,** 1177 (2012).
24. Son, J. S. *et al.* Large-scale soft colloidal template synthesis of 1.4 nm thick CdSe nanosheets. *Angew. Chem. Inter. Ed.* **121,** 6993–6996 (2009).
25. Sun, Z. *et al.* Generalized self-assembly of scalable two-dimensional transition metal oxide nanosheets. *Nat. Commun.* **5,** 3813 (2014).
26. Tan, C. & Zhang, H. Wet-chemical synthesis and applications of non-layer structured two-dimensional nanomaterials. *Nat. Commun.* **6,** 7873 (2015).
27. Xu, M., Liang, T., Shi, M. & Chen, H. Graphene-like two-dimensional materials. *Chem. Rev.* **113,** 3766–3798 (2013).
28. Liu, R. *et al.* Sodium chloride template synthesis of cubic tin dioxide hollow particles for lithium ion battery applications. *ACS Appl. Mater. Interfaces* **4,** 1537–1542 (2012).
29. Wang, B. *et al.* Synthesis of NaCl single crystals with defined morphologies as templates for fabricating hollow nano/micro-structures. *RSC Adv.* **5,** 5072–5076 (2015).
30. Li, N. *et al.* Free-standing and transparent graphene membrane of polyhedron box-shaped basic building units directly grown using a nacl template for flexible transparent and stretchable solid-state supercapacitors. *Nano Lett.* **15,** 3195–3203 (2015).
31. Tang, Y. *et al.* Efficient Ag@AgCl cubic cage photocatalysts profit from ultrafast plasmon-induced electron transfer processes. *Adv. Func. Mater.* **23,** 2932–2940 (2013).
32. Takenaka, S., Miyake, S., Uwai, S., Matsune, H. & Kishida, M. Preparation of metal oxide nanofilms using graphene oxide as a template. *J. Phys. Chem. C* **119,** 12445–12454 (2015).
33. Lee, J. M., Kim, I. Y., Han, S. Y., Kim, T. W. & Hwang, S. J. Graphene nanosheets as a platform for the 2D ordering of metal oxide nanoparticles:

mesoporous 2D aggregate of anatase TiO_2 nanoparticles with improved electrode performance. *Chemistry* **18,** 13800–13809 (2012).
34. Cao, H., Zhou, X., Zheng, C. & Liu, Z. Two-dimensional porous micro/nano metal oxides templated by graphene oxide. *ACS Appl. Mater. Interfaces* **7,** 11984–11990 (2015).
35. Kuykendall, T. *et al.* Crystallographic alignment of high-density gallium nitride nanowire arrays. *Nat. Mater.* **3,** 524–528 (2004).
36. Lee, W. C. *et al.* Graphene-templated directional growth of an inorganic nanowire. *Nat. Nanotechnol.* **10,** 423–428 (2015).
37. Zheng, L., Xu, Y., Jin, D. & Xie, Y. Novel metastable hexagonal MoO_3 nanobelts: synthesis, photochromic, and electrochromic properties. *Chem. Mater.* **21,** 5681–5690 (2009).
38. Tsumura, T. & Inagaki, M. Lithium insertion/extraction reaction on crystalline MoO_3. *Solid State Ionics* **104,** 183–189 (1997).
39. Augustyn, V. *et al.* High-rate electrochemical energy storage through Li^+ intercalation pseudocapacitance. *Nat. Mater.* **12,** 518–522 (2013).
40. Lin, M. C. *et al.* An ultrafast rechargeable aluminium-ion battery. *Nature* **520,** 325–328 (2015).
41. Brezesinski, T., Wang, J., Tolbert, S. H. & Dunn, B. Ordered mesoporous α-MoO_3 with iso-oriented nanocrystalline walls for thin-film pseudocapacitors. *Nat. Mater.* **9,** 146–151 (2010).
42. Conway, B. E. Transition from 'supercapacitor' to 'battery' behavior in electrochemical energy storage. *J. Electrochem. Soc.* **138,** 1539–1548 (1991).

Acknowledgements

This work was financially supported by the National Natural Science Foundation of China (grants 51322210, 61434001, 61274055 and 61322401) and Director Fund of WNLO. J.Z. wishes to thank the National Program for Support of Top-notch Young Professionals for financial support. J.T. wishes to thank the National 1,000 Young Talents project for financial support. Y.G. was supported by the US Department of Energy, Office of Electricity, Energy Storage Systems Program through Sandia National Laboratory. We wish to thank Kathleen Maleski at Drexel University for helpful comments on the manuscript. We also wish to thank Dr D. Yong and Professor Z.L. Wang from Georgia Institute of Technology for helpful discussions on the growth mechanism. In addition, we wish to thank facility support from the Center for Nanoscale Characterization & Devices, WNLO of Huazhong University of Science and Technology (HUST) and the Analytical and Testing Center of HUST.

Author contributions

J.Z. and J.T. conceived the idea. X.X., H.S., J.Z., J.T. and Y.G. designed the experiments. X.X., H.S., S.L., X.Z. and J.S. carried out the experiments. X.X., H.S., J.Z., J.T., and Y.G. analysed the data. Y.Z. and B.Y. assisted in analysis. Z.H. assisted with the X-ray diffraction measurement. T.L. assisted in the analysis of the electrochemical performance. Q.Z. assisted with the AFM measurements. L.J. provided helpful suggestions for analysis and manuscript revision. X.X., H.S., J.Z., J.T. and Y.G. wrote the manuscript. All of the authors commented on the manuscript.

Additional information

Competing financial interests: The authors declare no competing financial interests.

27

Giant photoluminescence enhancement in tungsten-diselenide-gold plasmonic hybrid structures

Zhuo Wang[1,2,3], Zhaogang Dong[4], Yinghong Gu[5], Yung-Huang Chang[6], Lei Zhang[5], Lain-Jong Li[7], Weijie Zhao[2], Goki Eda[2,8,9], Wenjing Zhang[2,10], Gustavo Grinblat[3], Stefan A. Maier[3], Joel K.W. Yang[4,11], Cheng-Wei Qiu[1,5] & Andrew T.S. Wee[1,2,9]

Impressive properties arise from the atomically thin nature of transition metal dichalcogenide two-dimensional materials. However, being atomically thin limits their optical absorption or emission. Hence, enhancing their photoluminescence by plasmonic nanostructures is critical for integrating these materials in optoelectronic and photonic devices. Typical photoluminescence enhancement from transition metal dichalcogenides is 100-fold, with recent enhancement of 1,000-fold achieved by simultaneously enhancing absorption, emission and directionality of the system. By suspending WSe_2 flakes onto sub-20-nm-wide trenches in gold substrate, we report a giant photoluminescence enhancement of ~20,000-fold. It is attributed to an enhanced absorption of the pump laser due to the lateral gap plasmons confined in the trenches and the enhanced Purcell factor by the plasmonic nanostructure. This work demonstrates the feasibility of giant photoluminescence enhancement in WSe_2 with judiciously designed plasmonic nanostructures and paves a way towards the implementation of plasmon-enhanced transition metal dichalcogenide photodetectors, sensors and emitters.

[1] NUS Graduate School for Integrative Sciences & Engineering (NGS), National University of Singapore, 28 Medical Drive, Singapore 117456, Singapore. [2] Department of Physics, National University of Singapore, 2 Science Drive 3, Singapore 117542, Singapore. [3] Department of Physics, Imperial College London, London SW7 2AZ, UK. [4] Institute of Materials Research and Engineering, A*STAR (Agency for Science, Technology and Research), 3 Research Link, Singapore 117602, Singapore. [5] Department of Electrical and Computer Engineering, National University of Singapore, 4 Engineering Drive 3, Singapore 117583, Singapore. [6] Department of Electrophysics, National Chiao Tung University, Hsinchu 30010, Taiwan. [7] Physical Science and Engineering Division, King Abdullah University of Science and Technology, Thuwal 23955-6900, Kingdom of Saudi Arabia. [8] Department of Chemistry, National University of Singapore, 3 Science Drive 3, Singapore 117543, Singapore. [9] Centre for Advanced 2D Materials, National University of Singapore, 2 Science Drive 3, Singapore 117542, Singapore. [10] SZU-NUS Collaborative Innovation Center for Optoelectronic Science & Technology, Key Laboratory of Optoelectronic Devices and Systems of Ministry of Education and Guangdong Province, College of Optoelectronic Engineering, Shenzhen University, Shenzhen 518060, China. [11] Singapore University of Technology and Design, 8 Somapah Road, Singapore 487372, Singapore. Correspondence and requests for materials should be addressed to J.K.W.Y. (email: joel_yang@sutd.edu.sg) or to C.-W.Q. (email: chengwei.qiu@nus.edu.sg) or to A.T.S.W. (email: phyweets@nus.edu.sg).

Crystalline monolayers of transition metal dichalcogenides (TMDCs) are direct bandgap two-dimensional (2D) material semiconductors that are promising as light-active materials for optoelectronic applications[1,2]. Recently, TMDCs have shown great potential in ultrafast and ultrasensitive photodetectors and as ultrathin light absorbers and emitters[3–5]. However, their application in photonic devices is limited by their low absolute photoluminescence (PL) caused by low quantum efficiency and weak absorption[6]. Fortunately, significant enhancement is afforded by integrating noble metal nanostructures supporting localized surface plasmons[7–13] and propagating surface plasmon polaritons (SPPs)[14,15].

Tungsten diselenide is a promising material in creating electrically excited light-emitting diodes[1] and heterojunctions with MoS_2 (ref. 5). It proves to be a promising emitter at ~750 nm wavelength with a quantum yield that is 2 orders of magnitude larger than that of MoS_2 (refs 16,17). However, PL enhancements from WSe_2-plasmonic hybrid nanostructure have yet to be investigated. Recently reported enhancements in MoS_2 are potentially close to the ultimate limit that can be achieved from this material[9,18]. Although it is generally more challenging to achieve large enhancements from systems with a higher quantum yield, it is important to explore the ultimate limit in the PL enhancements that can be achieved in WSe_2.

Here we report the surprisingly large PL enhancement factor (EF) of ~20,000-fold from a single crystal monolayer of WSe_2. WSe_2-gold plasmonic hybrid nanostructures were created by suspending the monolayer over sub-20-nm-wide trenches that support lateral gap plasmons. By systematically tuning the gap plasmon resonances to the 633-nm pump laser, a close correlation was observed between the local near-field intensity, the Purcell factor and the measured PL enhancements. This paper sheds light on the utility of lateral gap plasmons in promoting excitation at the pump laser wavelength and emission at 750 nm, providing an effective way to obtain giant PL in TMDCs. Distinct from previous plasmon-enhanced structures[8,9,12,13], our hybrid plasmon-enhanced design provides full access to the top surface of WSe_2, for example, for layering of other 2D materials, electrical top contacts, chemical doping or optical waveguiding.

Results

WSe_2-plasmonic hybrid nanostructures. A schematic of the investigated system is shown in Fig. 1a. We transferred chemical vapour deposition (CVD)-grown WSe_2 monolayer flakes (Supplementary Note 1) onto gold substrates onto which trenches as narrow as sub-20 nm have been patterned. This sample enables a direct comparison of PL emission from the same single crystal WSe_2 flake, where regions over unpatterned gold were taken as the base reference. The gold substrate was prepared by a recently reported template-stripping method based on nano-patterned silicon templates (Supplementary Fig. 1 and Supplementary Note 2)[19]. These structures have advantages over other commonly used gold nano-bowties, nano-rods or nano-discs made by electron-beam lithography (EBL)[8,12,13], where template-stripping produces ultra-smooth surfaces, critical for reducing damping losses due to the reduction of plasmon scattering[20,21]. Moreover, we do template-stripping right before the transfer process of WSe_2 so as to reduce contamination from adsorbed air-borne molecular species and preserve the hydrophilic nature of freshly stripped gold[22].

The trenches in gold support strong field enhancements as shown in the finite-difference time-domain (FDTD) simulation results in Fig. 1b, where a monolayer WSe_2 flake is placed over the trench. The resonances are in the form of lateral gap plasmons in the sub-20 nm trenches with **E**-fields predominantly parallel to the plane of WSe_2 to promote strong light absorption[23]. This resonance can be tuned to be matched with the pump laser wavelength by varying the pitch of the structures, which are critical for plasmon coupling with light to achieve optimal field confinement[24,25]. Figure 1c presents a scanning electron micrograph (SEM) image of a triangular WSe_2 single crystal monolayer as transferred onto the gold substrate with a trench width of ~12 nm.

Observation of PL enhancement in WSe_2. Here we show that the lateral gap plasmons are able to enhance the PL emission significantly. A template-stripped gold substrate with a large pitch (760 nm) and 532-nm pump laser were chosen for this purpose so that the enhanced PL due to lateral gap plasmons in the trench could be spatially resolved. Figure 2a,b presents the SEM image and the corresponding PL mapping across the same WSe_2 flake, so that it enables us to have a self-consistent comparison of PL enhancement from a single flake, thus avoiding potential variations in PL emission between different flakes. The PL experiment was carried out using a 532-nm pump laser with a power of 30 μW and the intensity value at each pixel of Fig. 2b was obtained by integrating the PL spectrum across the spectrum window of 700–820 nm. This PL mapping shows that the sub-20 nm trenches on gold substrate were able to enhance the PL emission considerably.

Figure 2c presents a quantitative comparison of PL spectra from WSe_2 on patterned gold nanostructures (that is, Point A), unpatterned gold film (that is, Point B) and on sapphire. In this particular sample with a pitch of 760 nm, the 9-nm-wide trenches enhance the PL emission from WSe_2 up to 37-fold as compared with the emission from WSe_2 on sapphire. Moreover, PL emission from WSe_2 on unpatterned gold film is enhanced by sevenfold as compared with the emission from WSe_2 on sapphire (Fig. 2d), which might be due to the substrate-induced doping so as to reduce the non-radiative decay rates of the excitonic transitions[26]. In addition, the change in doping level could be induced from the downshift of Raman peak in WSe_2 when transferred from sapphire onto gold substrate[27] (Supplementary Fig. 2c). The weaker PL enhancement at Point C is due to the tear defect in the flake that is observable in the SEM image.

PL enhancement mechanism. The physical processes involved in the PL enhancement are investigated next. The PL intensity of emitters is determined by its excitation rate (that is, absorption of the pump laser) and its emission efficiency (that is, Purcell factor). When the plasmon resonance of WSe_2-gold nanostructures matches the wavelength of the pump laser, the excitation rate of WSe_2 will be enhanced[28], that is, $\gamma_{exc} \sim |\mathbf{E}_{NF}|^2$ (near-field intensity enhancement). In addition, the plasmon resonance at the emission wavelength enhances the radiative decay rate and quantum efficiency via the Purcell effect[29,30].

The largest PL enhancement was achieved at the maximum of the product between the field EF at the excitation wavelength and the Purcell factor at the emission wavelength. Figure 3a presents the relative reflectance spectra as measured from gold nanostructures with different pitches, as normalized to the unpatterned gold film (see Methods). The plasmon resonance at the dips of the spectra red-shifted with increasing pitch. FDTD simulations were done for each nanostructure pitch, with nanostructure geometry adjusted to match the SEM images (Fig. 3c). The simulations show an excellent agreement with the measured spectra, as seen in Fig. 3a. Moreover, the dependence of the plasmon resonance on the pitch size is summarized in Fig. 3b, where the plasmon resonance red-shifts from 507 to 633 nm as the pitch size is increased from 60 to 200 nm. Resonance

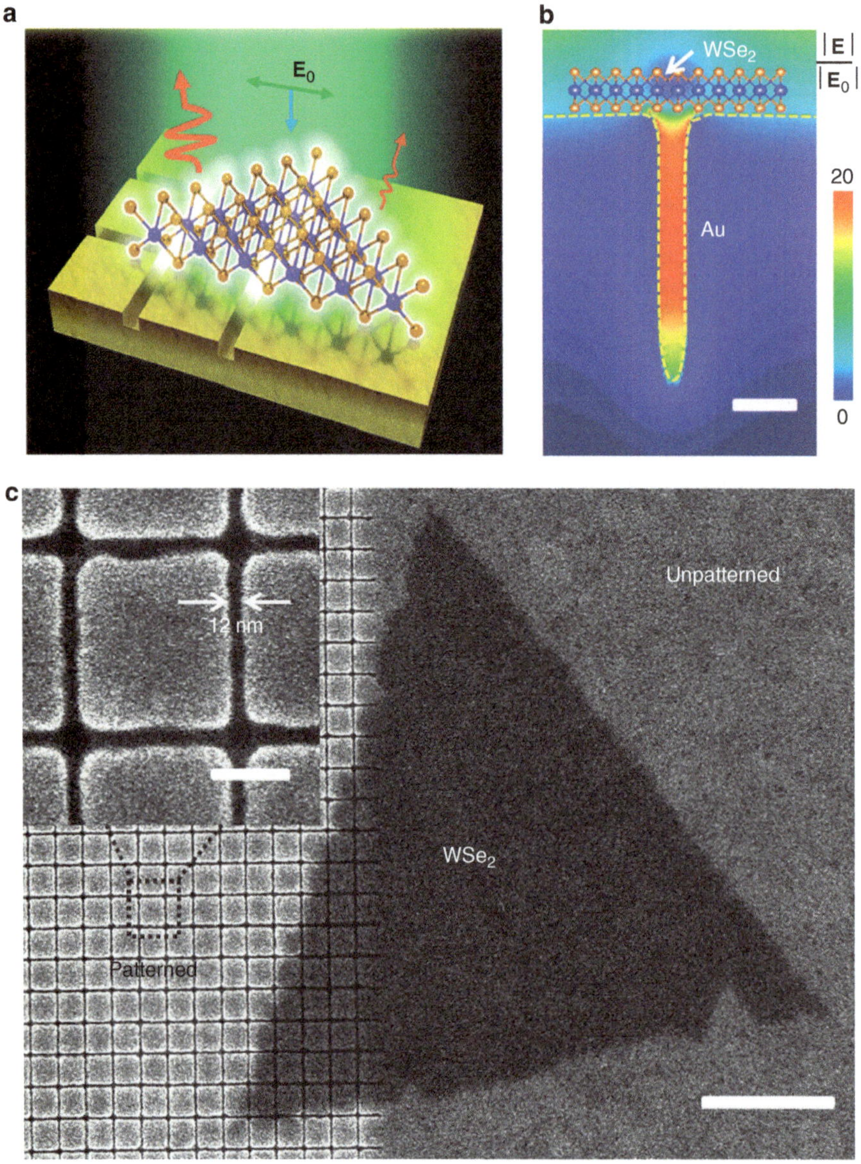

Figure 1 | Schematic of WSe$_2$-gold plasmonic hybrid structure with strong optical absorption. (**a**) Schematic of PL emission from a single crystal monolayer of WSe$_2$ flake on a gold substrate. Part of the triangular flake rests on the patterned region of the substrate consisting of sub-20-nm-wide trenches. (**b**) Representative simulation of the electric field distribution of the lateral gap plasmons with a WSe$_2$ monolayer flake suspended over a single trench. The polarization of the incident laser field is across the gap. The dashed yellow line denotes the boundary between air and gold. The scale bar, 20 nm. (**c**) Representative Scanning electron micrograph (SEM) image of WSe$_2$ on square arrays with a trench width of 12 nm (Patterned) and unpatterned gold film (Unpatterned). The scale bars in the main figure and the inset, 1 μm and 100 nm, respectively.

wavelengths beyond 633 nm were not obtained in our patterned nanostructures as the resonance wavelength for nanostructures with pitches >200 nm exhibited a blue shift due to a shallower trench depth as caused by imperfections in the fabrication process (see Supplementary Fig. 3 and Supplementary Note 3). Figure 3c presents the SEM images of the patterned trenches of different pitches. Well-defined square structures are observed in the patterned gold nanostructures for large pitch sizes of 140–200 nm, while the shape evolves into rounded disks as the pitch was reduced to 60 nm. This effect is due to the nanofabrication process, that is, proximity effects in EBL and micro-loading effects during etching that are more pronounced when the pitch is reduced.

Both 532-nm and 633-nm pump lasers were used to investigate the effect of plasmon-enhanced PL emission from monolayer WSe$_2$. Figure 4a,b presents the corresponding PL spectra as extracted from the PL mapping. We note that the 633-nm pump laser gives a higher PL emission as compared with the 532-nm pump laser. Despite the higher intrinsic absorption at 532 nm as compared with that at 633 nm in WSe$_2$, excitation at 532 nm resulted in low field enhancements (Supplementary Fig. 4a,b) and low PL enhancements (Supplementary Fig. 4d), due to inter-band transitions of electrons in gold at 532 nm (ref. 31). Figure 4c presents the experimental PL EF (corrected by the trench area fraction) excited by the 633-nm laser and the product of the simulated EF of near-filed intensity and Purcell factor (see Methods) as a function of pitch size. Despite the higher density of trenches on the smaller pitch structures, on-resonance excitation with the pitch of 200 nm results in the strongest near-field intensity and Purcell factor enhancements, which were computed 1 nm above the gold substrate, that is, at the expected z-plane of WSe$_2$. The s.d. in Fig. 4c was obtained by statistically analysing 10 spectra

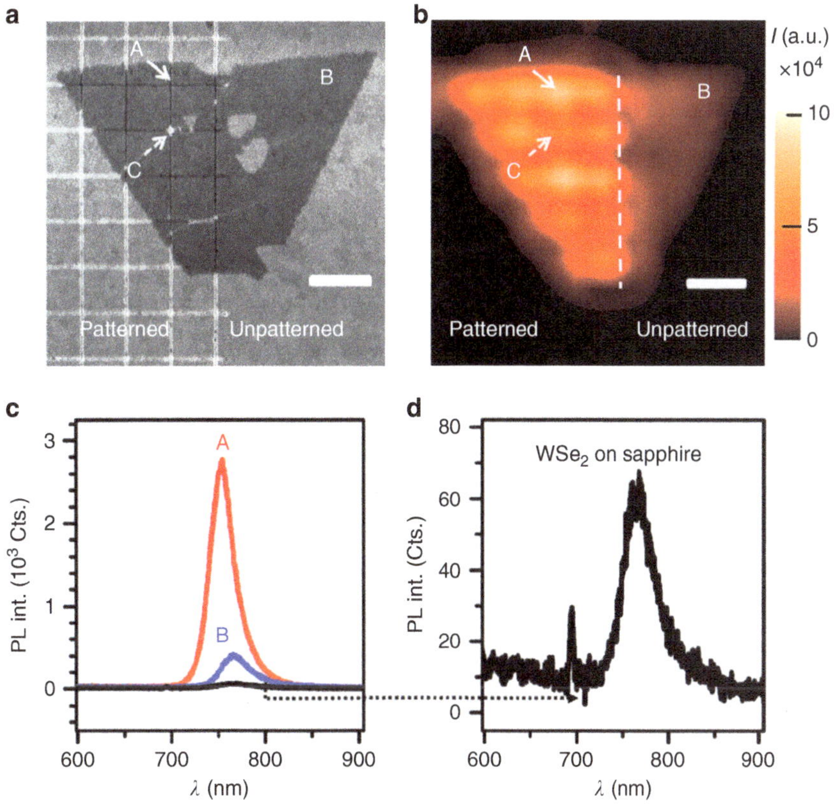

Figure 2 | Characterization results of WSe_2-gold plasmonic hybrid structure. (**a**) SEM image of a crystalline WSe_2 monolayer flake transferred onto a template-stripped substrate with a pitch of 760 nm. 'A' points to a portion of WSe_2 suspended above an intersection of two underlying trenches, while 'B' corresponds to the reference point, that is, WSe_2 on unpatterned smooth gold. 'C' points to a tear defect in the monolayer. The scale bar, 1 μm. (**b**) PL intensity (I) mapping on the WSe_2-gold plasmonic hybrid structure showing larger signals from patterned regions and resolvable modulations in intensity. The intensity value at each pixel was obtained by integrating the PL spectrum across the spectrum window of 700–820 nm. A 532-nm pump laser was chosen here for a fine PL mapping resolution. (**c**) PL spectra from WSe_2 on patterned gold nanostructures (A), unpatterned gold film (B) and the one from as-grown WSe_2 on sapphire. Their PL peak energy (full-width-at-half-maximum) are 1.65 eV (63 meV), 1.63 eV (78 meV) and 1.61 eV (83 meV), respectively. (**d**) Zoom-in PL spectrum of WSe_2 on sapphire. The laser power is 30 μw and integration time is 0.5 s.

of WSe_2 on each patterned gold nanostructure. Variations could be attributed to the slight trench size and shape variations of the patterned gold nanostructures, or optical non-uniformity among different triangular single crystals of WSe_2.

As seen in Fig. 4c, the maximum PL enhancement was achieved at WSe_2 on a nanostructure with the pitch of 200 nm. With the gap plasmon resonance tuned to the 633-nm pump laser wavelength, the integrated PL intensity from WSe_2 on this gold nanostructure was enhanced up to 1,810-fold (without correction by the trench area fraction) compared with the reference on unpatterned gold film (see Supplementary Fig. 5a,b). Due to the small pitch size of 200 nm, the modulation of WSe_2 PL intensity by the array of trenches could no longer be resolved in the PL mapping (see the inset of Fig. 4c). From Figs 1b and 5b as shown later on, we assume that the enhancements are still localized to WSe_2 at the trenches, where this assumption will be supported by the PL mappings as shown in Fig. 5d,f later on. As measurements here were done with the pump laser polarized along the x-axis, fields are confined only within the trenches along the y-axis. Correcting for the small area occupied by these trenches, we obtain the maximum effective PL EF of ~20,000 in WSe_2 over the trench using the formula[9]:

$$<\mathrm{PL\,EF}> = \frac{I_{\mathrm{patterned}}}{I_{\mathrm{unpatterned}}}\frac{A_0}{A_{\mathrm{gap}}}, \quad (1)$$

where $I_{\mathrm{patterned}}$ is the PL intensity from WSe_2 on the patterned gold nanostructure and $I_{\mathrm{unpatterned}}$ is the PL intensity from WSe_2 on the unpatterned gold film. A_0 represents the excitation area of the laser spot size ($\pi \times 600^2\,\mathrm{nm}^2$) and A_{gap} represents the area of the trenches ($1.0 \times 10^5\,\mathrm{nm}^2$) perpendicular to the polarization direction of the laser within the laser spot (see detailed calculation in Supplementary Fig. 6 and Supplementary Note 4). To the best of our knowledge, this magnitude of enhancement has not been previously observed from 2D materials[7–15].

Such a giant EF for the measured PL emission is due to the plasmon-enhanced excitation process and the plasmon-enhanced emission process. To be more specific, as shown in Supplementary Fig. 4b, the plasmonic nanostructure is able to enhance the excitation intensity by ~310-fold, at the pump laser wavelength of 633 nm, when the pitch size is 200 nm. Furthermore, as shown in Supplementary Fig. 4c, the Purcell factor on this 200-nm-pitch nanostructure is able to be enhanced by ~307-fold, as compared with the one on unpatterned gold film. Therefore, the experimental EF of PL emission could be predicted by the product of these two EFs, which is calculated to be more than 9×10^4, as shown in Fig. 4c for the pitch size of 200 nm. In addition, we would like to mention that it is unlikely that phase transition of WSe_2 is at play here, although there is direct contact between WSe_2 and gold plasmonic structures. The semiconducting 2H to metallic 1T phase transition, which has only been observed in high-vacuum and low-temperature conditions[32], has associated Raman and PL quenching effects[33]. However, we observed strong PL enhancement instead of quenching.

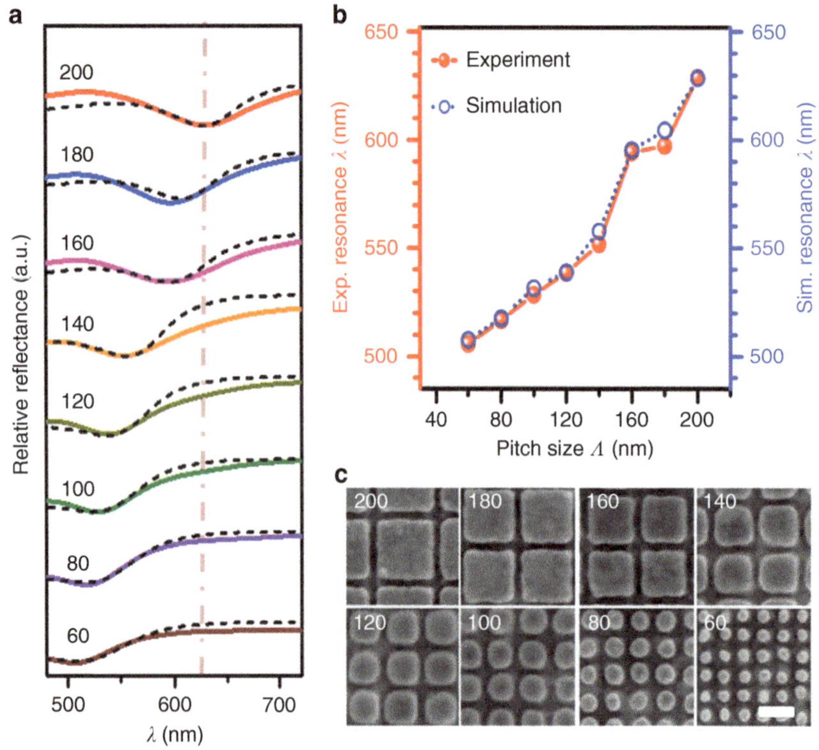

Figure 3 | Optical characteristics of patterned gold nanostructures with different pitch sizes. (**a**) Relative reflectance spectra of patterned gold nanostructures with pitch sizes of 60–200 nm with respect to the unpatterned gold film. The solid and dashed curves present the experimental and simulated spectra, respectively. Legends denote the pitch size in units of nm for each pattern. The dash dot line indicates the position of the pump laser of 633 nm. (**b**) Experimental (Exp.) and simulated (Sim.) resonance wavelength of the patterned gold nanostructures as a function of pitch size. (**c**) SEM images of the template-stripped gold nanostructures with respective pitch sizes. The inset numbers denote the pitch sizes in the unit of nm. The scale bar in the SEM images, 100 nm.

Polarization dependence of PL emission. Next, we investigate the polarization-dependent characteristics of the PL emission from WSe_2 on patterned gold nanostructures. To resolve the PL-enhanced regions, we chose a 920-nm-pitch substrate and a 532-nm pump laser to do the PL mapping. Here we choose the 532-nm pump laser to do the PL mapping simply because the 532-nm pump laser has a smaller spot size due to the shorter wavelength, as compared with the 633-nm pump laser. As a result, the imaging resolution of the PL mapping as obtained by the 532-nm pump laser is much clearer to show the enhanced PL emissions from trench regions. Figure 5a shows an SEM image of the sample. Figure 5b,c presents the simulated electric field distributions of lateral gap plasmons at the intersection of two trenches with 0°- and 45°-polarizations for the excitation. The simulated electric field distributions of this nanostructure agree with the experimental PL mappings in Fig. 5d–f. The maximum field enhancement occurs in trenches perpendicular to the polarization direction (Fig. 5b) and at the intersection of the trenches under 45°-polarization (Fig. 5c). A plot of PL intensity intersecting several trenches indicates a laser beam waist of ~600 nm (see Supplementary Fig. 7). These results provide further evidence that the highest PL enhancements are from WSe_2 directly above the trenches, where the optical fields are localized. It is noted that when the pitch size is larger than the pump laser wavelength, both lateral gap plasmons at the trenches and SPP on the flat squares are excited. However, the intensity of the SPP is only 1/6 of that of the lateral gap plasmon (see Fig. 5b,c). Thus, the contribution to the PL enhancement of WSe_2 still dominantly stems from the lateral gap plasmons supported by the trenches. Figure 5g,h presents the PL and Raman spectra of WSe_2 on the intersection and on the square centre, respectively. Both Raman and PL intensities of WSe_2 at the intersection are found to be larger than the ones at the square centre. The un-shifted Raman peak of WSe_2 over trenches and square centre suggests that no strain is induced by the trench and thus it also confirms that the observed PL enhancement was not induced by strain[34].

Discussion

Furthermore, the WSe_2-gold plasmonic hybrid nanostructure allows us to conveniently evaluate a recent claim in the literature[35], which states that suspended 2D materials are expected to have larger PL than the counterparts on a substrate, because the substrate depopulates the density of states in 2D materials through charger transfer. Nevertheless, as shown in Fig. 5d,f, the gaps parallel to the pump laser polarization do not exhibit any stronger PL over its surrounding, although WSe_2 is suspended here. Therefore, it suggests that the suspension nature of WSe_2 by the trenches contributes little to the PL enhancement.

It should be noted here that the square geometry chosen has the advantage of being independent of the incident polarization. In other words, if the incident polarization is either purely *x*-polarized or *y*-polarized, we will achieve the same PL EF. Therefore, for any other polarizations in-between, we always could decompose the incident optical field into the respective *x*-polarized component and *y*-polarized component. Since the PL emission intensity is linear with the excitation intensity, the PL EF remains the same.

Last, we would like to mention that there were no observable local heating effects in our experiments. In hybrid structure of

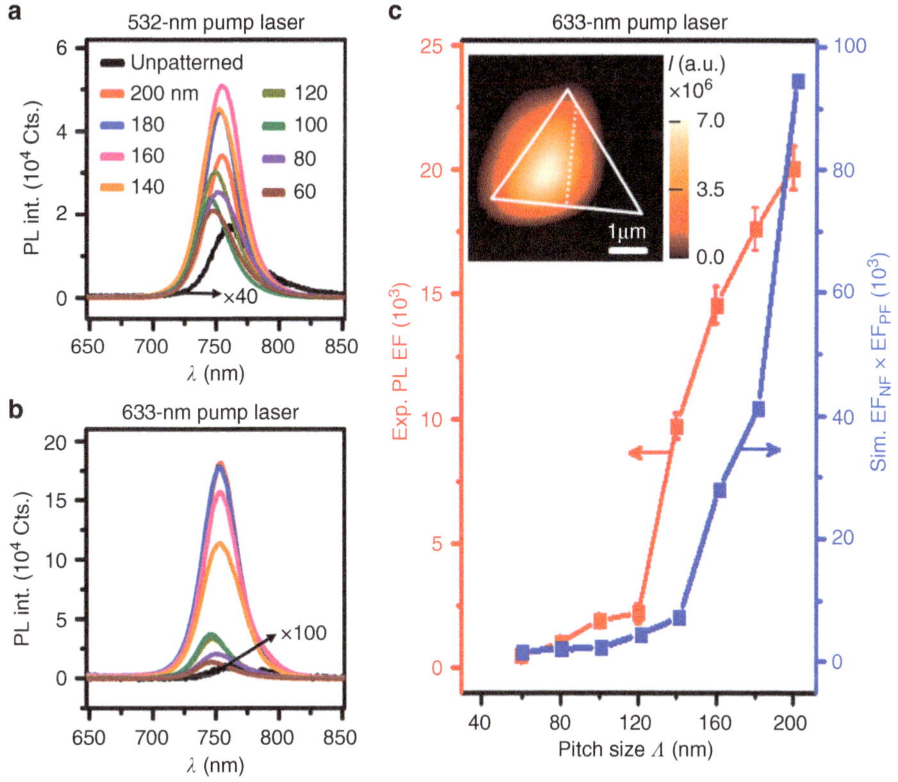

Figure 4 | Photoluminescence of WSe_2 on patterned gold nanostructures with different pitch sizes. (**a**) 532-nm pump laser. (**b**) 633-nm pump laser. (**c**) Experimental PL enhancement factor (EF) (left axis) excited by 633-nm laser (after correcting for the trench area fraction) and the simulated EF_{NF} (at 633 nm) × EF_{PF} (at 750 nm) (right axis), respectively. The data points and corresponding error bars are obtained, respectively, by calculating the mean value and s.d. of PL EF based on 10 experimental PL spectra for each nanostructure. The inset shows the PL mapping of a WSe_2 monolayer flake sited on patterned gold nanostructure with a pitch of 200 nm (on the left of the dashed line) and unpatterned gold film (on the right of the dashed line). The intensity value at each pixel was obtained by integrating the PL spectrum across the spectrum window of 700–820 nm. The solid line represent the shape of the flake and the dashed line represents the boundary between the patterned and unpatterned substrate. The laser power was 30 μW and integration time was 0.5 s.

MoS_2-gold nanoantenna, an increased absorption of the laser light at the nanoantenna may cause local heating so as to red shift and broaden the MoS_2 PL emission peak. Simultaneously, the local heating might cause a reduction in PL intensity of MoS_2 (ref. 12). In our experiment, the PL peak position and its full-width-at-half-maximum (FWHM) of WSe_2 on both patterned and unpatterned gold film have no power dependence (Supplementary Fig. 8a–d). Hence, for the range of laser powers used, the integrated PL and peak intensity increase linearly with respect to the laser power (Supplementary Fig. 8e,f), showing no signs of local heating effect.

In summary, concentrating and manipulating the electromagnetic field in sub-20 nm trenches can achieve unprecedented PL enhancement of WSe_2. Tuning the resonance of lateral gap plasmons to match with the pump laser wavelength effectively boosts the light–matter interactions in WSe_2, thus enhancing the light emission efficiency of WSe_2. A giant PL EF of ~20,000 was observed in WSe_2 when using the 633-nm pump laser. We expect further PL enhancements to be achieved through the design of double-resonance structures with directionality control. This work demonstrates an important method to enhance the PL of TMDCs since it could enable high-efficiency and high-quality photodetectors and sensors, where photon absorption and emission dictate the device performance. The incorporation of gold arrays with sub-20-nm-wide trenches onto WSe_2 monolayers opens up a new platform for investigating novel electrical/optical properties in 2D materials, such as electroluminescence and second harmonic generation, speeding up the applications of novel optoelectronic devices.

Methods

Material growth and quality of WSe_2. The triangular-shape monolayer WSe_2 was grown on sapphire by the CVD method (Supplementary Note 1). This technique yields high-quality monolayer WSe_2 with a crystal size of ~4.5 μm (Supplementary Fig. 2a). The typical PL spectrum of a pristine monolayer WSe_2 (Supplementary Fig. 2b) shows one pronounced emission peak at 768 nm (1.61 eV), that is, the A direct excitonic transition[36–38]. The Raman spectrum excited by 532-nm laser is shown in the inset of Supplementary Fig. 2b, where the two characteristic peaks for monolayer WSe_2 on sapphire are at 248 cm^{-1} and 259 cm^{-1}, which are from the degenerate E'/A_1' mode and 2LA(M) mode, respectively[39].

Fabrication of gold substrate and transferring of WSe_2. The gold substrate consisting of patterned gold structure and unpatterned gold film was prepared by a template-stripping method (Supplementary Note 2). WSe_2 flake was then transferred onto the gold substrate by a wet transfer approach (Supplementary Note 1). In comparison with the direct patterning of metallic nanostructures onto TMDCs by EBL, transferring WSe_2 on top of plasmonic nanostructure has several advantages. First, it allows for direct incident laser irradiation on WSe_2, instead of attenuated transmitted light irradiation on WSe_2 through the plasmonic structures[13]. Second, it avoids the potential undesired damage or doping of WSe_2 due to electron bombardment during electron-beam lithography process[40]. Last, the final structure will also be suitable for the direct access to the WSe_2 flake, for example, for subsequent patterning or for contact to external leads.

Characterizations. Atomic force microscope (AFM) images were measured by a BRUKER Dimension FastScan equipment. The optical reflectance spectra of the sample were then measured by using a CRAIC UV-VIS-NIR micro-spectrophotometer model

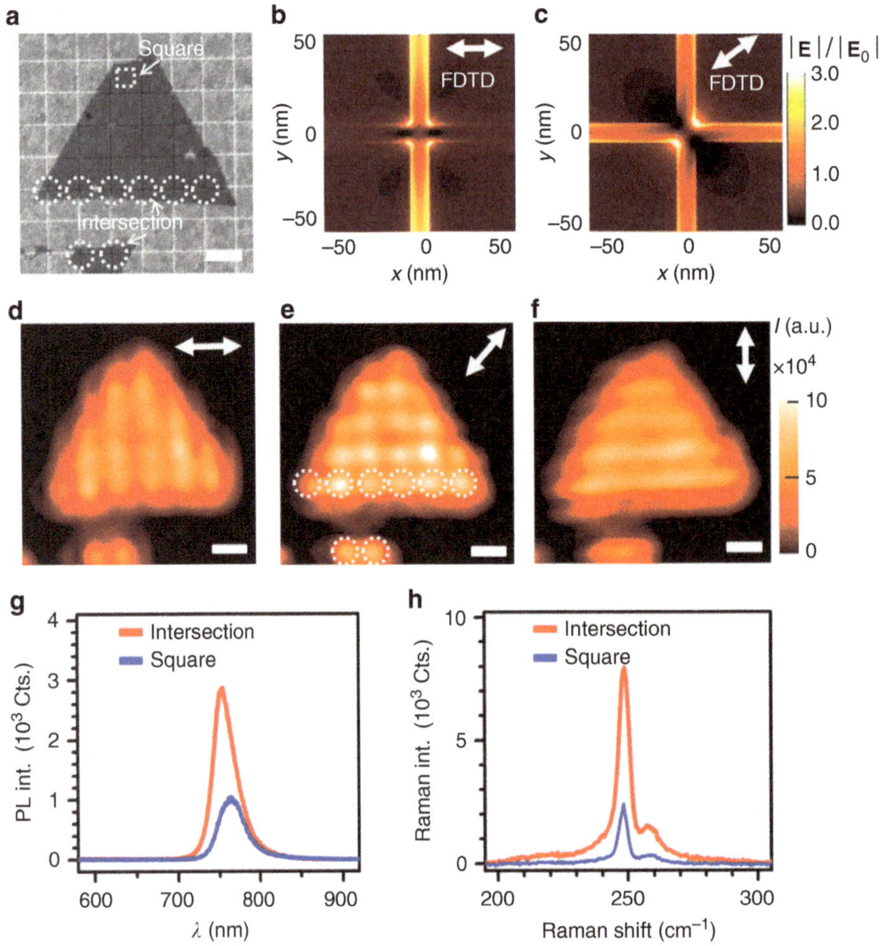

Figure 5 | Polarization-dependent characteristics of PL emission from WSe_2 flake on gold nanostructures with a pitch size of 920 nm and a pump laser of 532 nm. (**a**) SEM image of WSe_2 flake on the gold nanostructures. The white dashed circles denote the region where the trenches in the gold film intersect and the dashed square denotes the centre of the square region. (**b,c**) Simulated electric field distribution at a plane 1 nm above the surface of patterned gold nanostructures with 0°- and 45°-polarizations, respectively. (**d–f**) Corresponding PL intensity mappings with polarization angles of 0°, 45° and 90°, respectively. The intensity value at each pixel was obtained by integrating the PL spectrum across the spectrum window of 700–820 nm. The white arrows show the polarization directions of the pump laser. (**g,h**) PL and Raman spectra of WSe_2 taken at the intersection of gold trenches and square center, respectively. All the scale bars, 1 μm. The laser power and integration time for measuring PL (Raman) is 30 μW and 0.5 s, respectively (450 μW and 15 s).

QDI 2010 (equipped with a × 36 objective lens with numerical aperture (NA) = 0.5). Moreover, SEM images were measured by FEI Verios 460. A confocal micro-PL set-up was used to excite the sample by using a 532-nm or 633-nm CW pump laser focused by a × 100 microscope objective lens (NA = 0.65). PL emission was then collected by the same objective and detected by a nitrogen-cooled charge coupled device. The spot sizes of the 532-nm and 633-nm pump lasers were ~600 nm and 1.2 μm, respectively. The PL mappings in Figs 2 and 5 were measured by the 532-nm laser as the smaller laser spot size of this gave a better PL mapping resolution. The background PL signals of pure gold nanostructures (Supplementary Figs 2e and 5c) were subtracted from all the PL spectra of WSe_2 on patterned gold nanostructures. The PL EF is defined by equation (1).

Numerical simulations. A commercial software package, Lumerical FDTD Solutions, was used to simulate the optical field distributions and Purcell factors. Periodic boundary conditions were used along x- and y-axes, while perfect-matched layer was used along z-axis. The finest mesh size was set to be 0.5 nm in the structure. The geometries of the metal nanostructures in the simulations were designed to match with the SEM images (Fig. 3c), with a trench depth of 75 nm. This fine adjustment in the geometry was done for every pitch, which explains the discontinuous jumps in the numerical data points (Fig. 3b and Supplementary Fig. 4). The simulated EF of near-filed intensity is defined as the ratio of the simulated $|\mathbf{E}|^2$ on top of the trench to the simulated $|\mathbf{E}|^2$ on the unpatterned gold film at pump laser wavelength. To estimate the emission enhancement, we first calculated the Purcell factors for dipoles located on top of the trenches of gold nanostructures and for dipoles on top of the unpatterned gold film[41]. The emission enhancement was then determined by the ratio of these Purcell factors, which can be obtained directly from the FDTD analysis. Note that the classical calculations of Purcell factors do take into consideration both radiative and non-radiative rates for the dipole in the different environments[42]. Classically, the non-radiative decay rates are due to Ohmic losses in the metallic structures adjacent to the dipole.

References

1. Ross, J. S. *et al.* Electrically tunable excitonic light-emitting diodes based on monolayer WSe_2 pn junctions. *Nat. Nanotechnol.* **9,** 268–272 (2014).
2. Radisavljevic, B., Radenovic, A., Brivio, J., Giacometti, V. & Kis, A. Single-layer MoS_2 transistors. *Nat. Nanotechnol.* **6,** 147–150 (2011).
3. Zhang, W. *et al.* High gain phototransistors based on a CVD MoS_2 monolayer. *Adv. Mater.* **25,** 3456–3461 (2013).
4. Koperski, M. *et al.* Single photon emitters in exfoliated WSe_2 Structures. *Nat. Nanotechnol.* **10,** 503–506 (2015).
5. Chiu, M. H. *et al.* Determination of band alignment in the single-layer MoS2/WSe2 heterojunction. *Nat. Commun.* **6,** 7666 (2015).
6. Tonndorf, P. *et al.* Photoluminescence emission and Raman response of monolayer MoS_2, $MoSe_2$, and WSe_2. *Opt. Express* **21,** 4908–4916 (2013).
7. Lee, K. C. *et al.* Plasmonic gold nanorods coverage influence on enhancement of the photoluminescence of two-dimensional MoS_2 monolayer. *Sci. Rep.* **5,** 16374 (2015).
8. Lee, B. *et al.* Fano resonance and spectrally modified photoluminescence enhancement in monolayer MoS_2 integrated with plasmonic nanoantenna array. *Nano Lett.* **15,** 3646–3653 (2015).
9. Akselrod, G. M. *et al.* Leveraging nanocavity harmonics for control of optical processes in 2D semiconductors. *Nano Lett.* **15,** 3578–3584 (2015).

10. Zhou, K., Zhu, Y., Yang, X., Zhou, J. & Li, C. Demonstration of photoluminescence and metal-enhanced fluorescence of exfoliated MoS_2. *Phys. Chem. Chem. Phys.* **13,** 699–702 (2012).
11. Sobhani, A. *et al.* Enhancing the photocurrent and photoluminescence of single crystal monolayer MoS_2 with resonant plasmonic nanoshells. *Appl. Phys. Lett.* **104,** 031112 (2014).
12. Najmaei, S. *et al.* Plasmonic pumping of excitonic photoluminescence in hybrid MoS_2-Au nanostructures. *ACS Nano* **8,** 12682–12689 (2014).
13. Butun, S., Tongay, S. & Aydin, K. Enhanced light emission from large-area monolayer MoS_2 using plasmonic nanodiscarrays. *Nano Lett.* **15,** 2700–2704 (2015).
14. Goodfellow, K. M., Beams, R., Chakraborty, C., Novotny, L. & Vamivakas, A. N. Integrated nanophotonics based on nanowire plasmons and atomically thin material. *Optica* **1,** 149–152 (2014).
15. Lee, H. S. *et al.* Efficient exciton-plasmon conversion in Ag nanowire/ monolayer MoS_2 hybrids: direct imaging and quantitative estimation of plasmon coupling and propagation. *Adv. Opt. Mater.* **3,** 943–947 (2015).
16. Mouri, S. *et al.* Nonlinear photoluminescence in atomically thin layered WSe_2 arising from diffusion-assisted exciton-exciton annihilation. *Phys. Rev. B* **90,** 155449 (2014).
17. Wu, S. *et al.* Monolayer semiconductor nanocavity lasers with ultralow thresholds. *Nature* **520,** 69–72 (2015).
18. Matin, A. *et al.* Near-unity photoluminecescence quantum yield in MoS_2. *Science* **350,** 1065–1068 (2015).
19. Dong, Z. *et al.* Second-harmonic generation from sub-5-nm gaps by directed self-assembly of nanoparticles onto template-stripped gold substrates. *Nano Lett.* **15,** 5976–5981 (2015).
20. Nagpal, P., Lindquist, N. C., Oh, S. H. & Norris, D. J. Ultrasmooth patterned metals for plasmonics and metamaterials. *Science* **325,** 594–597 (2009).
21. Liu, H. *et al.* Enhanced surface plasmon resonance on a smooth silver film with a seed growth layer. *ACS Nano* **4,** 3139–3146 (2010).
22. Chai, L. & Klein, J. Large area, molecularly smooth (0.2 nm rms) gold films for surface forces and other studies. *Langmuir* **23,** 7777–7783 (2007).
23. Chen, X. *et al.* Atomic layer lithography of wafer-scale nanogap arrays for extreme confinement of electromagnetic waves. *Nat. Commun.* **4,** 2361 (2013).
24. Siegfried, T., Ekinci, Y., Martin, O. & Sigg, H. Gap plasmons and near-field enhancement in closely packed sub-10 nm gap resonators. *Nano Lett.* **13,** 5449–5453 (2013).
25. Tan, S. F. *et al.* Quantum plasmon resonances controlled by molecular tunnel junctions. *Science* **343,** 1496–1499 (2014).
26. Mak, K. F., Lee, C., Hone, J., Shan, J. & Heinz, H. F. Atomically thin MoS_2: a new direct-gap semiconductor. *Phys. Rev. Lett.* **105,** 136805 (2010).
27. Mukherjee, B. *et al.* Raman analysis of gold on WSe_2 single crystal film. *Mater. Res. Express* **2,** 065009 (2015).
28. Wientjes, E. *et al.* Strong antenna-enhanced fluorescence of a single light-harvesting complex shows photon antibunching. *Nat. Commun.* **5,** 017402 (2014).
29. Purcell, E. Spontaneous emission probabilities at radio frequencies. *Phys. Rev.* **69,** 681 (1946).
30. Salehzadeh, O. *et al.* Optically pumped two-dimensional MoS_2 lasers operating at room-temperature. *Nano Lett.* **15,** 5302–5306 (2015).
31. Hu, H., Duan, H., Yang, J. & Shen, Z. X. Plasmon-modulated photoluminescence from gold nanostructures and its dependence on plasmon resonance, excitation energy, and band structure. *Opt. Express* **23,** 5547–5564 (2015).
32. Li, Z. *et al.* Active Light control of the MoS_2 monolayer exciton binding energy. *ACS Nano* **9,** 10158–10164 (2015).
33. Ma, Y. *et al.* Reversible semiconducting-to-metallic phase transition in chemical vapor deposition grown monolayer WSe_2 and applications for devices. *ACS Nano* **9,** 7383–7391 (2015).
34. Desai, S. B. *et al.* Strain-induced indirect to direct bandgap transition in multilayer WSe_2. *Nano Lett.* **14,** 4592–45978 (2014).
35. Jin, W. *et al.* Substrate interactions with suspended and supported monolayer MoS_2: angle-resolved photoemission spectroscopy. *Phys. Rev. B* **91,** 121409 (2015).
36. Huang, J. K. *et al.* Large-area synthesis of highly crystalline WSe_2 monolayers and device applications. *ACS Nano* **8,** 923–930 (2013).
37. Zhou, H. *et al.* Large area growth and electrical properties of p-type WSe_2 atomic layers. *Nano Lett.* **15,** 709–713 (2015).
38. Srivastava, A. *et al.* Valley Zeeman effect in elementary optical excitations of monolayer WSe_2. *Nat. Phys.* **11,** 141–147 (2015).
39. Terrones, H. *et al.* New first order Raman-active modes in few layered transition metal dichalcogenides. *Sci. Rep.* **4,** 4215 (2014).
40. Siller, G. A., Kurasch, S., Sedighi, M., Lehtinen, O. & Kaiser, U. The pristine atomic structure of MoS_2 monolayer protected from electron radiation damage by graphene. *Appl. Phys. Lett.* **103,** 203107 (2013).
41. Zhou, W. *et al.* Lasing action in strongly coupled plasmonic nanocavity arrays. *Nat. Nanotechnol.* **8,** 506–511 (2013).
42. Anger, P., Bharadwaj, P. & Novotny, L. Enhancement and quenching of single-molecule fluorescence. *Phys. Rev. Lett.* **96,** 113002 (2006).

Acknowledgements

Z. W. acknowledges scholarship support from NUS Graduate School for Integrative Sciences & Engineering (NGS). Z. W. and A.T.S.W. acknowledge the funding support from MOE Tier 1 grant R 144-000-321-112 and facility support from NUS Center for Advanced 2D Materials. Z.D. and J.K.W.Y. acknowledge the funding support from the Agency for Science, Technology and Research (A*STAR) Young Investigatorship (grant number 0926030138), SERC (grant number 092154099), the National Research Foundation (grant number NRF-CRP 8-2011-07), and A*STAR-JCO under project number 1437C00135. C.-W.Q acknowledges the financial support from A*STAR Pharos Programme (Grant No. 152 70 00014, with Project No. R-263-000-B91-305). S.A.M. acknowledges the EPSRC Reactive Plasmonics Programme Grant (EP/M013812/1), the Royal Society, and the Lee-Lucas Chair in Physics. L.-J.L. acknowledges support from KAUST (Saudi Arabia) and Taiwan Consortium of Emergent Crystalline Materials (TCECM). G.E. acknowledges Singapore National Research Foundation, Prime Minister s Office, Singapore, for funding the research under its Medium-sized Centre program as well as NRF Research Fellowship (NRF-NRFF2011-02).

Author contributions

Z.W., Z.D., J.K.W.Y., A.T.S.W., C.-W.Q. and S.A.M. conceived the experiments; Z.W. transferred WSe_2 on gold substrate and characterized the OM, AFM, SEM, Reflectance, Raman and PL of WSe_2, gold substrate and hybrid structure. Z.D. prepared the gold substrate. Y.G. and C.-W.Q. contributed to the theoretical simulation and modelling, G.E. and W.J.Zhao. provided the PL/Raman system and guidance, Y.-H.C. and L.-J.L. grew the WSe_2 monolayers samples, S.A.M., G.G., W.J.Zhang and L.Z. provided expertise in data analysis and interpretation. Z.W. and Z.D. contributed equally to this work. The paper was drafted by Z.W. with inputs from Z.D., A.T.S.W., J.K.W.Y., C.-W.Q., S.A.M., G.G., G.E., W.J.Zhang., W.J.Zhao., Y.G., L.Z. and L.-J.L.

Additional information

Competing financial interests: The authors declare no competing financial interests.

Permissions

All chapters in this book were first published in NC, by Nature Publishing Group; hereby published with permission under the Creative Commons Attribution License or equivalent. Every chapter published in this book has been scrutinized by our experts. Their significance has been extensively debated. The topics covered herein carry significant findings which will fuel the growth of the discipline. They may even be implemented as practical applications or may be referred to as a beginning point for another development.

The contributors of this book come from diverse backgrounds, making this book a truly international effort. This book will bring forth new frontiers with its revolutionizing research information and detailed analysis of the nascent developments around the world.

We would like to thank all the contributing authors for lending their expertise to make the book truly unique. They have played a crucial role in the development of this book. Without their invaluable contributions this book wouldn't have been possible. They have made vital efforts to compile up to date information on the varied aspects of this subject to make this book a valuable addition to the collection of many professionals and students.

This book was conceptualized with the vision of imparting up-to-date information and advanced data in this field. To ensure the same, a matchless editorial board was set up. Every individual on the board went through rigorous rounds of assessment to prove their worth. After which they invested a large part of their time researching and compiling the most relevant data for our readers.

The editorial board has been involved in producing this book since its inception. They have spent rigorous hours researching and exploring the diverse topics which have resulted in the successful publishing of this book. They have passed on their knowledge of decades through this book. To expedite this challenging task, the publisher supported the team at every step. A small team of assistant editors was also appointed to further simplify the editing procedure and attain best results for the readers.

Apart from the editorial board, the designing team has also invested a significant amount of their time in understanding the subject and creating the most relevant covers. They scrutinized every image to scout for the most suitable representation of the subject and create an appropriate cover for the book.

The publishing team has been an ardent support to the editorial, designing and production team. Their endless efforts to recruit the best for this project, has resulted in the accomplishment of this book. They are a veteran in the field of academics and their pool of knowledge is as vast as their experience in printing. Their expertise and guidance has proved useful at every step. Their uncompromising quality standards have made this book an exceptional effort. Their encouragement from time to time has been an inspiration for everyone.

The publisher and the editorial board hope that this book will prove to be a valuable piece of knowledge for researchers, students, practitioners and scholars across the globe.

List of Contributors

Ying-Shuang Fu
School of Physics andWuhan National High Magnetic Field Center, Huazhong University of Science and Technology,Wuhan 430074, China
RIKEN Center for Emergent Matter Science, Wako, Saitama 351-0198, Japan

T. Hanaguri and M. Kawamura
RIKEN Center for Emergent Matter Science, Wako, Saitama 351-0198, Japan

M.S. Bahramy
RIKEN Center for Emergent Matter Science, Wako, Saitama 351-0198, Japan
226-8503, Japan
Department of Applied Physics, University of Tokyo, Bunkyo-ku, Tokyo 113-8656, Japan

T. Sasagawa and K. Igarashi
Materials and Structures Laboratory, Tokyo Institute of Technology, Yokohama, Kanagawa

Sungmin Park, Seok Young Son, Inhee Cho and Youngrok Cho
Department of Electrical and Computer Engineering, Seoul National University, Seoul 08826, South Korea

Hyomin Lee
Department of Electrical and Computer Engineering, Seoul National University, Seoul 08826, South Korea
Institute of Advanced Machines and Design, Seoul National University, Seoul 08826, South Korea

Sung Jae Kim
Department of Electrical and Computer Engineering, Seoul National University, Seoul 08826, South Korea
Big Data Institute, Seoul National University, Seoul 08826, South Korea
Inter-university Semiconductor Research Center, Seoul National University, Seoul 08826, South Korea

Yeonsu Jung
Department of Mechanical and Aerospace Engineering, Seoul National University, Seoul 08826, South Korea

Ho-Young Kim
Department of Mechanical and Aerospace Engineering, Seoul National University, Seoul 08826, South Korea
Institute of Advanced Machines and Design, Seoul National University, Seoul 08826, South Korea
Big Data Institute, Seoul National University, Seoul 08826, South Korea

Arun Devaraj
Physical and Computational Sciences Directorate, Pacific Northwest National Laboratory, 902 Battelle Boulevard, Richland, Washington 99354, USA

Vineet V. Joshi and Curt Lavender
Energy and Environment Directorate, Pacific Northwest National Laboratory, 902 Battelle Boulevard, Richland, Washington 99354, USA

Ankit Srivastava
Department of Materials Science and Engineering, Texas A&M University, 3003 College Station, Texas 77843, USA

Sandeep Manandhar
Environmental Molecular Sciences Laboratory, Pacific Northwest National Laboratory, 902 Battelle Boulevard, Richland, Washington 99354, USA

Vladimir Moxson and Volodymyr A. Duz
Advance Materials Products Inc. (ADMA), 1890 Georgetown Road, Hudson, Ohio 44236, USA

Xinyong Tao
College of Materials Science and Engineering, Zhejiang University of Technology, Hangzhou 310014, China
Department of Materials Science and Engineering, Stanford University, Stanford, California 94305, USA

JianguoWang and Qiuxia Cai
College of Chemical Engineering, Zhejiang University of Technology, Hangzhou 310014, China

Chong Liu, HaotianWang, Hongbin Yao, Guangyuan Zheng, ZhiWei Seh, Weiyang Li, Guangmin Zhou and Chenxi Zu
Department of Materials Science and Engineering, Stanford University, Stanford, California 94305, USA

Yi Cui
Department of Materials Science and Engineering, Stanford University, Stanford, California 94305, USA
Stanford Institute for Materials and Energy Science, SLAC National Accelerator Laboratory, Menlo Park, California 94025, USA

Yunlong Zi, Jie Wang, Sihong Wang, Shengming Li, Zhen Wen and Hengyu Guo
School of Materials Science and Engineering, Georgia Institute of Technology, Atlanta, Georgia 30332, USA

Zhong Lin Wang
School of Materials Science and Engineering, Georgia Institute of Technology, Atlanta, Georgia 30332, USA
Beijing Institute of Nanoenergy and Nanosystems, Chinese Academy of Sciences, Beijing 100083, China

Mo-Rigen He and Gyuseok Kim
Department of Materials Science and Engineering, University of Pennsylvania, Philadelphia, Pennsylvania 19104, USA

Saritha K. Samudrala, Peter J. Felfer, Andrew J. Breen and Julie M. Cairney
Australian Centre for Microscopy and Microanalysis, School of Aerospace, Mechanical and Mechatronic Engineering, University of Sydney, Sydney, New South Wales 2006, Australia

Daniel S. Gianola
Department of Materials Science and Engineering, University of Pennsylvania, Philadelphia, Pennsylvania 19104, USA
Department of Materials, University of California Santa Barbara, Santa Barbara, California 93106, USA

R.J. Coles, D.M. Price, J.E. Dixon, B. Royall, P. Kok, M.S. Skolnick, A.M. Fox and M.N. Makhonin
Department of Physics and Astronomy, University of Sheffield, Hicks Building, Sheffield S3 7RH, UK

E. Clarke
EPSRC National Centre for III-V Technologies, Department of Electronic and Electrical Engineering, University of Sheffield, Sheffield S1 3JD, UK

Haiming Yu
Physik Department E10, Technische Universität München, James-Franck-Strasse 1, D-85748 Garching bei München, Germany
Fert Beijing Institute, School of Electronic and Information Engineering, Beihang University, Xueyuan Road 37, Beijing 100191, China

O. d' Allivy Kelly, V. Cros, R. Bernard, P. Bortolotti and A. Anane
Unité Mixte de Physique, CNRS, Thales, Univ Paris-Sud, Université Paris-Saclay, 91767 Palaiseau, France

F. Brandl and F. Heimbach
Physik Department E10, Technische Universität München, James-Franck-Strasse 1, D-85748 Garching bei München, Germany

D. Grundler
Physik Department E10, Technische Universität München, James-Franck-Strasse 1, D-85748 Garching bei München, Germany
Laboratory of Nanoscale Magnetic Materials and Magnonics, Institute of Materials, School of Engineering, École Polytechnique Fédérale de Lausanne, STI-IMX-LMGN, Station 17, CH-1015 Lausanne, Switzerland

Chenhong Lang
College of Textiles, Donghua University, Shanghai 201620, China
Institute for Frontier Materials, Deakin University, Geelong, Victoria 3216, Australia

Xin Ding
College of Textiles, Donghua University, Shanghai 201620, China

Jian Fang, Hao Shao and Tong Lin
Institute for Frontier Materials, Deakin University, Geelong, Victoria 3216, Australia

M. Taupin, M. Meschke and J.P. Pekola
Low Temperature Laboratory, Department of Applied Physics, Aalto University School of Science, P.O. Box 13500, FI-00076 Aalto, Finland

I.M. Khaymovich
Low Temperature Laboratory, Department of Applied Physics, Aalto University School of Science, P.O. Box 13500, FI-00076 Aalto, Finland
Institute for Physics of Microstructures, Russian Academy of Sciences, GSP-105, 603950 Nizhni Novgorod, Russia

A.S. Mel'nikov
Institute for Physics of Microstructures, Russian Academy of Sciences, GSP-105, 603950 Nizhni Novgorod, Russia
Lobachevsky State University of Nizhni Novgorod, 23 Prospekt Gagarina, 603950 Nizhni Novgorod, Russia

Gi Seok Jeong
Department of Biomedical Engineering, College of Health Science, Korea University, Anam-ro 145, Seongbuk-gu, Seoul 02841, Korea
Biomedical Engineering Research Center, Asan Institute for Life Sciences, Asan Medical Center, 88 olympic-Ro, Songpa-gu, Seoul 05505, Korea

Da Yoon No
Department of Biomedical Engineering, College of Health Science, Korea University, Anam-ro 145, Seongbuk-gu, Seoul 02841, Korea
Department of Bioengineering, School of Engineering and Medicine, Stanford University, 443 Via Ortega, Stanford, California 94305, USA

JaeSeo Lee
KU-KIST Graduate School of Converging Science and Technology, Korea University, Seoul 02841, Korea

Junghyo Yoon and Seok Chung
School of Mechanical Engineering, Korea University, Anam-ro 145, Seongbuk-gu, Seoul 02841, Korea

Sang-Hoon Lee
Department of Biomedical Engineering, College of Health Science, Korea University, Anam-ro 145, Seongbuk-gu, Seoul 02841, Korea
KU-KIST Graduate School of Converging Science and Technology, Korea University, Seoul 02841, Korea

Tian Zhou, Dmitri Barbash, Christopher Y. Li and Hao Qi
Department of Materials Science and Engineering, Drexel University, Philadelphia, Pennsylvania 19104, USA

Lin Han
School of Biomedical Engineering, Science and Health Systems, Drexel University, Philadelphia, Pennsylvania 19104, USA

Borja Cirera, Francisco Martínez-Peña, Alberto Martin-Jimenez, Ana M. Pizarro and David Ecija
IMDEA Nanoscience, c/Faraday 9, Cantoblanco, 28049 Madrid, Spain

Roberto Otero
IMDEA Nanoscience, c/Faraday 9, Cantoblanco, 28049 Madrid, Spain
Departamento de Física de la Materia Condensada, Universidad Autónoma de Madrid, c/Francisco Tomás y Valiente 7, Cantoblanco, 28049 Madrid, Spain

Nelson Giménez-Agulló
Institute of Chemical Research of Catalonia, Barcelona Institute of Science and Technology, Avinguda Països Catalans 16, Tarragona 43007, Spain

Jonas Björk
Department of Physics, Chemistry and Biology, IFM, Linköping University, 58183 Linköping, Sweden

Jonathan Rodriguez-Fernandez
Departamento de Física de la Materia Condensada, Universidad Autónoma de Madrid, c/Francisco Tomás y Valiente 7, Cantoblanco, 28049 Madrid, Spain

José M. Gallego
Instituto de Ciencia de Materiales de Madrid, c/ Sor Juana Inés de la Cruz 3, Cantoblanco, 28049 Madrid, Spain

Pablo Ballester and José R. Galan-Mascaros
Institute of Chemical Research of Catalonia, Barcelona Institute of Science and Technology, Avinguda Països Catalans 16, Tarragona 43007, Spain
Catalan Institution for Research and Advanced Studies, Passeig Lluis Companys 23, Barcelona 08010, Spain

Eli Sutter
Department of Mechanical and Materials Engineering, University of Nebraska-Lincoln, Lincoln, Nebraska 68588, USA

Peter Sutter
Department of Electrical and Computer Engineering, University of Nebraska-Lincoln, Lincoln, Nebraska 68588, USA

Alexei V. Tkachenko
Center for Functional Nanomaterials, Brookhaven National Laboratory, Upton, New York 11973, USA

Roman Krahne, Milena Arciniegas and Liberato Manna
Department of Nanochemistry, Istituto Italiano di Tecnologia (IIT), via Morego 30, Genova IT-16163, Italy

Joost de Graaf
Faculty 8: Mathematics and Physics, Institute for Computational Physics (ICP), University of Stuttgart, Allmandring 3, Stuttgart 70569, Germany

Giancarlo Soavi and Daniele Viola
Dipartimento di Fisica, Politecnico di Milano, Piazza Leonardo Da Vinci 32, Milano 20133, Italy

Stefano Dal Conte and Cristian Manzoni
Istituto di Fotonica e Nanotecnologie, CNR, Piazza Leonardo Da Vinci 32, Milano 20133, Italy

Giulio Cerullo
Dipartimento di Fisica, Politecnico di Milano, Piazza Leonardo Da Vinci 32, Milano 20133, Italy
Istituto di Fotonica e Nanotecnologie, CNR, Piazza Leonardo Da Vinci 32, Milano 20133, Italy

Klaus Müllen, Akimitsu Narita, Yunbin Hu and Xinliang Feng
Max Planck Institute for Polymer Research, Ackermannweg 10, Mainz 55128, Germany

Ulrich Hohenester
Institute of Physics, University of Graz, Universitätsplatz Graz 8010, Austria

Elisa Molinari
Dipartimento di Scienze Fisiche, Informatiche e Matematiche, Universitá di Modena e Reggio Emilia, Modena 41125, Italy
Istituto Nanoscienze, CNR, via G. Campi 213/a, Modena 41125, Italy

Deborah Prezzi
Istituto Nanoscienze, CNR, via G. Campi 213/a, Modena 41125, Italy

Haoxue Han and Sebastian Volz
Laboratoire EM2C, CNRS, CentraleSupélec, Université Paris-Saclay, Grande Voie des Vignes, 92295 Châtenay-Malabry, France

Yong Zhang and Johan Liu
SMIT Center, School of Automation and Mechanical Engineering and Institute of NanomicroEnergy, Shanghai University, 20 Chengzhong Road, Shanghai 201800, China
Electronics Materials and Systems Laboratory, Department of Microtechnology and Nanoscience, Chalmers University of Technology, Kemivägen 9, SE-41296 Gothenburg, Sweden

Nan Wang, Majid Kabiri Samani, Michael Edwards and Murali Murugesan
Electronics Materials and Systems Laboratory, Department of Microtechnology and Nanoscience, Chalmers University of Technology, Kemivägen 9, SE-412 96 Gothenburg, Sweden

Yuxiang Ni
Department of Mechanical Engineering, University of Minnesota, 111 Church Street SE, Minneapolis, Minnesota 55455, USA

Zainelabideen Y. Mijbil
Quantum Technology Center, Physics Department, Lancaster University, Lancaster LA1 4YB, UK
Science Department, Veterinary Medicine College, Al-Qasim Green University, Babylon, Iraq

Shiyun Xiong
Max Planck Institute for Polymer Research, Ackermannweg 10, D-55128 Mainz, Germany

Kimmo Sääskilahti
Department of Biomedical Engineering and Computational Science, Aalto University, FI-00076 Aalto, Finland

Yifeng Fu
Electronics Materials and Systems Laboratory, Department of Microtechnology and Nanoscience, Chalmers University of Technology, Kemivägen 9, SE-412 96 Gothenburg, Sweden, SHT Smart High Tech AB, Ascherbergsgatan 46, SE-411 33 Gothenburg, Sweden

Lilei Ye
SHT Smart High Tech AB, Ascherbergsgatan 46, SE-411 33 Gothenburg, Sweden

Hatef Sadeghi, Steven Bailey and Colin J. Lambert
Quantum Technology Center, Physics Department, Lancaster University, Lancaster LA1 4YB, UK

Yuriy A. Kosevich
Laboratoire EM2C, CNRS, CentraleSupélec, Université Paris-Saclay, Grande Voie des Vignes, 92295 Châtenay-Malabry, France
Department of Polymers and Composite Materials, Semenov Institute of Chemical Physics, Russian Academy of Sciences, Kosygin Street 4, 119991 Moscow, Russia

Subhajit Biswas and Jessica Doherty
Materials Chemistry & Analysis Group, Department of Chemistry, Tyndall National Institute, University College Cork, Cork T12 YF78, Ireland

Dzianis Saladukha and Tomasz Ochalski
Department of Photonics, Tyndall National Institute, University College Cork, Cork T12 R5CP, Ireland
CAPPA, Cork Institute of Technology, Cork T12 T66T, Ireland

Quentin Ramasse
SuperSTEM Laboratory, SciTech Daresbury Campus, DaresburyWA4 4AD, UK

Dipanwita Majumdar and Achintya Singha
Department of Physics, Bose Institute, Kolkata 700009, India

Moneesh Upmanyu
Group for Simulation and Theory of Atomic-Scale Material Phenomena (stAMP), Department of Mechanical and Industrial Engineering and Department of Bioengineering, Northeastern University, Boston, Massachusetts 02115, USA

Michael A. Morris
AMBER, CRANN, Trinity College Dublin, Dublin D02 R590, Ireland

Justin D. Holmes
Materials Chemistry & Analysis Group, Department of Chemistry, Tyndall National Institute, University College Cork, Cork T12 YF78, Ireland
AMBER, CRANN, Trinity College Dublin, Dublin D02 R590, Ireland

Yueqi Li, Limin Xiang and Julio L. Palma
Center for Bioelectronics and Biosensors, Biodesign Institute, Arizona State University, Tempe, Arizona 85287-5801, USA
School of Molecular Sciences, Arizona State University, Tempe, Arizona 85287-5801, USA

Yoshihiro Asai
Research Center for Computational Design of Advanced Functional Materials (CD-FMat), National Institute of Advanced Industrial Science and Technology (AIST), Central 2, Umezono 1-1-1, Tsukuba, Ibaraki 305-8568, Japan

Nongjian Tao
Center for Bioelectronics and Biosensors, Biodesign Institute, Arizona State University, Tempe, Arizona 85287-5801, USA
School of Chemistry and Chemical Engineering, Nanjing University, Nanjing, Jiangsu 210093 China

Wanbin Li, Congyang Zhang, Zehai Xu, Pengcheng Su, Zheng Fan, Lei Qin and Guoliang Zhang
Institute of Oceanic and Environmental Chemical Engineering, State Key Lab Breeding Base of Green Chemical Synthesis Technology and Collaborative Innovation Center of Membrane Separation and Water Treatment of Zhejiang Province, Zhejiang University of Technology, Chaowang Road 18#, Hangzhou 310014, China

Qingbiao Li
Department of Chemical and Biochemical Engineering, College of Chemistry and Chemical Engineering, wwNational Laboratory for Green Chemical Productions of Alcohols, Ethers and Esters, Key Lab for Chemical Biology of Fujian Province, Xiamen University, Xiamen 361005, China

Yufan Zhang
Department of Materials Science and Engineering, College of Engineering, University of California, Berkeley, California 94720, USA

Qin Meng and Chong Shen
Department of Chemical and Biological Engineering, College of Chemical and Biological Engineering, State Key Laboratory of Chemical Engineering, Zhejiang University, Hangzhou 310027, China

Jesús Ferrando-Soria, Eufemio Moreno Pineda, Antonio Fernandez, Samantha A. Magee, Iñigo J. Vitorica-Yrezabal, Floriana Tuna, Grigore A. Timco, Eric J.L. McInnes and Richard E.P. Winpenny
School of Chemistry and Photon Science Institute,The University of Manchester, Oxford Road, Manchester M13 9PL, UK

Alessandro Chiesa, Stefano Carretta and Paolo Santini
Dipartimento di Fisica e Scienze della Terra, Universitádi Parma, viale delle Scienze 7/a, Parma 43123, Italy

Jonathan Rivnay, Sahika Inal, Eleni Stavrinidou, Xenofon Strakosas and George G. Malliaras
Department of Bioelectronics, Ecole Nationale Supérieure des Mines, CMP-EMSE, MOC, Gardanne 13541, France

Brian A. Collins
Material Science and Engineering Division, National Institute of Standards and Technology (NIST), Gaithersburg, Maryland 20899, USA
Department of Physics and Astronomy, Washington State University, Pullman, Washington 99164, USA

Michele Sessolo
Instituto de Ciencia Molecular, Universidad de Valencia, Paterna 46980, Spain

Christopher Tassone
Stanford Synchrotron Radiation Lightsource (SSRL), SLAC National Accelerator Laboratory, Menlo Park, California 94025, USA

Dean M. Delongchamp
Material Science and Engineering Division, National Institute of Standards and Technology (NIST), Gaithersburg, Maryland 20899, USA

Momchil T. Mihnev, Charles J. Divin and Theodore B. Norris
Department of Electrical Engineering and Computer Science, University of Michigan, Ann Arbor, Michigan 48109, USA
Center for Ultrafast Optical Science, University of Michigan, Ann Arbor, Michigan 48109, USA

Faris Kadi, Torben Winzer and Andreas Knorr
Institut für Theoretische Physik, Nichtlineare Optik und Quantenelektronik, Technische Universität Berlin, D-10623 Berlin, Germany

Seunghyun Lee
Department of Electrical Engineering and Computer Science, University of Michigan, Ann Arbor, Michigan 48109, USA
Department of Electronics and Radio Engineering, Kyung Hee University, Gyeonggi 446-701, South Korea

Che-Hung Liu and Zhaohui Zhong
Department of Electrical Engineering and Computer Science, University of Michigan, Ann Arbor, Michigan 48109, USA

Claire Berger
School of Physics, Georgia Institute of Technology, Atlanta, Georgia 30332, USA
Institut Neel, CNRS UJF-INP, Grenoble 38042, Cedex 6, France

Walt A. de Heer
School of Physics, Georgia Institute of Technology, Atlanta, Georgia 30332, USA King Abdulaziz University, Jeddah 22254, Saudi Arabia

Ermin Malic
Institut für Theoretische Physik, Nichtlineare Optik und Quantenelektronik, Technische Universität Berlin, D-10623 Berlin, Germany
Department of Applied Physics, Chalmers University of Technology, SE-412 96 Gothenburg, Sweden

Hyeon-Ho Jeong
Max Planck Institute for Intelligent Systems, Heisenbergstrasse 3, 70569 Stuttgart, Germany
Institute of Materials, École Polytechnique Fédérale de Lausanne (EPFL), CH-1015 Lausanne, Switzerland

Andrew G. Mark and Peter Oswald
Max Planck Institute for Intelligent Systems, Heisenbergstrasse 3, 70569 Stuttgart, Germany

Mariana Alarcón-Correa, Insook Kim and Peer Fischer
Max Planck Institute for Intelligent Systems, Heisenbergstrasse 3, 70569 Stuttgart, Germany
Institute for Physical Chemistry, University of Stuttgart, Pfaffenwaldring 55, 70569 Stuttgart, Germany

Tung-Chun Lee
Max Planck Institute for Intelligent Systems, Heisenbergstrasse 3, 70569 Stuttgart, Germany
UCL Institute for Materials Discovery and Department of Chemistry, University College London, Christopher Ingold Building, 20 Gordon Street, London WC1H 0AJ, UK

G. Zeevi, M. Shlafman, T. Tabachnik, Z. Rogachevsky, S. Rechnitz, I. Goldshtein, S. Shlafman, N. Gordon, G. Alchanati, M. Itzhak, Y. Moshe, E.M. Hajaj, H. Nir, Y. Milyutin, T.Y. Izraeli, A. Razin, O. Shtempluck, V. Kotchtakov and Y.E. Yaish
Andrew and Erna Viterbi Faculty of Electrical Engineering, Technion, Haifa 32000, Israel

Xu Xiao, Huaibing Song, Shizhe Lin, Ying Zhou, Xiaojun Zhan, Zhimi Hu, Jiyu Sun, Bo Yang, Tianqi Li, Jun Zhou and Jiang Tang
Wuhan National Laboratory for Optoelectronics and School of Optical and Electronic Information, Huazhong University of Science and Technology, Wuhan, Hubei 430074, China

Qi Zhang and Liying Jiao
Key Lab of Organic Optoelectronics & Molecular Engineering, Department of Chemistry, Tsinghua University, Beijing 100084, China

Yury Gogotsi
Department of Materials Science and Engineering and A.J. Drexel Nanomaterials Institute, Drexel University, Philadelphia, Pennsylvania 19104, USA

Zhuo Wang
NUS Graduate School for Integrative Sciences & Engineering (NGS), National University of Singapore, 28 Medical Drive, Singapore 117456, Singapore
Department of Physics, National University of Singapore, 2 Science Drive 3, Singapore 117542, Singapore
Department of Physics, Imperial College London, London SW7 2AZ, UK

Zhaogang Dong
Institute of Materials Research and Engineering, A*STAR (Agency for Science, Technology and Research), 3 Research Link, Singapore 117602, Singapore

Yinghong Gu and Lei Zhang
Department of Electrical and Computer Engineering, National University of Singapore, 4 Engineering Drive 3, Singapore 117583, Singapore

Yung-Huang Chang
Department of Electrophysics, National Chiao Tung University, Hsinchu 30010, Taiwan

Lain-Jong Li
Physical Science and Engineering Division, King Abdullah University of Science and Technology, Thuwal 23955-6900, Kingdom of Saudi Arabia

Weijie Zhao
Department of Physics, National University of Singapore, 2 Science Drive 3, Singapore 117542, Singapore

Goki Eda
Department of Physics, National University of Singapore, 2 Science Drive 3, Singapore 117542, Singapore
Department of Chemistry, National University of Singapore, 3 Science Drive 3, Singapore 117543, Singapore
Centre for Advanced 2D Materials, National University of Singapore, 2 Science Drive 3, Singapore 117542, Singapore

Wenjing Zhang
Department of Physics, National University of Singapore, 2 Science Drive 3, Singapore 117542, Singapore
SZU-NUS Collaborative Innovation Center for Optoelectronic Science & Technology, Key Laboratory of Optoelectronic Devices and Systems of Ministry of Education and Guangdong Province, College of Optoelectronic Engineering, Shenzhen University, Shenzhen 518060, China

Gustavo Grinblat and Stefan A. Maier
Department of Physics, Imperial College London, London SW7 2AZ, UK

Joel K.W. Yang
Institute of Materials Research and Engineering, A*STAR (Agency for Science, Technology and Research), 3 Research Link, Singapore 117602, Singapore
Singapore University of Technology and Design, 8 Somapah Road, Singapore 487372, Singapore

Cheng-Wei Qiu
NUS Graduate School for Integrative Sciences & Engineering (NGS), National University of Singapore, 28 Medical Drive, Singapore 117456, Singapore
Department of Electrical and Computer Engineering, National University of Singapore, 4 Engineering Drive 3, Singapore 117583, Singapore

Andrew T.S. Wee
NUS Graduate School for Integrative Sciences & Engineering (NGS), National University of Singapore, 28 Medical Drive, Singapore 117456, Singapore
Department of Physics, National University of Singapore, 2 Science Drive 3, Singapore 117542, Singapore
Centre for Advanced 2D Materials, National University of Singapore, 2 Science Drive 3, Singapore 117542, Singapore

Index

Q

S

T

U

X

Z